Lecture Notes in Computer Science 824

Edited by G. Goos and J. Hartmanis

Advisory Board: W. Brauer   D. Gries   J. Stoer

Erik M. Schmidt Sven Skyum (Eds.)

Algorithm Theory –
SWAT '94

4th Scandinavian Workshop
on Algorithm Theory
Aarhus, Denmark, July 6-8, 1994
Proceedings

Springer-Verlag
Berlin Heidelberg New York
London Paris Tokyo
Hong Kong Barcelona
Budapest

Erik M. Schmidt   Sven Skyum (Eds.)

Algorithm Theory – SWAT '94

4th Scandinavian Workshop
on Algorithm Theory
Aarhus, Denmark, July 6-8, 1994
Proceedings

Springer-Verlag

Berlin Heidelberg New York
London Paris Tokyo
Hong Kong Barcelona
Budapest

Series Editors

Gerhard Goos
Universität Karlsruhe
Postfach 69 80
Vincenz-Priessnitz-Straße 1
D-76131 Karlsruhe, Germany

Juris Hartmanis
Cornell University
Department of Computer Science
4130 Upson Hall
Ithaca, NY 14853, USA

Volume Editors

Erik M. Schmidt
Sven Skyum
Computer Science Department, Aarhus University
Ny Munkegade, Building 540, DK-8000 Aarhus C, Denmark

CR Subject Classification (1991): F.1-2, E.1-2, G.2, I.3.5

ISBN 3-540-58218-5 Springer-Verlag Berlin Heidelberg New York
ISBN 0-387-58218-5 Springer-Verlag New York Berlin Heidelberg

CIP data applied for

© Springer-Verlag Berlin Heidelberg 1994
Printed in Germany

Typesetting: Camera-ready by author
SPIN: 10472576   45/3140-543210 - Printed on acid-free paper

Foreword

The papers in this volume were presented at SWAT'94, the Fourth Scandinavian Workshop on Algorithm Theory. The workshop, which is really a conference, continues the tradition of SWAT'88, SWAT'90 and SWAT'92, and of the Workshops on Algorithms and Data Structures (WADS'89, WADS'91 and WADS'93), and is intended as an international forum for researchers in the area of design and analysis of algorithms. The SWAT conferences are coordinated by the SWAT steering committee, which consists of B. Aspvall (Bergen), S. Carlsson (Luleå), H. Hafsteinsson (Reykjavík), R. Karlsson (Lund), A. Lingas (Lund), E. M. Schmidt (Aarhus) and E. Ukkonen (Helsinki).

The call for papers sought contributions in algorithms and data structures, in all areas, including combinatorics, computational geometry, data bases, parallel and distributed computing, and graphics. A total of 100 papers were submitted and the program committee selected 31 for presentation. In addition, invited lectures were presented by Michael Fredman (Rutgers), Johan Håstad (Stockholm) and Ketan Mulmuley (Chicago).

SWAT'94 was held in Århus, July 6-8, 1994 and was organized by an organizing committee consisting of L. Arge, G. S. Frandsen, K. Kjær Møller, E. M. Schmidt (chairman) and S. Skyum, all from the Computer Science Department of Aarhus University.

We wish to thank all referees who aided in evaluating the papers. We also wish to thank the Danish Natural Science Research Council (SNF), the Centre for Basic Research in Computer Science of Aarhus University (BRICS), and the Aarhus University for financial support.

Århus, May 1994

Erik M. Schmidt
Sven Skyum

Program Committee

R. Freivalds (University of Latvia)
H. Hafsteinsson (University of Iceland)
T. Hagerup (Max-Planck-Institut, Saarbrücken)
G. F. Italiano (IBM T.J.Watson Research Center, Yorktown Heights)
M. Jerrum (Edinburgh University)
R. Karlsson (Lund University)
R. Klein (Fernuniversität, Hagen)
D. Kozen (Cornell University)
I. Munro (University of Waterloo)
S. Skyum (Aarhus University), chairman
G. Tel (Utrecht University)
E. Ukkonen (University of Helsinki)
E. Welzl (Freie Universität, Berlin)

Referees for SWAT'94

Table of Contents

Computing Depth Orders and Related Problems*

Pankaj K. Agarwal[1], Matthew J. Katz[2], Micha Sharir[3]

[1] Department of Computer Science, Duke University
[2] School of Mathematical Sciences, Tel Aviv University
[3] School of Mathematical Sciences, Tel Aviv University, and Courant Institute of
Mathematical Sciences, New York University

Abstract. Let $\mathcal{K}$ be a set of n non-intersecting objects in 3-space. A depth order of $\mathcal{K}$, if exists, is a linear order $<$ of the objects in $\mathcal{K}$ such that if $K, L \in \mathcal{K}$ and K lies vertically below L then $K < L$. We present a new technique for computing depth orders, and apply it to several special classes of objects. Our results include: (i) If $\mathcal{K}$ is a set of n triangles whose xy-projections are all 'fat', then a depth order for $\mathcal{K}$ can be computed in time $O(n \log^6 n)$. (ii) If $\mathcal{K}$ is a set of n convex and simply-shaped objects whose xy-projections are all 'fat' and their sizes are within a constant ratio from one another, then a depth order for $\mathcal{K}$ can be computed in time $O(n \lambda_s^{1/2}(n) \log^4 n)$, where s is the maximum number of intersections between the xy-projections of the boundaries of any pair of objects in $\mathcal{K}$.

1 Introduction

We describe a general technique for solving problems of the following form: Let $\mathcal{C} = \{c_1, \ldots, c_n\}$ be a set of n objects in the plane, and let $\prec$ be a partial binary anti-symmetric relation over $\mathcal{C}$ such that for any pair of objects $c_i, c_j \in \mathcal{C}$, if $c_i \cap c_j \neq \emptyset$ then either $c_i \prec c_j$ or $c_j \prec c_i$, and if $c_i \cap c_j = \emptyset$ then c_i and c_j are incomparable. Moreover, we assume that for any pair of intersecting objects in $\mathcal{C}$ it is possible to determine in constant time which of the two possibilities holds, and that the transitive closure of $\prec$ is acyclic, i.e., that it is a partial order. We wish to compute a linear extension of this partial order. We refer to this problem as the *2-dimensional linear-extension problem*.

The main motivation for studying this problem comes from the problem of computing a *depth order* in three dimensions. Specifically, we are given a set $\mathcal{K}$ of n non-intersecting, convex and simply-shaped objects in 3-space (by 'simple shape' we mean that each object can be described by a constant number of

---

* Work on this paper by the first author has been supported by National Science Foundation Grant CCR–93–01259 and an NYI award. Work on this paper by the third author has been supported by NSF Grant CCR-91-22103, by a Max-Planck Research Award, and by grants from the U.S.-Israeli Binational Science Foundation, the Fund for Basic Research administered by the Israeli Academy of Sciences, and the G.I.F., the German-Israeli Foundation for Scientific Research and Development.

polynomial equalities and inequalities of constant maximum degree). For a pair of objects $K, L \in \mathcal{K}$, we say that K lies *below* L (and L lies *above* K) if there exists a vertical line λ that intersects both K and L and $\lambda \cap K$ lies fully below $\lambda \cap L$ (the convexity of K, L implies that this relation is anti-symmetric). We denote this relation by $K \prec L$. Assuming that the transitive closure of $\prec$ is an acyclic relation, any linear extension of it is called a *depth order* of $\mathcal{K}$. It is then clear that the set of the xy-projections of the objects in $\mathcal{K}$ is a proper input for the 2-dimensional linear-extension problem, as formulated above, where for any pair of projections K^*, L^* of two respective objects K, L of $\mathcal{K}$, we have $K^* \prec L^*$ if and only if $K \prec L$. Hence, the problem of computing depth orders in 3-space can be reduced to the 2-dimensional linear-extension problem. Computing depth orders is a preliminary step of many algorithms for *hidden surface removal* in computer graphics, and of several other algorithms; see [5] for more details.

We note that the 2-dimensional linear-extension problem arises in some other applications as well. For example, suppose that each $c_i \in \mathcal{C}$ designates some area of the plane to which we want to apply a certain process (e.g., spraying it with some substance, painting it, etc.), and that whenever two such areas overlap, there is some order in which the two respective processes must be scheduled. The problem is then to find a global scheduling order for the processes so that they are executed in the correct order for each point in the plane. In fact, the painter's algorithm for hidden surface removal (see e.g. [10]) is an instance of this more general problem.

de Berg et al. [7] presented an algorithm for computing a depth order of a collection of n arbitrary non-intersecting triangles in 3-space; their algorithm runs in time $O(n^{4/3+\varepsilon})$, for any $\varepsilon > 0$. It is conceptually fairly simple, but it uses rather involved range searching methods, and it does not extend to sets of more general objects.

We present here a different approach to the problem (actually, to the 2-dimensional linear-extension problem), which is simpler than the approach of [7]. The algorithm applies to collections of objects that satisfy certain 'fatness' properties, so it does not apply to arbitrary triangles. We call the objects in a collection $\mathcal{K}$ as above α-*fat* if the xy-projection of any object in $\mathcal{K}$ has the property that it is contained in some axis-parallel square S^+ and contains another axis-parallel square S^-, so that the ratio between the edge lengths of S^+ and S^- is at most α. (Note that the notion of fatness depends also on the 3-D orientations of the objects in $\mathcal{K}$ and not just on their shapes; for example, if $\mathcal{K}$ is a collection of disks in 3-space, then these disks are α-fat if and only if the angles between their normals and the z-axis are all smaller than some fixed angle $\delta = \delta(\alpha)$.) We say that the objects in $\mathcal{K}$ have the α-*ratio* property if the xy-projections K^*, L^* of any pair of objects in $\mathcal{K}$ have the property that K^* is contained in some axis-parallel square S^+, L^* contains another axis-parallel square S^-, and the ratio between the edge lengths of S^+ and S^- is at most α (and a symmetric condition holds when interchanging K^* and L^*).

Our main results are:

Theorem 1. *A depth order of a collection of n non-intersecting α-fat triangles* *in 3-space, for any $\alpha > 1$, can be computed (if it exists) in time $O(n \log^6 n)$, where the constant of proportionality depends on α.*

Theorem 2. *Let $\mathcal{K}$ be a collection of n non-intersecting convex simply-shaped α-fat objects in 3-space with the α-ratio property. Then a depth order for $\mathcal{K}$ can be computed (if it exists), in an appropriate model of computation, in time $O(n \lambda_s^{1/2}(n) \log^4 n)$, where s is the maximum number of intersections between the boundaries of any pair of projections of objects in $\mathcal{K}$.*[4]

Theorem 3. *The 2-dimensional linear-extension problem for a collection $\mathcal{C}$ of n disks can be solved in $O(n^{7/5+\varepsilon})$ time, for any $\varepsilon > 0$. Moreover, if $\mathcal{C}$ has the α-ratio property, this problem can be solved in $O(n \log^7 n)$ time.*

Thus, for the case of fat triangles, we obtain a solution that is much faster and simpler than the algorithm of [7]. This result is especially significant since it enables us to perform hidden surface removal for a collection of fat triangles with unknown depth order, using the algorithm of [12], whose running time is close to linear. Prior to our result, one had to spend close to $O(n^{4/3})$ time on computing the depth order, before proceeding to apply the algorithm of [12], or, alternatively, one could apply the hidden surface removal algorithm of [6], which does not require a depth order to be given, but still runs in time close to $O(n^{4/3})$. The running time of our algorithm for general collections of objects, as in Theorem 2, which is close to $O(n^{3/2})$, is somewhat worse than the running time of the algorithm of [7], but it is nevertheless (significantly) subquadratic, and it applies to many cases where no previous subquadratic algorithm was known, such as the case of disks in 3-space (satisfying the conditions of Theorem 2).

We still do not know how to extend our technique to detect cycles in the 'above-below' relation $\prec$ (the algorithm of [7] can detect such cycles). We are currently studying this problem.

The paper is organized as follows. In Section 2 we describe the general technique for the 2-dimensional linear-extension problem. In Section 3, we specialize this technique to the case of fat triangles in the plane, to obtain Theorem 1, and, in Section 4, we specialize it to the case of general fat objects in the plane with the α-ratio property, to obtain Theorem 2. For lack of space, we omit the proof of Theorem 3 in this version.

2 Outline of the Method

We focus here on the 2-dimensional linear-extension problem. The algorithm consists of two stages. In the first stage we compute a collection of pairs of subsets of $\mathcal{C}$, $\mathcal{F} = \{(\mathcal{R}_1, \mathcal{B}_1), (\mathcal{R}_2, \mathcal{B}_2), \ldots, (\mathcal{R}_m, \mathcal{B}_m)\}$, such that

(i) For each intersecting pair of objects $c_1, c_2 \in \mathcal{C}$, there is an $1 \leq i \leq m$ such that either $c_1 \in \mathcal{R}_i$ and $c_2 \in \mathcal{B}_i$, or $c_2 \in \mathcal{R}_i$ and $c_1 \in \mathcal{B}_i$.

---

[4] $\lambda_s(n)$ is the maximum length of (n, s) Davenport-Schinzel sequences [9].

(ii) For each $i = 1, \ldots, m$, all the objects in $\mathcal{B}_i$ have a common point (we say that $\mathcal{B}_i$ has the *common point* property).

In the second stage we construct a directed graph G over the set $\mathcal{C}$ of size $O(\sum_{i=1}^{m}(|\mathcal{R}_i| + |\mathcal{B}_i|))$, such that any permutation of $\mathcal{C}$ that is obtained by topologically sorting G is a linear extension of the underlying relation $\prec$. Initially, the edge set of G is empty. We process each pair $(\mathcal{R}, \mathcal{B})$ of the above collection separately, and obtain from it $O(|\mathcal{R}| + |\mathcal{B}|)$ edges of G, as follows. By property (ii), the set $\mathcal{B}$ is a chain of the relation $\prec$, and we can therefore sort $\mathcal{B}$ into a linear list, denoted as $b_1 \prec b_2 \prec \cdots \prec b_q$. We add the $q - 1$ edges (b_i, b_{i+1}), for $i = 1, \ldots, q - 1$, to the graph G. Next we 'merge' the elements of $\mathcal{R}$ into the list $\mathcal{B}$. That is, we want to compute, for each element r of $\mathcal{R}$, two pointers to its immediate predecessor and successor (if they exist) in the list $(b_1, \ldots, b_q)$; these are denoted, respectively, as $b_{pred(r)}$ and $b_{succ(r)}$. We then add, for each $r \in \mathcal{R}$, the (at most) two corresponding edges $(b_{pred(r)}, r)$ and $(r, b_{succ(r)})$ to G.

We claim that, after this step is repeated for all pairs $(\mathcal{R}, \mathcal{B})$, the relation $\prec$ is contained in the transitive closure of G; that is, for each pair $c \prec c'$ of objects of $\mathcal{C}$, there exists a directed path in G from c to c'. (Note that G is acyclic, by assumption.) Indeed, if c, c' is such a pair then $c \cap c' \neq \emptyset$, and therefore there exists a pair $(\mathcal{R}, \mathcal{B})$ in the collection computed in the first stage, such that one of the objects, say c, belongs to $\mathcal{B}$ and the other belongs to $\mathcal{R}$. In the second stage we find an object $c'' \in \mathcal{B}$ such that c'' is the immediate predecessor of c' in $\mathcal{B}$, and the edge (c'', c') is added to G (note that c'' exists, since c is a predecessor of c' in $\mathcal{B}$). If $c'' = c$ then the claim is obvious, and if $c'' \neq c$ then necessarily $c \prec c''$, so G contains a directed path from c to c', as claimed. The case where $c \in \mathcal{R}$ and $c' \in \mathcal{B}$ is completely symmetric.

Suppose we have an algorithm $\mathcal{A}$ that, given a subset $\mathcal{R}' \subseteq \mathcal{R}$ and a subset $\mathcal{B}' \subseteq \mathcal{B}$, determines for each object $r \in \mathcal{R}'$ whether r intersects any object of $\mathcal{B}'$ (which is the same as asking whether r intersects the region $\bigcup_{b \in \mathcal{B}'} b$), and, if so, it also computes a 'witness' object $b \in \mathcal{B}'$ that intersects r. Using $\mathcal{A}$ as a subroutine, we can perform the merge step for a pair of sets $(\mathcal{R}, \mathcal{B})$, with $\mathcal{B}$ already sorted, as follows. Let w be a point in $\bigcap \mathcal{B}$. Since the objects of $\mathcal{C}$ are assumed to be convex, the union of any subset of $\mathcal{B}$ is a star-shaped region with respect to w. Let T be a minimum-height binary tree over the objects $b_1, \ldots, b_q$, where these objects are stored at the leaves of T from left to right in increasing order. Denote by $\mathcal{B}_v$ the set of objects stored at the leaves of the subtree of T rooted at v. Associate with each internal node v of T the region $U_v = \bigcup_{b \in \mathcal{B}_v} b$. We process the nodes of T by levels, beginning at the root level. Initially, we assign a set $\mathcal{R}_{root} = \mathcal{R}$ to the root. Upon completion of the processing of the j-th level, we assign to each node v of the $(j + 1)$-th level an appropriate subset $\mathcal{R}_v$ of $\mathcal{R}$. After completing the processing at the root, we form the sets $\mathcal{R}_{left(root)}$ and $\mathcal{R}_{right(root)}$, as follows. For each object $r \in \mathcal{R}_{root} = \mathcal{R}$, if r does not intersect U_{root} we can ignore r, since it is $\prec$-incomparable with all objects of $\mathcal{B}$. If r does intersect U_{root} and the witness b is greater than r (i.e., $r \prec b$), then, if b belongs to $\mathcal{B}_{left(root)}$, add r to $\mathcal{R}_{left(root)}$, otherwise (i.e., $b \in \mathcal{B}_{right(root)}$) add r to both sets. The case where the witness b is smaller than r is treated symmetrically.

In general, we maintain the following invariant property: After processing each level, each object r of $\mathcal{R}$ is present in at most four $\mathcal{R}$-subsets, $\mathcal{R}_{v_1}$, $\mathcal{R}_{v_2}$, $\mathcal{R}_{v_3}$, $\mathcal{R}_{v_4}$, so that the largest predecessor and the smallest successor of r (if they exist) are stored at one or two of these nodes. After we finish the processing of all nodes at the current level, we form the $\mathcal{R}$-subsets for the next level by distributing the elements of the sets $\mathcal{R}_v$, for the current level nodes v, among their children. This is done by inspecting, for each object $r \in \mathcal{R}$, the (at most) four j-level nodes that store it, by identifying (at most) two of these nodes (the rightmost of these nodes for which a predecessor witness of r was found, and the leftmost of these nodes for which a successor witness was found) whose children might still need to store r, and by distributing r to these (at most four) children, based on a criterion similar to that used at the root. The validity of the invariant property easily follows by induction. At the leaf level, each object r of $\mathcal{R}$ is assigned to at most four leaves, and the desired objects $b_{pred(r)}$, $b_{succ(r)}$ (if they exist) are two of these leaves, and can thus be determined in constant time.

This concludes the outline of the algorithm. In the subsequent sections we will specialize this technique to a set of fat triangles (Section 3), and to a set of general fat objects (Section 4).

3   Fat Triangles

In this section we apply the general technique outlined above to the case where the objects of $\mathcal{C}$ are all δ-fat triangles, that is, all the angles of each triangle in $\mathcal{C}$ are at least δ, for some fixed constant δ. (For triangles, this is an equivalent definition of fatness.) Fat triangles have been investigated in [13], where it was shown that the combinatorial complexity of the union of m δ-fat triangles is $O(m \log \log m)$, where the constant of proportionality depends on δ (see also [2, 8, 16]). Moreover, such a union can be computed in time $O(m \log^2 m)$ by a deterministic algorithm [13], or in expected time $O(m 2^{\alpha(m)} \log m)$ by a randomized algorithm [15].

3.1   The First Stage

The first stage of the algorithm is implemented in this case as follows. We first fix a family $\mathcal{O}$ of $O(1/\delta)$ orientations, evenly spaced in $[0, 2\pi)$, so that the angle between any pair of successive orientations is $\leq \delta/3$. We then represent each triangle $\Delta \in \mathcal{C}$ as the union of three overlapping triangles, such that each subtriangle Δ' is 'semi-canonical', i.e., it has two sides at orientations belonging to $\mathcal{O}$; we refer the reader to Figure 1 and to [13] for the easy details. This allows us to partition the collection of new triangles into a constant number of subfamilies, so that all triangles within any subfamily $\mathcal{C}'$ are 'almost-homothetic', that is, two sides of each triangle of $\mathcal{C}'$ are at two fixed orientations, and the orientation of the third side lies in some narrow range of orientations (of size, say, $\delta/3$).

Recall that in the first stage we need to compute a collection of pairs of subsets of $\mathcal{C}$ satisfying properties (i) and (ii) of Section 2. We do this by computing

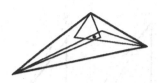

Fig. 1. Representing a fat triangle Δ as the union of three 'semi-canonical' subtriangles (the middle point is the center of the inscribed circle of Δ)

such a collection for each pair of (not necessarily distinct) subfamilies C' and C''. That is, we compute a family $\mathcal{F} = \{(\mathcal{R}_1, \mathcal{B}_1), \ldots, (\mathcal{R}_m, \mathcal{B}_m)\}$ of pairs such that

(i) For each i, $\mathcal{R}_i \subseteq C'$ and $\mathcal{B}_i \subseteq C''$. Moreover, for each intersecting pair of objects $(c_1, c_2) \in C' \times C''$, there is an i such that either $c_1 \in \mathcal{R}_i$ and $c_2 \in \mathcal{B}_i$, or $c_2 \in \mathcal{R}_i$ and $c_1 \in \mathcal{B}_i$.
(ii) Each $\mathcal{B}_i$ has the common point property.

The union of these collections over all pairs of subfamilies (after replacing each semi-canonical triangle in $\mathcal{R}_i, \mathcal{B}_i$ by its original containing triangle) constitutes an appropriate collection for C.

We now describe how to construct $\mathcal{F}$ for a fixed pair of subfamilies C' and C''. First assign to each triangle $\Delta \in C' \cup C''$ the radius $\rho(\Delta)$ of the smallest circle containing Δ, and sort the triangles in $C' \cup C''$ in increasing order of these radii. Let $\mathcal{L}'$ (resp. $\mathcal{H}'$) denote the subset of triangles of C' that lie in the smaller (resp. larger) half of the sorted sequence. We define $\mathcal{L}'', \mathcal{H}'' \subseteq C''$ in complete analogy. Our strategy is to interact $\mathcal{L}'$ with $\mathcal{H}''$, in a manner that will be described in a moment, then interact $\mathcal{L}''$ with $\mathcal{H}'$, in the same manner, and finally handle recursively the sets $\mathcal{L}' \cup \mathcal{L}''$ and $\mathcal{H}' \cup \mathcal{H}''$.

Next, we describe how to compute an appropriate collection for the pair $\mathcal{L}', \mathcal{H}''$. Associate with a semi-canonical triangle $\Delta = uvw$ two triangles, $\Delta^+ \supseteq \Delta$ and $\Delta^- \subseteq \Delta$, and a trapezoid Δ^*, as follows (see Figure 2 for an illustration). The triangle Δ^+ contains Δ and is obtained from Δ by replacing the side vw whose orientation is not fixed by another side. The new side has a vertex v in common with the former side, its orientation is in $\mathcal{O}$ and is greater than the orientation of the former side, and the angle between the former side and the new side is at most $\delta/3$. The triangle Δ^- is contained in Δ and is obtained by replacing the new side of Δ^+ by a parallel side through the other vertex w. Finally, $\Delta^* \equiv \Delta^+ - \Delta^-$.

Lemma 4. *If $\Delta_1 \cap \Delta_2 \neq \emptyset$, where $\Delta_1 \in \mathcal{L}'$ and $\Delta_2 \in \mathcal{H}''$, then at least one of the following conditions must hold:*

1. *One of the two sides of Δ_2 of fixed orientation intersects a side of Δ_1^+.*
2. *A vertex of Δ_2 lies inside Δ_1^+.*
3. *$\Delta_1^+ \cap \Delta_2^- \neq \emptyset$.*

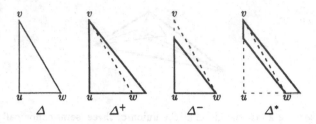

Fig. 2. A triangle Δ, the corresponding triangles Δ^+, Δ^-, and the trapezoid Δ^*

4. A vertex of Δ_1^+ lies inside Δ_2^*.

Proof. Since $\Delta_1 \cap \Delta_2 \neq \emptyset$, and since $\Delta_1^+ \supseteq \Delta_1$, we must also have $\Delta_1^+ \cap \Delta_2 \neq \emptyset$. This is easily seen to imply at least one of the above conditions. $\qquad\square$

Reporting intersections of type 1: Let us denote by $\mathcal{L}'^+$ the set of expanded triangles Δ_1^+, for $\Delta_1 \in \mathcal{L}'$. Fix one of the three possible orientations, θ_1, of the sides of expanded triangles in $\mathcal{L}'^+$ and one of the two orientations, θ_2, of the sides of triangles in $\mathcal{H}''$ at fixed orientations. Let S_1 (resp. S_2) denote the collection of all sides at orientation θ_1 (resp. θ_2) of triangles in $\mathcal{L}'^+$ (resp. in $\mathcal{H}''$). Since the segments in S_1 are all parallel, and so are the segments of S_2, it is straightforward, using standard orthogonal range searching techniques (such as 2-dimensional *range trees* [14]), to report all intersecting pairs in $S_1 \times S_2$, as a collection of complete bipartite graphs $\{\mathcal{R}_j \times \mathcal{B}_j\}$, so that $\sum_j |\mathcal{R}_j|, \sum_j |\mathcal{B}_j| = O(m \log^2 m)$, where m is the combined size of $\mathcal{L}'$ and $\mathcal{H}''$; the cost of this procedure is also $O(m \log^2 m)$. Repeating this step six times, for all possible pairs θ_1, θ_2, we obtain a collection of pairs $\{(\mathcal{R}_j, \mathcal{B}_j)\}$, so that:

(i) $\mathcal{R}_j \subseteq \mathcal{L}'$ and $\mathcal{B}_j \subseteq \mathcal{H}''$ for each j.
(ii) For each pair $(\mathcal{R}_j, \mathcal{B}_j)$, each pair of triangles $\Delta_1 \in \mathcal{R}_j$, $\Delta_2 \in \mathcal{B}_j$ is such that $\rho(\Delta_1) \leq \rho(\Delta_2)$.
(iii) For each pair $(\mathcal{R}_j, \mathcal{B}_j)$, each pair of triangles $\Delta_1 \in \mathcal{R}_j$, $\Delta_2 \in \mathcal{B}_j$ is such that Δ_1^+ and Δ_2 have nonempty intersection.
(iv) $\sum_j |\mathcal{R}_j|, \sum_j |\mathcal{B}_j| = O(m \log^2 m)$.

Next we need to refine the pairs $(\mathcal{R}_j, \mathcal{B}_j)$, so as to enforce the common point property for one of the sets in each pair. Let $(\mathcal{R}, \mathcal{B})$ be one of the pairs produced above, and let Δ_0 be a fixed triangle in $\mathcal{R}$. Each triangle in $\mathcal{B}$ must contain, by property (iii), a point inside the expanded triangle Δ_0^+. Let z be the center of the circumscribed circle of Δ_0^+, and let $\Delta_0^{++} = z + 2(\Delta_0^+ - z)$; that is, Δ_0^{++} is an expansion of Δ_0^+ about z by a factor of 2. It is easy to verify that for each triangle $\Delta \in \mathcal{B}$ we have $area(\Delta \cap \Delta_0^{++}) \geq \beta \cdot area(\Delta_0^{++})$, for some constant fraction β (this follows from the fatness of the triangles and from the fact that Δ_0 is 'smaller' than the triangles in $\mathcal{B}$); see Figure 3 for an illustration. This however

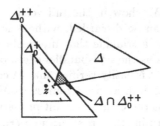

Fig. 3. The area of $\Delta \cap \Delta_0^{++}$ must be 'large'

is easily seen to imply that there exist a constant number (depending on δ) of points within Δ_0^{++} so that each triangle of $\mathcal{B}$ contains at least one of these points. Thus, by splitting $\mathcal{B}$ into a constant number of subfamilies $\mathcal{B}_1, \dots, \mathcal{B}_l$, and by replacing the pair $(\mathcal{R}, \mathcal{B})$ by the l corresponding pairs $(\mathcal{R}, \mathcal{B}_1), \dots, (\mathcal{R}, \mathcal{B}_l)$, we ensure the common point property for all the new sets $\mathcal{B}_j$, without affecting the other properties listed above.

Reporting intersections of type 2: Next we find all pairs of triangles $\Delta_1^+ \in \mathcal{L}'^+$, $\Delta_2 \in \mathcal{H}''$, such that a vertex of Δ_2 lies inside Δ_1^+. Since the orientations of the edges of the triangles Δ_1^+ are all fixed, this condition can be easily expressed in terms of orthogonal range searching queries: We need to determine, for each vertex-triangle test, whether the vertex lies on the appropriate sides of the three lines containing the edges of the triangle; this is easily achieved, using, e.g., 3-level range trees. We can thus collect all these pairs of triangles, represented by an appropriate collection of complete bipartite graphs, as above, where the total size of all the graphs, and the computation time, are both now $O(m \log^3 m)$. We enforce the common point property as in the previous type of intersection.

Reporting intersections of type 3: We now find all pairs of intersecting triangles (Δ_1^+, Δ_2^-), for $\Delta_1 \in \mathcal{L}'$, $\Delta_2 \in \mathcal{H}''$. Since all edges of each of these triangles have fixed orientations, all these intersections can be computed and represented using an appropriate combination of the techniques used in the two previous cases; the total size of the graphs, and the computation time, are both $O(m \log^3 m)$.

Reporting intersections of type 4: We are left with the task of finding intersections between triangles Δ_1^+, for $\Delta_1 \in \mathcal{L}'$, and trapezoids Δ_2^*, for $\Delta_2 \in \mathcal{H}''$, where a vertex of Δ_1^+ lies inside Δ_2^*. Again, since the orientations of all four edges of these trapezoids are fixed, and assume only three values, it is easy to report all such vertex-trapezoid containments, using a 3-level orthogonal range tree structure, as above. This produces a collection of pairs of triangles $(\mathcal{R}_i, \mathcal{B}_i)$ that satisfies the above properties (where the overall size of these sets is $O(m \log^3 m)$).

Let $(\mathcal{R}, \mathcal{B})$ denote one of the canonical pairs of sets of triangles produced in this case, where $\mathcal{R}$ consists of 'small' triangles and $\mathcal{B}$ of 'large' triangles. The orientations of the edges of all the trapezoids Δ^*, for $\Delta \in \mathcal{B}$, are fixed, so the intersection $\tau = \bigcap_{\Delta \in \mathcal{B}} \Delta^*$ is a nonempty trapezoid whose sides have the same three fixed orientations. We show in the full version that all triangles $\Delta \in \mathcal{B}$ that are disjoint from τ can be discarded, without losing any intersecting pair of triangles. Let $\Delta \in \mathcal{B}$ be a triangle that intersects τ. Since two of the edges of Δ have the same slopes as the non-parallel edges of τ and since they do not intersect the interior of τ, it is easily verified that Δ contains one of the vertices of τ. Hence, by splitting $\mathcal{B}$ into at most four subcollections, we can guarantee that each subcollection has the common point property.

To sum up, after performing the procedures described above, we end up with a collection of pairs $\{(\mathcal{R}, \mathcal{B})\}$ of sets of triangles, so that the union of the sets $\mathcal{R} \times \mathcal{B}$ contains all intersecting pairs of $\mathcal{L}' \times \mathcal{H}''$, the total size of all these sets is $O(m \log^3 m)$, and for each pair $(\mathcal{R}, \mathcal{B})$, the set $\mathcal{B}$ has the common point property. We repeat the same procedure for the pair $\mathcal{L}''$ and $\mathcal{H}'$. Taking into account the recursive handling of $\mathcal{L}' \cup \mathcal{L}''$ and of $\mathcal{H}' \cup \mathcal{H}''$, we obtain an overall collection of pairs of sets that covers all intersecting pairs of triangles in $\mathcal{C}' \times \mathcal{C}''$, where the total size of all these sets is $O(n \log^4 n)$. This completes the processing for the pair $\mathcal{C}', \mathcal{C}''$, and we repeat this procedure for the (constant number of) other pairs of subfamilies.

3.2 The Second Stage

We now describe the implementation of the second stage of the algorithm. Let $(\mathcal{R}, \mathcal{B})$ be one of the canonical pairs of sets of triangles produced in the first stage. Since $\mathcal{B}$ has the common point property, we can fully sort it, according to $\prec$, in $O(n \log n)$ time. Next we follow the scheme described in the preceding section. Each subproblem that arises there is of the following type: We have two subsets $\mathcal{B}_v \subseteq \mathcal{B}$, $\mathcal{R}_v \subseteq \mathcal{R}$, and our goal is to find all triangles $\Delta \in \mathcal{R}_v$ that intersect the union U_v of the triangles in $\mathcal{B}_v$, and for each such triangle Δ, to report a witness triangle of $\mathcal{B}_v$ that intersects Δ. Since $\mathcal{B}$ has the common point property, U_v is a star-shaped polygon, with respect to the common point z, whose complexity is linear in $n_v = |\mathcal{B}_v|$ (this latter property follows from the fatness of the triangles, as is easily checked); U_v can be easily constructed in time $O(n_v \log n_v)$. (More details concerning this construction can be found in the following section.)

We can thus preprocess ∂U_v, in $O(n_v)$ time, into a data structure of linear size, for logarithmic-time ray-shooting queries, as in [3, 4, 11]. Then, for each $\Delta \in \mathcal{R}_v$, we pick a vertex w of Δ, and test, in $O(\log n_v)$ time, whether $w \in U_v$; this can be done by a binary search with the orientation of zw through the list of orientations of the segments connecting z to the vertices of U_v. If this is the case, we report that Δ intersects U_v, and the above binary search also easily yields a triangle in $\mathcal{B}_v$ containing w (e.g., we can take the triangle whose boundary appears along ∂U_v in the direction zw from z). Otherwise, w lies outside U_v, and we perform three ray shooting queries along the sides of Δ (where the first two are with the sides adjacent to w), to determine whether Δ intersects U_v; if an intersection is found then a witness triangle can also be easily produced.

Hence the whole procedure at node v of the tree takes time $O((|\mathcal{R}_v| + |\mathcal{B}_v|) \log |\mathcal{B}_v|)$, and, summing this over all nodes of the tree, the total running time of the second stage is $O((|\mathcal{R}| + |\mathcal{B}|)\log^2 |\mathcal{B}|)$. Putting everything together, we have shown:

Theorem 5. *The 2-dimensional linear-extension problem for a collection of n fat triangles can be solved in time $O(n\log^6 n)$. Hence, within the same time bound, we can also compute a depth order (if it exists) for a collection of n non-intersecting α-fat triangles in 3-space, for any $\alpha > 1$, (with the constant of proportionality depending on α).*

4   General Fat Objects

In this section we apply our technique to a collection $\mathcal{C} = \{c_1, \ldots, c_n\}$ of n convex, simply-shaped α-fat objects with the α-ratio property, as defined in the introduction. For each $c_i \in \mathcal{C}$, let s_i^+ be a smallest axis-parallel square containing c_i and let s_i^- be a largest axis-parallel square contained in c_i; denote the edge lengths of s_i^+, s_i^-, respectively, as ρ_i^+, ρ_i^-. By assumption, we have $\rho_i^+ \leq \alpha\rho_j^-$, for any pair of (not necessarily distinct) objects $c_i, c_j \in \mathcal{C}$.

Put $\mathcal{S}^+ = \{s_1^+, \ldots, s_n^+\}$. Clearly, for any pair of objects $c_i, c_j \in \mathcal{C}$, $c_i \cap c_j \neq \emptyset$ implies that $s_i^+ \cap s_j^+ \neq \emptyset$. Moreover, two axis-parallel squares intersect if and only if a vertex of one of them lies inside the other. This observation allows us, using orthogonal range searching techniques as in the preceding section, to obtain all pairs of intersecting squares in $\mathcal{S}^+$ as a collection of complete bipartite graphs $\{\mathcal{S}_i' \times \mathcal{S}_i''\}$, where the total size of their vertex sets is $O(n\log^2 n)$, and where one set of each such pair has the common point property (all squares in that set contain a vertex of some square in the second set of the pair). This representation can be constructed in time $O(n\log^2 n)$.

Now replace each such pair $(\mathcal{S}_i', \mathcal{S}_i'')$ by the corresponding pair $(\mathcal{R}_i, \mathcal{B}_i)$ of subsets of $\mathcal{C}$. Property (i) of Section 2 clearly holds for the new collection, but property (ii) (the common point property) may fail to hold. This can be enforced as follows. Let p be a point that lies in, say $\bigcap \mathcal{S}_i''$. Then all the squares in $\mathcal{S}_i''$, and thus all the objects in $\mathcal{B}_i$, are contained in an axis-parallel square s centered at p whose edge length is $2\rho_{max}$ (where $\rho_{max} = \max_j \rho_j^+$). Partition s into a constant number of small squares of edge length less than $\rho_{min}/2$, where $\rho_{min} = \min_j \rho_j^-$. This induces a partitioning of $\mathcal{B}_i$ into a constant number of subsets $\mathcal{B}_{i,1}, \ldots, \mathcal{B}_{i,l}$, so that each of these subsets has the common point property, since all the objects in such a subset contain entirely one of those small squares. We replace the pair $(\mathcal{R}_i, \mathcal{B}_i)$ by the l pairs $(\mathcal{R}_i, \mathcal{B}_{i,1}), \ldots, (\mathcal{R}_i, \mathcal{B}_{i,l})$. Note that this increases the overall size of the vertex sets of our bipartite graphs by only a constant factor.

We next consider the implementation of the second stage. As described in Section 2, we need to solve efficiently the following subproblem: We are given two sets, $\mathcal{R}$, $\mathcal{B}$, of objects of $\mathcal{C}$, so that $\mathcal{B}$ has the common point property. For each $r \in \mathcal{R}$, we want to determine whether it intersects any object in $\mathcal{B}$, and, if so, find a witness $b \in \mathcal{B}$ that intersects r. Put $p = |\mathcal{R}|$, $q = |\mathcal{B}|$.

We form the union U of all objects in $\mathcal{B}$. Since all these objects contain a common point, z, and are convex, we can represent each $b \in \mathcal{B}$ as a function

$f_b = f_b(\theta)$, in polar coordinates about z, where $f_b(\theta)$ is the distance from z to the point of intersection of ∂b with the ray emerging from z in direction θ. Then the boundary of the union U is the graph, in polar coordinates, of the upper envelope $f_U = f_U(\theta) = \max_{b \in \mathcal{B}} f_b(\theta)$ of these functions. If s is the maximum number of intersections between the boundaries of any pair of objects in $\mathcal{B}$, then the number of connected portions of the boundaries of the objects of $\mathcal{B}$ which constitute the boundary of U is $O(\lambda_s(q))$, where $\lambda_s(n)$ is the maximum length of (n, s) Davenport-Schinzel sequences, and, under an appropriate model of computation, f_U (and thus U) can be computed in time $O(\lambda_s(q) \log q)$. (Note that here we did not make use of the fact that the given objects are fat; we only needed the common point property.)

We next partition $\mathcal{R}$ into $p/\sqrt{\lambda_s(q)}$ subsets, each of size at most $\sqrt{\lambda_s(q)}$. For each subset $\mathcal{R}^{(j)}$, we perform a line-sweeping procedure on the collection of all arcs constituting ∂U and all arcs constituting the boundaries ∂r of the objects $r \in \mathcal{R}^{(j)}$. The goal of this procedure is to detect intersections between ∂U and the boundaries ∂r, and also to detect containments of boundaries ∂r within U. When the sweep encounters an intersection between two boundaries ∂r, $\partial r'$, it processes this intersection in a standard manner (there are only $O(\lambda_s(q))$ such intersections). However, when the sweep reaches an intersection between an arc of ∂U and an arc of ∂r, or when it reaches the leftmost point of an arc of ∂r and finds that it lies inside the union U, the algorithm first reports that r intersects the union, and finds a witness object of $\mathcal{B}$ that r intersects (as in the case of fat triangles described in Section 3), but then it removes the arcs of ∂r from the collection of arcs being processed. This guarantees that the total number of events processed by the sweep is only $O(\lambda_s(q))$, and the whole procedure thus takes $O(\lambda_s(q) \log q)$ time. We repeat this procedure for all sets $\mathcal{R}^{(j)}$, and thus solve this subproblem of the second stage in time $O((p/\sqrt{\lambda_s(q)} + 1)\lambda_s(q) \log q)$. (Similar ideas were used in [1].)

Recall that in the second stage of the algorithm we need to solve the above subproblem for each pair of subsets $(\mathcal{R}_v, \mathcal{B}_v)$ at each node v of the tree constructed for each of the pairs $(\mathcal{R}_i, \mathcal{B}_i)$ generated in the first stage. Summing the cost of the above procedure over all these subproblems, we conclude that the overall cost of stage 2 is

$$\sum_{(\mathcal{R}_i, \mathcal{B}_i)} O((|\mathcal{R}_i|/\sqrt{\lambda_s(|\mathcal{B}_i|)} + 1)\lambda_s(|\mathcal{B}_i|) \log^2 n) = O(n\lambda_s^{1/2}(n) \log^4 n) .$$

Theorem 6. *The 2-dimensional linear-extension problem for a collection of n convex, simply-shaped, α-fat objects with the α-ratio property, can be solved, in an appropriate model of computation, in time $O(n\lambda_s^{1/2}(n) \log^4 n)$, where s is the maximum number of intersections between the boundaries of any pair of objects in $\mathcal{C}$. Hence, within the same time bound, we can also compute a depth order (if it exists) for a collection of n non-intersecting convex objects in 3-space, whose xy-projections have the above properties.*

Remark. We also consider a related problem. Given a collection of objects in 3-space as in Theorem 6, preprocess it for *vertical ray shooting* queries, that is, preprocess it so that for a given query point, the object lying immediately under it (if exists) can be found efficiently. We present a solution with $O(n \log^a n)$ preprocessing time and $O(\log^b n)$ query time, where a and b are small constants. Thus this problem seems to be easier than the problem of computing a depth order.

References

1. P.K. Agarwal and M. Sharir, Red-blue intersection detection algorithms, with applications to motion planning and collision detection, *SIAM J. Computing* 19 (1990), 297–321.

2. H. Alt, R. Fleischer, M. Kaufmann, K. Mehlhorn, S. Näher, S. Schirra and C. Uhrig, Approximate motion planning and the complexity of the boundary of the union of simple geometric figures, *Algorithmica* 8 (1992), 391–406.

3. B. Chazelle, H. Edelsbrunner, M. Grigni, L. Guibas, J. Hershberger, M. Sharir and J. Snoeyink, Ray shooting in polygons using geodesic triangulations, to appear in *Algorithmica*.

4. B. Chazelle and L. Guibas, Visibility and intersection problems in plane geometry, *Discrete Comput. Geom.* 4 (1989), 551–581.

5. M. de Berg, *Ray Shooting, Depth Orders and Hidden Surface Removal*, Lecture Notes in Computer Science, vol. 703, Springer-Verlag, 1993.

6. M. de Berg, D. Halperin, M.H. Overmars, J. Snoeyink and M. van Kreveld, Efficient ray shooting and hidden surface removal, *Proc. 7th ACM Symp. on Computational Geometry*, 1991, 21–30.

7. M. de Berg, M. Overmars and O. Schwarzkopf, Computing and verifying depth orders, *SIAM J. Computing* 23 (1994), 437–446.

8. A. Efrat, G. Rote and M. Sharir, On the union of fat wedges and separating a collection of segments by a line, *Comp. Geom. Theory and Appls* 3 (1993), 277–288.

9. S. Hart and M. Sharir, Nonlinearity of Davenport–Schinzel sequences and of generalized path compression schemes, *Combinatorica* 6 (2) (1986), 151–177.

10. D. Hearn and M.P. Baker, *Computer Graphics*, Prentice-Hall International, 1986.

11. J. Hershberger and S. Suri, A pedestrian approach to ray shooting: Shoot a ray, take a walk, *Proc. 4th ACM-SIAM Symp. Discrete Algorithms*, 1993, 54–63.

12. M.J. Katz, M.H. Overmars, M. Sharir, Efficient hidden surface removal for objects with small union size, *Comp. Geom. Theory and Appls* 2 (1992), 223–234.

13. J. Matoušek, J. Pach, M. Sharir, S. Sifrony and E. Welzl, Fat triangles determine linearly many holes, *SIAM J. Computing* 23 (1994), 154–169.

14. K. Mehlhorn, *Data Structures and Algorithms 3: Multidimensional Searching and Computational Geometry*, Springer-Verlag, Berlin, 1984.

15. N. Miller and M. Sharir, Efficient randomized algorithms for constructing the union of fat triangles and of pseudodiscs, manuscript, 1991.

16. M. van Kreveld, On fat partitioning, fat covering, and the union size of polygons, *Proc. 3rd Workshop on Algorithms and Data Structures*, 1993, 452–463.

Selection in Monotone Matrices and Computing k^{th} Nearest Neighbors *

Pankaj K. Agarwal[†] Sandeep Sen[‡]

Abstract

We present an $O((m + n)\sqrt{n} \log n)$ time algorithm to select the kth smallest item from an $m \times n$ totally monotone matrix for any $k \le mn$. This is the first subquadratic algorithm for selecting an item from a totally monotone matrix. Our method also yields an algorithm for *generalized row selection* in monotone matrices of the same time complexity. Given a set $S = \{p_1, \ldots, p_n\}$ of n points in convex position and a vector $\mathbf{k} = \{k_1, \ldots, k_n\}$, we also present an $O(n^{4/3} \log^{O(1)} n)$ algorithm to compute the k_i^{th} nearest neighbor of p_i for every $i \le n$; c is an appropriate constant. This algorithm is considerably faster than the one based on a row-selection algorithm for monotone matrices. If the points of S are arbitrary, then the k_i^{th} nearest neighbor of p_i, for all $i \le n$, can be computed in time $O(n^{7/5} \log^c n)$, which also improves upon the previously best-known result.

1 Introduction

An $m \times n$ matrix $\mathcal{A} = (a_{ij})$, $1 \le i \le m$ and $1 \le j \le n$, is called a *totally monotone* matrix if for all i_1, i_2, j_1, j_2, satisfying $1 \le i_1 < i_2 \le m, 1 \le j_1 < j_2 \le n$,

$$a_{i_1 j_1} < a_{i_1 j_2} \implies a_{i_2 j_1} < a_{i_2 j_2}. \tag{1}$$

Totally monotone matrices were originally introduced by Aggarwal et al. [4]. They presented an $O(m+n)$ time algorithm for computing the leftmost maximal element in each row of an $m \times n$ totally monotone matrix, and applied it to a number of problems in computational geometry and VLSI routing, including all-farthest neighbors of convex polygons. Since the publication of their paper, several new applications of monotone matrices have been discovered, e.g., dynamic programming, largest area (or perimeter) triangle in a set of points, economic

---

*Pankaj Agarwal has been supported by an NYI award and National Science Foundation Grant CCR–93–01259.

†Department of Computer Science, Duke University, PO Box 90129, Durham, NC 27708-0129, USA.

‡Department of Computer Science, Indian Institute of Technology, New Delhi, INDIA.

lot problems, etc.; see [8, 5, 6, 7, 9, 20, 21] for some of these applications.[1]

Most of the papers cited above, however, consider only the problem of computing the maximal or minimal elements of each row. Surprisingly, very little is known about more general selection problems for monotone matrices. There are two natural selection problems for monotone matrices.

(i) *Array selection:* Given an $m \times n$ totally monotone matrix $\mathcal{A}$, and a positive integer $k \leq mn$, find the kth smallest element, x_k, of $\mathcal{A}$, i.e.,

$$|\{a_{ij} \mid a_{ij} < x_k, a_{ij} \in \mathcal{A}\}| < k \quad \text{and} \quad |\{a_{ij} \mid a_{ij} \leq x_k, a_{ij} \in \mathcal{A}\}| \geq k.$$

(ii) *Row selection:* Given an $m \times n$ totally monotone matrix $\mathcal{A}$ and a positive integer $k \leq n$, find the kth smallest element of each row of $\mathcal{A}$.

One can, of course, use a well-known linear time selection algorithm [12] to solve both of these problems in $O(mn)$ time, so the challenge is to develop an $o(mn)$ time algorithm by exploiting the properties of monotone matrices. Selecting the kth-smallest element from a set possessing certain properties has been studied by several researchers [18, 19]. Frederickson and Johnson [18] presented an $O(m \log(2n/m))$ time algorithm to select the kth-smallest item from an $m \times n$ matrix each of whose rows and columns is sorted in nondecreasing order. Kravets and Park [20] showed that for an $m \times n$ totally monotone matrix, whose transpose is also totally monotone, the kth smallest element can be computed in time $O(m + n + k \log(mn/k))$. Their algorithm is optimal for $k = O(1)$, but its running time is $\Omega(mn)$ when k is large. Moreover, their algorithm does not work if $\mathcal{A}^T$ is not totally monotone.

Kravets and Park also presented an $O(k(m + n))$ time algorithm for the row selection problem. Again, their problem is quite inefficient for large values of k. Recently Mansour et al. [21] developed another algorithm whose running time is $O(n\sqrt{m \log n} + m)$. A drawback of both of these algorithms is that they require the value of k to be the same for all rows of $\mathcal{A}$. One can formulate a generalized row-selection problem as follows: Given an $m \times n$ totally monotone matrix $\mathcal{A}$ and a vector $\mathbf{k} = \langle k_1, \ldots, k_m \rangle$ of length m, where $0 < k_i \leq n$, compute the k_i^{th}-smallest element of the ith row of $\mathcal{A}$, for all $1 \leq i \leq m$. Alon and Azar [10] proved that $O(n\sqrt{m} \log n \sqrt{\log m})$ comparisons are sufficient to solve the generalized row selection problem, but their approach falls short of giving an efficient algorithm.

In this paper we present an $O((m + n)\sqrt{n} \log n)$ time algorithm for the array selection problem. Unlike Kraverts-Park algorithm, it does not require $\mathcal{A}^T$ to be totally monotone. As far as we know, this is the first subquadratic algorithm for the array selection problem. A natural variant of our algorithm solves the generalized row selection problem also in time $O((m + n)\sqrt{n} \log n)$.

---

[1]In order to obtain an $o(mn)$ algorithm, we have to assume that $\mathcal{A}$ is not represented explicitly. We assume that any particular entry $a_{ij} \in \mathcal{A}$ can be computed in constant time. If the time required to compute a_{ij}, for a given pair i, j, is $t(m, n)$, then we have to multiply the running time of the algorithm by a factor of $t(m, n)$.

In the second part of this paper, we consider the *all k^{th}-nearest neighbor* problem for a convex set: Given a set S of n points in the plane in convex position and a parameter $k \leq n$, find the kth nearest neighbor of every point in S. As shown in [20, 21], this problem can be reduced to the row selection problem. However, there are known algorithms which solve the all k^{th}-nearest neighbor problem for an arbitrary set of points in the plane directly. Using k^{th}-order Voronoi diagrams, the kth-nearest neighbor of every point in an arbitrary planar point set can be computed in time $O(k^{3/2}n \log n)$. The running time has been improved by Vaidya [25] to $O(nk \log n)$, and recently by Callahan and Kosaraju [13] to $O(n \log n + nk)$. An undesirable feature of these algorithms is that their running time depends on k, and is $\Omega(n^2)$ for $k = \Omega(n)$. Agarwal and Matoušek presented an $O(n^{3/2+\varepsilon})$ time algorithm for the all k^{th}-nearest neighbor problem, for any $\varepsilon > 0$ [2].

Here, we present an $O(n^{4/3} \log^c n)$ time algorithm, for some constant $c > 0$, for the all-k^{th}-nearest neighbor problem of a convex set. In fact, we consider a more general problem. We show that an arbitrary set S of n points and a convex polygon P in the plane can be preprocessed in time $O(n^2 \log n)$ into a data structure of size $O(n^2)$, so that the kth nearest neighbor of a query point p lying on ∂P, the boundary of P, can be computed in time $O(\log n)$. This data structure can also count the number of points of S lying in a disk whose center lies on P. (If the center of a query disk lies anywhere in the plane, then the best known data structure with $O(\log n)$ query time requires $O(n^3)$ space.) Combining this data structure with the range searching data structure described in [3], we can preprocess S into a data structure of size $O(s)$, for any s ($n \leq s \leq n^2$), so that the kth-nearest neighbor query, as described above, can be answered in time $O(\frac{n}{\sqrt{s}} \log^c n)$ time, where $c \geq 1$ is some constant. By setting $s = n^{4/3}$ and querying the data structure with each point of S, the all k^{th}-nearest for neighbor problem points for a convex set can be solved in time $O(n^{4/3} \log^c n)$. For an arbitrary set of points in the plane, the running time is $O(n^{7/5} \log^c n)$.

2 Selection in Monotone Matrices

In this section we present algorithms for the array selection and the row selection problems. Let $\mathcal{A} = (a_{ij})$, $1 \leq i \leq m, 1 \leq j \leq n$, be an $m \times n$ totally monotone matrix, and let $k \leq mn$ be a positive integer. The algorithm works in two steps. In the first step, it partitions $\mathcal{A}$ into a family $\mathcal{L} = \{L_1, \dots, L_t\}$ of $t = O(m\sqrt{n})$ lists, such that each list $L_i \in \mathcal{L}$ satisfies the following two properties.

(i) All elements of L_i belong to the same row. Let r_i denote the row of $\mathcal{A}$ containing the elements of L_i.

(ii) The elements of each L_i are in nondecreasing order.

After computing the family $\mathcal{L}$, it selects the kth smallest element among the elements of lists of $\mathcal{L}$ in time $O(|\mathcal{L}| \log n) = O(m\sqrt{n} \log n)$, using a simplified version of the Frederickson-Johnson algorithm [18]. We now describe each of the two steps in detail.

2.1 Computing the family of lists

We first describe the algorithm for computing the family $\mathcal{L}$. We also compute a collection $\Pi = \{\pi_1, \ldots, \pi_s\}$ of arrays, which we refer to as *index-arrays*. Each list $L_i \in \mathcal{L}$ is represented implicitly, as follows. We store the value of r_i, the row that contains L_i, and a pointer to an array $\pi_j \in \Pi$ that contains the indices of elements in L_i; the size of π_j is the same as that of L_i, and that the lth item of π_j is k if the lth item of L_i is $a_{r_i k}$. We will refer to π_j as $\varphi(L_i)$.

We will use the following properties of monotone matrices, which are quite easy to prove; see, e.g., [20, 21] for proofs.

Lemma 2.1 *(i) If $\mathcal{A}$ is a totally monotone matrix, then any submatrix of $\mathcal{A}$ is also a totally monotone matrix.*
(ii) For any three integers i, j, k, satisfying $1 \leq i \leq m$, $1 \leq j < k \leq n$,
   (a) $a_{ij} \geq a_{ik} \Longrightarrow a_{i'j} \geq a_{i'k}$ for all $i' \leq i$, and
   (b) $a_{ij} \leq a_{ik} \Longrightarrow a_{i'j} \leq a_{i'k}$ for all $i' \geq i$.

We also use a result of Erdős and Szekeres [17]. Let $U = \langle u_1, u_2, \ldots, u_m \rangle$ be a sequence of real numbers. A subsequence $U' = \langle u_{i_1}, u_{i_2}, \ldots, u_{i_k} \rangle$ of U is called *monotone* if either $u_{i_1} \leq u_{i_2} \leq \cdots \leq u_{i_k}$ or $u_{i_1} \geq u_{i_2} \geq \cdots \geq u_{i_k}$.

Lemma 2.2 (Erdős-Szekeres) *Given a sequence U of n real numbers, there always exists a monotone subsequence U of length at least $\lceil \sqrt{n} \rceil$.*

Dijkstra [15] gave a simple $O(n \log n)$ time algorithm for computing the longest monotone subsequence U. Recently, Bar Yehuda and Fogel [11] presented an $O(n^{3/2})$ algorithm for partitioning U into a collection of at most $2\sqrt{n}$ monotone subsequences.

We are now in position to describe the algorithm for computing $\mathcal{L}$. At each step of the algorithm we have a $u \times v$ submatrix M of $\mathcal{A}$; M consists of a contiguous block of u rows of $\mathcal{A}$, say $\alpha+1, \ldots, \alpha+u$, and a subsequence $C(M)$ of columns of $\mathcal{A}$, i.e., $M = (a_{ij})$, $\alpha+1 \leq i \leq \alpha+u, j \in C(M)$. By Lemma 2.1 (i), M is a totally monotone matrix. Let $\mathcal{L}' = \{L_1, \ldots, L_\xi\}$ and $\Pi' = \{\pi_1, \ldots, \pi_\zeta\}$ be the families of lists and index-arrays, respectively, that we have computed so far. If M consists of a single column, i.e., $C(M) = \langle z \rangle$, then we set $\Pi' = \Pi' \cup \{\pi_{\zeta+1}\}$ and $\mathcal{L}' = \mathcal{L}' \cup \{L_{\xi+1}, \ldots, L_{\xi+u}\}$, where $\pi_{\zeta+1} = \langle z \rangle$ and, for $1 \leq i \leq u$, $L_{\xi+i} = \langle a_{(\alpha+i)z} \rangle$, $r_{\xi+i} = \alpha+i$, $\varphi(L_{\xi+i}) = \pi_{\zeta+1}$.

Next, assume that $v > 1$ (M consists of more than one column). Set $w = \lceil u/2 \rceil$, and let $U = \langle a_{(\alpha+w)j} \mid j \in C(M) \rangle$ be the wth row of M. Using Dijkstra's algorithm, we compute the longest monotone subsequence $\sigma = \langle a_{(\alpha+w)j_1}, a_{(\alpha+w)j_2}, \ldots, a_{(\alpha+w)j_k} \rangle$ of U. Suppose σ is a monotonic non-increasing subsequence, then for all x, $\alpha+1 \leq x \leq \alpha+w$, the subsequence $\langle a_{xj_1}, \ldots, a_{xj_k} \rangle$ is also monotonic non-increasing (cf. Lemma 2.1 (ii)). We set $\Pi' = \Pi' \cup \{\pi_{\zeta+1}\}$ and $\mathcal{L}' = \mathcal{L}' \cup \{L_{\xi+1}, \ldots, L_{\xi+w}\}$, where $\pi_{\zeta+1} = \langle j_k, j_{k-1}, \ldots, j_1 \rangle$, and $L_{\xi+i} =$

$\langle a_{(\alpha+i)j_k}, a_{(\alpha+i)j_{k-1}}, \ldots, a_{(\alpha+i)j_1} \rangle$, $r_{\xi+i} = \alpha + i$ and $\varphi(L_{\xi+i}) = \pi_{\zeta+1}$ for $1 \leq i \leq w$.

We decompose the remaining entries of M, the ones that do not belong to any list L_i, into two matrices M_1, M_2, and recursively decompose them into sorted lists. $M_1 = (a_{ij})$ is composed of the bottom $u - w$ rows of M and all the columns of $\mathcal{A}$, i.e., $\alpha + w + 1 \leq i \leq \alpha + u$ and $C(M_1) = C(M)$. $M_2 = (a_{ij})$ is composed of the top w rows of M and those columns of M that are not in σ, i.e., $\alpha + 1 \leq i \leq \alpha + w$ and $C(M_2) = C(M) - \langle j_1, \ldots, j_k \rangle$.

The case when σ is monotonical increasing subsequence can be handled similarly. Next, let us analyze the size of $\mathcal{L}$ and the total time spent by the algorithm. Let $\psi(u, v)$ be the maximum number of lists generated by the above algorithm. By Lemma 2.2, the longest subsequence has length at least $\lceil \sqrt{v} \rceil$, so we get the following recurrence

$$\psi(u, v) \leq \psi\left(\left\lfloor \frac{u}{2} \right\rfloor, v\right) + \psi\left(\left\lceil \frac{u}{2} \right\rceil, v - \lceil \sqrt{v} \rceil\right) + \left\lceil \frac{u}{2} \right\rceil.$$

The solution of this recurrence is $\psi(u, v) \leq 2u\sqrt{v}$. Since we spend $O(v \log v)$ time in computing the monotone subsequence and $O(u)$ time in generating the lists plus the two submatrices M_1, M_2, we can prove that the total time spent in computing $\mathcal{L}$ is $O(u\sqrt{v} + v^{3/2} \log v \log u)$, which can be improved to $O(u\sqrt{v} + v^{3/2} \log \log v)$ using the algorithm of Bar Yehuda and Fogel [11]; see the full version for details. Hence, we can conclude

Lemma 2.3 *Any $m \times n$ totally monotone matrix can be decomposed in time $O(m\sqrt{n} + n^{3/2} \log \log n)$ into $O(m\sqrt{n})$ non-decreasing lists.*

2.2 Selecting the kth element

Let $\mathcal{L} = \{L_1, \ldots, L_t\}$ be a set of t lists, each sorted in nondecreasing order, and let us assume that, for any $i \leq t, j \leq |L_i|$, the jth item of L_i can be obtained in $O(1)$ time. We present an algorithm, which is a simplified version of the algorithm by Fredrickson and Johnson [18] for selecting the kth smallest item in $\bigcup_{1 \leq i \leq t} L_i$.

Let $p = \lceil \log_2(\max_{L \in \mathcal{L}} |L|) \rceil \leq \lceil \log_2 n \rceil$. Without loss of generality assume that each list $L_i \in \mathcal{L}$ has exactly 2^p elements; if $|L_i| \leq 2^p$, then we assume that $2^p - |L_i|$ ∞'s are padded at the end of L_i. The algorithm works in p phases. In the beginning of the ith phase, we have a set Λ of at most $2t$ lists, each of size $m_i = 2^{p-i}$ (including the padded ∞'s), and a parameter $s \leq k$. Each list in Λ is a contiguous sublist of some list in $\mathcal{L}$, and the sth-element of $\bigcup \Lambda$ is the same as the kth-element of $\bigcup \mathcal{L}$. Initially, $\Lambda = \mathcal{L}$ and $s = k$. The goal is to find the sth-element in the set $\bigcup \Lambda$.

In the ith step we first divide each list $\lambda \in \Lambda$ into two lists λ_1, λ_2; λ_1 consists of the first $m_i/2$ elements of Λ, and λ_2 consists of the next $m_i/2$ elements of Λ. Next, we discard some of the lists that are guaranteed not to contain the sth element of $\bigcup \Lambda$.

Recall that we are not storing the lists of Λ explicitly, so the divide-step is somewhat involved. For each list $\lambda \in \Lambda$, we store the following information (i)

The list $L_j \in \mathcal{L}$ that contains λ, and (ii) the index of the first item of λ in L_j. When we split λ into λ_1, λ_2, this information for λ_1, λ_2 can be computed in $O(1)$ time. With a slight abuse of notation, let Λ also denote the set of lists after the divide-step.

The prune-step works in two stages. Let $\alpha = \lceil s/m_i \rceil + t$, and let $\lambda(i)$ denote the ith-element of λ. We select the αth smallest element, x_α, from the set $A = \{\lambda(1) \mid \lambda \in \Lambda\}$; A can be obtained in $O(|\Lambda|)$ time and x_α can be computed in another $O(|\Lambda|)$ time. If $\alpha < |\Lambda|$, we discard all $|\Lambda| - \alpha$ lists of Λ whose minimum item is not smaller than x_α. Next, let $\beta = \lfloor s/m_i \rfloor - t$. If $\beta > 0$, we select the βth element, x_β, from the set $B = \{\lambda(m_i) \mid \lambda \in \Lambda\}$, and discard all the lists whose maximum element is not larger than x_β. Set $s = s - \beta m_i$.

After the pth step each list in Λ has at most one item. We then use the linear time selection algorithm to find the sth item in another $O(t)$ time. Notice that after each phase, at most $2t$ lists are retained in Λ, so the running time is obviously $O(tp) = O(m\sqrt{n}\log n)$. The correctness of the algorithm follows from the fact that in the ith step, the sth element of $\bigcup \Lambda$ cannot lie in one of the discarded lists; see the original paper [18] for the proof. Plugging this algorithm to Lemma 2.3, we can conclude

Theorem 2.4 *Let $\mathcal{A}$ be an $m \times n$ totally monotone matrix and let $k \leq mn$ be some positive integer. The kth smallest item of $\mathcal{A}$ can be computed in time $O(m\sqrt{n}\log n + n^{3/2}\log\log n)$.*

We can use the same idea to solve the generalized row selection problem: Let $\mathbf{k} = \langle k_1, k_2, \ldots, k_m \rangle$ be a vector of length m, where each $k_i \leq n$ is a positive integer. The goal is to select the k_i^{th}-smallest item from the ith row of $\mathcal{A}$. Let $\mathcal{L}_i = \{L_j \mid r_j = i\}$ be the set of lists belonging to the ith row of $\mathcal{A}$. We run the above selection algorithm for each $\mathcal{L}_i$ separately and select the k_i^{th} item of $\mathcal{L}_i$. The running time for each $\mathcal{L}_i$ is $O(|\mathcal{L}_i|\log n)$, thereby implying that the overall running time is again $O(m\sqrt{n}\log n + n^{3/2}\log\log n)$. Hence, we have

Theorem 2.5 *Let $\mathcal{A}$ be an $m \times n$ totally monotone matrix and let $\mathbf{k}$ be a vector of length m as defined above. The k_i^{th} smallest item of the ith row of $\mathcal{A}$, for all $1 \leq i \leq n$, can be computed in time $O(m\sqrt{n}\log n + n^{3/2}\log\log n)$.*

3   Computing All k^{th}-Nearest Neighbors

Let S be a set of n points in the plane in convex position and $\mathbf{k} = \langle k_1, k_2, \ldots, k_n \rangle$ be a vector of length n, where each $k_i \leq n$ is a positive integer. The goal is to compute the k_i^{th} nearest neighbor of p_i for $i \leq n$. As mentioned in the Introduction, we present an algorithm for the kth-nearest neighbor searching: Let S be a set of n points and P a convex polygon in the plane. Preprocess S and P into a data structure so that, for a query point $p \in \partial P$ and a positive integer $k \leq n$, we can compute the kth-nearest neighbor of p in S efficiently. We first present a data structure for the circular range searching — given a disk D whose center lies on ∂P, compute $|D \cap S|$ — and then apply parametric searching to answer a kth-nearest neighbor query.

3.1 Circular range searching

In this subsection we show that if the center of the query disk always lies on the boundary of a convex polygon P, then a circular range query can be answered in $O(\log n)$ time using only $O(n^2)$ space. We then combine it with the data structure of Agarwal and Matoušek [3] to obtain a space/query-time tradeoff.

Data structure with $O(\log n)$ query time: Let S be a set of n points in the plane. For a pair of points $p_i, p_j \in S$, let ℓ_{ij} denote the perpendicular bisector of p_i and p_j, and let $\mathcal{L} = \{\ell_{ij} \mid 1 \leq i < j \leq n\}$. For a point q in the plane, let $\sigma(q)$ denote the sequence of points of S sorted in nondecreasing order of their distances from q, i.e., if $d(q, p_{i_1}) \leq d(q, p_{i_2}) \leq \cdots \leq d(q, p_{i_n})$, then $\sigma(q) = \langle p_{i_1}, p_{i_2}, \ldots, p_{i_n} \rangle$. We will refer to $\sigma(q)$ as the *distance-ordering* of S with respect to q. It can be easily verified that, for any two points q, q' lying in the same face of $\mathcal{A}(\mathcal{L})$, $\sigma(q) = \sigma(q')$. Abusing the notation slightly, let $\sigma(f)$ denote the distance-ordering with respect to any point in f.

Lemma 3.1 *If f and f' are two adjacent faces in $\mathcal{A}(\mathcal{L})$ separated by the line ℓ_{kl}, then p_k, p_l are adjacent in $\sigma(f)$ and $\sigma(f')$. Moreover, if $\sigma(f) = \langle p_{i_1}, \ldots, p_{i_m}, p_k, p_l, p_{i_{m+3}}, \ldots, p_{i_n} \rangle$, then $\sigma(f') = \langle p_{i_1}, \ldots, p_{i_m}, p_l, p_k, p_{i_{m+3}}, \ldots, p_{i_n} \rangle$.*

If we know the face f of $\mathcal{A}(\mathcal{L})$ that contains the center of a disk D and the distance ordering $\sigma(f)$, then $|S \cap D|$ can be computed by a binary search on $\sigma(f)$ — find the first element in $\sigma(f)$ whose distance from the center of D is more than the radius of D. However, we cannot afford to store the entire arrangement $\mathcal{A}(\mathcal{L})$, because there are $\Omega(n^4)$ faces in $\mathcal{A}(\mathcal{L})$. This is where we use the fact that the center of D lies on ∂P.

Partition ∂P into maximal connected intervals that do not intersect any line of $\mathcal{L}$, so that each interval lies within a face of $\mathcal{A}(\mathcal{L})$. Let $\mathcal{I} = \langle I_1, \ldots, I_m \rangle$ be a sequence of the resulting intervals sorted along ∂P in clockwise direction; $m = O(n^2)$, for any line intersects ∂P in at most two points. Since each interval I_j lies within a single face of $\mathcal{A}(\mathcal{L})$, the distance ordering of S with respect to all points of I_j remains the same, and let $\sigma(I_j)$ denote this ordering. We store each $\sigma(I_j)$ in a binary search tree T_j, whose i^{th} leftmost node v stores the index i, the rank of v, and the ith element of $\sigma(I_j)$. (Instead of storing i, we can also store the number of nodes in the subtree rooted at i.) For a query disk D, whose center lies on ∂P, $|S \cap D|$ can be computed in a straight-forward manner by traversing a path of T_j. Since $\sigma(I_j)$ and $\sigma(I_{j+1})$ differ only in two adjacent positions, and $\sigma(I_{j+1})$ can be obtained from $\sigma(I_j)$ by swapping these two adjacent elements, we can use a persistent data structure to store all T_j's [24]. However, we cannot use the update procedures described in [24] directly, because we are storing the rank of v (or the number of nodes in the subtree rooted at v) at each node v of T_j; see [24] a more detailed discussion. To circumvent this problem we will use a slight variant of their data structure, tailored to our application. Since the insertions and deletions are not arbitrary in our case, the update procedures are somewhat simpler than the ones described in [24].

Each node in the persistent data structure, described by Sarnak and Tarjan [24], has an extra pointer (other than the standard 'left' and 'right' pointers). Whenever we want to modify the information stored at any node v, we make a new copy v' of v and store the updated information there. If v is the root of the tree, we are done. Otherwise, let v' be the parent of v. If the extra pointer of v' is free, it now points to v'. If the extra pointer of v' is not free, v', too, is copied. Thus, copying propagates through successive ancestors until the root is copied or a node with the free extra pointer is reached; see [24] for details.

We now describe how to modify the update procedures to suit our application. Let $\mathcal{D}$ denote the overall data structure; $\mathcal{D}$ is a dag which implicitly stores all T_j's in the sense that T_j is a subgraph of $\mathcal{D}$, and that, given a node $v \in T_j$, we can find the children of v in T_j in $O(1)$ time. We also maintain an array A of length m whose j^{th} entry stores a pointer to a root of $\mathcal{D}$ that corresponds to the root of T_j. Each node $v \in \mathcal{D}$ stores the following information.

(i) A point of S, denoted as $val(v)$,

(ii) an integer, called $rank(v)$, which is defined below.

(iii) three pointers — *left, right* and *extra*. Initially, when v is created, *extra* (v) pointer is free. As the algorithm progresses, the extra pointer is assigned to one of its children.

(iv) An integer, denoted as $ver(v)$, which is 0 if the extra pointer of v is free; $ver(v) = j$, if the extra pointer was assigned while processing I_j.

(v) a bit, denoted as $\xi(v)$; $\xi(v) = 0$ if the extra pointer is assigned to the left child of v, and $\xi(v) = 1$ if the extra pointer is assigned as the right child of v.

$\mathcal{D}$ is constructed incrementally by processing the intervals of $\mathcal{I}$ one by one. Let $\mathcal{D}_j$ denote the data structures storing $T_1, \ldots, T_j$. $\mathcal{D}_1 = T_1$ is a minimum height binary tree storing $\sigma(I_1)$. The ith leftmost node v_i of $\mathcal{D}_1$ stores the ith element of $\sigma(I_1)$, $rank(v_i) = i$, and $ver(v) = 0$. Roughly speaking, $\mathcal{D}_{j+1}$ can be obtained from $\mathcal{D}_j$ as follows: let ℓ_{kl} be the line separating I_j and I_{j+1}, and let u, v be the nodes corresponding to T_j that store p_k, p_l, respectively. We make new copies u', v' of u and v, and store p_k at v' and p_l at u' (i.e., swap p_k and p_l), and propagate the copying of nodes to the ancestors of u and v, as sketched above. We omit the details from this version.

Let D be a query disk with center $c \in \partial P$ and radius r. We compute $|D \cap S|$ as follows. We first find in $O(\log n)$ time the interval $I_j \in \mathcal{I}$ that contains c; $A[j]$ gives the pointer to the root z of T_j. We search in $\mathcal{D}$ with the triple (c, r, j), starting from z. Suppose we are at node v. Let p_k be the point stored at v. If $d(c, p_k) = r$, we return $rank(v)$. If $d(c, p_k) > r$ (resp. $d(c, p_k) < r$), we descend to the right (resp. left) child of v in T_j. If v does not have the right (resp. left) child, then we return $rank(v)$ (resp. $rank(v) - 1$). The left and right children of v in T_j can be computed in $O(1)$ time using $ver(v), \xi(v)$, and the pointers stored at v. The total query time is thus $O(\log n)$ time. Notice that the same procedure can also compute the kth-nearest neighbor of a point $p \in P$.

Theorem 3.2 *Let S be a set of n points in the plane and let P be a convex polygon. S and P can be preprocessed in time $O(n^2 \log n)$ into a data structure of size $O(n^2)$, so that the number of points lying a query disk whose center lies on ∂P can be counted in $O(\log n)$ time. The same data structure can also compute the kth-nearest neighbor of a query point $p \in P$ in time $O(\log n)$.*

Space/query-time tradeoff: We can obtain a space/query-time tradeoff by combining Theorem 3.2 with the data structure proposed by Agarwal and Matoušek [3]. They construct a partition tree T on S each of whose node v is associated with a subset $S_v \subseteq S$ and a 'pseudo-trapezoid' τ_v such that $S_v \subseteq \tau_v$; τ_v has at most four sides — the left and right sides are vertical segments and the top and bottom sides are portions of quadratic curves (see [3] for details). The degree of each node is r for some constant r. If $w_1, \ldots, w_r$ are the children of v, then $\max_{1 \le i \le r} |S_{w_i}| \le 2|S_v|/r$ and $\bigcup_{i=1}^r S_{w_i} = S_v$, so the height of T is $O(\log n)$. If v is a leaf, then $|S_v| = O(1)$. We say that a circle C crosses a node v if τ_v intersects C and $\tau_v \not\subseteq C$. T has the property that every circle crosses at most $O(\sqrt{r} \log^{3/2} r)$ children of any node of T. At each leaf v we store S_v itself and at each internal node $v \in T$ we store $|S_v|$ and τ_v.

Let $n \le s \le n^2$ be some fixed parameter. Suppose we want to construct a data structure of size s. Delete all those nodes of T whose parents are associated with less than $\lceil rs/n \rceil$ points. Let T' denote the resulting tree. Each leaf of T' contains at most s/n points. For each leaf w of T', we preprocess S_w using Theorem 3.2 and replace the resulting structure with w. Let Ψ denote the overall data structure. We will refer to T' as the 'top-structure' and to the structures stored at the leaves of T' as the 'bottom-structures'. Let L be the set of leaves of T'. Then for any $v \in L$ $|S_v| \le s/n$ and $\sum_{v \in L} |S_v| \le n$. The total size of the data structure is at most $O(n) + \sum_{v \in L} |S_v|^2 = O(s)$. The preprocessing time is easily seen to be $O(s \log n)$.

Let D be a query disk with center c and radius r. We visit Ψ in a top-down fashion, starting from the root. We maintain a global variable *count*, which is initially set to 0. At each step, we visit a node v. If v is leaf of T', we compute $|D \cap S_v|$ using the algorithm of Theorem 3.2, and add this quantity to *count*. Otherwise, we do the following. If $\tau_v \cap D = \emptyset$, we ignore v. If $\tau_v \subseteq D$, we add $|S_v|$ to *count* and do not visit any children of v. Finally, if D crosses τ_v, we recursively visit all the children of v. It can be shown that the maximum query time is $O(n^{1+\varepsilon}/\sqrt{s})$ which can be improved to $O(\frac{n}{\sqrt{s}} \log^c n)$, for some constant $c > 0$, by choosing $r = n^\delta$, for some sufficiently small $\delta > 0$. We leave out the details from here. Hence, we can conclude

Theorem 3.3 *Let S be a set of n points in the plane, let P be a convex polygon, and let $n \le s \le n^2$ be a parameter. S can be preprocessed in time $O(s \log n)$ into a data structure of size $O(s)$, so that the number of points lying a query disk whose center lies on ∂P can be counted in $O(\frac{n}{\sqrt{s}} \log^c n)$ time for some constant $c > 0$.*

As mentioned in the Introduction, if the center of a query disk lies anywhere in the plane, then $|D \cap S|$ can be computed in time $O(\log n)$ using $O(n^3)$ space. Plugging this data structure, instead of Theorem 3.2, into T', we obtain

Theorem 3.4 *Let S be a set of n points in the plane, and let $n \leq s \leq n^3$ be a parameter. S can be preprocessed in time $O(s \log n)$ into a data structure of size $O(s)$, so that the number of points lying a query disk can be counted in $O((n^3/s)^{1/4} \log^c n)$ time, for some constant $c > 0$.*

3.2 Applying parametric searching

We now apply the parametric searching technique, due to Megiddo [23], to the above data structure for answering queries of the following form: Given a point $q \in P$ and a parameter $k \leq n$, determine its kth nearest neighbor, $\varphi_k(q)$, in S. The basic idea is the same as described in [2], but we have to modify their technique a little.

Let $r_k = d(q, \varphi_k(p))$, and let D_k be the disk of radius r_k centered at q. We will query the data structure with D_k. We, of course, do not know the value of r_k, so we will simulate the query answering procedure, described above, without knowing the exact value of r_k.

We search through Ψ level-by-level. Let V_i be the set of nodes that we visit in the ith step. If a node $v \in V_i$ belongs to the bottom structure, we need to determine the relation between r_k and $d(q, p(v))$ in order to determine whether the left or the right child of v has to be visited in the next step; let $r_v = d(q, p(v))$. If v is a node of the top structure, then we want to determine whether D_k crosses τ_v, or D_k contains τ_v, or D_k is disjoint from τ_v. Let $r_v^- = \min_{p \in \tau_v} d(q, p)$ and $r_v^+ = \max_{p \in \tau_v} d(q, p)$. D_k crosses τ_v if and only if $r_v^- < r_k < r_v^+$, and D_k contains τ_v if and only if $r_v^+ \leq r_k$. The numbers r_v^-, r_v^+ can be computed in $O(1)$ time. Let

$$R_i = \{r_v^-, r_v^+ \mid v \in V_i \text{ and } v \text{ is a node of the top structure}\} \cup$$
$$\{r_v \mid v \in V_i \text{ and } v \text{ is a node of a bottom structure}\}.$$

We sort R_i and, by a binary search on the sorted list, we compute ρ_i, the largest element of R_i which is less than or equal to r_k. Each step of the binary search computes $|D(q, r) \cap S|$ for some $r \in R_i$. If $|D(q, r) \cap S| < k$, then $r < r_k$. If $|D(q, r) \cap S| = k$ and ∂D contains a point of S, $r = r_k$. Otherwise $r > r_k$. By Theorem 3.3, each step of the binary search requires $O(\frac{n}{\sqrt{m}} \log^c n)$ time. If $\rho_i = r_k$, we already know the value of r_k, so we stop right away. Next, assume that $\rho_i < r_k$. For a node $v \in V_i$ of the bottom structure, if $r_v \leq \rho_i$ (resp. $r_v > \rho_i$), i.e., $r_v < r_k$ (resp. $r_v > r_k$), then we visit the right (resp. left) child of v in the next step. For a node $v \in V_i$ of the top structure, if $r_v^- > \rho_i$ (i.e., $r_v^- > r_k$), then τ_v and D_k are disjoint, so we ignore v. If $r_v^+ \leq \rho_i$ (i.e., $r_v^+ < r_k$), $\tau_v \subseteq D_k$, so we add $|S_v|$ to *count*. If none of these two conditions are satisfied, then D_k crosses v and we visit all the children of v in the next step. By repeating this step for all nodes in V_i, we have at our disposal all the nodes that we have to visit in the next step. It is known that during the simulation of the algorithm, the outcome of one of the oracle calls (computation of $|D(q,r)| \cap S|$) will be $r = r_k$, so the algorithm will return the value of r_k before completing the simulation. Since Ψ has $O(\log n)$ levels, and we spend $O(\frac{n}{\sqrt{s}} \log^c n)$ time at each level, we can conclude

Theorem 3.5 *Let S be a set of n points in the plane, let P be a convex polygon as defined above, and let $n \le s \le n^2$ be a parameter. S can be preprocessed in time $O(s \log n)$ into a data structure of size $O(s)$, so that the kth nearest neighbor for any point $q \in \partial P$ can be counted in $O(\frac{n}{s} \log^c n)$ time.*

Going back to the problem of computing the k_i^{th} nearest neighbor of p_i, we set $s = n^{4/3}$, preprocess S as in Theorem 3.5, and query it with p_i, k_i for each $1 \le i \le n$.

Theorem 3.6 *Let S be a set of points in the plane in convex position, and let $\mathbf{k} = \langle k_1, \ldots, k_n \rangle$ be a vector of length n as defined earlier. Then the k_i^{th} nearest neighbor of p_i, for all $i \le n$, can be computed in time $O(n^{4/3} \log^c n)$. If the points in S are arbitrary, then the running time is $O(n^{7/5} \log^c n)$.*

References

[1] P.K. Agarwal, B. Aronov, M. Sharir and S. Suri, Selecting distances in the plane, *Algorithmica* 9 (1993), 495–514.

[2] P.K. Agarwal and J. Matoušek, Ray shooting and parametric search, *SIAM J. Computing* 22 (1993), 794–806.

[3] P.K. Agarwal and J. Matoušek, Range searching with semi-algebraic sets, *Proc. 17th Symp. Mathematical Foundations of Computer Science*, (Lecture Notes in Computer Science 629), Springer-Verlag 1992, pp. 1–13. (Also to appear in *Discrete and Computational Geometry*.)

[4] A. Aggarwal, M. Klawe, S. Moran, P. Shor, and R. Wilber, Geometric applications of a matrix-searching algorithm, *Algorithmica* 2 (1987), 195–208.

[5] A. Aggarwal and J. Park, Parallel searching in multidimensional monotone arrays, to appear in *J. Algorithms*.

[6] A. Aggarwal and J. Park, Sequential searching in multidimensional monotone arrays, to appear in *J. Algorithms*.

[7] A. Aggarwal and J. Park, Improved algorithm for economic lot size problems, *Operations Research* 41 (1993), 549–571.

[8] A. Aggarwal, D. Kravets, J. Park, and S. Sen, Parallel searching in generalized Monge arrays with applications, *Proc. 2nd ACM Symp. Parallel Algorithms and Architectures*, 1990, pp. 259–268.

[9] A. Aggarwal and S. Suri, Computing the longest diagonal of a simple polygon, *Information Processing Letters* 35 (1990), 13–18.

[10] N. Alon and Y. Azar, Comparison-sorting and selecting in totally monotone matrices, *Proceedings 3rd Annual ACM-SIAM Symposium on Discrete Algorithms*, 1992, pp. 403–408.

[11] R. Bar Yehuda and S. Fogel, Variations on ray shooting, *Algorithmica* 11 (1994), 133–145.

[12] M. Blum, R. Floyd, V. Pratt, R. Rivest, and R. Tarjan, Time bounds for selection, *J. Computer and Systems Sciences* 7 (1973), 448–461.

[13] P. Callahan and S. Kosaraju, Faster algorithms for some geometric graph problems in higher dimensions, *Proceedings 4th Annual ACM-SIAM Symposium on Discrete Algorithms*, 1993, pp. 291–300.

[14] B. Chazelle, R. Cole, F. Preparata and C. Yap, New upper bounds for neighbor searching, *Information and Control* 68 (1986), 105–124.

[15] E. Dijkstra, *A Discipline of Programming*, Prentice-Hall, Englewood Cliff, NJ, 1976.

[16] H. Edelsbrunner, *Algorithms in Combinatorial Geometry*, Springer-Verlag, Berlin, 1987.

[17] P. Erdős and G. Szekeres, A combinatorial problem in geometry, *Compositio Math.* 2 (1935), 463–470.

[18] G. Frederickson and D. Johnson, Generalized selection and ranking: sorted matrices, *SIAM J. Computing* 13 (1984), 14–30.

[19] D. Johnson and T. Mitzoguchi, Selecting the kth element in $X + Y$ and $X_1 + X_2 + \cdots + X_m$, *SIAM J. Computing* 7 (1978), 147–153.

[20] D. Kravets and J. Park, Selection and sorting in totally monotone arrays, *Mathematical Systems Theory* 24 (1991), 201–220.

[21] Y. Mansour, J. Park, B. Schieber and S. Sen, Improved selection in totally monotone arrays, *International J. of Computational Geometry and Applications* 3 (1993), 115–132.

[22] J. Matoušek and E. Welzl, Good splitters for counting points in triangles, *J. Algorithms* 13 (1992), 307–319.

[23] N. Megiddo, Applying parallel computation algorithms in the design of serial algorithms, *J. ACM* 30 (1983), 852–865.

[24] N. Sarnak, and R. E. Tarjan, Planar point location using persistent search trees, *Comm. ACM* 29 (1986), 609–679.

[25] P. Vaidya, An $O(n \log n)$ algorithm for the all-nearest-neighbors problem, *Discrete and Computational Geometry* 4 (1989), 101–115.

New On-Line Algorithms for the Page Replication Problem

Susanne Albers*   Hisashi Koga**

Abstract. We present improved competitive on-line algorithms for the page replication problem and concentrate on important network topologies for which algorithms with a constant competitive factor can be given. We develop an optimal randomized on-line replication algorithm for trees and uniform networks; its competitive factor is approximately 1.58. Furthermore we consider on-line replication algorithms for rings and present general techniques that transform large classes of c-competitive algorithms for trees into $2c$-competitive algorithms for rings. As a result we obtain a randomized on-line algorithm for rings that is 3.16-competitive. We also derive two 4-competitive on-line algorithms for rings which are either deterministic or memoryless.

1 Introduction

This paper deals with problems that arise in the memory management of large multiprocessor systems. Such multiprocessing environments typically consist of a network of processors, each of which has its local memory. A global shared memory is modeled by distributing the physical pages among the local memories. Accesses to the global memory are then accomplished by accessing the local memories. Suppose a processor p wants to read a memory address from page A. If A is stored in p's local memory, then this read operation can be accomplished locally. Otherwise, p determines a processor q holding the page and sends a request to q. The desired information is then transmitted from q to p, and the communication cost incurred thereby is proportional to the distance from q to p. If p has to access page A frequently, it may be worthwhile to move or copy A from q to p because subsequent accesses will become cheaper. However, transmitting an entire page incurs a high communication cost proportional to the page size times the distance from q to p.

If a page is writable, it is reasonable to store only one copy of the page in the entire system. This avoids the problem of keeping multiple copies of the page consistent. The *migration problem* is to decide in which local memory the single copy of the page should be stored so that a sequence of memory accesses can be processed at low cost. On the other hand, if a page is read-only, it is possible to keep several copies of the page in the system, i.e. a page may be copied from one local memory to another. In the *replication problem* we have to determine which local memories should contain copies of the page. Finding efficient migration

---

* Max-Planck-Institut für Informatik, 66123 Saarbrücken, Germany. E-mail: albers@mpi-sb.mpg.de.
** Department of Information Science, The University of Tokyo, Tokyo 113, Japan. E-mail: nwa@is.s.u-tokyo.ac.jp.

and replication strategies is an important problem that has been studied from a practical and theoretical point of view [DF82, SD89, BS89, BFR92, W92, ABF93, CLRW93, K93]. In this paper we study on-line algorithms for the page replication problem. We analyze the performance of on-line algorithms using *competitive analysis* [ST85], the worst case ratio of cost incurred by an on-line algorithm and the cost incurred by an optimal off-line algorithm.

Awerbuch *et al.* [ABF93] have presented a deterministic on-line replication strategy for general graphs that achieves an optimal competitive ratio of $O(\log n)$, where n is the number processors. However, for many important topologies, this bound is not very expressive. Black and Sleator [BS89] have proposed an optimal deterministic on-line algorithm for trees and uniform networks which is 2-competitive. A uniform network is a complete graph in which all edges have the same length. Recently Koga [K93] has developed a randomized replication algorithm for trees that is 1.71-competitive against oblivious adversaries. Bartal *et al.* [BFR92] have presented a randomized $2(2 + \sqrt{3})$-competitive replication algorithm for the case that the network topology forms a ring. The competitive ratio holds against adaptive on-line adversaries. Koga has proposed a randomized algorithm for rings which is 4-competitive against adaptive on-line adversaries. However, both algorithms for the ring use a large amount of randomness, namely one random number for each read operation. Using the 4-comptitive algorithm by Koga, one can construct a deterministic 16-competitive replication algorithm for the ring, see [BBKTW94]. However, that algorithm is very complicated and not useful in practical applications.

In this paper we develop a number of new deterministic and randomized on-line replication algorithms. We concentrate on network topologies that are important in practice and for which on-line algorithms with a constant competitive factor can be developed. In Section 4 we present a randomized on-line replication algorithm for trees and uniform networks, called GEOMETRIC, which is $(\frac{\rho^r}{\rho^r-1})$-competitive. Here $\rho = \frac{r+1}{r}$ and r is the page size factor. For large values of r, which occur in practice, GEOMETRIC's competitiveness is approximately $\frac{e}{e-1} \approx 1.58$. We also show that GEOMETRIC is optimal. Specifically we prove that no randomized on-line replication algorithm can be better than $(\frac{\rho^r}{\rho^r-1})$-competitive. Interestingly, our algorithm GEOMETRIC uses only one random number during an initialization phase and runs completely deterministically thereafter. Lund *et al.* [LRWY94] have independently developed the same results for trees and uniform networks. In Section 5 we consider replication algorithms for rings. We present a deterministic technique that transforms a large class of c-competitive algorithms for trees into $2c$-competitive algorithms for rings. As a result we obtain a randomized $(\frac{2\rho^r}{\rho^r-1})$-competitive algorithm for rings. Note that the competitive ratio is approximately 3.16 and beats the previously best ratio of 4. We also derive two 4-competitive algorithms for rings which are either deterministic or memoryless. Our 4-competitive deterministic algorithm is very simple and improves the competitive factor of 16 mentioned above. Finally we present a randomized version of our deterministic technique for constructing ring algorithms and prove that this variant achieves the same performance. All our randomized competitive factors hold against the oblivious adversary.

2 Problem statement and competitive analysis

Formally, the page replication problem can be described as follows. We are given an undirected graph G. Each node in G corresponds to a processor and the edges represent the interconnection network. Associated with each edge is a *length* that is equal to the distance between the connected processors. We assume that the edge lengths satisfy the triangle inequality. In the page replication problem we generally concentrate on one particular page. We say that *a node v has the page* if the page is contained in v's local memory. A *request at a node v* occurs if v wants to read an address from the page. The request can be satisfied at zero cost if v has the page. Otherwise the request is served by accessing a node w holding the page and the incurred cost equals the distance from v to w. Immediately after a request at a node v, the page may be replicated into v's local memory. The cost incurred by this replication is r times the distance from v to w. Here r denotes the page size factor. In practical applications, r is a large value, usually several hundred or thousand. We study the page replication problem under the assumption that a node having the page never drops it. A page replication algorithm is usually presented with an entire sequence of requests that must be served with low total cost. A page replication algorithm is *on-line* if it serves every request without knowledge of any future requests.

We analyze the performance of on-line page replication algorithms using *competitive analysis* [ST85]. In a competitive analysis, the cost incurred by an on-line algorithm is compared to the cost incurred by an *optimal off-line algorithm*. An optimal off-line algorithm knows the entire request sequence in advance and can serve it with minimum cost. Let $C_A(\sigma)$ and $C_{OPT}(\sigma)$ be the cost of the on-line algorithm A and the optimal off-line algorithm OPT on request sequence σ. In the context of page replication an on-line algorithm A is called *c-competitive* if $C_A(\sigma) \leq c \cdot C_{OPT}(\sigma)$ for all request sequences σ. If A is a randomized algorithm, then $C_A(\sigma)$ must be replaced by the expected cost incurred by A, where the expectation is taken over the random choices made by A. In this paper we evaluate randomized on-line algorithms only against the *oblivious adversary*, see [BBKTW94]. The oblivious adversary has to generate a request sequence in advance and is not allowed to see the random choices made by the on-line algorithm. The *competitive factor* of an on-line algorithm A is the infimum of all c such that A is c-competitive.

3 Basic definitions and techniques

A substantial part of this paper deals with on-line replication algorithms for trees. Even when considering uniform networks and rings, we will reduce the algorithms and their analyses to the case that the underlying topology forms a tree. For this reason we introduce some basic definitions for trees. The root of the given tree is generally denoted by s. We assume that initially, only s has the page. Consider an undirected edge $e = \{v, w\}$ in the tree. The node in $\{v, w\}$ that is farther away from the root is called the *child node* of e. The length of e is denoted by $l(e)$.

In the following we will always assume that if an algorithm (on-line or off-line) replicates the page from a node v to a node w, then the page is also replicated to all nodes on the path from v to w. This does not incur extra cost. Thus, the nodes with the page always form a connected component of the given tree. Note that if a node v does not have the page, then the closest node w with the page lies on the path from v to the root, and all paths from v to a node with the page pass through w. Therefore, we may assume without loss of generality that a replication algorithm always serves requests at a node not holding the page by accessing the closest node with the page. This cannot increase the total cost incurred in serving the whole request sequence.

We present a technique that we will frequently use to analyze on-line replication algorithms for trees. Let T be a tree and σ be a request sequence for T. We usually analyze an on-line replication algorithm A by partitioning the cost that is incurred by A and by OPT into parts that are incurred by each edge of the tree. Suppose an algorithm serves a request at a node v. Then an edge e incurs a cost equal to the length of e if e belongs to the path from v to the closest node with the page. If e does not belong to that path, then e incurs a cost of zero. An edge also incurs the cost of a replication across it. Let $C_A(\sigma, e)$ denote the cost that is incurred by edge e when A serves σ. Analogously, let $C_{OPT}(\sigma, e)$ be the cost that is incurred by e when OPT serves σ. (If A is a randomized algorithm, then $C_A(\sigma, e)$ is the expected cost incurred by e.) The performance of an on-line algorithm A is generally evaluated by comparing $C_A(\sigma, e)$ to $C_{OPT}(\sigma, e)$ for all edges e of the tree. In order to analyze $C_A(\sigma, e)$, we use some notation. Let $\sigma = \sigma(1), \sigma(2), \ldots, \sigma(m)$ be a request sequence of length m and let $\sigma(t)$, $1 \leq t \leq m$, be the request at time t. Suppose $\sigma(t)$ is a request at node v. We set

$$a_\sigma(e, \sigma(t)) = 1$$

if e belongs to the path from v to the root. Otherwise we set

$$a_\sigma(e, \sigma(t)) = 0.$$

If $a_\sigma(e, \sigma(t)) = 1$, we say that $\sigma(t)$ causes an access at edge e. Let

$$a_\sigma(e) = \sum_{t=1}^{m} a_\sigma(e, \sigma(t)),$$

i.e. $a_\sigma(e)$ is the number of requests that cause an access at edge e. The following simple lemma is crucial in our analyses.

Lemma 1. *Let A be an on-line replication algorithm that, given an arbitrary tree T and a request sequence σ for T, satisfies*

$$C_A(\sigma, e) \leq c \cdot \min\{a_\sigma(e), r\} \cdot l(e) \tag{1}$$

for all edges e. Then the algorithm A is c-competitive. (Again, if A is a randomized algorithm, then $C_A(\sigma, e)$ is the expected cost incurred by e.)

Proof. We prove that for any edge e, $C_{OPT}(\sigma, e) = \min\{a_\sigma, r\} \cdot l(e)$. By inequality (1), this implies $C_A(\sigma, e) \leq c \cdot C_{OPT}(\sigma, e)$ for all edges e, and hence A is c-competitive. If $a_\sigma(e) < r$, then OPT does not replicate the page across e and e incurs a cost of $a_\sigma(e)l(e)$. Hence $C_{OPT}(\sigma, e) = a_\sigma(e) \cdot l(e) = \min\{a_\sigma, r\}l(e)$. On the other hand, if $a_\sigma(e) \geq r$, then OPT replicates the page across e, and e incurs a cost of $rl(e)$. Thus $C_{OPT}(\sigma, e) = r \cdot l(e) = \min\{a_\sigma, r\}l(e)$. $\square$

4 An optimal algorithm for trees and uniform networks

First we describe and analyze a randomized on-line algorithm for trees. This algorithm can be applied to uniform networks, too. Then we prove that the competitive factor of our algorithm is optimal for all values of r. Throughout this section let $\rho = \frac{r+1}{r}$.

Algorithm GEOMETRIC (for trees): The algorithm first chooses a random number from the set $\{1, 2, \ldots, r\}$. Specifically, the number i is chosen with probability $p_i = \alpha \cdot \rho^{i-1}$, where $\alpha = \frac{\rho-1}{\rho^r-1}$. While processing the request sequence, the algorithm maintains a count on each edge of the tree. Initially, all counts are set to 0. If there is a request at a node v that does not have the page, then all counts along the path from v to the closest node with the page are incremented by 1. When a count reaches the value of the randomly chosen number, the page is replicated to the child node of the corresponding edge.

Before we analyze the performance of GEOMETRIC, we mention a few observations and remarks. The algorithm is called GEOMETRIC because $p_{i+1}/p_i = \rho$ is constant for all $i = 1, 2, \ldots, r-1$. It is easy to verify that $\sum_{i=1}^{r} p_i = 1$. Suppose that GEOMETRIC processes a request sequence σ. We can easily prove by induction on the number of requests processed so far that the counts on a path from the root to a node v are monotonically non-increasing. Furthermore, after each request, a node has the page if and only if it is the child node of an edge whose count is equal to the value of the randomly chosen number.

Theorem 2. *For any tree, the algorithm GEOMETRIC is $(\frac{\rho^r}{\rho^r-1})$-competitive.*

Note that $\frac{\rho^r}{\rho^r-1}$ goes to $\frac{e}{e-1} \approx 1.58$ as r tends to infinity. Furthermore, GEOMETRIC uses only one random number during an initialization phase and runs completely deterministically thereafter. The proof of the theorem follows from Lemma 1 and Lemma 3 below. For a given tree T and an arbitrary request sequence σ on T, let $E[C_G(\sigma, e)]$ denote the expected cost incurred by edge e when GEOMETRIC serves σ.

Lemma 3. *Given any tree T and request sequence σ for T, the algorithm GEOMETRIC satisfies*

$$E[C_G(\sigma, e)] \le \left(\frac{\rho^r}{\rho^r - 1}\right) \cdot \min\{a_\sigma(e), r\} \cdot l(e)$$

for all edges e of T.

Proof. Consider an arbitrary tree T and a request sequence σ for T. Let e be an edge of the tree. Furthermore, let $k = a_\sigma(e)$ and $\sigma(t_1), \sigma(t_2), \ldots, \sigma(t_k)$ be the requests that cause an access at the edge e. Note that the algorithm GEOMETRIC increases the count of e exactly at the requests $\sigma(t_1), \sigma(t_2), \ldots, \sigma(t_k)$, provided that the page has not been replicated across e so far.

First, assume that $k > r$. Since $\sum_{i=1}^{r} p_i = 1$, GEOMETRIC has replicated the page across e before request $\sigma(t_{r+1})$. Thus the edge e incurs the same cost as if we had $k = r$. For this reason it suffices to consider the case that k satisfies $1 \le k \le r$ and show $E[C_G(\sigma, e)] \le c \cdot k \cdot l(e)$, where $c = \frac{\rho^r}{\rho^r-1}$. This proves the lemma.

So suppose we have $1 \leq k \leq r$. The algorithm GEOMETRIC first chooses a random number i from the set $\{1, 2, \ldots, r\}$. If i satisfies $i \leq k$, the edge e incurs a cost of $r + i$. Otherwise e incurs a cost of k. Thus

$$E[C_G(\sigma, e)] = l(e)(\sum_{i=1}^{k}(r+i)p_i + \sum_{i=k+1}^{r} kp_i)$$

$$= l(e)(\sum_{i=1}^{k} r\alpha\rho^{i-1} + \sum_{i=1}^{k} i\alpha\rho^{i-1} + \sum_{i=k+1}^{r} k\alpha\rho^{i-1})$$

$$= \alpha l(e)(\frac{r(\rho^k - 1)}{\rho - 1} + \frac{k\rho^{k+1} - (k+1)\rho^k + 1}{(\rho - 1)^2} + \frac{k(\rho^r - \rho^k)}{\rho - 1}).$$

We have $\rho - 1 = \frac{1}{r}$. Thus

$$E[C_G(\sigma, e)] = \frac{\alpha l(e)}{\rho - 1}(r(\rho^k - 1) + k\rho^k - r(\rho^k - 1) + k(\rho^r - \rho^k))$$

$$= \frac{\alpha l(e)}{\rho - 1}(k\rho^r)$$

$$= \frac{\rho^r}{\rho^r - 1} \cdot k \cdot l(e). \quad \square$$

The algorithm GEOMETRIC is easily applied to uniform networks. Consider an arbitrary uniform network and let s be the node that has the page initially. Since all edges in the graph have the same length, we may assume without loss of generality that a replication algorithm (on-line or off-line) serves requests and replicates the page only along edges $\{s, v\}$. Hence the network can be reduced to a tree by neglecting the edges $\{v, w\}$ with $v \neq s$, $w \neq s$. Run on this tree, the algorithm GEOMETRIC is $(\frac{\rho^r}{\rho^r - 1})$-competitive.

We now prove that GEOMETRIC's competitive factor is optimal for all values of r.

Theorem 4. *Let A be a randomized on-line replication algorithm. Then A cannot be better than $(\frac{\rho^r}{\rho^r - 1})$-competitive, even on a graph consisting of two nodes.*

Proof of Theorem 4: Let s and t be two nodes that are connected by an edge of length 1. We assume that initially, only node s has the page. We will construct a request sequence σ consisting of requests at node t such that the expected cost incurred by A is at least $\frac{\rho^r}{\rho^r - 1}$ times the optimal off-line cost.

For $i = 1, 2, \ldots$, let q_i be the probability that A replicates the pages from s to t after exactly i requests, given a request sequence that consists only of requests at node t. In the following we compare the algorithm A to the algorithm GEOMETRIC. Let $E[C_A(\sigma)]$ and $E[C_G(\sigma)]$ denote the expected cost incurred by A and GEOMETRIC on a request sequence σ. Furthermore, for $i = 1, 2, \ldots, r$, let $p_i = \alpha \cdot \rho^{i-1}$. We consider two cases.

Case 1: There exists an l, where $1 \leq l \leq r$, such that $\sum_{i=1}^{l} q_i \geq \sum_{i=1}^{l} p_i$. Let k be the smallest number satisfying the above inequality, i.e. $\sum_{i=1}^{k} q_i \geq \sum_{i=1}^{k} p_i$ and $\sum_{i=1}^{j} q_i < \sum_{i=1}^{j} p_i$ for all j with $1 \leq j < k$. Let σ be the request sequence that consists of k requests at node t. We show that the inequality $E[C_A(\sigma)] - E[C_G(\sigma)] \geq 0$ holds. This implies $E[C_A(\sigma)] \geq E[C_G(\sigma)] = \frac{\rho^r}{\rho^r-1} \cdot k$ and A cannot be better than $(\frac{\rho^r}{\rho^r-1})$-competitive because the optimal off-line cost on σ equals k. We have

$$E[C_A(\sigma)] = \sum_{i=1}^{k}(r+i)q_i + \sum_{i \geq k+1} kq_i = \sum_{i=1}^{k}(r+i)q_i + k(1 - \sum_{i=1}^{k}q_i)$$

$$E[C_G(\sigma)] = \sum_{i=1}^{k}(r+i)p_i + \sum_{i \geq k+1} kp_i = \sum_{i=1}^{k}(r+i)p_i + k(1 - \sum_{i=1}^{k}p_i).$$

Hence $\quad E[C_A(\sigma)] - E[C_G(\sigma)] = \sum_{i=1}^{k} i(q_i - p_i) + (r-k)\sum_{i=1}^{k}(q_i - p_i).$

Since $\sum_{i=1}^{k}q_i \geq \sum_{i=1}^{k}p_i$ and $r-k \geq 0$, we obtain

$$E[C_A(\sigma)] - E[C_G(\sigma)] \geq \sum_{i=1}^{k} i(q_i - p_i) = \sum_{i=1}^{k}(\sum_{j=i}^{k} q_i - \sum_{j=i}^{k} p_i).$$

For $i = 2, 3, \ldots, k$ we have $\sum_{j=1}^{i-1}q_i < \sum_{j=1}^{i-1}p_i$ and hence $\sum_{j=i}^{k}q_i - \sum_{j=i}^{k}p_i > \sum_{j=1}^{k}q_i - \sum_{j=1}^{k}p_i$. We conclude

$$E[C_A(\sigma)] - E[C_G(\sigma)] \geq \sum_{i=1}^{k}(\sum_{j=i}^{k} q_i - \sum_{j=i}^{k} p_i) \geq \sum_{i=1}^{k}(\sum_{j=1}^{k} q_i - \sum_{j=1}^{k} p_i) \geq 0.$$

Case 2: For all $k = 1, 2, \ldots, r$, the inequality $\sum_{i=1}^{k}q_i < \sum_{i=1}^{k}p_i$ is satisfied. Let σ be the request sequence that consists of $2r$ requests at node t. Let A' be the on-line algorithm with $q_i' = q_i$, for $i = 1, 2, \ldots, r-1$, and $q_r' = \sum_{i \geq r} q_i$. It is easy to prove that $E[C_A(\sigma)] \geq E[C_{A'}(\sigma)]$. Since $\sum_{i=1}^{r}q_i' = \sum_{i=1}^{r}p_i = 1$ and $\sum_{i=1}^{j}q_i' < \sum_{i=1}^{j}p_i$ for all j with $1 \leq j < r$, Case 1 immediately implies $E[C_A(\sigma)] \geq E[C_{A'}(\sigma)] \geq E[C_G(\sigma)] = \frac{\rho^r}{\rho^r-1} \cdot r$, and A cannot be better than $(\frac{\rho^r}{\rho^r-1})$-competitive because the optimal off-line cost equals r. $\square$

5 Algorithms for the ring

In this section we assume that the given net of processors forms a ring. We will present techniques that transform a large class of c-competitive algorithms for trees into $2c$-competitive algorithms for rings.

We assume that initially, only one node of the ring, say s, has the page. Let n be the number of nodes in the ring and let $v_1, v_2, \ldots, v_n$ be the nodes if we scan the ring in clockwise direction starting from s, i.e. $v_1 = s$. For $i = 1, 2, \ldots, n$, let $e_i = \{v_i, v_{i+1}\}$ be the undirected edge from v_i to v_{i+1}. Naturally, v_{n+1} equals v_1. Let x and y be any two points on the ring; x and y need not necessarily be processor nodes. We denote by (x, y) the arc of the ring that is obtained if we start in x and go to y in clockwise direction. Let $l(x, y)$ be the length of the arc (x, y).

Algorithm RING: Let P, $P \neq s$, be the point on the ring satisfying $l(s, P) = l(P, s)$, i.e. P is the point "opposite" to s. The algorithm first cuts the ring at P. It regards the resulting structure as a tree T with root $s = v_1$. The arc (s, P) represents one branch of the tree and the arc (P, s) represents another branch of the tree (see Figure 1). We assume that the point P becomes part of the arc (s, P). This is significant if P coincides with one of the nodes v_i. The algorithm RING then uses an on-line replication algorithm A for trees in order to serve a request sequence σ. That is, RING assumes that σ is a request sequence for T and serves the request sequence using the tree algorithm A.

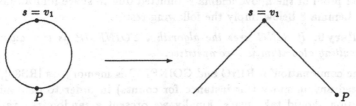

Fig. 1: A cut of the ring

Theorem 5. *Let A be an on-line replication algorithm that, given an arbitrary tree T and a request sequence σ for T, satisfies*

$$C_A(\sigma, e) \leq c \cdot \min\{a_\sigma(e), r\} l(e) \tag{2}$$

for all edges e of the tree. ($C_A(\sigma, e)$ is the expected cost incurred by e if A is a randomized algorithm.) If the algorithm RING uses A as tree algorithm, then the resulting algorithm is $2c$-competitive.

Before we prove this theorem, we mention some important implications. Lemma 3 immediately implies the following result.

Corollary 6. *If RING uses the algorithm GEOMETRIC as tree algorithm, then the resulting algorithm is c-competitive, where $c = \frac{2\rho^r}{\rho^r - 1}$.*

We observe that c goes to $\frac{2e}{e-1} \approx 3.16$ as r tends to infinity. Also note that if RING uses the GEOMETRIC algorithm, then only one random number is used during an initialization phase. Next we consider the deterministic replication algorithm for trees proposed by Black and Sleator [BS89]. The algorithm achieves an optimal competitive factor of 2.

Algorithm DETERMINISTIC_COUNT: The algorithm works in the same way as the algorithm GEOMETRIC. However DETERMINISTIC_COUNT does not choose a random number in order to determine when a replication should occur. Rather it replicates the page to the child node of an edge when the corresponding count reaches r.

It is easy to see that, given an arbitrary tree T and a request sequence σ, $C_{DC}(\sigma, e) \leq 2 \cdot \min\{a_\sigma(e), r\} \cdot l(e)$ for all edges e. Here $C_{DC}(\sigma, e)$ denotes the cost that is incurred by edge e when DETERMINISTIC_COUNT serves σ.

Corollary 7. *If the algorithm RING uses DETERMINISTIC_COUNT as tree algorithm, then the resulting algorithm is 4-competitive.*

We remark that the combination of RING and DETERMINISTIC_COUNT runs completely deterministically. Another interesting on-line replication algorithm for trees was presented by Koga [K93].

Algorithm COINFLIP: If there is a request at a node v without the page, the algorithm serves the request by accessing the closest node u with the page. Then with probability $\frac{1}{r}$, the algorithm replicates the page from u to v.

Lemma 8. *Let T be an arbitrary tree and σ be arbitrary request sequence for T. For any edge e in T, let $E[C_{CF}(\sigma, e)]$ be the expected cost COINFLIP incurs on e. Then $E[C_{CF}(\sigma, e)] \leq 2 \cdot \min\{a_\sigma(e), r\} \cdot l(e)$.*

The proof of the above lemma is omitted due to space limitations. Theorem 5 and Lemma 8 below imply the following result.

Corollary 9. *If RING uses the algorithm COINFLIP as tree algorithm, then the resulting algorithm is 4-competitive.*

The combination of RING and COINFLIP is memoryless [RS89], i.e. it does not need any memory (for instance for counts) in order to determine when a replication should take place. Finally we present a randomized variant of the algorithm RING and a statement analogous to Theorem 5.

Algorithm RING(RANDOM): The algorithm works in the same as the algorithm RING. However, instead of cutting the ring at the point opposite to s, the algorithm RING(RANDOM) chooses a point P uniformly at random on the ring and cuts the ring at that point P.

Theorem 10. *Let A be an on-line replication algorithm that, given an arbitrary tree T and a request sequence σ for T, satisfies*

$$C_A(\sigma, e) \leq c \cdot \min\{a_\sigma(e), r\}l(e) \tag{3}$$

for all edges e of the tree. ($C_A(\sigma, e)$ is the expected cost incurred by e if A is a randomized algorithm.) If the algorithm RING(RANDOM) uses A as tree algorithm, then the resulting algorithm is 2c-competitive.

Theorem 10 implies that statements analogous to Corollaries 1 - 3 hold. Note, however, that a combination of RING(RANDOM) and DETERMINISTIC_COUNT is not a purely deterministic algorithm. It remains to prove the two main theorems. In the following we present a proof of Theorem 5. Theorem 10 can be proved using similar techniques.

Proof of Theorem 5: Let $\sigma = \sigma(1), \sigma(2), \ldots, \sigma(m)$ be a request sequence for the ring. We start with some observations on how OPT serves σ. Consider the state of the ring after OPT has served σ. Let v_a and v_b be the nodes farthest from s to

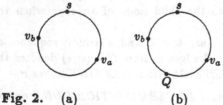

which OPT has replicated the page in clockwise and counter-clockwise direction, respectively. Figure 2(a) illustrates this situation. We may assume without loss of generality that OPT replicates the page from s to

Fig. 2. (a)   (b)

v_a and from s to v_b at the beginning of the request sequence, before any requests are served. This does not incur a higher cost as if the replication is done while requests are processed.

Any request at a node that belongs to (s, v_a) or (v_b, s) can then be served at zero cost. Let Q be the point on (v_a, v_b) which satisfies $l(v_a, Q) = l(Q, v_b)$, see Figure 2(b). Any request at a node v_i that belongs to (v_a, Q) is served by accessing v_a and the incurred cost equals $l(v_a, v_i)$. Any request at a node v_i that belongs to (Q, v_b) is served by accessing v_b, and the incurred cost equals $l(v_i, v_b)$. Let $I = \{1, 2, \ldots, n\}$. Let $\sigma(t_1), \sigma(t_2), \ldots, \sigma(t_k)$ be the requests in σ which request a node that does not belong to (s, v_a) or (v_b, s). For $j = 1, 2, \ldots, k$, let $v_{\mu(t_j)}$ be the node requested by $\sigma(t_j)$. The cost incurred by OPT in serving σ equals

$$C_{OPT}(\sigma) = rl(s, v_a) + rl(v_b, s) + \sum_{j=1}^{k} \min\{l(v_a, v_{\mu(t_j)}), l(v_{\mu(t_j)}, v_b)\}.$$

Note that also OPT implicitly uses a tree in order to serve the requests sequence. This tree is obtained if the ring is cut at the point Q. Let $C_R(\sigma)$ be the cost incurred by RING in serving σ. In the following we show that

$$C_R(\sigma) \leq 2c \cdot C_{OPT}(\sigma). \tag{4}$$

This implies the theorem.

For the analysis of $C_R(\sigma)$ we need some more notation. Let T be the tree that is obtained if the ring is cut at point P. Let $i \in \{1, 2, \ldots, n\}$ and $t \in \{1, 2, \ldots, m\}$ be arbitrary. We denote by $v_{\mu(t)}$ the node requested at time t. We set

$$a_\sigma(e_i, t) = 1$$

if in the tree T, e_i is on the path from $v_{\mu(t)}$ to the root. Otherwise we set $a_\sigma(e_i, t) = 0$. If $a_\sigma(e_i, t) = 1$, we say that $\sigma(t)$ causes an access at the edge e_i in the tree T. We set

$$a_\sigma(e_i) = \sum_{t=1}^{m} a_\sigma(e_i, t).$$

By inequality (2), $RING$ incurs a total cost of

$$C_R(\sigma) = \sum_{j=1}^{n} c \min\{a_\sigma(e_i), r\} l(e_i).$$

In the following we investigate two cases. First we will consider the case that P belongs either to (s, v_a) or to (v_b, s). Then we will study the case that P belongs neither to (s, v_a) nor to (v_b, s).

Case 1: Suppose that P belongs either to (s, v_a) or to (v_b, s), see Figure 3. Then

$$\sum_{i \in I} l(e_i)$$
$$\leq 2 \max\{l(s, v_a), l(v_b, s)\}$$
$$\leq 2(l(s, v_a) + l(v_b, s))$$

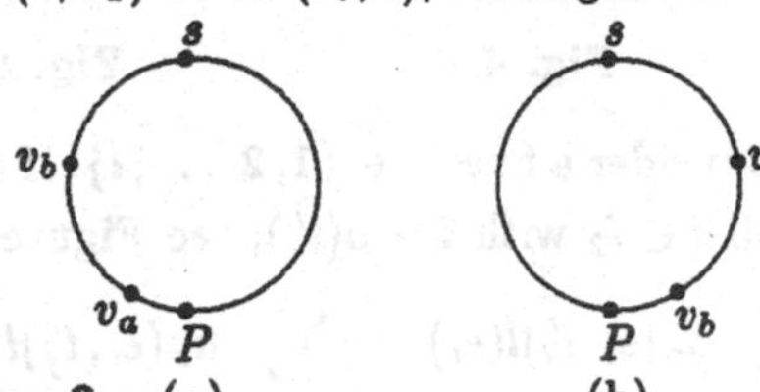

Fig. 3. (a)   (b)

and we can prove inequality (4).

$$C_R(\sigma) \leq \sum_{i \in I} c \min\{a_\sigma(e_i), r\} l(e_i) \leq cr \sum_{i \in I} l(e_i) \leq 2cr(l(s, v_a) + l(v_b, s))$$
$$\leq 2c \cdot C_{OPT}(\sigma)$$

Case 2: Now suppose that P belongs neither to (s, v_a) nor to (v_b, s).
We only consider the case that $l(s, v_a) \geq l(v_b, s)$. The case $l(v_b, s) \geq l(s, v_a)$
is symmetric. Let w be the point on the arc (P, s) such that $l(w, s) = l(s, v_a)$,
see Figure 4. We may assume without loss of generality that w coincides with a
processor node. Define $I_1 = \{i \in I | e_i$ belongs to the arc $(w, v_a)\}$ and $I_2 = \{i \in I | e_i$ belongs to the arc $(v_a, w)\}$. We have

$$C_R(\sigma) = c \sum_{i \in I} \min\{a_\sigma(e_i), r\} l(e_i) \leq c \sum_{i \in I_1} r l(e_i) + c \sum_{i \in I_2} a_\sigma(e_i) l(e_i).$$

Furthermore, $\sum_{i \in I_1} l(e_i) = 2 \cdot l(s, v_a) \leq 2(l(s, v_a) + l(v_b, s))$ and hence

$$C_R(\sigma) \leq 2c(r l(s, v_a) + r l(v_b, s)) + c \sum_{i \in I_2} a_\sigma(e_i) l(e_i)$$

$$\leq 2c(r l(s, v_a) + r l(v_b, s) + \sum_{i \in I_2} a_\sigma(e_i) l(e_i)).$$

In the following we show

$$\sum_{i \in I_2} a_\sigma(e_i) l(e_i) \leq \sum_{j=1}^{k} \min\{l(v_a, v_{\mu(t_j)}), l(v_{\mu(t_j)}, v_b)\}. \tag{5}$$

Note that a request $\sigma(t)$ can only cause an access at an edge e_i, $i \in I_2$, if $v_{\mu(t)}$
belongs to (v_a, w) and $v_{\mu(t)} \neq v_a$, $v_{\mu(t)} \neq w$. Let $\sigma(t_1'), \sigma(t_2'), \ldots, \sigma(t_l')$ be all the
requests in σ satisfying these properties. Then we have

$$\sum_{i \in I_2} a_\sigma(e_i) l(e_i) = \sum_{i \in I_2} \sum_{j=1}^{l} a_\sigma(e_i, t_j') l(e_i)$$

$$= \sum_{j=1}^{l} \left(\sum_{\substack{i \in I_2 \\ i < \mu(t_j')}} a_\sigma(e_i, t_j') l(e_i) + \sum_{\substack{i \in I_2 \\ i \geq \mu(t_j')}} a_\sigma(e_i, t_j') l(e_i) \right).$$

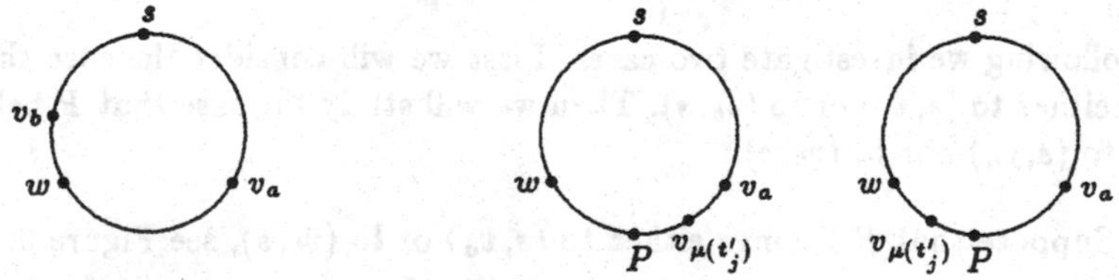

Fig. 4.   **Fig. 5.** (a)   (b)

Consider a fixed $j \in \{1, 2, \ldots, l\}$. If $v_{\mu(t_j')}$ belongs to (s, P), then $a_\sigma(e_i, t_j') = 0$
for all $i \in I_2$ with $i \geq \mu(t_j')$, see Figure 5(a). Thus

$$\sum_{\substack{i \in I_2 \\ i < \mu(t_j')}} a_\sigma(e_i, t_j') l(e_i) + \sum_{\substack{i \in I_2 \\ i \geq \mu(t_j')}} a_\sigma(e_i, t_j') l(e_i) =$$

$$\sum_{\substack{i \in I_2 \\ i < \mu(t_j')}} a_\sigma(e_i, t_j') l(e_i) = \sum_{\substack{i \in I_2 \\ i < \mu(t_j')}} l(e_i) = l(v_a, v_{\mu(t_j')}) \leq \min\{l(v_a, v_{\mu(t_j')}), l(v_{\mu(t_j')}, w)\}.$$

If $v_{\mu(t'_j)}$ belongs to (P, s) and $v_{\mu(t'_j)} \neq P$, then $a_\sigma(e_i, t'_j) = 0$ for all $i \in I_2$ with $i < \mu(t'_j)$, see Figure 5(b). As above we can show

$$\sum_{\substack{i \in I_2 \\ i < \mu(t'_j)}} a_\sigma(e_i, t'_j) l(e_i) + \sum_{\substack{i \in I_2 \\ i \geq \mu(t'_j)}} a_\sigma(e_i, t'_j) l(e_i) \leq \min\{l(v_a, v_{\mu(t'_j)}), l(v_{\mu(t'_j)}, w)\}.$$

We obtain $C_R(\sigma) \leq 2c(rl(s, v_a) + rl(v_b, s) + \sum_{j=1}^l \min\{l(v_a, v_{\mu(t'_j)}), l(v_{\mu(t'_j)}, w)\})$.
Since $l(v_{\mu(t'_j)}, w) \leq l(v_{\mu(t'_j)}, v_b)$ and $\sigma(t'_1), \sigma(t'_2), \ldots, \sigma(t'_l)$ is a subsequence of $\sigma(t_1), \sigma(t_2), \ldots, \sigma(t_k)$ we have

$$C_R(\sigma) \leq 2c(rl(s, v_a) + rl(v_b, s) + \sum_{j=1}^k \min\{l(v_a, v_{\mu(t_j)}), l(v_{\mu(t_j)}, v_b)\} = 2c \cdot C_{OPT}(\sigma)$$

and inequality (4) is proved. □

References

[ABF93] B. Awerbuch, Y. Bartal and A. Fiat. Competitive distributed file allocation. In *Proc. 25th Annual ACM Symposium on Theory of Computing*, pages 164-173, 1993.

[BFR92] Y. Bartal, A. Fiat and Y. Rabani. Competitive algorithms for distributed data management. In *Proc. 24th Annual ACM Symposium on Theory of Computing*, pages 39-50, 1992.

[BBKTW94] S. Ben-David, A. Borodin, R.M. Karp, G. Tardos and A. Wigderson. On the power of randomization in on-line algorithms. *Algorithmica*, special issue on on-line algorithms, 11(1):2-14, 1994.

[BS89] D.L. Black and D.D. Sleator. Competitive algorithms for replication and migration problems. Technical Report Carnegie Mellon University, CMU-CS-89-201, 1989.

[CLRW93] M. Chrobak, L.L. Larmore, N. Reingold and J. Westbrook. Page migration algorithms using work functions. In *Proc. 4th International Annual Symposium on Algorithms and Complexity*, Springer LNCS Vol. 762, pages 406-415, 1993.

[DF82] D. Downey and D. Foster. Comparative models of the file assignment problem. *Computing Surveys*, 14(2):287-313, 1982.

[K93] H. Koga. Randomized on-line algorithms for the page replication problem. In *Proc. 4th International Annual Symposium on Algorithms and Complexity*, Springer LNCS Vol. 762, pages 436-445, 1993.

[LRWY94] C. Lund, N. Reingold, J. Westbrook and D. Yan. On-line distributed data management. Manuscript, submitted, 1994.

[RS89] P. Raghavan and M. Snir. Memory versus randomization in on-line algorithms. In *Proc. 16th International Colloquium on Automata, Languages and Progamming*, Springer LNCS Vol. 372, pages 687-703, 1989.

[SD89] C. Scheurich and M. Dubois. Dynamic page migration in multiprocessors with distributed global memory. *IEEE Transactions on Computers*, 38(8):1154-1163, 1989.

[ST85] D.D. Sleator and R.E. Tarjan. Amortized efficiency of list update and paging rules. *Communication of the ACM*, 28:202-208, 1985.

[W92] J. Westbrook. Randomized Algorithms for the multiprocessor page migration. In *Proc. of the DIMACS Workshop on On-Line Algorithms*, AMS, pages 135-149, 1992.

Serving Requests with On-line Routing*

Giorgio Ausiello[1], Esteban Feuerstein[1], Stefano Leonardi[1], Leen Stougie[2] and
Maurizio Talamo[1]

[1] Dipartimento di Informatica e Sistemistica, Università di Roma "La Sapienza",
via Salaria 113, 00198-Roma, Italia.
e-mail: {ausiello,esteban,leonardi,talamo}@dis.uniroma1.it.
[2] Department of Operations Research, University of Amsterdam,
Roetersstraat 11, 1018WB Amsterdam, The Netherlands.
e-mail: leen@sara.nl.

Abstract. In this paper we consider the on-line version of the routing problem with release times. Formally, it consists in a metric space M with a distinguished point o (the origin), plus a sequence of triples $< t_i, p_i, r_i >$ where p_i is a point of M, r_i specifies the first moment in which the request is ready to be served, and t_i represents the moment in which the request is known. A server located at point o at time 0 that moves at constant unit speed must serve the sequence of requests trying to minimize the total time till all the requests are served.

An on-line server follows a solution computed on the basis of the set of requests presented in the past; this solution may be updated each time some new piece of information is known. We show that no on-line algorithm, neither deterministic nor randomized, can achieve a competitive factor lower than 2. We give a 5/2-competitive exponential algorithm and a 3-competitive polynomial algorithm for the plane, and a 7/3-competitive algorithm for the line.

1 Introduction

The problem of routing and scheduling a server in a network has a great quantity of applications from robotics to several transportation problems. The most general problem considers a server that must serve a set of requests in some prescribed locations of a metric space, minimizing some objective function, typically the total distance traveled or the completion time.

Several versions of the problem have been studied, in which additional constraints are imposed and particular metric spaces are considered. For example in [11] each request can be served only after a certain *release time*. In that work it has been shown that if the metric space is a line the optimal solution may be found in quadratic time, while it was conjectured to be NP-hard for some other simple metric spaces. In [8] the metric space is restricted to be a tree and each request has, besides a release time, an associated *handling time* that is the time needed to serve it. The problem is shown to

* This work was partly supported by ESPRIT BRA Alcom II under contract No.7141, and by Italian Ministry of Scientific Research Project 40% "Algoritmi, Modelli di Calcolo e Strutture Informative".

be NP-hard in that context, and a 2-approximate solution is given. In general these problems are called *routing and scheduling with time window constraints*. Sometimes more than one server are considered, and other restrictions are given by imposing that requests must be served before a specified *deadline* (see for example [14, 15]). In general metric spaces all these problems are at least NP-complete, since they contain the Hamiltonian Path problem as a particular case. Other related works are [1, 4, 5, 6].

All the previously cited works consider off-line problems, in the sense that the entire input of the problem must be available before the server starts its work and hence, the solution (whether optimal or approximated) can be computed once and be accomplished by the server.

In many applications the requests are not known in advance, but are presented in an on-line fashion possibly by independent sources. In this context it would be necessary to wait till the last request is presented to compute a solution; but this could not be feasible if we do not know which request is the last one. Besides, the time spent till the last request is presented could be used to do part of the work. The lack of knowledge about the future introduces an additional source of difficulty that cannot be overcome even using a non-polynomial algorithm. This means that in many cases an optimal solution on any instance cannot be obtained also disregarding computational complexity. Examples of applications of this problem are given by a robot that must provide a service to a set of independent clients located in a metric space, a repairman with a cellular phone that must decide his route, and allocation of shared resources in a distributed network. To the best of our knowledge, there are not previous references concerned with the on-line version of the problem, even if it is very natural and realistic.

In this paper we consider the on-line version of the routing problem with release times. Formally, it consists in a metric space M with a distinguished point o (the origin), plus a sequence of triples $< t_i, p_i, r_i >$ where p_i is a point of M and t_i and r_i are real numbers, $0 \leq t_i \leq t_j$ for every $i < j$. The r_i are called the *release times* of the requests, and specify the first moment in which the request is ready to be served, while t_i represents the moment in which the request is known. The relation $t_i \leq r_i$ holds for every i, with the meaning that requests may be presented at any moment not later than their release times. A server located at point o at time 0 that moves at constant unit speed must serve the sequence of requests trying to minimize the total time till all the requests are served.

An on-line algorithm for this problem determines the behavior of the server in a certain moment t as a function of all requests $< t_i, p_i, r_i >$ such that $t_i \leq t$. The server follows a solution computed on the basis of the set of requests presented in the past; this solution may be updated each time some new piece of information (as for example a new request or simply the passing of time) is known. We suppose that there is no way of knowing if a certain request is the last one of the sequence.

One of the most widely studied on-line problems is the *k-server problem* ([10]) in which it must be decided which one of k servers moves to cover a request in a metric space, with the goal of minimizing the total distance traveled by the k servers. Our problem differs from the trivial 1-server problem because our task is to decide the order in which the requests are served, while in the 1-server problem the order must be precisely the order in which the requests are presented.

Another related work ([9]) considers the problem of visiting the whole set of vertices of an unknown graph, when the set of edges leaving a node is revealed only once the node is visited. In our case, the metric space is completely known from the beginning, but what is revealed in an on-line way is the set of locations that must be visited.

An on-line algorithm will in general produce a non optimal solution. The quality of a certain on-line strategy is measured by the worst-case ratio between the time needed by the on-line algorithm for a sequence of requests and the optimal time needed to serve it by an algorithm who knows the entire sequence in advance. In the case of on-line algorithms that *approximation ratio* is called the *competitive ratio* of the algorithm. Competitive analysis provides a more appropriate analysis tool for on-line algorithms than, for example, worst-case or average-case analysis. It has been introduced in [12], and has been widely used for analyzing the performance of on-line strategies for a great variety of on-line algorithms, ranging from scheduling to financial decision taking, and for many data structure problems (among the wide literature on the subject, see for example [3, 10, 13]).

As we do not know which request will be the last one, the simple idea of waiting till the last request is presented and then serving all the requests in an optimal way can not be applied. Had this been possible, we would have a 2-competitive algorithm, and we will see later that such a result is the best one can expect, even if we allow the strategies the capability of computing optimal hamiltonian paths. As a result of this fact, we will see that "good" strategies must start working even if the work done may happen to reveal as "wasted" in front of future requests.

In this work we study the on-line routing and scheduling problem in the cases in which the metric space is the euclidean plane or the real line. We provide lower and upper bounds for the competitiveness of on-line algorithms. In particular we show that no on-line algorithm, neither deterministic nor randomized, can achieve a competitive factor lower than 2. We give a 5/2-competitive exponential algorithm and a 3-competitive polynomial algorithm for the plane, and a 7/3-competitive algorithm for the line.

The remainder of this paper is organized as follows: Section 2 is dedicated to the case in which the metric space is the euclidean space. In Section 3 we study the particular case of the real line, and we provide an algorithm achieving a competitiveness ratio nearer to the lower bound. Finally, Section 4 depicts conclusions and open problems.

2 The on-line routing problem in the plane

As was stated above, we restrict our attention to the case in which the metric space is the euclidean plane. In this section we show a lower bound on the competitiveness of any on-line algorithm for this problem as well as some strategies with their respective competitive ratio.

For simplicity of notation, we will say for a triple $< t, p, r >$ that "a request in point p with release time r is presented at time t". When the release time is not mentioned, we mean that $t = r$. All the algorithms presented in this paper ignore every request with release time greater than the moment in which it is presented till its release time. Hence we can consider all requests as presented exactly at their release time.

Given an on-line algorithm and a sequence of requests σ we denote as T_σ^* the time required by the algorithm to serve the sequence of requests. Conversely, t_σ^* denotes the optimal off-line time needed if the sequence is completely known before time 0. An algorithm is said to be *c-competitive* if for every sequence σ, $T_\sigma^* \leq ct_\sigma^* + d$, where d is a constant that does not depend on σ. The subscript σ will be omitted each time this does not cause confusion. As it is usually done in competitive analysis, we compare the behavior of an on-line algorithm with that of an *adversary* that starts at the same time as the on-line algorithm and who knows (in fact, who chooses) the sequence of requests and, hence, serves it optimally.

2.1 Lower bounds

We show that no on-line algorithm can achieve a competitive ratio smaller than 2. With this aim, we provide a family of sequences of requests for which no algorithm can finish within less than twice the optimal off-line time.

The lower bound we show is constructed on the real line. In fact, we use real numbers to denote the points p of the metric space and the initial position o is at point 0.

Theorem 2.1 *No deterministic algorithm for the on-line routing problem in the plane can be c-competitive with $c < 2$.*

Proof We consider a family of sequences defined in the following way: Let $t_0 = 0$ and $t_i = 3^{i-1}, i = 1, 2, \ldots$ Let σ_n be the sequence consisting in making two requests in points $-t_i, t_i$ at time t_i, for $i = 1, \ldots, n$. Intuitively, each sequence σ_n is obtained from σ_{n-1} by adding two new symmetric requests at a distance from the origin equal to the optimal time needed to serve σ_{n-1}.

Assume the position of the on-line server at each time t_j with $j \leq i$ is 0 and suppose the on-line server is at point p at time $t_{i+1} = 3t_i$; without loss of generality we suppose $p \geq 0$. It is easy to see that if $p > 0$ the server has not been to $-t_i$ yet, while if $p = 0$ without loss of generality we assume that it has not been in $-t_i$ since he could not have visited both the extremes. At time t_{i+1}, the adversary may be in point t_i, having served all the requests, so, to be 2-competitive, the on-line algorithm must finish serving all the requests before time $2t_{i+1} = 6t_i$. It is clear that at a certain moment the on-line server will have to move left towards $-t_i$. Hence, the adversary can continue moving rightward until the position of the on-line server is exactly the center of the interval defined by $-t_i$ and the adversary's position $t_i + x$. If at that moment a request is presented in position $t_i + x$, the total time required by the adversary will be exactly $t_i + x$, while the on-line algorithm will have still to travel three halves of an interval of length $2t_i + x$, the competitive ratio being $\frac{(3t_i+x)+3/2(2t_i+x)}{3t_i+x} \geq 2$. Note that if $p = 0$ then $x = 0$ and the ratio is exactly 2.

Note that the lower bound tends asymptotically to 2 even if the time passed between 0 and the moment in which the first request is presented is not charged to the on-line algorithm. Otherwise, the much simpler sequence that we use in the proof of the next theorem could be used also for deterministic algorithms, obtaining the same lower bound of 2.

Next theorem refers to randomized algorithms for this problem. The competitive ratio of a randomized algorithm is defined in terms of the expected value of the ratio T^*/t^* (see [2]).

Theorem 2.2 *No randomized algorithm for the on-line routing problem in the plane is c-competitive, with $c < 2$.*

Proof. Suppose there are no requests till time 1, then the expected position of the server must be the origin, as if that is not the case and its expected position is point $p \neq 0$, then the adversary may make a request in point $-|p|/p$ (that is in -1 if $p > 0$ and 1 if $p < 0$). The cost for serving that sequence of requests is clearly 1 for the adversary, while the on-line algorithm will have an expected cost of $1 + |p| + 1 = 2 + p$ and hence the algorithm is not 2-competitive. But if at time 1 the expected position of the on-line algorithm is the origin, then the expected cost for the sequence in which a request is given in point 1 at time 1 will be greater or equal than 2, and hence, as the cost for the adversary is exactly 1, the algorithm is at most 2-competitive.

2.2 The greedy algorithm

The greedy algorithm basically follows at every moment the shortest Hamiltonian path starting from the current position of the server and visiting all points where there are requests not yet served. A request such that its release time is greater than the moment in which it is presented is simply ignored by the algorithm till its release time. Hence, we may suppose that requests are presented exactly at their release times. The route is re-computed each time a new request arrives. Note that each computation of a Hamiltonian path may take exponential time.

The greedy algorithm achieves a competitive ratio of $5/2$, as we show in the following. Let S be the set of points where requests have been posed and not yet served by the algorithm, let p be the current position of the algorithm. Let T be the optimal path that visits S starting from p. Let $\mathcal{T}$ be the optimal path on the set S of requests presented until now, and $\mathcal{C}$ be the convex hull built on S and the starting point. Notice that T and $\mathcal{T}$ do not consider the release times of the requests. The greedy algorithm consists in following the path T that is recomputed every time a new request is presented. Notice that the current position of the greedy algorithm is always inside $\mathcal{C}$, and the following property holds.

Lemma 2.3 *The distance from any point inside $\mathcal{C}$ to the nearest extreme of $\mathcal{T}$ is at most half of the length of $\mathcal{T}$.*

Proof. Let a and b be the extreme points of $\mathcal{T}$, and c any point inside $\mathcal{C}$. Consider the perpendicular to the line containing segment $\overline{ab}$ that passes through c and let c' be its intersection with $\mathcal{C}$ in the semi-plane containing c. Let p and q be the vertices of the edge of $\mathcal{C}$ containing c', such that the path $\mathcal{T}$ passes through p before q going from a to b (see Figure 1). It is simple to see that:

$$\overline{ac} + \overline{cb} \leq \overline{ac'} + \overline{c'b} \leq \overline{ap} + \overline{pc'} + \overline{c'q} + \overline{qb} \leq \overline{ap} + \overline{pq} + \overline{qb} \leq |\mathcal{T}|$$

Then the minimum among $\overline{ac}$ and $\overline{cb}$ is no more than the half of $|\mathcal{T}|$.

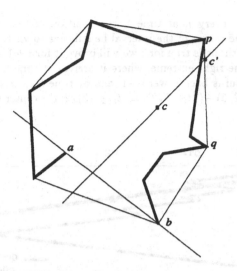

Fig. 1. The convex hull and the optimal hamiltonian path of a set of points

Theorem 2.4 *The greedy algorithm is 5/2-competitive.*

Proof. The length of T is no more that the time required to serve the remaining requests in the order specified by $|\mathcal{T}|$ followed starting at its nearest extreme towards the furthest. By Lemma 2.3 that amount of time is less than 3/2 times $|\mathcal{T}|$. Given that each time a new request is presented the optimal time is greater or equal than the current time, we have that the total time needed by the greedy algorithm is at most 5/2 the time required by the optimal off-line strategy.

The following example shows that the factor of 5/2 is asymptotically tight. Note that if all the requests fall in the real line, the greedy strategy translates in going always to the nearest extreme of the smallest interval containing the requests that have yet to be visited. We suppose that ties are broken in an arbitrary way. This is not a limitation because any choice can be forced by moving requests a negligible distance. The example is illustrated in figure 2. Consider a sequence starting at time 1 with two requests in -1 and 1, and suppose without loss of generality that the greedy algorithm goes towards 1. At time 3 it will be back in 0 and the adversary may put a request at point 1 again. Let $t_0 = 3, p_0 = 1$ and in general $p_i = t_i - 2$, and $t_i = \frac{5}{3}t_{i-1} - 2/3$, for $i = 1, 2, 3, \ldots, n$. The sequence continues with a request in point p_i at time t_i. The adversary's cost for this sequence will be exactly t_n, since it can start going to -1 and then going always to the right arriving at each point p_i when the request is presented. As for the cost of the greedy algorithm, we will show that at time t_n it is in the middle of the interval $[-1, p_n]$, and hence he must still travel 3/2 times the length of the interval, as it has not yet served the requests in the extremes. The total cost charged will then be equal to $t_n + 3/2(p_n + 1) = t_n + 3/2(t_n - 1)$, and hence the ratio between the costs of the greedy algorithm and that of the adversary's tends to 5/2 as n tends to infinity.

Let's see that for every i, at time t_i the position of the greedy algorithm is $(t_i - 3)/2$, that is, the center of the interval he still has to visit. This is obvious for $i = 0$. Assuming the thesis is true for i we will prove it for $i + 1$. At time t_i it leaves $(t_i - 3)/2$ towards the right extreme, where it arrives at time $t_i + (t_i - 3)/2 + 1 = \frac{3}{2}t_i - 1/2$. Then, it turns back towards -1, and at time $t_{i+1} = \frac{5}{3}t_i - 2/3$ will be at point $t_i - 2 - [(\frac{5}{3}t_i - 2/3) - (\frac{3}{2}t_i - 1/2)] = (t_{i+1} - 3)/2$, the center of the new interval.

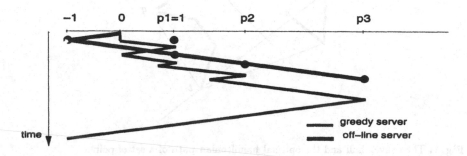

Fig. 2. A worst-case sequence for the greedy algorithm

2.3 The behavior of polynomial algorithms

If we restrict on-line algorithms to be polynomial, it is clear that the competitive ratio achievable is lower-bounded by the approximation ratio that may be obtained by the best approximate algorithm, as a particular instance could consist in a set of requests with their relative release times presented together at time 0.

One known approximation algorithm for the Euclidean Traveling Salesman Problem (ETSP) is the 2-approximate Minimum Spanning Tree heuristic (for details refer to [7]). The MST heuristic constructs a 2-approximate tour by traversing all the edges of a minimum spanning tree over the set of cities. This heuristic also provides a 2-approximation to the Hamiltonian Path Problem, since the size of a minimum spanning tree is less than the total length of an optimal hamiltonian path.

An algorithm that uses the MST heuristic can be described as follows: Each time a new request is presented, arrive to the destination point (if not already in a point). Then follow the 2-approximate path given by the minimum spanning tree of the requests that still have to be served.

Theorem 2.5 *Algorithm MST is 3-competitive.*

Proof. Let a denote the last visited point when a new set of requests B' are presented at a certain time t. Let B denote the set of requests that had to be served before the arrival of the new set of requests. Finally, let S be the set containing all requests presented, and T the minimum hamiltonian path over S starting from the origin. We first note that

$$MST(\{a\} \cup B \cup B') \leq MST(S \cup \{o\}) \leq t^*$$

where $MST(.)$ represents the total length of a minimum spanning tree over a set of points. At the time when the new set of requests is available the on-line server is on an edge leading from a to a point in B. By Lemma 2.6 this edge is always included in a 2-approximate tour on $\{a\} \cup B \cup B'$. Hence, the remaining distance to be traveled by the on-line server is less than $2MST(\{a\} \cup B \cup B')$. Hence, the total cost of the on-line server will be less than $t + 2MST(S \cup \{o\})$ that is less than 3 times the cost of the optimal off-line algorithm, since $t \leq t^*$.

Lemma 2.6 *For every pair of points a and b in a set S, there exists a 2-approximate tour on S including the segment $\overline{ab}$.*

Proof. If the segment $\overline{ab}$ is part of the MST, then the thesis holds trivially. Otherwise, a legal 2-approximate tour starts in a and visits at each node first the subtrees not containing the path leading to b till b is reached, then the subtrees rooted at b not yet visited, and finally the segment $\overline{ab}$.

We finish this section giving an example in the line that shows that the bound of 3 on the competitiveness of the MST heuristic is tight. At time 0 a request at point 1 arrives. At time ϵ a request at point 0 arrives. The MST contains the edge $\overline{01}$, and the on-line server is at point ϵ and continues to follow the segment $\overline{01}$ until 1. At time $1 + \epsilon$ the on-line server is at position $1 - \epsilon$ and a new request is given at point 1. Now the on-line goes towards 0 and afterwards it goes back to 1. The total time is 3, while the optimal solution takes $1 + \epsilon$. The example is illustrated in Figure 3.

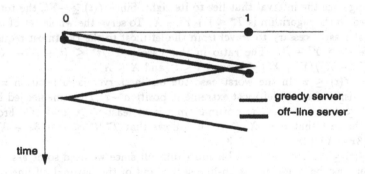

Fig. 3. A worst-case sequence for the MST algorithm

An important feature of algorithm MST is that it can be applied (obtaining the same competitive ratio) not only to the euclidean plane but to *any metric space*. It is interesting to note that the competitive factor of 3 obtained by this strategy is exactly the sum of the 2-factor of the approximation rate of the heuristic plus 1, the same that would be obtained following the approximate path after the last request is presented. As we said before, such a strategy can not be considered because no information about which is the last request is given to the on-line algorithms.

3   On-line routing on the line

In this section we give an algorithm for the particular case in which the metric space is the real line that achieves a competitive factor of 7/3, and hence is better than any of the algorithms we proposed for the plane. We recall that the lower bound of 2 from the previous section was obtained in the line. As we did before, we will consider the origin o in point 0, and requests are specified by real numbers.

Let I be the smallest interval containing the requests not yet served The algorithm, which we call SNB, for "Start near the beginning", consists in going always to the extreme of I that is nearer to the origin.

Theorem 3.1 *Algorithm SNB is 7/3-competitive.*

Proof. We analyze the situation at a generic time t in which a new request is presented. We consider only the case in which I includes the origin. The case in which I is either to the left or to the right of 0 is easier to consider.

Without loss of generality we suppose that the leftmost extreme point is the one nearest to the origin.

Then, our interval is $I = [-x, X]$ with $x \le X$. Moreover, say $-Y'$ and X' the leftmost and the rightmost requests in the past, and $f(t)$ the position of the on-line algorithm at time t. Clearly, at time t the following hold: $t \le t^*$, $x \le Y'$, $X \le X'$ and $-Y' \le f(t) \le X'$. We consider four cases depending on the position in the line of the on-line server:

1. $-Y' \le f(t) \le -x$. The on-line server is to the left of $-x$ and will finish its work visiting once the interval that lies to its right. Since $f(t) \ge -Y'$, the total time needed to the algorithm is $T^* \le t + Y' + X$. To serve the whole set of requests it is at least necessary to travel from the leftmost to the rightmost request and hence $t^* \ge Y' + X'$. The ratio in this case is $T^*/t^* \le (t + Y' + X)/t^* \le 1 + (Y' + X)/(Y' + X') \le 2$, since $t \le t^*$ and $X \le X'$.

2. $-x \le f(t) \le x$. In the worst case the on-line server is in position x and it must visit before the leftmost extreme in position $-x$. The time needed by SNB is $T^* \le t + 3x + X$. The optimal time is at least $t^* \ge 2x + X'$. From this and the fact that $x \le X \le X'$ it follows that $T^*/t^* \le (t + 3x + X)/t^* \le 1 + (3x + X)/(2x + X') \le 7/3$.

3. $x \le f(t) \le X$. This case is a bit more difficult since we need some assumptions on the past behavior of the on-line server and of the optimal off-line solution. We consider two different cases:

   - The optimal strategy ends to the left of the origin. Then we have $t^* \ge 2X' + x$. The on-line server must at most cover twice the interval if it is very close to the rightmost extreme X. Then it will finish by $T^* \le t + 2x + 2X$ time and the ratio is $T^*/t^* \le 1 + (2x + 2X)/(x + 2X') \le 7/3$.
   - The optimal strategy ends to the right of the starting point. We have two more cases:

     • The optimal off-line strategy visits X' before $-x$. It follows that $t^* \ge 2x + 2X'$ since in the best case the optimal algorithm has visited X', then it has gone to $-x$ and has finished to the right of the starting point. In this case $T^* \leq t + 2x + 2X$ and the ratio is $T^*/t^* \leq 1 + (2x + 2X)/(2x + 2X') \leq 2$.

- The optimal off-line strategy visits X' after having visited $-x$. Assume that the optimal off-line algorithm visits for the last time $-x$ at time d (with $d \geq x$ since it has traveled from 0 to $-x$ before time d). Then, at time d it has still to travel at least from $-x$ to X', and $t^* \geq d + x + X'$. If $d \geq t$ we have that $T^*/t^* \leq (d + 2x + 2X)/(d + x + X') \leq 2$. Otherwise, if $d \leq t$, the following two facts hold:

Fact 3.2 *At every time t', $d \leq t' \leq t$, the position of the on-line server is $f(t') \geq x$.*

Proof. We show this by contradiction. The release time of the request in position $-x$ is less than d since at time d the optimal off-line algorithm has already served $-x$. Suppose SNB was in $f(t') \leq -x$ at time t'. Then the request in $-x$ would have already been served at time t since the on-line server was in position $f(t) \geq x$ at time t, a contradiction. Otherwise, if SNB was in $-x \leq f(t') \leq x$ at time d, then it must have crossed position x towards right before time t. By definition of SNB this could only be possible if at a certain moment the nearest extreme is to the right of x. But, given that at time t the request in $-x$ is the leftmost one, and that anything to the right of x is further from the origin than $-x$, this is a contradiction.

Fact 3.3 *Starting at time d the on-line server moves towards left until time t.*

Proof. From the previous fact we know that the on-line server is always in position $f(t') \geq x$ with $d \leq t' \leq t$. It is sufficient to show that between d and t the extreme point nearest to the origin is in the interval $[-x, x]$ in which case the on-line server always moves to the left. This follows immediately from the observation that during this period $-x$ always remains the leftmost point not yet served.

If at time d the on-line server starts from a position $f(d) \geq x$ to travel towards left and at time t it is still to the right of x we have that $T^* \leq d + X' + 2x + X$ leading the ratio to be $T^*/t^* \leq (d + X' + 2x + X)/(d + x + X') \leq 2$.

4. $X \leq f(t) \leq X'$. The on-line server is to the right of X and it will finish its work by $T^* \leq t + X' + x$. On the other hand, we have that $t^* \geq Y' + X'$. The ratio is $T^*/t^* \leq (t + X' + x)/t^* \leq 1 + (x + X')/(Y' + X') \leq 2$.

We finally prove that there is a sequence of requests for which SNB achieves a ratio of $7/3$.

At time 1 two requests in -1 and $1/2$ are presented. At time 1 the on-line algorithm leaves 0 towards $1/2$ and arrives at the starting point at time 2 when a new request is presented in $1 - \epsilon$. Again, the on-line server goes to the right and arrives in $1 - \epsilon$ at time $3 - \epsilon$. At that time a new request is given in $1 + \epsilon$ and the on-line algorithm goes to the left since the extreme point in -1 is nearer to the starting point than $1 + \epsilon$. Finally the on-line server takes $7 - \epsilon$ to serve the requests, while the optimal off-line solution takes $3 + \epsilon$. Then the ratio tends to $7/3$ as ϵ tends to 0. The example is illustrated in Figure 4.

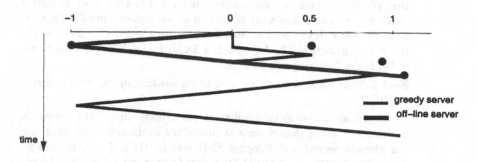

Fig. 4. A worst-case sequence for the SNB algorithm

4 Conclusions and open problems

We have studied a classical scheduling problem from a new point of view that seems very natural and realistic, given the great quantity of applications in which work must be started before having complete information about the input, and in which things can not always be done optimally.

The main open questions regards the gap between lower and upper bounds for the problems in the plane and in the line. We conjecture that 2-competitive algorithms should exist in both cases. Was this conjecture true, it would imply that the use of randomization does not help to improve the performance of on-line algorithms for this problem.

During our research to establish the conjecture, we have considered a great quantity of algorithms, more or less intuitive, specially in the case of the line. All of them have been ruled out for having counterexamples in which their competitive ratio was not less than $7/3$. Among them we may cite the strategy that always follows the optimal off-line solution computed over the set of known requests; another strategy similar to SNB but that considers as the distance from the origin the maximum between the geometric distance of the point and the release time of the request; an algorithm that waits till the optimal off-line time before starting to move, and many others. However, all the efforts done trying to obtain a lower bound bigger than 2 have also failed.

As for polynomial algorithms for the plane, it would be interesting to use a better heuristic, for instance Christofides' heuristic for ETSP that approximates the optimal solution up to a factor of $3/2$. However, a property such as that of Lemma 2.6 does not hold for it, and hence it has not been possible to use it to obtain a better competitive ratio for our problem.

A challenging problem is that in which more than one server are used. Another interesting research direction regards the consideration of different metric spaces.

The lower bound of 2 we have proved in this work is not necessarily valid for every metric space. We think that the factor of 2 could be beaten under certain conditions. Finally, a different problem to consider is that in which the server must go back to the origin after serving all the requests. In this case the lower bound of 2 does not hold either. Even if very similar to our problem, the differences with the latter make it deserve study on its own.

References

1. M. Atallah and S. Kosaraju, Efficient solutions for some transportation problems with application to minimizing robot arm travel, *SIAM J. on Computing*, 17 (1988), pp. 849-869.
2. S. Ben-David, A. Borodin, R.M. Karp, G. Tardos and A. Widgerson, On the power of randomization in on-line algorithms, *Proc. of the 22nd Annual ACM Symposium on Theory of Computing*, (1990), pp. 379-386.
3. R. El-Yaniv, A. Fiat, R.M. Karp and G. Turpin, Competitive analysis of financial games, *Proc. 33rd Annual Symposium on Foundations of Computer Science* (1992), pp. 327-333.
4. G. Frederickson, A note on the complexity of a simple transportation problem, *SIAM J. on Computing*, 22-1 (1993), pp. 57-61.
5. G. Frederickson and D. Guan, Preemptive ensemble motion planning on a tree, *SIAM J. on Computing*, 21-6 (1992), pp. 1130-1152.
6. G. Frederickson, M. Hecht and C. Kim, Approximation algorithms for some routing problems, *SIAM J. on Computing*, 7-2 (1978), pp. 178-193.
7. M. Garey, and D. Johnson, *Computers and intractability: a guide to the theory of NP-completeness*, Freeman, San Francisco (1979).
8. Y. Karuno, H. Nagamochi and T. Ibaraki: Vehicle scheduling on a tree with release times and handling times, *Proc. 4th. International Symposium on Algorithms and Computation ISAAC'93*, LNCS 762 (1993), Springer-Verlag, pp. 486-495.
9. B. Kalyanasundaram and K.R. Pruhs, Constructing competitive tours from local information, *Proc. 20th International Colloquium on Automata, Languages and Programming*, LNCS 700 (1993), Springer-Verlag.
10. M. Manasse, L.A. McGeoch and D. Sleator, Competitive algorithms for server problems, *Journal of Algorithms* 11 (1990), pp. 208-230.
11. H. Psaraftis, M. Solomon, T. Magnanti and T. Kim, Routing and scheduling on a shoreline with release times, *Management Science* 36-2 (1990), pp. 212-223.
12. D. Sleator, R. Tarjan, Amortized efficiency of list update and paging algorithms, *Comm. ACM* 28 (1985), pp. 202-208.
13. D. Sleator, R. Tarjan, Self-adjusting binary search trees, *Journal of the ACM*, 32 (1985), pp. 652-686.
14. M. Solomon, Algorithms for the vehicle routing and scheduling problem with time window constraints, *Operations Research*, 35-2 (1987), pp. 254-265.
15. M. Solomon and J. Desrosiers, Time window constrained routing and scheduling problems: a survey, *Transportation Science*, 22 (1988), pp. 1-13.

A New Algorithm for the Construction of Optimal B-Trees

Peter Becker

Wilhelm-Schickard-Institut für Informatik,
Universität Tübingen
Sand 13, 72076 Tübingen, Germany
E-Mail: becker@informatik.uni-tuebingen.de
Fax: +49 7071 / 295958

ABSTRACT

In this paper the construction of optimal B-trees for n keys, n key weights, and $n+1$ gap weights, is investigated. The best algorithms up to now have running time $O(k\,n^3\log n)$, where k is the order of the B-tree. These algorithms are based on dynamic programming and use step by step construction of larger trees from optimal smaller trees. We present a new algorithm, which has running time $O(k\,n^\alpha)$, with $\alpha=2+\log 2/\log(k+1)$. This is a substantial improvement to the former algorithms. The improvement is achieved by applying a different dynamic programming paradigm. Instead of step by step construction from smaller subtrees a decison model is used, where the keys are placed by a sequential decision process in such a way into the tree, that the costs become optimal and the B-tree constraints are valid.

1. Introduction

In this article the problem of constructing an optimal B-tree for n keys, n key weights, and $n+1$ gap weights is considered. The best algorithms up to now have running time $O(k\,n^3\log n)$, where k is the order of the B-tree, see [7, 3]. These algorithms are adaptations of the well known algorithm of Knuth [4] for the construction of optimal binary search trees with n key weights and $n+1$ gap weights. Furthermore, in [7, 2] algorithms for the construction of optimal multiway trees of order t are presented. These algorithms also adapt Knuth's dynamic programming scheme and have running time $O(t\,n^3)$. All these algorithms use step by step construction of larger trees from optimal smaller trees, that means an optimal tree for the set $\{k_i, \ldots, k_j\}$ is constructed from optimal trees for $\{k_i, \ldots, k_{b-1}\}$ and $\{k_{b+1}, \ldots, k_j\}$ for some b $(i \leq b \leq j, i \neq j)$.

In this article a new algorithm for the problem of constructing an optimal B-tree is presented. The improvement is achieved by applying a different dynamic programming paradigm. Instead of step by step construction from smaller subtrees a decison model is used, where the keys are placed by a sequential decision process in such a way into the tree, that the search costs become optimal and the B-tree constraints are valid. This new approach results in a running time of $O(k n^\alpha)$ with $\alpha = 2 + \log 2/\log(k+1)$, which is a substantial improvement to the former algorithms, especially for k of realistic size. Common values of k in database applications are between 20 and 50. This implies values of 2.23 resp. 2.18 for the exponent α.

The rest of the paper is structured in the following way: in section 2 a formal description of the problem is given. Section 3 reviews the best algorithms up to now. In section 4 the new approach is presented: the decision model is explained, the attached dynamic program is formulated and the solution algorithm is stated. Section 5 gives the complexity results and section 6 presents some ideas about further improvements.

2. The problem

First, we review the definition of a B-tree, cf. [1].

Definition 1: A *B-tree* of order k is a multiway search tree that satisfies the following conditions:

(1) Each node has at most $2k$ keys.

(2) Each node, except the root, has at least k keys.

(3) The root has at least one key.

(4) A nonleaf node with d keys has exactly $d+1$ sons.

(5) All leafs are on the same level.

The following well known theorem (cf. [1]) is crucial for our complexity results in section 5:

Theorem 1: Let h_u be the maximum height of a B-tree of order k with n keys and let h_l be its minimum height. For the height h of a B-tree of order k with n keys the following equation is valid:

$$ h_l = \left\lceil \frac{\log(n+1)}{\log(2k+1)} \right\rceil \le h \le \left\lfloor 1 + \log\frac{n+1}{2}/\log(k+1) \right\rfloor = h_u $$

□

Now we give the problem formulation. We have keys $k_1 < k_2 \cdots < k_n$, an order k and positive weights $q_0, p_1, q_1, p_2, \ldots, p_n, q_n$. p_i are the *key weights* and q_j are the *gap weights*. We define the sum of all weights as

$$ w := \sum_{i=1}^{n} p_i + \sum_{i=0}^{n} q_j $$

We may interpret the value $\alpha_i := p_i/w$ as the probability, that key k_i is requested and the value $\beta_j := q_j/w$ as the probability, that a search is made for a key d with $k_j < d < k_{j+1}$. We assume that we have artificial keys $k_0 = -\infty$ and $k_{n+1} = \infty$. We define $pq(i,j) := (q_i, p_{i+1}, q_{i+1}, \ldots, p_j, q_j)$ as the sequence of gap and key weights from i to j and $w(i,j) := \sum_{v=i}^{j} q_v + \sum_{v=i+1}^{j} p_v$ is the weight of such a sequence. For a B-tree b with n

keys and height h we define the *weighted path length* wpl(b) by

$$\text{wpl}(b) := \sum_{i=1}^{n} p_i \, \text{level}(k_i) + h \sum_{i=0}^{n} q_i$$

where level(k_i) is the level of the node which contains key k_i. The level of the root is defined to be one. The weighted path length is the expected number of visits in a search multiplied by w. The problem is now to find a B-tree b of order k for n keys, that minimizes the weighted path length wpl(b). Such a tree is defined to be an *optimal B-tree*.

3. Review of existing algorithms

The best algorithms known so far for the problem defined in section 2 are adaptations of the corresponding algorithms to construct optimal multiway search trees, cf. [7, 2]. These algorithms themselves are extensions of Knuth's algorithm for the construction of optimal binary search trees.

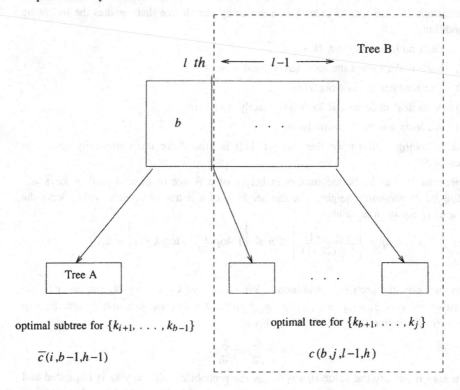

Figure 1. Construction principle for optimal trees

Figure 1 shows the basic principle underlying all these algorithms. Trees for larger sets of keys are constructed by joining trees for smaller sets. An optimal tree for the set $\{k_i, \ldots, k_j\}$ with exactly l keys in the root is constructed by joining an optimal tree for the set $\{k_{b+1}, \ldots, k_j\}$ with exactly $l-1$ keys in the root and an optimal tree for the set $\{k_i, \ldots, k_{b-1}\}$ in an optimal way. That means, we have to find a value for b, that minimizes the weighted path length of the resulting tree.

If we want to use this principle for B-trees, we have to notice some additional constraints. First, subtrees of any node must have the same height. This leads to the following restriction: if tree B of figure 1 has height h we can use for tree A only trees of height $h-1$. Second, optimal B-trees may not have a fully occupied root in contrast to multiway trees. For them it can be proven, that there exists always an optimal tree that has a fully occupied root. Third, an optimal B-tree does not necessarily have minimal height. So we have to take into consideration every possible height of a B-tree of order k with n keys, that means every possible value between h_l and h_u.

Taking the basic principle shown in figure 1 and paying attention to the additional constraints leads to the following recursive formulas. With $c(i,j,l,h)$ we denote the weighted path length of an optimal B-tree for the set $\{k_{i+1}, \ldots, k_j\}$ that has height h and exactly l keys in the root. This is the weighted path length for the sequence $pq(i,j)$. $\overline{c}(i,j,h)$ is the weighted path length of an optimal subtree with height h for $\{k_{i+1}, \ldots, k_j\}$. Subtrees must have at least k keys in their root, see point (2) in Definition 1. We define $\hat{c}(h)$ as the weighted path length of an optimal B-tree for $\{k_1, \ldots, k_n\}$ with height h. $\tilde{c}$ is the weighted path length of an optimal B-tree for $\{k_1, \ldots, k_n\}$. We have the following formulas (cf. [7]):

(i) $c(i,i,l,1) = \infty$, $0 \leq i \leq n$, $1 \leq l \leq 2k$

(ii) $c(i,j,l,1) = \begin{cases} w(i,j) & \text{if } l = j-i \\ \infty & \text{otherwise} \end{cases}$ $0 \leq i < j \leq n$, $1 \leq l \leq 2k$

(iii) $\overline{c}(i,j,h) = \min_{k \leq l \leq 2k} c(i,j,l,h) + w(i,j)$, $0 \leq i < j \leq n$, $1 \leq h \leq h_u$

(iv) $c(i,j,1,h) = \min_{i < b < j} (\overline{c}(i,b-1,h-1) + p_b + \overline{c}(b,j,h-1))$, $0 \leq i < j \leq n$, $2 \leq h \leq h_u$

(v) $c(i,j,l,h) = \min_{i < b < j} (\overline{c}(i,b-1,h-1) + p_b + c(b,j,l-1,h))$, $0 \leq i < j \leq n$, $2 \leq l \leq 2k$

(vi) $\hat{c}(h) = \min_{1 \leq l \leq 2k} c(0,n,l,h)$, $1 \leq h \leq h_u$

(vii) $\tilde{c} = \min_{h_l \leq h \leq h_u} \hat{c}(h)$

Formulas (i) and (ii) initialize the recursion. Formulas (iv) and (v) state the construction principle. Formula (iii) computes the weighted path length of an optimal subtree for $\{k_i, \ldots, k_j\}$. We have to add $w(i,j)$, because the level of each key and gap weight in the subtree will be increased by one, if we use an optimal tree as subtree of another tree. (vi) and (vii) are closely related to the definitons of $\hat{c}$ and $\tilde{c}$. To get the minimum weighted path length in (vii), we have to examine all possible heights between h_l and h_u.

Now we give the algorithm (cf. [7, 3]):

Algorithm 1:

> **for** $i \leftarrow 1$ **to** n **do**
> $\quad w(i,i) \leftarrow q_i$
> $\quad c(i,i,l,1) \leftarrow \infty$
> $\quad$**for** $j \leftarrow i+1$ **to** n **do**
> $\quad\quad w(i,j) \leftarrow w(i,j-1)+p_j+q_j$
> $\quad\quad$**for** $l \leftarrow 1$ **to** $2k$ **do**
> $\quad\quad\quad$**if** $l = j-i$ **then**
> $\quad\quad\quad\quad c(i,j,l,1) \leftarrow w(i,j)$
> $\quad\quad\quad\quad r[i,j,l,1] \leftarrow i+1$
> $\quad\quad\quad$**else**
> $\quad\quad\quad\quad c(i,j,l,1) \leftarrow \infty$
> $\hat{c}(1) \leftarrow \min_{1\leq l \leq 2k} c(0,n,l,1)$
> $h_u \leftarrow 1+\log((n+1)/2)/\log(k+1)$
> **for** $h \leftarrow 2$ **to** h_u **do**
> $\quad$**for** $i \leftarrow n$ **downto** 0 **do**
> $\quad\quad$**for** $j \leftarrow i$ **to** n **do**
> $\quad\quad\quad \bar{c}(i,j,h-1) \leftarrow \min_{k\leq l \leq 2k} c(i,j,l,h-1)+w(i,j)$
> $\quad\quad\quad$**forall** $i<b<j$ **do**
> $\quad\quad\quad\quad$*compute b which minimizes*
> $\quad\quad\quad\quad \bar{c}(i,b-1,h-1)+p_b+\bar{c}(b,j,h-1)$
> $\quad\quad\quad r[i,j,1,h] \leftarrow b$
> $\quad\quad\quad$**for** $l \leftarrow 2$ **to** $2k$ **do**
> $\quad\quad\quad\quad$**forall** $i<b<j$ **do**
> $\quad\quad\quad\quad\quad$*compute b which minimizes*
> $\quad\quad\quad\quad\quad \bar{c}(i,b-1,h-1)+p_b+c(b,j,l-1,h)$
> $\quad\quad\quad\quad c(i,j,l,h) \leftarrow \bar{c}(i,b-1,h)+p_b+c(b,j,l-1,h)$
> $\quad\quad\quad\quad r[i,j,l,h] \leftarrow b$
> $\quad\hat{c}(h) \leftarrow \min_{1\leq l \leq 2k} c(0,n,l,h)$
> $\tilde{c} \leftarrow \min_{h_l \leq h \leq h_u} \hat{c}(h)$

After the termination of the algorithm the multidimensional array r contains the information to construct the optimal B-tree. $r[i,j,l,h]$ defines the leftmost key in the root of an optimal tree of height h for $\{k_{i+1}, \ldots, k_j\}$ with exactly l keys in the root. From the nesting of the loops it is clear that the algorithm has a time complexity of $O(k\, n^3 \log n)$.

4. The new approach

As mentioned before the construction principle underlying the algorithm of the last section has originally been used for optimal multiway trees. In contrast to multiway trees the height of B-trees is bounded by $O(\log n)$ and their structure is much more restricted than the structure of multiway trees. These aspects are not considered by algorithm 1. We present a sequential decision approach where keys are placed directly on some level l ($1 \leq l \leq h_u$) of the tree. Using this approach yields a linear iteration over the keys instead of a cubic one (i, j, b) as in algorithm 1. The crucial question is whether the number of trees that have to be examined for an optimal assignment of k_i is bounded drastically enough by the B-tree restrictions, so that we overall get an improved running time.

We now model the process of constructing an optimal B-tree as a decision problem with n stages. For every key k_i we have to decide, on which level this key should be placed. Whether placing on some level is feasible, depends on the former decisions for the keys k_1 to k_{i-1}, which define a certain state in the decision process. Then placing the key k_i on any level results in an increasing weighted path length and a new state. The amount of increasing as well as the new state depend on our decision.

Using this approach, the optimal B-tree is the result of a sequence of optimal decisions starting in a unique initial state. This leads to a dynamic program DP of the form $DP = (S_v, A_v, D_v, T_v, c_v, C_{n+1})$, where n is the number of the stages of DP, S_v is the *state set* of stage v, $1 \leq v \leq n+1$, and A_v is the *decision set* of stage v, $1 \leq v \leq n$. The sets $D_v \subseteq S_v \times A_v$ define the *feasible decisions* for the states of stage v. It holds: $(s, a) \in D_v$, if and only if a is feasible in state s on stage v. The set $D_v(s) := \{a \in A_v \mid (s, a) \in D_v\}$ contains all feasible decisions for state s on stage v. $T_v: D_v \to S_{v+1}$ is the *transition function*. Making decision a in state s on stage v results in state $T_v(s, a)$ on stage $v+1$. $c_v: D_v \to \mathbb{R}$ is the *cost function* of stage v. $c_v(s, a)$ gives the costs that arise if we decide to make decision a in state s on stage v. $C_{n+1}: S_{n+1} \to \mathbb{R}$ is the *terminal cost* function. $C_{n+1}(s)$ gives the costs that arise if our final state is s.

Now we have to define the components of the dynamic program in such a way that the decision process models the construction of a B-tree.

First we give the definition of the states. For a motivation take a look at figure 2. Suppose we have $k=2$. Then we have to place the first two keys in the first block. This results in (a) of figure 2. Now for k_3 we have two possibilities. We may place k_3 in the same block as the former keys (b) or we may create a new root, which leads to (c). In the latter case it is not feasible to place one of the following keys in the block of $\{k_1, k_2\}$. Instead we have to place k_4 and k_5 in a new block on leaf level. Doing this we get (d). Now there are again two choices for k_6. On the other hand in situation (e) we have to place k_8 in the root, because the leaf already contains $2k$ keys. For assigning k_{10} it doesn't matter if we have tree (f) or tree (g). For both trees we have the choices to create a new root with k_{10} as the first key or to place k_{10} on the actual root or leaf level. The rules that we have implicitly used in this example have led to trees that are valid B-trees with the exception of the rightmost path. Such trees will be filtered on the last stage ($n+1$) by the terminal cost function C_{n+1}. For instance tree (c) gets the terminal costs ∞ for $n=3$.

From the examples above we can deduce, that for a correct placing of a key in the partial tree only the occupation of the blocks in the rightmost path from the root to the

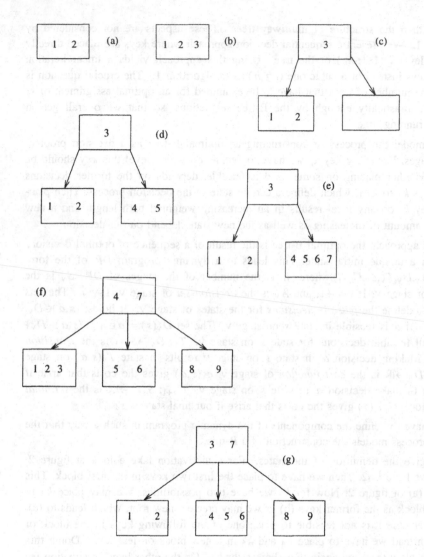

Figure 2. B-tree states in the construction process

leaf is relevant. Due to this fact we can represent a state $s \in S_v$ by a vector with h_s components, that means $s = (s_1, \ldots, s_{h_s})$ with $0 \le s_i \le 2k$ resp. $s \in \mathbb{N}_{2k}^h$. In the following we will use h_s as the denotation for the *component number* of a state. Each component s_i gives the number of keys in the block of level i in the rightmost path of the tree. For instance the state resulting from tree (d) in figure 1 is represented by (1,2) and the state resulting from tree (g) by (2,2). The set S_v is defined to be the set of all vectors that are possible after the assignment of $v-1$ keys. A more formal definition of the state sets S_v follows below.

A decision is characterized by the level on which a key is placed. So we define $A = A_v = \{0, \ldots, h_u\}$. Making decision $a=0$ means creating a new root, as in the transition from (a) to (c) of figure 2. $a \geq 1$ means, that the corresponding key is placed on level a. For instance, the tree (g) is constructed by the decision sequence $DS = (0,1,0,2,2,2,1,2,2)$, assuming that $s_1 = ()$ (i.e. $h_s = 0$) is the initial state. Why we start with $s_1 = ()$ and not with $s_1 = (0)$ is explained below.

Let $s = (s_1, \ldots, s_{h_s})$ be a state. A feasible decision a has to fulfill the following conditions:

    (i) $0 \leq a \leq h_s$

    (ii) $s_\mu \geq k$, $a+1 \leq \mu \leq h_s$

    (iii) $a = 0 \vee s_a < 2k$

Condition (ii) is due to (2) of definition 1 and (iii) is due to (1) of definition 1. (i) guarantees the consistency state s. So we can define $D_v = \{(s,a) \mid s \in S_v, a \text{ fulfills (i)}$ to (iii)$\}$. Observe that the feasible decisions of a state s are independent of the stage v. So we define $D(s) = \{a \in A \mid a \text{ fulfills (i) to (iii)}\}$ as the *set of feasible decisions for state s*. It is obvious that for every B-tree there exists a unique feasible decision sequence that constructs the tree. As an example see the decison sequence to construct tree (g) above. Using these definition each feasible decision sequence leads to trees that are valid B-trees with the exception of the rightmost path. Such trees are filtered by the terminal cost function C_{n+1}.

As explained before, making a decison a has two effects. First, the block on level a of the rightmost path gets one additional key and second, the blocks on the levels from $a+1$ to h_s become closed, that means we cannot assign following keys to these blocks. We get new empty blocks for these levels as in (c) of figure 2. So the definition for the transition function is

$$T_v(s,a) = T(s,a) = \begin{cases} (1,0, \ldots, 0) \in \mathbb{N}_{2k}^{h_s+1} & \text{if } a=0 \\ \\ (s_1, \ldots, s_{a-1}, s_a+1, 0, \ldots, 0) \in \mathbb{N}_{2k}^{h_s} & \text{if } 1 \leq a \leq h \end{cases}$$

With this definition of T_v the state sets S_v can be formally defined by:

    $S_1 = \{()\}$

    $S_{v+1} = T(D_v)$, $v=1, \ldots, n$

The cost functions c_v are defined by

$$c_v(s,a) = \begin{cases} h_s \, q_v + a \, p_v & \text{if } a>0 \\ \\ (h_s+1) q_v + p_v + w(0,v-1) & \text{if } a=0 \end{cases}$$

If we create a new root ($a=0$), the tree constructed so far becomes a subtree of the new root. In that case the levels of the previous keys and gaps increase by one. This results in an additional weighted path length of $w(0,v-1)$. Observe that by using $s_1=()$ the first gap weight q_0 is treated correctly.

The terminal costs C_{n+1} model whether our final state fulfills the B-tree conditions, especially condition (2) of definition 1. For instance, tree (c) of figure 2 is not a valid B-tree for $n=3$. So we have to verify whether, the rightmost path contains underfull nodes. We have

$$C_{n+1}(s) = \begin{cases} 0 & s_v \geq k, \ 2 \leq v \leq h_s \\ \infty & \text{otherwise} \end{cases}$$

Now the definition of the dynamic program DP is complete. Using this definition the optimization problem is

$$F := \sum_{v=1}^{n} c_v(s_v, a_v) + C_{n+1}(s_{n+1}) \rightarrow \min$$

subject to

$$s_1 = ()$$

$$a_v \in D(s_v) \quad 1 \leq v \leq n$$

$$s_{v+1} = T(s_v, a_v) \quad 1 \leq v \leq n$$

The value F of the objective function yields the minimum weighted path length. The tree is given by the optimal sequence of feasible decisions.

For the solution of this optimization problem we use a common dynamic programming algorithm, cf. [5].

Algorithm 2:

> forall $s \in S_{n+1}$
> $W(s) \leftarrow C_{n+1}(s)$
> for $v \leftarrow n$ **downto** 1 **do**
>   forall $s \in S_v$ **do**
>   forall $a \in D(s)$ **do**
>   *compute* $\overline{a}$ *that minimizes* $c_v(s,a) + W(T(s,a))$
> $V(s) \leftarrow c_v(s,\overline{a}) + W(T(s,\overline{a}))$
> $\pi_v(s) \leftarrow \overline{a}$
> $W \leftarrow V$
> $s \leftarrow ()$
> $\overline{c} \leftarrow 0$
> for $v \leftarrow 1$ **to** n **do**

$$a_v \leftarrow \pi_v(s)$$
$$\bar{c} \leftarrow \bar{c} + c_v(s, a_v)$$
$$s \leftarrow T(s, a_v)$$

The result of the algorithm is a sequence $DS = (a_1, \ldots, a_n)$ that defines the optimal tree. Having DS we are able to build the corresponding tree in linear time, as for each key k_v the level where k_v has to be placed is given by the sequence of the a_v.

5. Complexity results

Now we have to prove, that solving our DP using algorithm 2 yields the above mentioned running time. Cornerstones of this proof are bounds for the cardinalities of the state sets S_v and the sets D_v that define the feasible decisions.

Theorem 2: For all state sets S_v we have:

$$|S_v| = O(n^\beta) \text{ with } \beta = 1 + \frac{\log 2}{\log k + 1}, \quad v = 1, \ldots, n+1$$

Proof: Let $S := \{0, \ldots, 2k\}^{h_u}$ and let $\tilde{S} := \bigcup_{v=1}^{n+1} S_v$. The function $f : \tilde{S} \to S$ defined by

$$f(s) = (0, \ldots, 0, s_1, \ldots, s_{h_s}) \in \mathbb{N}_{2k}^{h_u}$$

is an injective mapping from $\tilde{S}$ to S. The vector $f(s)$ is simply constructed from s by adding leading $h_u - h_s$ zeros. Observe that the first component s_1 of a state $s \in \tilde{S}$ is always positive. Due to the injectivity of the mapping we get

$$|S_v| \leq |S| = (2k+1)^{h_u}, \quad v = 1, \ldots, n+1$$

Using theorem 1 we get

$$|S| \leq (2k+1)^{1+\log(\frac{n+1}{2})/\log(k+1)}$$

$$= (2k+1) \exp\left[\log(\frac{n+1}{2})\frac{\log 2k+1}{\log k+1}\right]$$

Using $\log(2k+1) \leq \log 2 + \log(k+1)$ we get

$$\leq (2k+1)\left[\frac{n+1}{2}\right]^{1+\log 2/\log(k+1)}$$

$\square$

The next theorem states that the cardinality of the feasible decisions is bounded by the same function as the cardinality of the states.

Theorem 3: Let $D := \bigcup_{v=1}^{n+1} D_v$. Then we have:

$$|D| = O(|S|)$$

Proof: We have $D \subset \tilde{S} \times A$ and

$$|D| = \sum_{s \in \tilde{S}} |D(s)|$$

Now we define a partition of S by

$$R_d := \{s \in S \mid k \leq s_{h_u-d+2}, \ldots, s_{h_u} \leq 2k \wedge (0 < s_{h_u-d+1} < k \vee h_u = d)\}, \, d=1, \ldots, h_u$$

i.e. R_d contains the vectors of s that have exactly $d-1$ components equal or greater than k at the backend. Observe that this is a necessary condition for a state to have d feasible decisions (see the definition of $D(s)$). Because the definition of R_d relaxes the conditions given in the definiton of $D_v(s)$ we have

$$(\tilde{s} \in \tilde{S} \wedge |D(\tilde{s})| = d \wedge f(\tilde{s}) \in R_e) \Rightarrow d \leq e$$

i.e. a state $s \in \tilde{S}$ which has exactly d feasible decisions is mapped to some vector $f(\tilde{s}) \in R_e$ with $d \leq e$. Using that the mapping is injective yields

$$|D| \leq \sum_{d=1}^{h_u} d \, |R_d|$$

Using the definition of R_d we get

$$|R_d| = \frac{k}{(2k+1)} \left[\frac{k+1}{2k+1}\right]^{d-1} |S|, \, d=1, \ldots, h_u-1$$

and $|R_{h_u}| = ((k+1)/(2k+1))^{h_u} |S|$. Extending the upper bound of the sum to ∞ yields

$$|D| \leq |S| \frac{k}{2k+1} \sum_{v=0}^{\infty} (v+1) \left[\frac{k+1}{2k+1}\right]^{v}$$

$1/2$ is an upper bound for the second term and $2/3$ is an upper bound for $\frac{k+1}{2k+1}$. To compute the sum we examine the power series $\sum_{v=0}^{\infty} (v+1) x^v$. We have

$$\sum_{v=0}^{\infty} (v+1) x^v = \frac{d}{dx} x \sum_{v=0}^{\infty} x^v = \frac{1}{(1-x)^2}$$

Using the upper bounds we get:

$$|D| \leq \frac{9}{2} |S|$$

$\square$

Now we consider algorithm 2. There are nested loops over v, S_v, and $D(s)$ for each $s \in S_v$. Using theorem 2 and 3 we get the result, that algorithm 2 will have running time $O(k \, n^{1+\beta})$ if we are able to compute the remaining statements efficiently enough. But this is simple, we map the states to components of an array AS. The function

$$\text{index}(s) = \sum_{v=1}^{h_u} s_v (2k+1)^{h_u-v}$$

defines a bijective mapping between S and the index domain of AS. Each component $AS(i)$ of AS represents a certain state s. Attached to this component is an array AD, that represents $D(s)$. Each component of AD contains a pointer to another state. This pointer implements the transition function $T(s,a)$. Due to this, we are able to compute

$W(T(s,a))$ in $O(1)$. Initialization of these data structures needs time $O(\mid D \mid)$. Moreover, it is not necessary to use exactly the set S_v on stage v. Because of the fact that the states $\overline{S}_v \setminus S_v$ can never be reached in the forward computation of algorithm 2, the algorithm remains correct if some superset $\overline{S}_v \supset S_v$ is used, for instance $\overline{S}_v := S$. For this reason we do not need any precomputation of the state sets S_v. But even if we did, we would not need more than $O(k\, n^{1+\beta})$ time by using the above described data structure. Putting all together yields:

Theorem 4: Algorithm 2 has running time $O(k\, n^\alpha)$ with $\alpha = 2 + \log 2 / \log(k+1)$.

6. Summary

We have presented a new algorithm to construct optimal weighted B-trees. The key to the improvement has been the formulation of a dynamic program, that models the construction of the tree in a decision oriented way. In the model we have to decide key by key, on which level the key should be placed. The B-tree conditions are included by additional constraints and a terminal cost function.

One idea for further improvements is to use an A^* algorithm in combination with a strong admissible estimation function, cf. [6]. In this case we would use a forward computation instead of the backward computation used in algorithm 2. A state would be modelled as in our model but with an additional component that gives the number of keys already assigned. That means a state would correspond to a state-stage pair of our model. Expanding a state then means adding a key to the tree that is modelled by the state. The main difference to the former approach is, that we may get a strong lower bound for the costs of the final tree by using an adequate admissible estimation function. In this way we may achieve a better performance, because a lot of states would never be expanded. Of course we can use $w(.,.)$ as estimate function, but the performance of $w(.,.)$ is poor, especially for increasing n.

References

1. R. Bayer and E. M. McCreight, "Organization and Maintenance of Large Ordered Indexes," *Acta Informatica*, vol. 1, pp. 173-189, 1972.

2. L. Gotlieb, "Optimal Multi-Way Search Trees," *SIAM Journal of Computing*, vol. 10, no. 3, pp. 422-433, 1981.

3. Shou-Hsuan Stephen Huang and Venkatraman Viswanathan, "On the Construction of Weighted Time-Optimal B-Trees," *Bit*, vol. 30, pp. 207-215, 1990.

4. D. E. Knuth, "Optimum Binary Search Trees," *Acta Informatica*, vol. 1, pp. 14-25, 1971.

5. K. Neumann and M. Morlock, *Operations Research*, Hanser, Munich, 1993.

6. J. Pearl, *Heuristics — Intelligent Search Strategies for Computer Problem Solving*, Addison-Wesley, 1984.

7. V. K. Vaishnavi, H. P. Kriegel, and D. Wood, "Optimum Multiway Search Trees," *Acta Informatica*, vol. 14, pp. 119-133, 1980.

New Results on Binary Space Partitions in the Plane (extended abstract)

Mark de Berg, Marko de Groot, and Mark Overmars *

Dept. of Computer Science, Utrecht University, P.O.Box 80.089, 3508 TB Utrecht, The Netherlands.

Abstract. We prove the existence of linear size binary space partitions for sets of objects in the plane under certain conditions that are often satisfied in practical situations. In particular, we construct linear size binary space partitions for sets of fat objects, for sets of line segments where the ratio between the lengths of the longest and shortest segment is bounded by a constant, and for homothetic objects. For all cases we also show how to turn the existence proofs into efficient algorithms.

1 Introduction

Problems where the input is a set of objects in the plane or in space are often solved by partitioning the space into subspaces, and then solving the problem on the subspaces recursively. A natural way to perform the partitioning is to make a linear cut of the space, that is, to split the space (and possibly some of the objects) with a hyperplane. The splitting process is repeated for each of the half-spaces with the corresponding sets of (fragments of) objects. This continues until there is at most one fragment of an object left in each of the subspaces. Such a partitioning scheme is called a *binary space partition*, or *bsp* for short. A binary space partition is naturally modeled as a tree structure: a *binary space partition tree*, or *bsp tree*. The nodes of the bsp tree store the splitting hyperplanes; the leaves correspond to the cells (subspaces) in the final partitioning and store (the fragment of) the object that is left in that cell.

Binary space partition trees are popular in many application areas. In computer graphics, for example, bsp trees are used for efficient implementations of the painter's algorithm [7]. In this algorithm one tries to "paint" the objects in a back-to-front order onto the screen. Thus objects in the front are painted on top of objects in the back, resulting in a correct view of the scene. Note that a depth order does not always exist, since there can be cyclic overlap among the objects. When the objects are stored in a bsp tree, however, then a back-to-front order for the object fragments in the tree can easily be obtained for any given viewing direction by traversal of the tree. Other uses of bsp trees in computer graphics include shadow generation [3]. In geometric modelling bsp trees have

---

* Supported by the Dutch Organisation for Scientific Research (N.W.O.) and by ESPRIT Basic Research Action No. 7141 (project ALCOM II: *Algorithms and Complexity*)

been used for the implementation of set operations on polyhedra [8, 14] and in robotics for (approximate) cell decomposition methods [1].

Obviously, the efficiency of algorithms that are based on binary space partitions depends strongly on the *size* of the bsp that is used, that is, on the number of cells in the final partitioning. Hence, it is important to choose the splitting hyperplanes in such a way that the fragmentation is kept as low as possible.

Several results in this direction have been obtained by Paterson and Yao [10, 11]. They proved that for any set of n line segments[2] in the plane there exists a bsp of size $O(n \log n)$, which was proved earlier by Preparata [12]; when the segments are orthogonal then there exists a bsp of $O(n)$ size. In both cases the bsp can be computed in $O(n \log n)$ time. For orthogonal objects the same result (with a slightly better constant for the combinatorial bound) was achieved by d'Amore and Franciosa [4]. Paterson and Yao have conjectured that any set of segments in the plane admits a bsp of linear size, but until now this conjecture is still open. Paterson and Yao also proved bounds on bsp trees in higher dimensions: they have shown that any set of $(d-1)$-simplices in d-space admits a bsp of size $O(n^{d-1})$, and any set of orthogonal rectangles admits a bsp of size $O(n^{d/(d-1)})$. In three-dimensional space they have given lower bound constructions which match their upper bounds, namely $\Omega(n^2)$ for the general case and $\Omega(n\sqrt{n})$ for the axis-parallel case.

In this paper we prove the existence of linear size binary space partitions for three classes of objects in the plane, namely for fat objects, for line segments where the ratio between the lengths of the longest and the shortest one is bounded by a constant, and for homothetic objects. We also give efficient algorithms for these three cases. For fat objects and for line segments with bounded length ratios our algorithms run in $O(n \log^2 n)$ time and for homothets we obtain an $O(n \log n)$ algorithm.

The methods for fat objects and for homothets both transform the problem to a problem on orthogonal line segments. The latter problem is then solved using the algorithm of Paterson and Yao [11]. In the case of homothets the transformation is relatively simple. For fat objects the transformation is more involved. Here we prove the following result which is of independent interest: for any set of fat objects in the plane there exists a linear size orthogonal subdivision such that any region of the subdivision contains a constant number of objects.

The strategy for segments with bounded length ratio is to first construct a partitioning where each segment is cut at least once. This implies that all segments inside any cell of this initial partitioning are attached to the boundary of that cell. We then show how to solve the subproblem inside a cell efficiently.[3]

The merit of this work is twofold. First of all, we prove that the conjecture of Paterson and Yao is true for several interesting special cases. The techniques that we use in our proofs—especially for the case of segments with bounded length ratios—might be useful to prove the general conjecture, although we have not

---

[2] Here and in the sequel we assume that the input objects are disjoint, since otherwise no bsp exists where each cell contains at most one object.

[3] For the last step Chazelle [2] has independently obtained similar results.

been able to do so yet. Secondly, although tight worst-case bounds are known on binary space partitions in 3-space, we feel that this case has not been solved satisfactorily from a practical point of view. Indeed, the quadratic lower bound example given in [10] is rather artificial. It would be very useful if one could establish better bounds for three-dimensional bsp trees under certain conditions on the input objects that are satisfied in practical situations. We believe that two of the conditions that we study in the planar case—fatness and bounded length ratios—are realistic for many three-dimensional applications and we hope that our planar results can be generalized to three dimensions.

The rest of the paper is organized as follows. We start with the case of fat objects in Section 2. Then we consider convex homothets in Section 3 and line segments with bounded length ratios in Section 4. We conclude the paper in Section 5, where we give some directions for further research.

2 Fat Objects

In this section we address the problem of finding a binary space partition for a set of n non-intersecting fat objects of constant complexity in 2-space. Intuitively, an object is called *fat* if it contains no extremely "skinny" parts. (See below for a more formal definition of fatness.) In practice, this is often the case. Our method for fat objects consists of three stages. In the first stage we transform the problem on non-intersecting fat objects to a problem on non-intersecting orthogonal line segments. The next stage solves the latter problem using the orthogonal partitioning algorithm of Paterson and Yao [11]. The final stage completes this orthogonal subdivision to an $O(n)$ size binary space partition for the set of fat objects. Fat objects can formally be defined as follows, see van der Stappen et al.[15].

Definition 1. Let $e \subseteq R^2$ be an object and let k be a positive constant. The object e is k-fat if for each disc D that has its center inside e and whose boundary intersects the boundary of e the following holds: $k \cdot area(e \cap D) \geqslant area(D)$. We call a set E of objects a set of fat objects if there is a constant k such that all objects in E are k-fat.

Let E be a set of n fat objects of constant complexity. The following property [15] of fat objects will be crucial to our solution.

Lemma 2. Let E be a set of non-intersecting fat objects in the plane and let $c > 0$ be a constant. Let $e \in E$ be an object and let δ be the diameter of its minimal enclosing circle. Then the number of objects $e' \in E$ with a minimal enclosing circle with diameter at least δ intersecting any rectangular box with side length $c \cdot \delta$ is bounded by a constant.

We now transform the set E of fat objects to a set $L = L(E)$ of axis-parallel line segments. First we sort the objects in order of increasing size of their minimal enclosing circle. Let $\{e_1, e_2, \ldots, e_n\}$ be the sorted set of objects and let b_i be the bounding box of e_i, that is, b_i is the smallest axis-parallel rectangle that contains e_i. Lemma 2 shows that only $O(1)$ objects from $\{e_{i+1}, \ldots, e_n\}$ will intersect the interior of b_i.

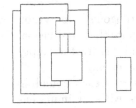

The set L consists of parts of the edges of the bounding boxes b_i. More precisely, for each edge s of b_i we add $s - \cup_{j=1}^{i-1} b_j$ to L. Intuitively, it is as if the boxes are added one by one, where b_i is added *behind* the bounding boxes that already have been added, see figure 1. The arrangement $\mathcal{A}(L)$ induced by the line segments in L has the following useful property.

Fig. 1. the arrangement

Lemma 3. *The interior of any cell C in $\mathcal{A}(L)$ is intersected by $O(1)$ objects $e \in E$.*

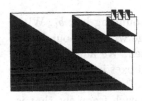

If we now apply the orthogonal partition algorithm of Paterson and Yao to the set L of orthogonal segments, then we create a linear size binary space partition such that any cell in this partition contains only a constant number of objects from E. Unfortunately the number of segments in L can be $O(n^2)$, as can be seen in Figure 2. Many cells in the arrangement $\mathcal{A}(L)$, however, appear to be empty, that is, their interiors do not intersect any object. Indeed, we prove that if we only consider the cells that are intersected by the boundary of at least one object then the complexity of the arrangement reduces to $O(n)$. To analyze the complexity of all such cells in $\mathcal{A}(L)$, we distinguish between *rectangular cells*, which have exactly four vertices, and *non-rectangular cells*, which have more than four vertices. The complexity of a non-rectangular cell is linear in the number of bounding box vertices on its boundary. This implies the following lemma.

Fig. 2. quadratic complexity

Lemma 4. *The complexity of all non-rectangular cells in $\mathcal{A}(L)$ is $O(n)$ in total.*

To bound the number of rectangular cells, we bound the number of intersection points that appear between the boundaries of the bounding boxes and the boundaries of the input objects. Using Lemma 2 we can show the following.

Lemma 5. *The total number of rectangular cells in $\mathcal{A}(L)$ that are intersected by the boundary of an object in E is $O(n)$.*

Let us discard from L all those segments that contribute neither to the boundary of any non-rectangular cell nor to the boundary of any rectangular cell that is intersected by the boundary of an object. Now we are left with an orthogonal subdivision $\mathcal{A}(L)$ of linear complexity in total. Each cell of the arrangement is intersected by only a constant number of object boundaries. We obtain the following lemma.

Lemma 6. *For any set E of n non-intersecting fat objects of constant complexity in the plane there exists an $O(n)$ size orthogonal subdivision $A(L)$ such that each cell C in $A(L)$ is intersected by a constant number of objects in E.*

Now that we have analysed the complexity of the orthogonal subdivision $A(L)$ while leaving out the empty cells, we show how to compute this subdivision. Notice that we cannot simply compute all cells of $A(L)$ and discard the empty cells afterwards because of the complexity of such an arrangement.

We first calculate the intersection points of the boundary of each bounding box b_i with the boundaries of the input objects $e_{i+1}, \cdots, e_n$ with minimum enclosing circle larger then the minimum enclosing circle of e_i. Lemma 3 shows there are $O(n)$ of these intersection points in total. Because the boundaries of the objects have constant complexity and therefore can be split into a constant number of axes-monotone curves, according to Overmars [9] finding these points will take us $O(n \log^2 n)$ time in total.

Let L initially only consist of the four edges of the bounding box b_1. With each edge of b_1 we store the intersection points of this edge with the boundaries of the input objects and we store the boundary of b_1 itself. We add fragments of edges of the bounding boxes b_i to L by processing the boxes in their order as follows. For each edge s of b_i we only add those fragments of $s - \cup_{j=1}^{i-1} b_j$ to L that give rise to a cell C for which $C \cap e_i$ is non-empty. With each fragment that is added to L we store the intersection points of this fragment with the boundaries of the larger input objects $e_{i+1}, \cdots, e_n$ and the boundary of b_i itself. Notice that in this way we compute an arrangement that is somewhat different from the one of which we analysed the complexity. There we added all fragments of $s - \cup_{j=1}^{i-1} b_j$ to L that gave rise to a cell C for which $C \cap E$ was non-empty. But the number of cells we compute in this way in the arrangement is still linear, since we add only fewer segments to L. It is less obvious to see that the number of object fragments within each cell C remains constant. However, because all objects $e_1, \cdots, e_{i-1}$ remain fully contained within $\cup_{j=1}^{i-1} b_j$, any cell induced by fragments of the edges of b_i will still only be intersected by the objects $e_{i+1}, \cdots, e_n$ and by e_i itself.

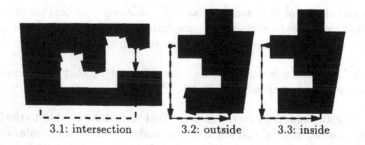

3.1: intersection   3.2: outside   3.3: inside

Let $U(L)$ denote the set of (fragments of) edges that lie on the boundary of the union of cells in $A(L)$ after b_{i-1} has been added. To compute the fragments of the edges of b_i that have to be added to L we start shooting rays *along* the edges of $\mathcal{U}(L)$ and along the edges of b_i. Each ray will hit an edge of $\mathcal{U}(L)$ or b_i along which the next ray can be shot. The rays initially start at the intersection points of the edges in $\mathcal{U}(L)$ with the boundary of e_i and at the vertices of b_i. We call these points the origins for ray shooting. Shooting the rays is guided by the following rules:

- Shooting along the edges of $\mathcal{U}(L)$ located inside b_i we continue until an edge of $\mathcal{U}(L)$ is hit that lies outside b_i, see figure 3.1. We calculate the intersection point of this ray shot and the boundary of b_i, using it as an origin for the next ray along the boundary of b_i outside $\cup_{j=1}^{i-1} b_j$.
- Shooting along the boundary of b_i outside $\cup_{j=1}^{i-1} b_j$ we continue until an edge of $\mathcal{U}(L)$ is hit. All edges traversed along the boundary of b_i are added to L. Again the intersection point of the last ray shot and the edge of $\mathcal{U}(L)$ hit is calculated. We continue ray shooting starting from this intersection point along the edges of $\mathcal{U}(L)$ located inside b_i, see figure 3.2. If we do not hit any edge of $\mathcal{U}(L)$, we can add all edges of b_i to $\mathcal{U}(L)$.
- Shooting along the boundary of b_i inside $\cup_{j=1}^{i-1} b_j$ we continue until an edge of $\mathcal{U}(L)$ is hit. None of the edges traversed along the boundary of b_i is added to L. Again the intersection point of the last ray shot and the edge of $\mathcal{U}(L)$ hit is calculated. We continue ray shooting starting from this intersection point along the boundary of b_i outside $\cup_{j=1}^{i-1} b_j$, see figure 3.3. If we do not hit any edge of $\mathcal{U}(L)$, we can discard b_i.

Observe that we can start shooting at an origin in two directions along an edge. The ray shooting is continued until no more origins are left to start a ray from. Note that in the case of the first item above, the part of $\mathcal{U}(L)$ that we traversed is no longer on the boundery of $\cup_{j=1}^{i-1} b_j$ after b_i is added.

According to de Berg[5] a structure exists for ray shooting into a fixed direction in a set of axis-parallel polygons with n vertices in total, such that queries can be answered in $O(\log n \log \log n)$ time with a structure that uses $O(n \log n)$ storage. An axis-parallel polygon of constant complexity can be inserted into or deleted from the structure in $O(\log n \log \log n)$ amortized time. Using this structure to dynamically maintain the datastructures for $\mathcal{U}(L)$, we can prove the following lemma:

Lemma 7. *For any set E of n non-intersecting fat objects of constant complexity in the plane there exists an $O(n)$ size orthogonal subdivision $\mathcal{A}(L)$ such that each cell C in $\mathcal{A}(L)$ is intersected by a constant number of objects in E and which can be computed in time $O(n \log^2 n)$ using $O(n \log n)$ space.*

Now we have available a scheme to construct an $O(n)$ size orthogonal subdivision $\mathcal{A}(L)$ for a set of non-intersecting fat objects of constant complexity, where each cell of the subdivision contains a constant number of objects. If we apply the orthogonal partition algorithm of Paterson and Yao to the set L of orthogonal segments, then we create a linear size binary space partition such that any cell in this partition contains only a constant number of objects from

E. If we can be sure that we need only $O(1)$ partition lines to further subdivide the space within each cell such that subspaces result containing one object fragment, we are able to construct an $O(n)$ size binary space partition for the set of input objects. Thus if the set of input objects is a set of non-intersecting convex fat objects of constant complexity or a set of non-intersecting polygonal fat objects of constant complexity we are able to construct an $O(n)$ size binary space partition. We obtain the following theorems.

Theorem 8. *For any set of n non-intersecting convex fat objects of constant complexity in the plane there exists an $O(n)$ size bsp, which can be computed in time $O(n \log^2 n)$ using $O(n \log n)$ space.*

Theorem 9. *For any set of n non-intersecting polygonal fat objects of constant complexity in the plane there exists an $O(n)$ size bsp, which can be computed in time $O(n \log^2 n)$ using $O(n \log n)$ space.*

3 Convex Homothets

In the previous section we constructed a linear size binary space partition for a set of fat objects in the plane by reducing the problem to finding a bsp for a set of orthogonal segments. It turns out that the same strategy can be applied to obtain a linear size bsp for a set of convex *homothets* as well, as we explain in this section. (A set of objects is a set of homothets if all objects are identical except for their scale and position in space. For example, a set of discs of varying radius in the plane is a set of homothets.) Homothets are fat, so we could use the results of the previous section. But the fact that they are identical except for their scale and position in space makes a much more simple strategy possible.

Fig. 4. homothets

Let H be the input set of n convex non-intersecting homothets. We assume that the homothets have constant complexity. The transformation consists thereof that we represent each homothet $h \in H$ by three non-intersecting orthogonal line segments, as follows. As the first line segment we choose the longest line segment contained in h. (If this segment is not uniquely defined then we take of all longest segments the one with smallest slope.) This line segment s cuts the homothet h into two convex halves h^- and h^+. For each of these halves we again choose a longest line segment $s^- \subseteq h^-$ and $s^+ \subseteq h^+$ contained in this half and such that s^- and s^+ are orthogonal to s. See figure 4 for an illustration. We denote the resulting set of $3n$ line segments by L_H. The following lemma is straightforward.

Lemma 10. *A set H of n non-intersecting convex homothets of constant complexity in the plane can be transformed in linear time into a set L_H of $3n$ non-intersecting line segments with the following properties:*

- *inside each homothet in H there are three segments in L_H*
- *there are two orthogonal axes such that each segment in L_H is parallel to one of them*
- *each line segment that is parallel to one of these two orthogonal axes and that intersects the boundary of a homothet in H twice intersects at least one segment in L_H.*

By applying the orthogonal partition algorithm of Paterson and Yao to the set L_H of line segments, we create an $O(n)$ size binary space partition $\mathcal{B}(L_H)$ for L_H with the property that no cell of $\mathcal{B}(L_H)$ contains a fragment of a line segment of L_H in its interior. Let C denote a single cell of the subdivision $\mathcal{B}(L_H)$.

Lemma 11. *At most 4 homothets intersect the interior of each cell C of the binary space partition $\mathcal{B}(L_H)$.*

Putting it all together we obtain the following theorem.

Theorem 12. *Given a set H of n non-intersecting convex homothets of constant complexity, a binary space partition of size $O(n)$ for H can be constructed in time $O(n \log n)$.*

Remark. In fact, we have proved the existence of a linear size bsp for arbitrary sets of convex non-intersecting homothets; the assumption that the homothets have constant complexity is only used to guarantee that we can compute the line segments representing the homothets in constant time per homothet. If the homothets do not have constant complexity, but are, for example, convex polygons with N vertices in total then computing the segments representing a homothet takes $O(N)$ time [13]. Furthermore, a line separating two homothets (which we need in the final stage of the algorithm) can be found in $O(\log N)$ time [6]. So in this case we can find a bsp of $O(n)$ size in $O(N + n \log N)$ time.

4 Line Segments with Bounded Length Ratios

Let S be a set of n disjoint line segments in the plane such that the ratio between the length of the longest segment and the length of the shortest one is bounded by a constant c. In this section we show that S admits a linear size binary space partition (with the constant of proportionality depending on c.)

Our method consists of two stages. In the first stage we construct a partitioning such that each segment is intersected at least once (but not too many times, of course). This implies that inside a cell of the partitioning all segments are connected to the boundary of that cell. The next stage constructs for each cell a linear size bsp tree on the segments inside the cell.

Stabbing the segments. To construct a bsp such that each segment is stabbed at least once we proceed as follows. Assume without loss of generality that the length of the shortest segment is one. First we add a number of vertical partition lines in the following way. Define $\ell(x^*)$ as the vertical line with x-coordinate x^*. The vertical lines that we use are the lines of the form $\ell(\frac{i}{2}\sqrt{2})$ that intersect at least one segment in S, for some integer i. In between any pair of lines (and to the left of the leftmost line, and to the right of the rightmost line) we next add horizontal partition lines in the same manner. That is, between two adjacent vertical lines ℓ and ℓ' we add all the horizontal partition lines—which are segments connecting ℓ and ℓ'—with y-coordinates $\frac{i}{2}\sqrt{2}$ for some integer i, provided that they intersect at least one segment in S.

As remarked earlier, the fact that any segment is intersected at least once implies that any segment inside a cell of the partitioning is *anchored*, that is, it intersects the boundary of that cell. It is straightforward to prove that the above partitioning scheme yields the following result.

Lemma 13. *Let S be a set of line segments in the plane such that the ratio between the length of the longest segment in S and the length of the shortest one is bounded by a constant c. Then there is a binary space partition of size $\sqrt{2}(c(n-1)+1)$ into cells $C_1, \ldots, C_k$ such that the fragments of the segments in S that lie in any cell C_i are anchored. Furthermore, $\sum_{i=1}^{k} |S(C_i)| \leqslant (c\sqrt{2}+3)n$, where $S(C_i)$ is the set of segment fragments inside C_i.*

Notice that the partitioning of Lemma 13 is, strictly speaking, not a bsp for S, since there will be more than one segment left in the cells of the partitioning.

The case of anchored segments. Let C be a convex cell and let $S(C)$ be a set of anchored segments. We start with a simple special case, where all segments are anchored at the same edge of C. Here we can obtain a linear size bsp by taking splitting lines containing the segments, starting with the segment that extends farthest from the edge.

Lemma 14. *Let C be a convex polygon, and let $S(C)$ be a set of segments inside C that are all anchored at the same edge of C. Then there exists a bsp for $S(C)$ inside C of size $|S(C)|$.*

We now consider the case where the segments are incident to different edges of the boundary of the cell. Our strategy is to partition C into a number of subcells in such a way that the new fragments that we generate by cutting segments lie in only one of the subcells. Moreover, all these fragments are incident to a single edge of that subcell so we can employ Lemma 14 to partition it. The remaining cells are partitioned recursively.

Let s be an arbitrary segment in $S(C)$. The segment s defines a so-called *successor sequence*, denoted by $S(s)$, which we use to obtain the partitioning into subcells. The successor sequence of s is a sequence $s_0, s_1, \ldots, s_m$ of segments in $S(C)$ defined as follows. The first segment s_0 of the sequence is s itself. Extend s_0 until it either hits another segment or it hits the boundary of C. The extension of s_0 (including s_0 itself) obtained in this way is denoted by $\text{ext}(s)$. If $\text{ext}(s)$ hits the boundary of C then $S(s) = s_0$ and we are ready. Otherwise $\text{ext}(s)$ hits another segment and this segment—the successor of s_0, denoted by $successor(s_0)$—is the next segment s_1 of $\mathcal{S}(s)$. The rest of $\mathcal{S}(s)$ is found in a similar manner: Suppose that the part of $\mathcal{S}(s)$ that we have found so far is $s_0, s_1, \ldots, s_i$. To find the next segment in the sequence we extend s_i until it either hits the boundary of C or it hits the extension $\text{ext}(s_j)$ of one of the already added segments, or it hits a new segment s'. In the first two cases the successor sequence has been completed. In the last case we add $successor(s_i) := s'$ as the next segment s_{i+1} to $\mathcal{S}(s)$ and repeat the process. See Figure 5.1 for an illustration. Observe that the process

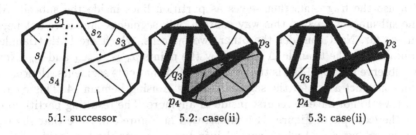

5.1: successor   5.2: case(ii)   5.3: case(ii)

must end after at most $|\mathcal{S}(C)|$ steps, because a segment is added at most once to the successor sequence.

We next describe how to use $\mathcal{S}(s) = s_0, s_1, \ldots, s_m$ to partition C into subcells. There are two cases to consider.

case (i): $\text{ext}(s_m)$ hits the boundary of C.

> In this case we can add the extensions in reverse order as partition lines, that is, we add the partition lines $\text{ext}(s_m), \text{ext}(s_{m-1}), \ldots, \text{ext}(s_0)$. We recurse on the resulting subcells.

case (ii): $successor(s_m) = s_j$ for some $0 \leqslant j < m$.

> This is the elaborate case. The problem is that none of the extensions $\text{ext}(s_i)$ crosses C completely. Somehow we have to "break" the cycle $s_j, \ldots, s_m$ by extending one of the segments even further. Then we can use that segment to start the partitioning process; the other extensions can then be added in reverse order, as in case (i). However, extending one of the segments further may cause a lot of fragmentation. If we would treat all subcells recursively then we cannot keep the fragmentation under control. Hence, we proceed in a slightly different fashion.
>
> For $0 \leqslant i \leqslant m$, let p_i be the endpoint of $\text{ext}(s_i)$ that is on the boundary of C and let q_i be the other endpoint. First we add $\overline{p_{m-1}p_m}$, the segment connecting p_{m-1} and p_m, as a partition line. (It may happen that p_m and p_{m-1} are already on the same edge of C, in which case the addition of $\overline{p_{m-1}p_m}$ can be omitted.) Next we extend $\text{ext}(s_{m-2})$ until it hits $\overline{p_{m-1}p_m}$ and add this extension as a partition line. We have now broken the cycle and we can add the extensions $\text{ext}(s_{m-3}), \text{ext}(s_{m-4}), \ldots \text{ext}(s_0), \text{ext}(s_m), \text{ext}(s_{m-1})$ as partition lines. This is illustrated in Figure 5.2 for the example in Figure 5.1.

Because of the way $S(s)$ is defined, $\overline{p_{m-1}p_m}$ and the extension of $\text{ext}(s_{m-2})$ are the only partition lines that can cut segments of $S(\mathcal{C})$ into fragments. Thus there are at most three subcells in the partitioning that contain fragments of segments, all other subcells contain only segments that have not been cut. These three subcells lie in the convex region enclosed by $\text{ext}(s_{m-1})$, $\overline{q_{m-1}p_m}$ and a part of the boundary of $\mathcal{C}$ connecting p_m to p_{m-1}. This region is depicted shaded in Figure 5.2. First consider the subcell enclosed by $\overline{p_{m-1}p_m}$ and the boundary of $\mathcal{C}$. All fragments in this subcell completely cross the subcell so we can make a *free cut* along each fragment, that is, we can use the fragments themselves as partition lines inside the subcell. All the subsubcells created this way contain only segments that have not been cut so far. Next consider the other two subcells, which are both triangles contained in the triangle Δ defined by the points p_{m-1}, q_{m-1} and p_m. Now we observe that $\overline{p_{m-1}p_m}$ is the only edge of Δ that cuts segments. Hence, we can construct a bsp for the segments inside Δ using Lemma 14. This means that we do not have to recurse inside Δ anymore. The resulting partitioning for the example of Figure 5.1 is illustrated in Figure 5.3. Note that the extension of $\text{ext}(s_{m-2})$ which cuts Δ into two also cuts the bsp inside Δ into two, since it already has been added. But this only doubles the size of the bsp.

Lemma 15. *The procedure described above yields a valid partitioning of $S(\mathcal{C})$ into subcells $\mathcal{C}_1, \ldots, \mathcal{C}_k$, with $k \geqslant 2$, such that $\sum_{i=1}^{k} |S(\mathcal{C}_i)| \leqslant |S(\mathcal{C})| - k/3$, where $S(\mathcal{C}_i)$ is the set of segments inside $\mathcal{C}_i$.*

To obtain the complete bsp for $S(\mathcal{C})$ inside $\mathcal{C}$ we apply the above procedure in $\mathcal{C}$ and recurse on the subcells $\mathcal{C}_i$ with $|\mathcal{C}_i| > 1$. Using Lemma 15 we can proof the following.

Lemma 16. *Let $\mathcal{C}$ be a convex polygon and let S be a set of segments inside $\mathcal{C}$ that are all anchored at the boundary of $\mathcal{C}$. Then there exists a bsp for $S(\mathcal{C})$ inside $\mathcal{C}$ of size at most $3|S(\mathcal{C})|$.*

Putting it all together. Summarizing, we obtain a linear size bsp for a set of segments in the plane with bounded length ratios in the following way. First we construct a partitioning such that each segment is intersected at least once. Then we solve the subproblem inside each cell using the method for anchored segments. We obtain the following result.

Theorem 17. *Let S be a set of n line segments in the plane such that the ratio between the length of the longest segment in S and the length of the shortest one is bounded by a constant c. Then there exists a bsp for S of size $O(n)$, which can be computed in $O(n \log^2 n)$ time using $O(n)$ space.*

5 Conclusion

We have proved the existence of linear size binary space partitions for three classes of objects in the plane, namely for fat objects, for homothets, and for segments with bounded length ratios. We also presented efficient algorithms for computing linear size binary space partitions for each of these classes. This extends a previous result by Paterson and Yao [10], who constructed linear size bsp's for orthogonal segments in the plane. We have, however, been unable to prove the conjecture of Paterson and Yao that any set of line segments in the plane admits a linear size bsp. This is still the most important problem in this area. We hope that our techniques provide some insight that is useful for proving this conjecture.

Another direction for further research is to study binary space partitions for fat objects in three-dimensional space or, in general, for objects with certain realistic properties. We believe that the study of those objects (and not only in the context of binary space partitions) is of great importance.

References

1. C. Ballieux. Motion planning using binary space partition. Report inf/src/93-25, Utrecht University, 1993.
2. B. Chazelle. personal communication, 1993.
3. N. Chin and S. Feiner. Near real time shadow generation using bsp trees. In *SIGGRAPH'90*, pages 99–106, 1990.
4. F. d'Amore and P. G. Franciosa. On the optimal binary plane partition for sets of isothetic rectangles. In *Proc. 4th Canad. Conf. Comput. Geom.*, pages 1–5, 1992.
5. M. de Berg. *Efficient algorithms for ray shooting and hidden surface removal*. Ph.D. dissertation, Dept. Comput. Sci., Univ. Utrecht, Utrecht, Netherlands, 1992.
6. H. Edelsbrunner. Computing the extreme distances between two convex polygons. *J. Algorithms*, 6:213–224, 1985.
7. H. Fuchs, Z. M. Kedem, and B. Naylor. On visible surface generation by a priori tree structures. *Comput. Graph.*, 14(3):124–133, 1980.
8. B. Naylor, J. Amanatides, and W. Thibault. Merging bsp trees yields polyhedral set operations. In *SIGGRAPH'90*, pages 115–124, 1990.
9. M. H. Overmars. Range searching in a set of line segments. In *Proc. 1st Annu. ACM Sympos. Comput. Geom.*, pages 177–185, 1985.
10. M. S. Paterson and F. F. Yao. Efficient binary space partitions for hidden-surface removal and solid modeling. *Discrete Comput. Geom.*, 5:485–503, 1990.
11. M. S. Paterson and F. F. Yao. Optimal binary space partitions for orthogonal objects. *J. Algorithms*, 13:99–113, 1992.
12. F. P. Preparata. A new approach to planar point location. *SIAM J. Comput.*, 10:473–482, 1981.
13. F. P. Preparata and M. I. Shamos. *Computational Geometry: an Introduction*. Springer-Verlag, New York, NY, 1985.
14. W. C. Thibault and B. F. Naylor. Set operations on polyhedra using binary space partitioning trees. In *Proc. SIGGRAPH'87*, pages 153–162, 1987.
15. A. van der Stappen, D. Halperin, and M. Overmars. The complexity of the free space for a robot moving amidst fat obstacles. *Computational Geometry: Theory and Applications*, 3:353–373, 1993.

A Nearly Optimal Parallel Algorithm for the Voronoi Diagram of a Convex Polygon

Piotr Berman　　　　Andrzej Lingas

Penn State University　　Lund University

Abstract

We present a parallel algorithm for the Voronoi diagram of the set of vertices of a convex polygon. The algorithm runs in time $O(\log n)$ and uses $O(n \log \log n / \log n)$ processors in the CRCW PRAM model. The concurrent write is used only by an integer sorting subroutine. We also obtain an $O(\log n)$-time and $O(n \log \log n / \log n)$-processor CRCW PRAM algorithm for the construction of the medial axis of a convex polygon. Our algorithms use the solution to the duration-unknown task scheduling problem due to Cole and Vishkin and the optimal parallel algorithm for the convex hull of a polygon due to Wagener. They are randomized in the sense that for any given $l > 0$ they terminate in time $O(\log n)$ with probability greater than $1 - n^{-l}$.

1 Introduction

The Voronoi diagram and its dual, the Delaunay triangulation, are the most fundamental structures in computational geometry, see [2, 13, 19]. Computing the diagram of n points in the Euclidean plane is well known to have $\Theta(n \log n)$-time sequential complexity. As for parallel complexity, a work optimal $O(\log^2 n)$-time algorithm in the CREW PRAM model is due to Cole, Goodrich and Ó Dúnlaing [5]. Its faster CRCW variant runs in time $O(\log n \log \log n)$ performing $O(n \log^2 n)$ work. A work optimal, $O(\log n)$-time randomized algorithm based on polling (a refinement of random sampling technique) is due to Reif and Sen [21]. Since the Voronoi diagram is so useful it is natural to ask if more efficient sequential or parallel algorithms are available for special site configurations.

In 1987, Aggarwal, Guibas, Saxe, and Shor [1] obtained a linear-time upper bound for the problem of computing the Voronoi diagram of the vertices or edges of a convex polygon. In [12, 15], it was shown that the method of Aggarwal et al. yields also a linear-time upper bound on computing the Voronoi diagram of more general site configurations, e.g., a monotone point sequence (i.e., for points sorted by their x-coordinates that have, in this order, monotone y-coordinates) or more generally, a point sequence for which a special Hamiltonian path in the dual of the Voronoi diagram is known in advance [15]. On the other hand, already in 1986 Chew presented an extremely simple linear-time randomized algorithm for the Voronoi diagram of a convex polygon [7] (for its generalization to include a monotone sequence of line segments see [17]). Devillers used the random sampling method to derive an $O(n \log^* n)$-time randomized algorithm for the so called medial axis or skeleton of a simple polygon [11], i.e., the Voronoi diagram of a simple polygon where the edges of the polygon are sites (see [2, 19]). Also, he showed that if a connected, spanning subgraph of the dual to the Voronoi diagram of a planar point set is known in advance then the diagram can be computed in $O(n \log^* n)$ time [11].

Surprisingly, there are no known fast parallel algorithms for the above special site configurations more work efficient than the aforementioned parallel algorithms for the general configurations (with the exception of the much simpler problem of computing all the nearest neighbours for vertices of a convex polygon [3, 4]). The randomized incremental method of Chew is inherently sequential. In the method of Aggarwal et al. several of the basic steps can be naturally parallelized. However the presence of the second recursive call which has to wait for the outcome of the first slows down time performance of such a direct parallelization to moderately sublinear (i.e., n^δ for a large $\delta < 1$) [15]. As in general one cannot sequentially merge the Voronoi diagrams of two halves of a convex polygon performing sublinear work, the parallel method of Cole et al. based on the standard bisection and efficient parallel merging is of no help here. On the other hand, the divide-and-conquer method based on random-sampling due to Clarkson [8, 9, 10] and its parallel version called polling due to Reif and Sen [21] require only linear work summed over recursion levels for putting solutions together. The bottleneck of this method is the cost of splitting into subproblems based on the locus approach [11]. We take advantage of the fact that in the case of a sequence of vertices or edges of a convex polygon, for each site the neighbours of the site in the subsequence are also neighbours in the Voronoi diagram of the subsequence. This allows us to bound the total cost of splitting to a linear one using the solution to the so called duration-unknown task scheduling problem due to Cole and Vishkin [6].

In effect, we obtain a nearly work optimal logarithmic-time algorithm for the Voronoi diagram of the vertices of a convex polygon. The algorithm works in time $O(\log n)$ and uses $O(n \log \log n / \log n)$ processors in the CRCW PRAM model. The concurrent write is used only by an integer sorting subroutine. Analogously, we also obtain a nearly optimal logarithmic-time algorithm for the Voronoi diagram of the edges of a convex polygon operating within the same asymptotic resource bounds.

2 Random Sampling and Polling

Random sampling has been widely used to design efficient sequential algorithms for several fundamental geometric problems including Voronoi diagrams, segment intersection and higher dimension convex hulls [8, 9, 10, 11]. The general idea of this method is to solve the problem on a randomly chosen subset R of the input S in order to partition the input into smaller problems. Clarkson proved that for many standard geometric problems, the expected size of each subproblem is $O(|S|/|R|)$ and the expected total size of the subproblems is $O(|S|)$ [9]. Reif and Sen observed that Clarkson's result is not sufficient in the design of efficient parallel recursive algorithms for the following reasons:

1. The sum of maximum subproblem sizes over the recursion levels decides about the parallel time performance.

2. A bound on the maximum subproblem size should be highly likely (i.e., it should hold with probability $1 - n^\alpha$ for an appropriate $\alpha > 0$) if there are n^β ($0 \le \beta \le 1$) subproblems on a given recursion level.

3. The total size of subproblems at any recursion level should be $O(n)$ in order to achieve the optimal processor-time product.

They overcame the aforementioned difficulties by introducing a refinement of random sampling called polling. Instead of a single random sample, $O(\log n)$ samples are choosen randomly. Their goodness with respect to the maximum problem size and total problem size is tested by "polling" a small fraction of the input. In this way, a good sample is found with high-likelihood.

Reif and Sen successfully applied their polling technique to the general problem of constructing the common intersection of n half-spaces in E^3. By duality and the paraboidal transformation [13], they could report logarithmic-time, work optimal randomized algorithms for the problems of constructing the convex hull in E^3 and the Voronoi diagram in E^2.

The rough idea of partitioning the intersection problem for the input set S of n half-spaces into smaller problems used by Reif and Sen is as follows. Suppose that a point p^* inside the intersection of the input half-spaces is known and that a "good" sample R of n^ϵ half-spaces is already found. The intersection I_R of the half-spaces in R is found by a simple parallel algorithm testing each intersection of three half-spaces for belonging to each half-space in R. In this way, the vertices of I_R, and then also the faces of I_R induced by triples of the vertices, can be found in logarithmic time and sublinear work assuming ϵ to be small enough. Next, the faces are triangulated and I_R is divided into cones (tetrahedrons) induced by the triangles and the point p^*. In the following crucial partitioning stage, for each cone the list of intersecting half-spaces in $S - R$ is computed. In the next filtering stage, several of the half-spaces on the list that don't contribute to the final intersection of S within the cone are removed (for the sake of keeping the total subproblem size linearly bounded in the sum of the input and output sizes). Now, the half-spaces remaining on the list together with the four half-spaces determining the cone form a half-space intersection subproblem which is solved recursively. The common intersection of the half-spaces in S is then computed as a simple union of the subproblem intersections.

A sample R of n^ϵ half-spaces from S is termed "good" in [21] if it satisfies two following conditions:

1. the total number of the intersections between the half-spaces in $S - R$ and the cones induced by R is less than $k_{tot}n$;

2. the maximum number of of half-spaces intersecting a cone induced by R is less than $k_{max}n^{1-\epsilon}\log n$,

where k_{tot} and k_{max} are constants.

Reif and Sen designed a polling procedure testing $O(\log n)$ random samples of n^ϵ half-spaces for goodness in [21]. Their analysis of the polling procedure (Section 4 in [21]) for our purpose can be summarized as follows.

Lemma 2.1 *[21] For any $l > 0$, there is a randomized CREW PRAM algorithm for finding a "good" sample of n^ϵ half-spaces which runs in time $O(\log n)$ with probability $> 1 - n^{-l}$. For sufficiently small ϵ, the algorithm performs $o(n)$ operations.*

3 Voronoi diagram for convex polygons

By the lifting mapping and duality the problem of constructing the Voronoi diagram of the vertices of a convex polygon is equivalent to the problem of constructing the common intersection I of n half-spaces H_1, H_2, ...H_n in E^3 for which we know a plane P intersecting I in such a way that the i-th edge of the convex polygon induced by $P \cap I$ lies on the plane bounding H_i for $i = 1, 2, ...n$ [21]. For simplicity, we may assume further without loss of generality that H_n is bounded by the plane P. (In the general case, we can extend the sequence of n-half-spaces by the appropriate half-space induced by P to obtain the part of I respectively over or below P.)

To solve this specific half-space intersection problem we shall follow the general parallel algorithm of Reif and Sen for half-space intersection in E_3 based on the polling technique (see Lemma 2.1) with the exception of the processor bottleneck steps of partitioning the current problem into subproblems and filtering out redundant half-spaces from the subproblems.

In the partitioning step, we divide the intersection I_R of a good sample R of n^ϵ half-spaces into cones touching a distinguished point p^* inside the polygon induced by $P \cap I$ (similarly as in [21]). We assume w.l.o.g. that $P \cap I_R$ induces a "closed" convex polygon. Next, for each cone, we report the half-spaces outside the sample, i.e., in $S - R$, intersecting the cone. To accomplish this task in logarithmic time and total linear work we split the partitioning step into two stages.

3.1 The first stage

In the first stage, by using the special assumption about the plane P, for all half-spaces H_j in $S - R$, we find a "starting" cone C_j intersected by H_j, in logarithmic time performing linear work, as follows.

By parallel prefix [18], for all half-spaces H_j in $S - R$, we find the half-spaces $H_{l(j)}$ in R such that for any $l(j) < l < j$, $H_l \notin R$ (w.l.o.g. $H_1 \in R$). It takes logarithmic time and linear work. On the other hand, to each $H_k \in R$, we assign the cone D_k that has faces co-planar with H_k and the plane P. It takes totally constant time and $O(n^\epsilon)$ processors. Since the convex polygon induced by $P \cap I_R$ includes that induced by $P \cap I$, H_j has to insert $D_{l(j)}$. Consequently, we set C_j to $D_{l(j)}$.

Note that the cones intersected by H_j form a connected subgraph of the 3-regular graph dual to the cone partition. Consider the following sequential job: given the starting cone C_j, report all the cones intersected by H_j. It can be easily done in time proportional to the subgraph size, e.g., by depth first search and testing the three cones adjacent to the current cone for intersection with H_j. Also, the total number of intersections between the cones and half-spaces is $O(n)$ by the goodness of the sample R (see Lemma 2.1). Therefore, for the jobs requiring only logarithmic work, we can use the solution to the following duration-unknown task scheduling problem due to Cole and Vishkin [6].

There are given n sequential independent jobs, that together take $O(n)$ steps, where a single job takes between 1 and $c \log n$ steps. The only way to test how many steps a job will take is to execute it. The goal is to execute all these jobs in time $O(\log n)$ using $O(n/\log n)$ processors in the EREW PRAM model (or, respectively, in the CREW PRAM model if the jobs may simultaneously read the same memory cells).

Hence, in logarithmic time and linear work in the CREW PRAM model, we can report:

1. for each half-space H that intersect less than $c \log n$ cones all of the intersected cones, and

2. for each remaining half-space only $c \log n$ intersected cones.

Note that the remaining half-spaces in $S - R$ can be easily identified by the reported $c \log n$ lower bound on the number of cones they intersect and the total number of the remaning half-spaces is $O(n/\log n)$ by Lemma 2.1.

3.2 The second stage

In the second stage, the intersected cones for the remaining $O(n/\log n)$ half-spaces are reported analogously as in [21] by reducing the problem to a point location in higher dimension. The parallel point location structure given in [21] can be built in logarithmic time using $O(m^7)$ processors. For a query half-space H_j it allows to find the list of all intersected cones in logarithmic time using a single processor. Since the number of queries is only $O(n/\log n)$ due to the first stage, all the queries take only linear work.

Now, using the lists and the optimal logarithmic-time randomized algorithm for integer sorting from [20], we can list for each cone the half-spaces in $S - R$ intersecting it in logarithmic-time, performing linear work in the randomized CRCW PRAM model.

4 Filtering

In the previous sections we have described how to divide our special version of the problem of finding the intersection of n half-spaces in E^3 (within a cone) into $n^{1-\epsilon}$ subproblems of total size $\leq cn$ for some constant c. If one applied this method recursively the problem size could increase by a constant factor at each of the recursion levels leading to an $O(n(\log n))^k$ total work.

To avoid this difficulty in the general case of half-space intersections, Reif and Sen apply a complicated method of filtering out few classes of redundant half-spaces which don't contribute to the final output in respective cones. In their method they construct among others the intersections of the final intersection of the n half-spaces with each of the three inner faces of each cone. They call these resulting convex polygons *contours*. For the purpose of the filtering, for each cone they also construct the intersection of all the half-spaces that contribute to the three contours with the cone, and call it the *skeletal hull* within the cone. In fact, to compute the skeletal hulls they are forced to call a simplified version of their algorithm as a subroutine. Fortunately, we have the following lemma.

Lemma 4.1 *Let U be a set of k half-spaces occurring on the three contours of a cone A. Divide the cone into two cones B and C along a plane passing through the apex p^* and cutting the outer face of the cone. The skeletal-hull of A is the union of the skeletal-hulls of B and C with respect to U.*

Proof: The contour on the dividing plane with respect to U divides the faces of the skeletal-hull of A into the following three categories: the faces occurring only in B, the faces cut by the plane, and the faces occurring only in C. Note that the half-planes corresponding to the faces of the first category (respectively, third category) don't contribute to the part of the skeletal-hull within A contained in C (respectively, in B). On the other hand, the above half-spaces contribute to the contours on the faces of B (respectively, C) by our assumptions. Finally, the half-spaces corresponding to the faces of the second category clearly contribute to the common contour of B and C. $\qquad\square$

Thus, by induction, if a cone is recursively divided into smaller cones, its skeletal hull H is the union of the skeletal hulls (with respect to the half-spaces contributing to H) within the smaller cones. Therefore, in the case of the construction of a skeletal hull, the filtering can be reduced to testing whether a half-space intersecting a cone contributes to at least one of its contours. This fact enormously facilitates filtering in our specific case which is indeed the skeletal hull case! Simply, one can easily enclose the final intersection into a tetrahedron based on P (recall that the intersection lies on one side of P by our assumption w.l.o.g. on H_n). Because all the half-spaces contribute to the contour on P, the intersection is easily seen to be a skeletal hull with respect to the bounding tetrahedron.

The construction of the three contours for each cone can also be done more efficient in our case than that in the general case in [21].

Lemma 4.2 *The sequence of points dual to the intersections of a subsequence $H_{j_1},...,H_{j_k}$ of the sequence of the half-spaces $H_1,...,H_n$ with a plane Q is a simple polygon.*

Proof: For the sake of the proof, consider an arbitrary quadruple of half-spaces H_{j_i}, $H_{j_{i+1}}$, H_{j_k}, $H_{j_{k+1}}$, where $i < k$, intersecting the plane Q. Consider the "small" contour formed just by these four half-spaces. It can have at most four edges. Note that the cyclic sequence of the half-spaces corresponding to the edges along the small contour will be a subsequence of the input sequence $H_1,...,H_n$ of half-spaces by the initial assumptions on the common intersection of the n half-spaces with P (*). Apply the dual transform to consider the points p_{j_i}, $p_{j_{i+1}}$, p_{j_k}, $p_{j_{k+1}}$ corresponding to the intersections of the half-spaces H_{j_i}, $H_{j_{i+1}}$, H_{j_k}, $H_{j_{k+1}}$ with the plane the face lies on. By duality and (*), the convex hull of this quadruple of points is $(p_{j_i}, p_{j_{i+1}}, p_{j_k}, p_{j_{k+1}})$ or it has only three vertices. Thus, in either case the segments $(p_{j_i}, p_{j_{i+1}})$, $(p_{j_k}, p_{j_{k+1}})$ cannot intersect! We conclude that the set of points dual to the intersections of the half-planes intersecting the face with the plane induced by the face given in the order of the indices of the half-spaces yields a simple polygon. □

Now, we can proceed as follows. After finding the intersection of each half-space intersecting the cone with each of its inner faces, we order the intersections on each inner face by the indices of the corresponding half-spaces and then construct the contours. Finding the intersections can be clearly done in logarithmic time using $O(n/\log n)$ processors [20]. As for ordering, we can apply here the optimal logarithmic-time CRCW algorithm for integer sorting from [20]. Since the construction of the contours is equivalent to the construction of the convex hull of a simple polygon, the contours can also be computed within the above resource bounds in the CREW PRAM model [22]. Thus, we can conclude with the following lemma.

Theorem 4.3 *For all cones T, we can remove all the half-spaces intersecting T that don't contribute to the final output within T in time $O(\log n)$ using $O(n/\log n)$ processors in the CRCW PRAM model.*

5 Final analysis

There are $O(\log\log n)$ recursion levels in our algorithm similarly as in that in [21]. By the preceding sections, the i-th recursion level takes $O(\log n^{\epsilon^i})$ time and $O(n/\log n^{\epsilon^i})$ processors in the CRCW PRAM model with very high likelihood. This would yield a logarithmic total time by $\sum_{i=1}^{\log\log n} \epsilon^i \log n = O(\log n)$ and unfortunately a linear number of processors by $\log n^{\epsilon^i} = O(1)$ for $i = \Omega(\log\log n)$. In this way, the time-processor product would be $\Omega(n \log n)$ in spite of the fact that each recursion level takes only linear work with very high likelihood.

Let k be the smallest natural number satisfying $\epsilon^k \leq 1/\log\log n$. To obtain a nearly optimal time-processor product we simply slow down the recursion levels

i where $k > i$ by the factor $\log n / \log n^{\epsilon^i} \log n \log n$ keeping their time-processor product still linear by Brent's principle. In result, each recursion level i where $i > k$ is implemented in time $O(\log n / \log n \log n)$ using $O(n \log \log n / \log n)$ processors. As there are $O(\log \log n)$ recursion levels the total time taken by the levels i where $i > k$ is $O(\log n)$ with very high likelihood. The remaining levels 1 through k take logarithmic time by $\sum_{i=k}^{\log \log n} \epsilon^i \log n = O(\log n)$ and use $O(n / \epsilon^k \log n)$,i.e., $O(n \log n \log n / n)$, processors.

Hence, we obtain our main result.

Theorem 5.1 *For any $l > 0$, the Voronoi diagram of a convex polygon can be computed in time $O(\log n)$ with probability $> 1 - n^{-l}$ using $O(n \log \log n / \log n)$ processors in the CRCW PRAM model. The concurrent write is used only by a subroutine for integer sorting.*

6 The medial axis of a convex polygon

The problem of computing the Voronoi diagram of the edges e_i, $i = 1, ..., n$ of a convex polygon in E^2 (called the medial axis or the skeleton of the polygon [2, 19]) immediately reduces to the intersection problem for half-spaces H_i, $i = 1, ..., n$, in E^3 such that (see [1]):

1. H_i includes the convex polygon lying on the plane XY;

2. e_i lies on the plane bounding H_i;

3. the plane bounding H_i forms a constant angle α, $0 < \alpha < \pi/2$, with the interior of the polygon on the plane XY.

The above reduction follows from the fact that the projections of the edges of the common intersection of the half-spaces on the plane XY form the skeleton.

The above problem of half-space intersection is clearly a special case of the specific half-space intersection problem considered in the previous sections. Also, the reduction as well as the computing of the projections can easily be done in logarithmic time and total linear work in the CREW PRAM model. Hence, we obtain our second main result.

Theorem 6.1 *For any $l > 0$, the medial axis of a convex polygon can be computed in time $O(\log n)$ with probability $> 1 - n^{-l}$ using $O(n \log \log n / \log n)$ processors in the CRCW PRAM model. The concurrent write is used only by a subroutine for integer sorting.*

7 Two final remarks

The number of random bits used for sampling and polling through the $O(\log \log n)$ recursion levels of our algorithms can be decreased to $O(\log^2 n)$ analogously as in [21].

We believe that the methods of this paper can be extended to derive nearly optimal fast parallel algorithms for the more general site configurations for which linear-time or almost linear-time sequential algorithms are known (see Introduction).

8 Acknowledgements

The authors are very grateful to Christos Levcopoulos for his careful reading and insightful comments on this work and to an unknown referee for her/his hint on improving the time bound by $\log \log n$ factor without affecting the asymptotic time-processor product in Theorems 5.1 and 6.1.

References

[1] A. Aggarwal, L.J. Guibas, J. Saxe, and P.W. Shor. A Linear-Time Algorithm for Computing the Voronoi Diagram of a Convex Polygon. Discrete and Computational Geometry 2, 1987, Springer Verlag.

[2] F. Aurenhammer. Voronoi Diagrams—A Survey. Tech. Rep., Graz Technical University, 1988.

[3] O. Berkman, D. Breslauer, Z. Galil, B. Schieber and U. Vishkin. Highly Parallelizable Problems. Proc. 21st ACM STOC, pp. 309-319.

[4] R. Cole and M.T. Goodrich. Optimal Parallel Algorithms for Polygon and Point-Set Problems. Proc. 4th ACM Symp. on Computational Geometry, 1988.

[5] R. Cole, M.T. Goodrich and C. Ó Dúnlaing. Merging Free Trees in Parallel for Efficient Voronoi Diagram Construction. Proc. 17th ICALP, LNCS 443, Springer Verlag, pp. 432-445.

[6] R. Cole and U. Vishkin. Approximate Parallel Scheduling. Part 1: The Basic Technique with Applications to Optimal Parallel List Ranking in Logarithmic Time. SIAM J. Comput. 17(1), 1988, pp. 128-142.

[7] P. Chew. Building Voronoi Diagrams for Convex Polygons in Linear Expected Time. Manuscript (1986).

[8] K.L. Clarkson. New applications of random sampling in computational geometry. Discrete and Computational Geometry, 1987, pp. 195-222.

[9] K.L. Clarkson. Applications of random sampling in computational geometry II. Proc. 4th ACM Symp. on Computational Geometry, 1988, pp. 1-11.

[10] K.L. Clarkson and P. Shor. Algorithms for diametral pairs and convex hulls that are optimal, randomized and incremental. Proc. 4th ACM Symp. on Computational Geometry, 1988, pp. 12-22.

[11] O. Devillers. Randomization yields simple $O(n \log^* n)$ algorithms for difficult $\Omega(n)$ problems. International Journal of Computational Geometry and Applications, Vol 2, No 1 (1992), pp. 97-111.

[12] H. Djidjev and A. Lingas. On Computing the Voronoi Diagram for Restricted Planar Figures. Proc. WADS'91, pp. 54-64, LNCS, Springer Verlag. To appear in IJCGA.

[13] H. Edelsbrunner. Algorithms in Combinatorial Geometry. EATCS Monographs on Theoretical Computer Science 10, 1987, Springer Verlag.

[14] L.J. Guibas and J. Stolfi. Primitives for the Manipulation of General Subdivisions and the Computation of Voronoi Diagrams. ACM Trans. Graphics 4, 1985, pp. 74–123.

[15] R. Klein and A. Lingas. A linear-time randomized algorithm for the bounded Voronoi diagram of a simple polygon. Proc. 9th ACM Symposium on Computational Geometry, San Diego, 1993.

[16] R. Klein and A. Lingas. Hamiltonian abstract Voronoi diagrams in linear time. Manuscript, 1993.

[17] R. Klein and A. Lingas. A note on generalizations of Chew's algorithm for the Voronoi diagram of a convex polygon. Proc. of the 5th CCCG, pp. 370-374.

[18] R. M. Karp and V. Ramachandran, Parallel Algorithms for Shared-Memory Machines. Handbook of Theoretical Computer Science, Edited by J. van Leeuwen, Volume 1, Elsevier Science Publishers B.V., 1990.

[19] F.P. Preparata and M.I. Shamos. Computational Geometry: An Introduction. Texts and Monographs in Theoretical Computer Science, Springer Verlag, New York, 1985.

[20] S. Rajasekaran and J.H. Reif. Optimal and Sublogarithmic Time Randomized Parallel Sorting Algorithms. SIAM Journal on Computing 18(3), pp. 594-607.

[21] J.H. Reif and S. Sen. Polling: A New Randomized Sampling Technique for Computational Geometry. Proc. 21st STOC, Seattle, 1989, pp. 394–404.

[22] H. Wagener. Parallel Computational Geometry: Exploiting polygonal order for optimally parallel algorithms. Ph.D. Thesis, Techn. Univ. Berlin, 1986.

On Triangulating Planar Graphs under the Four-Connectivity Constraint

Therese Biedl*   Goos Kant†   Michael Kaufmann‡

Abstract

Triangulation under constraints is a fundamental problem in the representation of objects. Graph augmentation and mesh generation are related keywords from the areas of graph algorithms and computational geometry. In this paper we consider the triangulation problem for planar graphs under the constraint that four-connectivity has to be satisfied.

Our first result states that triangulating embedded planar graphs without introducing new separating triangles can be solved in linear time and space. If the planar graph is not embedded then deciding whether there exists an embedding with at most K separating triangles is NP-complete. A linear time algorithm for this problem is presented, yielding a solution with at most twice the optimal number. Several related remarks and results are included as well.

1   Introduction

The problem of augmenting a graph to reach a certain connectivity requirement by adding edges has important applications in network reliability [7, 24] and fault tolerant computing. The general version of the augmentation problem is to augment the input graph to reach a given connectivity requirement by adding a smallest set of edges. Recent papers present linear time augmentation algorithms to admit the 2-connectivity constraint [6, 22, 13], and the 3-connectivity constraint [12]. With respect to 4-connectivity Kanevsky et al. [14] presented an $\mathcal{O}(n\alpha(m,n) + m)$ time algorithm for testing 4-connectivity, and Hsu presented an $\mathcal{O}(n\alpha(m,n) + m)$ time algorithm to compute the minimal set of edges to augment a 3-connected graph to a 4-connected graph [11] (here, $\alpha(m,n)$ is the functional inverse of Ackermann's function). Kant described several algorithms for the augmentation problem with the additional constraint of planarity [16]. He not only considered the 2- and 3-connectivity constraint, but also the problem of triangulating planar graphs while minimizing the maximum degree. This problem is NP-hard, and an approximation algorithm, working at most $\frac{3}{2}$ from optimal, is known. The augmentation problems with the additional constraint

*RUTCOR, Rutgers University, P.O. Box 5062, New Brunswick (USA), NJ 08903-5062, therese@rutcor.rutgers.edu

†Department of Computer Science, Utrecht University, Padualaan 14, 3584 CH Utrecht, the Netherlands, goos@cs.ruu.nl. Research was supported by the ESPRIT Basic Research Actions program of the EC under contract No. 7141 (project ALCOM II).

‡Wilhelm-Schickard-Institut für Informatik, Universität Tübingen, Sand 13, 72026 Tübingen, Germany, mk@informatik.uni-tuebingen.de.

of planarity have important applications in planar network design and graph drawing algorithms.

In this paper we consider the problem of triangulating a planar graph while achieving 4-connectivity. During the last years 4-connected planar graphs received new attention due to the important characteristics they have: every 4-connected planar graph is hamiltonian, it can be drawn as a visibility representation in a very compact way [17], and if it is triangular it can be represented by a rectangular dual [1, 18]. Visibility representations and rectangular duals are widely used drawing representations, e.g. in industrial environments where rectangular duals are used in floor-planning problems [19].

Unfortunately not every planar graph can be triangulated with the additional constraint of 4-connectivity. If a planar graph contains a cycle of length 3 which is not a face, this is called a *separating triangle*. No graph containing a separating triangle can be made 4-connected while maintaining planarity. Also the star graph does not contain a separating triangle, but any triangulation of it does. So this graph again cannot be made 4-connected.

We present a linear time and space algorithm that, given an embedded planar graph G which is does not contain a star in some sense, triangulates G without introducing new separating triangles. If the initial graph does not contain a separating triangle the output graph is a 4-connected triangular planar graph. Hence then we can apply the rectangular dual algorithm of [1, 18]. If the planar graph has k separating triangles a visibility representation of it can be drawn on a grid of size at most $(n+k-1) \times (n-1)$. When k is small this improves the general bound of $(\lfloor \frac{3}{2}n \rfloor - 3) \times (n - 1)$ [15].

If the planar graph is not embedded the number of initial separating triangles depends on the chosen embedding. We show that deciding whether a biconnected planar graph can be embedded such that the number of separating triangles is at most K is NP-complete. On the positive side, we present a linear time algorithm that embeds a planar graph such that the number of separating triangles is at most twice times the optimal.

The paper is organized as follows. In Section 2 we give some necessary definitions. In Section 3 we present the linear time algorithm for triangulating an embedded planar graph without introducing separating triangles. In Section 4 we consider the problem of embedding planar graphs such that the number of separating triangles is minimal. Section 5 deals shortly the problem of triangulating planar graphs while achieving 5-connectivity, and gives some concluding remarks.

2 Definitions

Let $G = (V, E)$ be a planar graph with n vertices. A graph is called *planar* if it can be drawn without crossing edges. A *planar embedding* is a representation of a planar graph in which the edges incident to a vertex are given in clockwise order with respect to a planar drawing. The embedding divides the plane into *faces*. The unbounded face is the *exterior face* or *outerface*. A cycle of length 3 is a *triangle*. A planar graph is *triangular* if every face is a triangle. A triangular planar graph has $3n - 6$ edges and adding any edge destroys planarity. A cycle C of G divides the plane into its interior and exterior region. If C contains at least one vertex in its interior and its exterior it is called a *separating cycle*. A graph G is called k-connected, if deleting any $k - 1$

vertices does not disconnect G. Such a set of vertices is called a *separating $(k-1)$-set*. Separating 1-sets and 2-sets of vertices are called *cutvertices* and *separation pairs*, respectively. It is well known that a planar graph is at most 5-connected and every triangular planar graph is at least 3-connected (also called *triconnected*).

The *star graph* is a graph where one vertex is adjacent to all other vertices. A graph is said to *contain a star* if it has a face and a central vertex w such that all vertices belonging to the face are either w or are adjacent to it; and there are at least four vertices $\neq w$ on the face.

Finally we shall describe an *SPQR-tree*, a versatile data structure that represents the decomposition of a biconnected graph into its triconnected components [4]. The triconnected components of a biconnected graph G are defined as follows: If G is triconnected it itself is the unique triconnected component. Otherwise let (u, v) be a separation pair of G. We partition G into two subgraphs G_1 and G_2 which have only vertices u and v in common. We continue the decomposition process recursively on $G'_1 = G_1 + (u, v)$ and $G'_2 = G_2 + (u, v)$ until no decomposition is possible. The added edges are called virtual edges. The resulting graphs are each either a triconnected simple graph, or a set of three multiple edges (triple bond), or a cycle of length 3 (triangle). The triconnected components of G are obtained from such graphs by merging the triple bonds into maximal sets of multiple edges (bonds), and the triangles into maximal simple cycles (polygons). The triconnected components are unique.

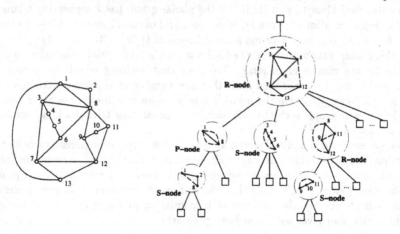

Figure 1: Example of a biconnected graph and the corresponding SPQR-tree (from [4]).

The SPQR-tree T is now defined as follows: for every triconnected component we create an R-node, for every polygon an S-node, for every bond a P-node, and for every edge a Q-node. The edges in T are defined as follows: Let u, v be nodes in T. If u is a Q-node then there is an edge between u and v if the edge represented by u belongs to the triconnected component represented by v. Otherwise there is an edge between u and v if and only if they contain the same virtual edge added in the same step of the decomposition process. In Figure 1 an example is given of a biconnected graph and the corresponding SPQR-tree.

3 Triangulating Embedded Planar Graphs

Assume in the following that G is an embedded biconnected planar graph. If G contains a star it cannot be triangulated without adding separating triangles. We will show that all other graphs can be triangulated without adding separating triangles. For the following algorithm we assume that G is given by its adjacency lists and does not contain a separating triangle.

Input: A planar biconnected graph G without separating triangle
Output: A triangulation of G without separating triangle if possible

while there are nontriangular faces
    do
        choose $v \in V$ such that v has maximum degree among all vertices
            which are part of a nontriangular face
        let F be one nontriangular face containing v
            with vertices $v = u_0, u_1, u_2, ..., u_p, u_{p+1} = v$ in clockwise order
(1)   **if** u_1 and u_p are not adjacent and have no common neighbor
        but v (resp. but v and u_2 if $p = 3$)
    then add (u_1, u_p)
    else
(2)   **if** u_1 and u_p are adjacent
        then determine minimal $j > 1$ such that u_j is not adjacent to u_p
            if such j does not exist
            then STOP, G contains a star with central vertex u_p
            else add edges $(u_j, u_{j-2}), ..., (u_j, u_1), (u_j, v)$
                determine maximal $k < p$ such that u_k is not adjacent to u_1
                add edges $(u_k, u_{k+2}), ..., (u_k, u_p)$
                if $k \neq j$ add edge (u_k, v)
(3)   **else**
            let w be common neighbor of nonadjacent u_1 and u_p
            determine minimal $j > 1$ such that u_j is not adjacent to w
            if such j does not exist
            then STOP, G contains a star with central vertex w
            if $j = p + 1$
            then add edges $(v, u_2), (v, u_3), ..., (v, u_{p-1})$
            else add edges $(u_j, u_{j-2}), ..., (u_j, u_1)$
                determine maximal $k < p$ such that u_k is not adjacent to w
                add edges $(u_k, u_{k+2}), ..., (u_k, u_p)$
                if $(u_p, u_j) \in E$ add edges $(u_k, u_1), (u_k, v)$
                else if $(u_1, u_k) \in E$ add edges $(u_j, u_p), (u_j, v)$
                    else add edges (u_j, v) and (u_k, v) (if different)
    od

The correctness of the triangulation is shown in the following

Lemma 3.1 *No separating triangles are introduced.*

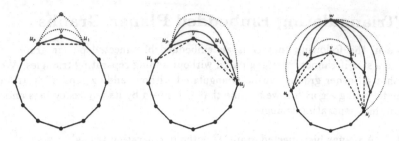

Figure 2: The three cases – nonexistent edges are dotted, added edges are dashed

Proof: We have three cases to consider. In the first case u_1 and u_p are not adjacent and have no neighbor in common. Introducing the edge between them cannot produce a separating triangle or a double edge.

The second and third case can be handled together. Let j and w be defined as above ($w = u_p$ in the second case). Assume that $j \leq p - 1$ and consider u_l, $1 \leq l \leq j - 2$. Adding edge (u_j, u_l) can only produce a separating triangle if these vertices have a common neighbor. For $1 < l < j - 2$ the only possible common neighbor is w which by definition is not adjacent to u_j. For $l = j - 2$ the vertices $\{u_l, u_{j-1}, u_j\}$ indeed induce a triangle, but it is a face and not a separating triangle. For $l = 1$ possible common neighbors are w and u_p. But $\{u_1, u_p, u_j\}$ do not form a triangle (in the second case $(u_p, u_j) \notin E$, in the third case $(u_1, u_p) \notin E$). Hence we introduce no separating triangle when adding (u_j, u_l).

Adding edge (v, u_j) happens only when u_p and u_j are not adjacent. The only possible common neighbors of v and u_j are u_p, w and u_1 (if $j = 2$). The vertices $\{v, u_1, u_2\}$ again form a face, and u_j is adjacent to neither w nor u_p, hence we do not add a separating triangle. If $j = p + 1$ then v is the only vertex of F not connected with w. Hence v has no common neighbor with u_l, $2 \leq l \leq p - 1$ but those where it forms faces, so we can add the edges between them.

The argument for k is symmetric. Here w is defined to be u_1 in the second case. $\square$

Lemma 3.2 *j can be found in $\mathcal{O}(j - 1)$ time.*

Proof: j is defined as "the minimum index $j > l$ such that a certain vertex w is not connected with u_j". When calling this procedure we know the edge (u_l, w) and (with a suitable implementation) its place in the neighborlist $N(w)$ of w.

Now if (w, u_{l+1}) is an edge it must be the next element in counterclockwise order in $N(w)$ (otherwise we had the separating triangle $\{w, u_l, u_{l+1}\}$). So if this element is not (w, u_{l+1}) we can return the value $l + 1$ and are done. Otherwise we can call the procedure again with vertices w and u_{l+1}. Each call of the procedure needs only $\mathcal{O}(1)$ time, completing the proof. $\square$

Lemma 3.3 *Finding the next v can be done in overall linear time.*

Proof: Throughout the algorithm we will keep a doubly linked list L of vertices. All vertices in L are possible candidates for the next v. The entries of L are sorted by descending degree. We also keep an array A where $A[i]$ indicates the first vertex in L of degree i.

We initialize L and A by performing bucket sort. This needs $\mathcal{O}(n)$ time. During the algorithm frequently the degree of a vertex will change. We will show how to maintain L and A in $\mathcal{O}(1)$ time when increasing one degree by 1. Hence overall we need $\mathcal{O}(\#\{\text{added edges}\}) = \mathcal{O}(n)$ time.

Assume vertex w has its degree increased from i to $i + 1$. The new place of w in L is before $A[i]$. Therefore we delete w from its old place and insert it before $A[i]$ (remark that if $w = A[i]$ then L remains unchanged). If $A[i + 1]$ is defined it remains unchanged, otherwise set $A[i + 1] = w$. If $A[i] \neq w$ it remains unchanged, otherwise let $A[i]$ be the successor of w in L, if this has degree i and undefined otherwise.

Our next candidate v is always the first element of L. When deleting v we have to set $A[deg(v)]$ as above to be the successor of v in L resp. to be undefined. All these operations need $\mathcal{O}(1)$ each, so overall we need $\mathcal{O}(n)$ time. □

Lemma 3.4 *The above algorithm works in linear time and space.*

Proof: Consider v and F to be fixed. To add edges in F while $v = u_0$ needs $\mathcal{O}(deg(u_1) + deg(u_p) + j + k)$ time. Remark that after doing this both edges (v, u_1) and (v, u_p) are contained in a triangular face which was not a triangular face before. Also by definition of v we know $deg(u_1), deg(u_p) \leq deg(v)$. As we add $\Omega(j + k)$ edges the time complexity for this loop is

$$\mathcal{O}\left(\sum_{e=(v_1,v_2)} \min\{deg(v_1), deg(v_2)\} + \#\{\text{added edges}\} \right),$$

where the sum runs over all edges where one face has been triangulated throughout the loop.

Every edge in the final graph $G^* = (V, E^*)$ was added at most once and belongs to two faces, hence is used at most twice for the above estimation. Also the degree of a vertex never decreases and we have an overall running time of

$$\mathcal{O}\left(|E^*| + \sum_{(v,w)\in E^*} \min\{deg(v), deg(w)\} \right) = \mathcal{O}(n),$$

where the last equality is due to Chiba and Nishizeki [2]. □

Assume that G is a biconnected graph which contains no separating triangle. We have to ensure that our algorithm does not transform G into a graph containing a star (unless it is unavoidable). So imagine we have a face f_1 and a vertex w such that all but one vertex u on f_1 are either w or are adjacent to it. Furthermore, u and w share another face f_2 which we are dealing with. As our algorithm avoids adding a separating triangle the danger of adding the edge (u, w) consists only if f_2 consists of w, u and the two neighbors of u in f_1 (say, x and y).

If f_1 contained at least five vertices but w then w is the vertex of highest degree in f_2, so our algorithm adds the edge (x, y) and the star has been avoided. Otherwise we have to force the algorithm to choose w to be the next v and can do this by adding the following lines:

> choose v and F as above
> **if** $deg(v) = 3$, $p = 3$ and $deg(u_2) = 3$
> > **if** $deg(u_1) = 3$ and $deg(u_3) = 2$ set $v = u_1$, renumber F
> > **if** $deg(u_3) = 3$ and $deg(u_1) = 2$ set $v = u_3$, renumber F

Assume that G contains separating triangles. Split G at its separating triangles (i.e., if $\{u, v, w\}$ form a separating triangle we split G at the three edges $(u, v), (v, w)$ and (w, u) and add these edges to both components). We repeat this process until there are no separating triangles anymore and apply the algorithm to the subgraphs. Using an initial list with pointers to the separating triangles in the graph this can be achieved in linear time and space (see also [15] for some details). This does not introduce a separating triangle iff none of the subgraphs contains a star. But this is the case iff G does not contains a star. Our algorithm (with the variation) never adds a star, so G is never made to contain a star throughout the algorithm.

We summarize these facts in the following theorem.

Theorem 3.5 *Let $G = (V, E)$ be a biconnected planar graph with a fixed embedding. Then G can be triangulated without introducing separating triangles if and only if G does not contains a star. The algorithm to do this works in linear time and space.*

If G is not biconnected we can add edges without adding separating triangles until it is biconnected by the algorithm of Read [21]: Let v be a cutvertex and let u and w be any two consecutive neighbors of v, belonging to different biconnected components. Add the edge (u, w) to G. Since u and w only share v among their neighbors this does not introduce a separating triangle. Also G has one fewer biconnected component than before and is still planar. Repeating this procedure yields a biconnected planar graph in linear time.

4   Triangulating Planar Graphs

4.1   Exact solutions are hard to find

In this section we consider the problem of triangulating a planar graph G where no embedding is given in advance. If G is triconnected the embedding is unique, so assume G is not triconnected. In this case we can change the embedding and the number of separating triangles can vary. More precisely we have the following theorem:

Theorem 4.1 *Given a planar graph G, deciding whether G can be embedded such that the number of separating triangles of G is at most K is NP-complete.*

Proof: The problem is in NP: compute an arbitrary planar embedding of G, and count the number of separating triangles in the embedding. Both can be done in polynomial time (e.g., see [3, 2]).

Let G be an arbitrary triangular planar graph. For every edge $(a, b) \in G$ we add a vertex x with edges to a and b. Let G' be the new graph. Clearly G' is biconnected and planar. Notice that if F was a face in G on vertices a, b, c, and we place x inside F then F is a separating triangle in G'. However, if (a, b) belong to face F and F' in

G then x must be embedded in either F or F', i.e. F or F' is a separating triangle in G'.

Let S be a minimum set of faces in G such that for every edge $(a, b) \in G$ at least one incident face belongs to S. It corresponds precisely to the minimum number of faces in G' which are separating (place x in the face which belongs to S). The set S also is a vertex cover in the dual graph G^* of G. Since G^* is a cubic triconnected planar graph and deciding whether there exists a vertex cover of size at most K is NP-hard for cubic triconnected planar graphs (cf. the following theorem), the problem of deciding whether a planar graph G can be embedded with at most K separating triangles is NP-complete. □

Theorem 4.2 *Vertex cover in cubic triconnected planar graphs is NP-hard.*

Proof: In the papers [8] and [9] it has already been shown that "vertex cover in planar graphs with maximum degree 3" is NP-hard. In the following we first show how to modify their graphs to transfer the result to cubic planar graphs. Then we give a construction how to increase the connectivity of the graph until it is 3-connected.

Lemma 4.3 *Vertex cover in cubic planar graphs is NP-hard.*

Proof: Given a planar graph G with maximum degree 3 and integer k we first construct a planar graph G' with degrees 2 and 3 and integer k' such that G' has a vertex cover of size k' if and only if G has a vertex cover of size k. We replace each vertex v of degree 1 with neighbor w by four vertices v_1, v_2, v_3, v_4 and the edges $(v_1, v_2), (v_2, v_3), (v_3, v_4), (v_4, v_1), (v_2, v_4)$ and (w, v_1).

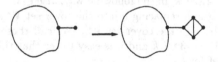

Figure 3: Removing degree 1.

Let p be the number of vertices of degree 1 in G and do the construction p times to get G'. Clearly G' is planar. We set $k' = k + 2p$.

Let V^* be a vertex cover for G of size k. We may assume that it does not contain a vertex v with degree 1. If such a vertex v exists in V^* we can replace v in V^* by its adjacent vertex w and obtain a new cover of not larger size. A vertex cover $V^{*'}$ for G' can be constructed by taking $V^{*'} = V^* \cup \bigcup_i \{v_{i2}, v_{i4}\}$ for all v_i in G with degree 1. The size of $V^{*'}$ is at most $k + 2p$.

Conversely let $V^{*'}$ be a vertex cover of size k'. We consider a subgraph with vertices v_1, v_2, v_3, v_4 which have been added to change G to G'. $V^{*'}$ contains either v_2 and v_4 or it contains v_1 and two more vertices, say v_2 and v_3. In the second case we change the cover by replacing v_1, v_2, v_3 by v_2, v_4, w where w is the other vertex adjacent to v_1. This transformation does not increase the size of vertex cover. Now we can transform G' back to G and obtain from $V^{*'}$ immediately a vertex cover for G. This new vertex cover has size $k' - 2p$.

With analogous arguments we can show how to replace vertices of degree 2 by a small subgraph of vertices with degree 3 such that the property of the vertex cover remain the same. In Figure 4 we demonstrate the replacement.

Figure 4: Removing degree 2.

This remark completes the proof of the first lemma.   □

Now we show how to make the graph for the reduction triconnected. Assume we are given any connected cubic planar graph G with n vertices without multiple edges and self-loops and an integer k. We replace each vertex as indicated in the next figure.

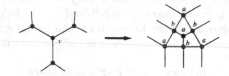

Figure 5: Achieving triconnectivity. Each vertex v is replaced by a cluster of 7 vertices, each edge by 3 edges.

Call the new graph G'. The degree of the vertices is either 3 or 4 and the number of vertices is seven times larger than in G.

We set $k' = 3n + k$. Let V^* be a vertex cover for G of size k. We construct a vertex cover $V^{*'}$ of G' of size k' in the following way: If $v \in V^*$ we include the vertices marked "a" in the subgraph replacing v into the cover set $V^{*'}$. If $v \notin V^*$ we include the vertices marked "b" into the cover set $V^{*'}$. We call those vertices side-vertices. Obviously the size of $V^{*'}$ is $3n + k$ and it is easy to see that this set is a cover set for G' if V^* was a cover set for G.

Conversely let $V^{*'}$ be a cover set of size k' for G'. We consider any subgraph of seven vertices that replaced a vertex v in G. If $V^{*'}$ contains only three vertices out of the seven they must be the side-vertices. In this case v will not be included in the cover set V^* for G. If $V^{*'}$ contains more than three vertices from the seven we can rearrange them such that they are the "a"-vertices. If we do that for all subgraphs we again get a vertex cover of size k'. In the second case v is included in the cover set for G. Now V^* contains at most k vertices and obviously is a vertex cover for G.

We conclude by giving the whole construction of the reduction:

We start with the graph by Garey and Johnson [8] with degree at most 3. We transform it to a cubic planar graph. If it is not triconnected yet we increase the connectivity by applying the technique described above. Then we make the graph cubic again. If it is still not triconnected we apply the technique once more and make the graph cubic afterwards. At that step we did obtain a cubic planar triconnected graph. Finding a vertex cover of prescribed size for planar triconnected graphs can therefore be reduced to solving the same problem for planar graphs of maximum degree 3 which in turn is reducible to the solvability of a 3-satisfiability problem.   □

4.2 An approximation

In the second part of this section we present a linear time algorithm to construct an embedding of a planar graph G with at most twice the optimal number of separating triangles. When the embedding is computed we can use the algorithm of Section 3 to get a triangulation of G without new separating triangles.

Following the proof of Theorem 4.1 we have to search for edges (u, v) in G where $\{u, v\}$ is forming a separation pair of a triconnected component. We call such edges *separation edges*. Finding them and changing the embedding by swapping the triconnected components to the other side of the separation edges is done by using the SPQR-tree defined in Section 2. Let T be the SPQR-tree, rooted at an arbitrary Q-node, and let all separation edges be marked in G. If (u, v) is a separation edge it follows that $G - \{u, v\}$ gives at least three connected components. Hence every separation edge corresponds to a P-node in T.

We visit the nodes of T in a bottom-up order and handle the corresponding triconnected components in this order. Separating triangles correspond to faces of length 3, therefore only R- and S-nodes of T need to be visited and of course only when the corresponding triconnected components contain faces of length 3. Let G' be such a triconnected component. We start with computing the dual graph H of G' in the following way: For every face F of length 3 in G' we add a vertex v_F in H. An edge is added between v_F and $v_{F'}$ if and only if F and F' share an edge e in G' which is a separation edge. Let G'' be the subgraph of G, represented by the edge e in G' (G'' is not necessarily triconnected). G'' has to be embedded in F or F'. This makes the corresponding face a separating triangle.

To decide whether G'' has to be embedded in F or F' we compute a vertex cover S in H. At least one endpoint of $(v_F, v_{F'})$, say v_F, will be part of S. As in Theorem 4.1, this will be the separating face, i.e. G'' will be embedded in face F in G'.

For the computation of the vertex cover we use a standard linear time algorithm, achieving at most twice the optimal size. We compute an arbitrary maximal matching M in H and for every edge $e \in M$ we add both endpoints to S. Since in every vertex cover at least one endpoint of e must be part of S it follows directly that the computed set S is at most twice optimal. We also use the following lemma:

Lemma 4.4 *The vertex cover in H corresponds to the set of separating triangles in G'.*

Proof: Let $e = (v_F, v_{F'})$ be an edge in H. Hence both F and F' are triangles in G'. Let (u, v) be the edge, shared by F and F'. Embedding the corresponding triconnected component in F or F' makes it a separating triangle.

If (u, v) is a separating edge belonging to faces F and F' in G' where F is not a triangle we place the corresponding triconnected components in F as F cannot become a separating triangle. Hence the set of separating triangles corresponds to the set of the vertex cover in H.   □

Theorem 4.5 *There is a linear time and space algorithm for computing an embedding of G which has at most twice the minimal number of separating pairs.*

Proof: We only have to show that the algorithm can be implemented to run in linear time and space. Di Battista and Tamassia [4] showed that the SPQR-tree can be computed in linear time and that the sum of all vertices and edges of the triconnected components is $\mathcal{O}(n)$. Let G' be a triconnected component with $n_{G'}$ vertices. Computing the dual graph H of G and its vertex cover S requires $\mathcal{O}(n_{G'})$ time. Embedding the corresponding subgraph G'' in face F or F' can be done in constant time, given the current planar embedding of G'' and G'. Hence the running time and space is $\mathcal{O}(n_{G'})$ for every triconnected component G' which implies a total time and space of $\mathcal{O}(n)$. □

5 Further Remarks and Open Questions

In this paper we considered the problem of triangulating a planar graph without introducing separating triangles. We showed that if the embedded planar graph has no separating triangles and does not contain star the resulting triangulation does not contain a separating triangle, i.e. the graph is 4-connected. This result has important applications in the area of visibility representations of planar graphs.

In order to apply visibility representations and rectangular duals the graph has to be 4-connected and triangular [15, 17]. In the general augmentation context the question arises how to find a mimimum set of edges whose addition 4-connects a graph without losing planarity. As far as we know this problem is open. Also the question whether testing 4-connectivity of planar graphs can be done in linear time is open.

We like to end the paper with some remarks on 5-connectivity. As mentioned in Section 2, every planar graph is at most 5-connected. Of course a triangular planar graph is 5-connected if it does not contain a separating 4-cycle (quadrangle). We can use this algorithm to test 5-connectivity of triangular planar graphs as follows: the algorithm of Chiba and Nishizeki [2] outputs the sets $(v, w, \{u_1, \ldots, u_p\})$, where $\{u_1, \ldots, u_p\}$ is the set of common neighbors of v and w, with $p \geq 2$. We assume that every vertex v has $deg(v) > 4$, otherwise the neighbors of v form a separating triangle or quadrangle. If $p > 2$ then u_1 and u_p are not consecutive in the adjacency list of v in which case there is a separating quadrangle on the vertices v, w, u_1, u_p. If $p = 2$ we test whether u_1 and u_2 are consecutive in the adjacency list of v. Using the adjacency lists of the embedded planar graph, this requires linear time and space. Hence testing 5-connectivity of triangular planar graphs can be done in linear time and space.

Triangulating planar graphs while admitting the 5-connectivity constraint is more difficult. Even triangulating a graph consisting of a cycle of length 5 introduces a separating quadrangle. Again the problem arises whether there is a precise class of planar graphs such that a planar graph can be augmented to a 5-connected triangular planar graph if and only if it does not belong to this class. Also the time complexity of this augmentation algorithm is an interesting problem for further research.

References

[1] Bhasker, J., and S. Sahni, A linear algorithm to check for the existence of a rectangular dual of a planar triangulated graph, *Networks* 7 (1987), pp. 307–317.

[2] Chiba, N., and T. Nishizeki, Arboricity and subgraph listing algorithms, *SIAM J. Comput.* 14 (1985), pp. 210–223.

[3] Chiba, N., T. Nishizeki, S. Abe and T. Ozawa, A linear algorithm for embedding planar graphs using PQ-trees, *J. of Computer and System Sciences* 30 (1985), pp. 54–76.

[4] Di Battista, G., and R. Tamassia, Incremental planarity testing, in: *Proc. 30th Annual IEEE Symp. on Foundations of Comp. Science*, 1989, pp. 436–441.

[5] Di Battista, G., Eades, P., and R. Tamassia, I.G. Tollis, *Algorithms for Automatic Graph Drawing: An Annotated Bibliography*, Brown Univ., Tech. Rep. , 1993.

[6] Eswaran, K.P., and R.E. Tarjan, Augmentation problems, *SIAM J. Comput.* 5 (1976), pp. 653–665.

[7] Frank, H. and W. Chou, Connectivity considerations in the design of survivable networks, *IEEE Trans. on Circuit Theory*, CT-17 (1970), pp. 486–490.

[8] Garey, M.R., D.S. Johnson, The Rectilinear Steiner Tree Problem is NP-complete, *SIAM J. Appl. Math.* **32** (1977), pp. 826-834.

[9] Garey, M.R., D.S. Johnson and L. Stockmeyer, Some simplified NP-complete graph problems, *Theoret. Comp. Science* 1 (1976), pp. 237–267.

[10] Kant, G., The Planar Triconnectivity Augmentation Problem, submitted to *Theoretical Computer Science*, 1993.

[11] Hsu, T.-S., On four-connecting a triconnected graph, in: *Proc. 33st Annual IEEE Symp. on Found. of Comp. Science*, 1992, pp. 70–79.

[12] Hsu, T.-S., and V. Ramachandran, A linear time algorithm for triconnectivity augmentation, *Proc. 32th Annual IEEE Symp. on Found. of Comp. Sc.*, 1991, pp. 548–559.

[13] Hsu, T.-S., and V. Ramachandran, On finding a smallest augmentation to biconnect a graph, in: W.L. Hsu and R.C.T. Lee (Eds.), *Proc. of the Second Annual Int. Symp. on Algorithms*, Lecture Notes in Comp. Science 557, Springer-Verlag, 1992, pp. 326–335.

[14] Kanevsky, A., R. Tamassia, G. Di Battista and J. Chen, On-line maintenance of the four-connected components of a graph, In *Proc. 32th Annual IEEE Symp. on Foundations of Comp. Science*, 1991, pp. 793–801.

[15] Kant, G., A More Compact Visibility Representation, in: J. van Leeuwen (Ed.), *Proc. 19th Intern. Workshop on Graph-Theoretic Concepts in Comp. Sc. (WG'93)*, LNCS, Springer 1994, to appear.

[16] Kant, G., *Algorithms for Drawing Planar Graphs*, PhD thesis, Utrecht University, 1993.

[17] Kant, G., and X. He, Two Algorithms for Finding Rectangular Duals of Planar Graphs, in: J. van Leeuwen (Ed.), *Proc. 19th Intern. Workshop on Graph-Theoretic Concepts in Comp. Sc. (WG'93)*, LNCS, Springer 1994, to appear.

[18] Koźmiński, K., and E. Kinnen, Rectangular dual of planar graphs, *Networks 5* (1985), pp. 145–157.

[19] Lengauer, Th., *Combinatorial Algorithms for Integrated Circuit Layout*, Wiley, 1990.

[20] Otten, R.H.J.M., and J.G. van Wijk, Graph representation in interactive layout design, in: *Proc. IEEE Int. Symp. on Circuits and Systems*, 1978, pp. 914–918.

[21] Read, R.C., A new method for drawing a graph given the cyclic order of the edges at each vertex, *Congr. Numer.* 56 (1987), pp. 31–44.

[22] Rosenthal, A., and A. Goldner, Smallest augmentation to biconnect a graph, *SIAM J. Comput* 6 (1977), pp. 55-66.

[23] Rosenstiehl, P., and R.E. Tarjan, Rectilinear planar layouts and bipolar orientations of planar graphs, *Discr. and Comp. Geometry* 1 (1986), pp. 343–353.

[24] Steiglitz, K., P. Weiner and D.J. Kleitman, The design of minimum-cost survivable networks, *IEEE Trans. on Circuit Theory*, CT-16 (1969), pp. 455-560.

[25] Tamassia, R., and I.G. Tollis, A unified approach to visibility representations of planar graphs, *Discr. and Comp. Geometry* 1 (1986), pp. 321–341.

Parallel and Sequential Approximation of Shortest Superstrings

Artur Czumaj[1][*][†] Leszek Gąsieniec[2][**] Marek Piotrów[1][***][†] Wojciech Rytter[2][**]

[1] Heinz Nixdorf Institut, Universität-GH-Paderborn,
Warburger Str. 100, 33098 Paderborn, Germany.
[2] Instytut Informatyki, Uniwersytet Warszawski, Banacha 2, 02-097 Warszawa, Poland.

Abstract. Superstrings have many applications in data compression and genetics. However the decision version of the shortest superstring problem is $\mathcal{NP}$-complete. In this paper we examine the complexity of approximating a shortest superstring. There are two basic measures of the approximations: the compression ratio and the approximation ratio. The well known and practical approximation algorithm is the sequential algorithm GREEDY. It approximates the shortest superstring with the compression ratio of $\frac{1}{2}$ and with the approximation ratio of 4. Our main results are:
(1) An $\mathcal{NC}$ algorithm which achieves the compression ratio of $\frac{1}{4+\epsilon}$.
(2) The proof that the algorithm GREEDY is not parallelizable, the computation of its output is P-complete.
(3) An improved sequential algorithm: the approximation ratio is reduced to 2.83. Previously it was reduced by Teng and Yao from 3 to 2.89.
(4) The design of an $\mathcal{RNC}$ algorithm with constant approximation ratio and an $\mathcal{NC}$ algorithm with logarithmic approximation ratio.

1 Introduction

Let $S = \{s_1, \ldots, s_n\}$ be a set of n strings over some alphabet Σ. A *superstring* of S is a string sp over Σ such that each string $s_i \in S$ appears as a substring (a consecutive block of characters) of sp. *The shortest superstring* problem is to find for a given set S the shortest superstring $ss(S)$. We use $opt(S)$ to denote the length of $ss(S)$. Assume further that no string $s_i \in S$ is a substring of any other $s_j \in S$.

It is known that the shortest superstring problem is $\mathcal{NP}$-hard [5, 6]. Because of its important applications in data compression practice [13] and DNA sequencing procedure [8, 12], it is of interest to find approximation algorithms with good performance guarantees. For example, a DNA molecule can be represented as a character string over a set of nucleotides $\{A, C, G, T\}$. Although a DNA string can have up

* Supported by DFG-Graduiertenkolleg "Parallele Rechnernetzwerke in der Produktionstechnik", ME 872/4-1. Email: artur@hni.uni-paderborn.de.
** Supported in part by the EC Cooperative Action IC 1000 Algorithms for Future Technologies "ALTEC". Email: {lechu,rytter}@mimuw.edu.pl.
*** Supported in part by Alexander von Humboldt-Stiftung and Volkswagen Stiftung. Permanent address: Instytut Informatyki, Uniwersytet Wrocławski, Przesmyckiego 20, 51-151 Wrocław, Poland. Email: marekp@uni-paderborn.de.
† Supported in part by the ESPRIT Basic Research Action No. 7141 (ALCOM II)

to 3×10^9 characters (for a human being), with current laboratory methods only small fragments of at most 500 characters can be determined at a time. Then from a huge number of these fragments, a biochemist should construct the superstring representing the whole molecule. Efficient superstring approximation algorithms are routinely used to cope with this job. In particular, good parallel algorithms would be useful in that context.

To evaluate how good is the obtained approximation, two kinds of measure are used. The first, most important in practice, is to find a superstring sp of S such that the ratio $\frac{|sp|}{opt(S)}$ is minimized. We will call this ratio to be the *approximation factor of a superstring*. The second approach is to find a superstring sp of S such that the ratio of the total compression obtained by sp and by $ss(S)$ is maximized. That is, we want to maximize $\frac{|S|-|sp|}{|S|-opt(S)}$, where $|S| = \sum_{1 \le i \le n} |s_i|$. We will call this ratio to be the *compression factor of a superstring*.

The algorithm GREEDY is a simple sequential approximation of a shortest superstring and appears to do quite well. It can be presented in the following way. Given a non-empty set of strings $S = \{s_1, \ldots, s_n\}$, repeat the following steps until S contains just one string (which is a superstring of S): Select a pair of strings $s', s'' \in S$ that maximizes overlap between s' and s''; Remove s' and s'' from S replacing them with the merge of s' and s''.

It was proved by Tarhio and Ukkonen [14] and Turner [16] that GREEDY achieves the compression factor of at least $1/2$. Other heuristics have been also considered by the authors, but $1/2$ is still the best obtained compression factor. The approximation factor of GREEDY was unknown for a long time. The first breakthrough was made by Blum et al. [2], where they proved that GREEDY achieves an approximation factor of 4. Furthermore, they showed a modified greedy algorithm that has an approximation factor of 3, and proved that the superstring problem is MAX-SNP-hard [11]. The recent result in [1] that MAX SNP-hard problems do not have polynomial time approximation scheme unless $\mathcal{P} = \mathcal{NP}$ implies that a polynomial time approximation scheme (that is, polynomial time algorithms with approximation factor of $1 + \varepsilon$ for any fixed $\varepsilon > 0$) for this problem is unlikely. Recently Teng and Yao [15] improved the result of Blum et al. [2] and presented an algorithm that achieved an approximation factor of 2.89. Our contribution is an algorithm whose approximation factor can be bounded by 2.83.

As far as we know, no parallel approximation algorithm for the superstring problem has been presented. In this paper, we give the following results concerning the parallel complexity of the problem:

1. An $\mathcal{NC}$ algorithm which achieves the compression ratio of $\frac{1}{4+\varepsilon}$.
2. The proof that the algorithm GREEDY is not parallelizable, the computation of its output is P-complete.
3. The design of an $\mathcal{RNC}$ algorithm with constant approximation factor and an $\mathcal{NC}$ algorithm with logarithmic approximation factor.

The open problem is to construct an $\mathcal{NC}$ algorithm with a constant approximation factor.

Below we introduce some necessary definitions.

For two strings s and t let v be the longest string such that $s = uv$ and $t = vw$ for some non-empty strings u and w. The *overlap* between two strings s and t is the length of the string v. We will denote it as $\text{ov}(s,t)$. The *prefix* of a string s with respect to a string t is the length of the string u. We will denote it as $\text{pref}(s,t)$. It will cause no confusion if sometimes we also call the string v to be the overlap $(\text{ov}(s,t))$ and the string u to be the prefix $(\text{pref}(s,t))$ of s and t. Define $s \circ t$ to be the string uvw, that is $s \circ t = \text{pref}(s,t)\,t$.

For a given set of strings $S = \{s_1, \ldots, s_n\}$ define an *overlap graph of* S to be the weighted digraph $OG(S) = (V, E, \text{ov})$ which has n vertices $V = \{1, \ldots, n\}$ and n^2 edges $E = \{(i,j) : 1 \le i, j \le n\}$. Here we take as weight function the overlap $\text{ov}(,)$: edge (i,j) has weight $\text{ov}(i,j) = \text{ov}(s_i, s_j)$.

Algorithm GREEDY can be also restated in terms of the overlap graph $OG(S)$. Repeat until selected edges do not form a Hamiltonian path in $OG(S)$: Scan the edges of $OG(S)$ in non-increasing order of weights and select an edge (i,j) if no edge of the form (i,p) or (q,j) has been previously selected and if the collection of paths constructed so far does not include a path from j to i. Obtained Hamiltonian path $(i_1, i_2), (i_2, i_3), \cdots, (i_{n-1}, i_n)$ defines us the superstring $s_{i_1} \circ s_{i_2} \circ s_{i_3} \circ \ldots \circ s_{i_{n-1}} \circ s_{i_n}$.

A weighted digraph G is an *overlap graph* if there exists a set S of strings such that the graph obtained from $OG(S)$ after removing all zero-weighted edges is isomorphic to G.

Due to space limitations some proofs and details are omitted here and will appear in the full version of the paper.

2 $\mathcal{NC}$-Approximation of Shortest Superstrings

Define a *cycle-cover* of a graph G to be a maximal collection of cycles in G such that each vertex is in at most one cycle. Let also a *path-cycle cover* be a collection of paths and cycles in G such that each vertex is in exactly one path or cycle. We call a path-cycle cover *maximal* if it can not be extended by any other edge in G.

Let $G = (V, E, w)$ be a complete weighted digraph without selfloops, where $V = \{1, \ldots, n\}$ is the set of vertices, $E = \{(i,j) : i \ne j \in V\}$ is the set of edges and $w : E \to \mathbb{R}_+$ is the (non-negative) weight function. Assume also that the unary weights are given. The *maximum cycle-cover* problem is to find a cycle-cover with the maximum weight (ie., the total weight of the cycles is maximized). It is known that this problem is reduced to the maximum-weighted matching problem in bipartite graphs [10]. Thus it can be solved in polynomial sequential time. However there is not known any $\mathcal{NC}$ algorithm for it. In this paper we are only focused on this problem for the case when all weights are given in unary. In this case there is known an $\mathcal{RNC}$ algorithm for the maximum-weighted matching problem in bipartite graphs, thus also for the maximum cycle-cover problem [17]. In what follows we show that there is an $\mathcal{NC}$ algorithm that finds an $(\frac{1}{2+\varepsilon})$-approximation of the maximum cycle-cover (and also for the maximum-weighted matching problem in bipartite graphs).

2.1 Sequential Approximation of a Maximum Cycle-Cover

We begin with a sequential $\frac{1}{2}$-approximation. The following is a simple greedy algorithm that finds a cycle-cover.

Algorithm CC-GREEDY :

Repeat until selected edges do not form a cycle-cover of G:

Scan the edges of G in non-increasing order of weight and select an edge (i, j) if no edge of the form (i, p) or (q, j) has been previously selected.

Lemma 1. *Algorithm* CC-GREEDY *finds a cycle-cover of weight that is at least half of the weight of a maximum cycle-cover.*

Proof. The proof follows the ideas of Turner's estimation of the superstring compression factor achieved by GREEDY [16]. Algorithm CC-GREEDY selects n edges in non-increasing order and let e_i be the i-th chosen edge. Let P_i be a maximum cycle-cover that includes edges $\{e_1, \ldots, e_i\}$ and let $C_i = P_i - \{e_1, \ldots, e_i\}$. We show that for $1 \le i \le n$, $w(C_{i-1}) \le w(C_i) + 2w(e_i)$. Since $w(C_0)$ is the weight of a maximum cycle-cover and $w(C_n) = 0$, this would imply the lemma.

Let $e_i = (p, q)$. Since e_i is the i-th edge chosen by CC-GREEDY , $w(e_i) \ge \max\{w(e) : e \in C_{i-1}\}$. There are at most two edges e' and e'' that are in $C_{i-1} - C_i$ and which share the head or the tail with e_i. Let $e' = (p, s)$, $e'' = (t, q)$ and when $s \ne t$, let us define $e^* = (t, s)$, Then we can obtain a cycle-cover $(P_{i-1} - \{e', e''\}) \cup \{e_i, e^*\}$ when $s \ne t$ or a cycle-cover $(P_{i-1} - \{e', e''\}) \cup \{e_i\}$ otherwise. In both cases we have:

$$
\begin{aligned}
w(C_i) + w(e_i) &\ge w((C_{i-1} - \{e', e''\}) \cup \{e_i\}) \\
&= w(C_{i-1}) - w(e') - w(e'') + w(e_i) \\
&\ge w(C_{i-1}) - w(e) - w(e) + w(e_i) \\
&= w(C_{i-1}) - w(e_i)
\end{aligned}
$$

which implies $w(C_{i-1}) \le w(C_i) + 2w(e_i)$.

Using well-known transformations (see eg. [10]) Lemma 1 can be restated as the following Corollary.

Corollary 2. *In a weighted bipartite graph the* GREEDY *matching algorithm finds a matching of weight that is at least half of the weight of a maximum-weighted matching.*

2.2 Parallel Approximation of a Maximum Cycle-Cover

In this section we describe an approximation of algorithm CC-GREEDY . Let $c > 1$, $G = (V, E, w)$ and for each edge e if $w(e) \in (c^{k-1}, c^k]$ then define c-level of e, LEVEL$_c(e)$, to be k (ie., LEVEL$_c(e) = \lceil \log_c w(e) \rceil$); additionally if $w(e) \le 1$ then LEVEL$_c(e) = 0$.

For a given path $v_1, v_2, \ldots, v_k$ its *contraction* is defined as follows. We remove vertices $v_2, \ldots, v_k$ together with the edges incident to them and change the edges (with their weights) outgoing from v_1 to be the edges outgoing previously from v_k. Vertex v_1 is called the *contraction* of the path $v_1, v_2, \ldots, v_k$. The *recontraction of a cycle or a path* is obtained by recursively replacing each contraction by its path.

Algorithm ACC-GREEDY

1. Let s be the maximum c-level of edges in G (ie., $s = \max_{e \in E}\{\text{LEVEL}_c(e)\}$) and let $\tilde{G} = G$ and $C = \emptyset$.
2. Repeat for $t = s$ downto $t = 0$
   (a) Let G_t be the (non-weighted) subgraph of $\tilde{G}$ induced by the edges from the t-th c-level
   (b) Find a maximal path-cycle cover of G_t
   (c) Remove all cycles in G_t from $\tilde{G}$ and after recontracting add them to C; contract all obtained paths
3. The set C contains a cycle-cover

Lemma 3. *For any $c > 1$ algorithm* ACC-GREEDY *finds a cycle-cover of weight at least $\frac{1}{2c}$ of the weight of a maximum cycle-cover.*

Proof. Let $\overline{G} = (V, E, \overline{w})$ be the graph with weights

$$\overline{w}(e) = \begin{cases} c^{\text{LEVEL}_c(e)} & \text{if ACC-GREEDY chooses } e \\ w(e) & \text{otherwise} \end{cases}$$

Note that $w(e) \leq \overline{w}(e) < c\,w(e)$ for every $e \in E$. Let $\text{MCC}(G)$ be the weight of a maximum cycle-cover in G and $w(AM(G))$ be the sum of weights of the edges chosen by algorithm ACC-GREEDY on G. Observe that algorithm ACC-GREEDY on the graph $\overline{G}$ is equivalent to CC-GREEDY assuming that both algorithms select the same edges in the case of equal weights. Thus $w(AM(\overline{G})) \geq \frac{1}{2}\text{MCC}(\overline{G}) \geq \frac{1}{2}\text{MCC}(G)$. On the other hand $w(AM(\overline{G})) < c\,w(AM(G))$. Thus finally we get $c\,w(AM(G)) \geq \frac{1}{2}\text{MCC}(G)$.

Now we want to show that this algorithm can be implemented to run in polylogarithmic time with polynomial number of processors. The main loop is executed $\lceil \log_c(\max_{e \in E}\{w(e)\}) \rceil$ times. Since the weights of the graph are given in unary, we only have to show that there is an $\mathcal{NC}$ algorithm that finds a maximal path-cycle cover. The proof of the following lemma will appear in the full version of the paper.

Lemma 4. *There is an $\mathcal{NC}$ algorithm that finds a maximal path-cycle cover of a digraph G.*

Lemma 3 and Lemma 4 immediately imply the following theorem.

Theorem 5. *There is an $\mathcal{NC}$ algorithm that finds a cycle-cover of a weighted digraph G with the cost of at least $\frac{1}{2+\varepsilon}$ of the weight of a maximum cycle-cover. Here $\varepsilon > 0$ is any arbitrary but fixed constant, and we assume that the weights of G are given in unary.*

The running time of this algorithm is either $O(\log^2 n \, \log_{1+\varepsilon}(\max_{e \in E}\{w(e)\}))$ with n^4 processors or $O(\log^3 n \, \log_{1+\varepsilon}(\max_{e \in E}\{w(e)\}))$ with n^2 processors.

2.3 Parallel Approximation of Shortest Superstrings

In this section we develop techniques presented in the previous section to design an $\mathcal{NC}$ algorithm that finds a superstring that has the overlap (the compression measure) at least $\frac{1}{4+\epsilon}$ that of a shortest superstring.

First we build the overlap graph $OG(S)$ for the set of strings S. We assume in this construction that there is no selfloop in $OG(S)$. Then we find a cycle-cover of OG using algorithm ACC-GREEDY from Section 2.2. We next remove from every cycle an edge with the minimum weight and join (by any edges) obtained paths to get a Hamiltonian path.

Lemma 6. *Obtained Hamiltonian path is of the weight at least $\frac{1}{4+\epsilon}$ of the weight of a maximum weight Hamiltonian path, for any $\epsilon > 0$.*

Proof. Any maximum cycle-cover MCC is of weight not smaller than the weight of a maximum Hamiltonian path MHP. Let C be a cycle-cover obtained by algorithm ACC-GREEDY and GS be a superstring obtained by the algorithm presented above. From Theorem 5 we get $w(C) \geq \frac{1}{2+\frac{1}{2}\epsilon}w(\text{MCC})$ for any $\epsilon > 0$. Since we remove the least weighted edge from every cycle in C, we get $w(GS) \geq w(C)/2$. Thus

$$w(GS) \geq w(C)/2 \geq \frac{1}{4+\epsilon}w(\text{MCC}) \geq \frac{1}{4+\epsilon}w(\text{MHP})$$

Theorem 7. *There is an $\mathcal{NC}$ algorithm that achieves compression factor of $\frac{1}{4+\epsilon}$. It runs either in $O(\log^2 n \cdot \log_{1+\epsilon} |S|)$ time with n^4 processors or in $O(\log^6 n \cdot \log_{1+\epsilon} |S|)$ time with n^2 processors.*

3 Algorithm GREEDY Is Not Parallelizable

Algorithm GREEDY appears to be very sequential in nature, since to select a current pair of strings with the largest overlap we need to know the results of previous merges. To formalize this observation we would like to prove that GREEDY applied to the superstring problem is $\mathcal{P}$-complete [3], what is commonly believed to mean: a hardly parallelizable algorithm.

We start with proving that the problem of finding the Hamiltonian path chosen by algorithm GREEDY is $\mathcal{P}$-complete. For a given Boolean circuit, a certain complete weighted digraph is introduced, in which a Hamiltonian path selected by GREEDY can simulate a computation of the circuit's value. Then we argue that the digraph is an overlap graph, i.e., a set of strings can be constructed whose overlap graph is isomorfic to the digraph.

Lemma 8. *The problem of finding the Hamiltonian path chosen by algorithm GREEDY is P-complete.*

Proof. Due to space limitations the construction of the graph is omitted here. From now on we will call the result digraph the *circuit-simulating graph*.

---

[3] To be more precise: a search problem of finding a superstring obtained by GREEDY is considered and proved to be P-complete

For a digraph G define its *skeleton* $\overline{G}$ to be an undirected graph with the vertex set the same as the vertex set of G and the edge set which is obtained from the edge set of G by removing directions.

In the following lemmas we would like to derive some sufficient conditions of a digraph to be an overlap graph. The first observation is that a positive-weighted edge in an overlap graph relates the beginning of one string to the end of another. Therefore, when we want to collect the related strings in a structure, we have to consider adjacent edges in alternating directions. This leads us to the following definitions:

An *alternating path* is a sequence of nodes and edges $v_1 e_1 v_2 e_2 \cdots v_{k-1} e_{k-1} v_k$ such that either $e_1 = (v_1, v_2), e_2 = (v_3, v_2), e_3 = (v_3, v_4), e_4 = (v_5, v_4) \cdots$, or $e_1 = (v_2, v_1), e_2 = (v_2, v_3), e_3 = (v_4, v_3), e_4 = (v_4, v_5) \cdots$. An *alternating tree* is a connected (ie., $\overline{AT}$ is connected) digraph $AT = (V_T, E_T)$ containing t nodes and $t-1$ edges such that each sequence $v_1 e_1 v_2 e_2 \cdots v_{l-1} e_{l-1} v_l$, where $v_i \in V_T$, $e_i \in E_T$ and $v_1 e_1 v_2 e_2 \cdots v_{l-1} e_{l-1} v_l$ is a path in $\overline{AT}$, is an alternating path. Assume that all edges in AT have weights defined by a function w. A weighted alternating tree AT is *monotone* if there exists a vertex r, called the *root* of AT , such that for each alternating path $r e_1 v_1 e_2 v_2 \cdots e_k v_k$ in AT , $w(e_1) < w(e_2) < \cdots < w(e_k)$ An *alternating cycle* in a digraph G is a cycle in $\overline{G}$ that can be transformed into an alternating path by splitting it in a node.

Lemma 9. *Every monotone alternating tree with positive, integer weights is an overlap graph.*

Lemma 10. *If a weighted digraph G with positive, integer weights can be edge-covered by a disjoint sum of monotone alternating trees in such a way that for each vertex in G all its incoming edges are in one tree and all its outgoing edges are in another tree, then G is an overlap graph.*

Proof. According to Lemma 9, each alternating tree AT in the cover C of G is an overlap graph. That is, strings can be assigned to nodes of AT to obtain the corresponding overlap graph. Let Σ_{AT} denote an alphabet of the strings. W.l.o.g. we can assume that alphabets Σ_{AT}, AT $\in C$, are pairwise disjoint. Let the incoming edges of a vertex v be in AT and its outgoing edges in AT '. Thus we have got two strings $in_v \in \Sigma_{AT}^*$ and $out_v \in \Sigma_{AT'}^*$. The result string for v we obtain by concatenating in_v with out_v. It can be easily checked that the overlap between two such strings is non-zero if and only if the corresponding nodes are joint by an edge.

Lemma 11. *If a digraph G does not contain alternating cycles, then it can be (uniquely) edge-covered by edge-disjoint alternating trees such that each vertex of G has all its incoming edges in one tree and all its outcoming edges in another tree.*

Theorem 12. *The problem of finding a superstring chosen by algorithm GREEDY is $\mathcal{P}$-complete.*

Proof. With respect to Lemma 8 we have only to show that a circuit-simulating graph G is an overlap graph. The gates in G can been designed in such a way that G contains no alternating cycles. By Lemma 11, G can be uniquely edge-covered by disjoint alternating trees. Moreover the trees are of size bounded by a constant (independent on the size of a circuit) and they are monotonic. Hence, by Lemma 10, G is an overlap graph.

4 Sequential Algorithm with 2.83 Approximation Factor

In this section we present a new sequential algorithm for the superstring problem that has an approximation factor of $2\frac{5}{6}$ and thus supersedes the algorithm of [15]. The later algorithm has the factor of $2\frac{8}{9}$ and is an improvement on algorithm TGREEDY [2] that has the factor of 3. We base on the ideas from both papers.

In this section, according to the previous papers, we will use the terms *assignment* and *cycle-cover* interchangeably. For a given set of strings $S = \{s_1, \ldots, s_n\}$ we consider two complete digraphs with S as a set of nodes: one $OG(S)$ weighted by $\text{ov}(\cdot, \cdot)$, the other $PG(S)$ weighted by $\text{pref}(\cdot, \cdot)$. We assume that both contain no selfloop. Let us notice that a minimum assignment in $PG(S)$ is also a maximum assignment in $OG(S)$, and vice versa. We will call such an assignment an *optimal assignment*. For a cycle c in an assignment C, let $d(c)$ denote the total $\text{pref}(\cdot, \cdot)$ weight of the edges in c. We refer to $d(c)$ as the *weight* of c.

For the sake of completeness we recall two crucial lemmas from [2] and [15]. Recall that a minimum assignment is called *canonical* if each string s is assigned to a cycle whose weight is the smallest among all cycles that s fits (see [15]).

Lemma 13. [2] *Let c_1 and c_2 be two cycles in a minimum weight assignment C with $s_1 \in c_1$ and $s_2 \subset c_2$. Then, the overlap between s_1 and s_2 is less than $d(c_1) + d(c_2)$.*

Lemma 14. (2-cycle Lemma, Teng and Yao [15])
Let c_1 and c_2 be two cycles in a canonical minimum assignment C with $r_1 \in c_1$ and $r_2 \in c_2$. Then $\text{ov}(r_1, r_2) + \text{ov}(r_2, r_1) < \max(|r_1|, |r_2|) + \min(d(c_1), d(c_2))$.

As the shortest superstring problem for S corresponds to the maximum Hamiltonian path problem in $OG(S)$ graph, approximation schemes start by computing an optimal assignment. Then the problem is how to join the cycles in the assignment to obtain a Hamiltonian path. Algorithm TGREEDY [2] opens each cycle by deleting the edge with the shortest overlap and joins the obtained strings by GREEDY . One can obtain the same approximation factor of 3 by the following procedure:

1. select a set R of cycles' representatives (one node s_i from each cycle c_i) and find an optimal assignment CC of R;
2. open each cycle in CC by deleting the shortest-overlap edge, and concatenate the obtained strings to form α;
3. split each cycle c_i in the selected node s_i (the result will begins and ends with s_i) and replace s_i in α by the results.

These stages form the basis of the algorithm presented by Teng and Yao [15]. The improvement on the approximation factor in [15] is obtained by making the assignment in Stage 1 canonical and by treating separately 2-cycles in Stage 2. Our further improvement is achieved by selecting "good" representatives of 2-cycles and 3-cycles in Stage 2, and by finding an optimal assignment on them.

Below there is an analog of Lemma 14 for a 3-cycle.

Lemma 15. (3-cycle Lemma)

Let c_1, c_2 and c_3 be cycles in a minimum assignment C with $r_1 \in c_1$, $r_2 \in c_2$ and $r_3 \in c_3$. Then $ov(r_1, r_2) + ov(r_2, r_3) + ov(r_3, r_1) \leq 2 \cdot \min(|r_1|, |r_2|, |r_3|) + d(c_1) + d(c_2) + d(c_3)$.

Proof. Without loss of generality we can assume that $|r_1| = \min(|r_1|, |r_2|, |r_3|)$. Then $ov(r_1, r_2) \leq |r_1|$, $ov(r_3, r_1) \leq |r_1|$. By Lemma 13, $ov(r_2, r_3) \leq d(c_2) + d(c_3)$.

The Algorithm

1. Find an optimal assignment C of S, and make C canonical.
2. Take an arbitrary string from each cycle of C to form a set R, and find an optimal assignment CC for R.
3. Select representative set RR containing one element for each 2-cycle in CC and one element for each 3-cycle in CC. From each 2-cycle take the longer string and from each 3-cycle take a string that is not in the pair with the longest overlap (i.e. when s_1, s_2 and s_3 are the strings of a 3-cycle, $ov(s_1, s_2) \geq ov(s_2, s_3)$ and $ov(s_1, s_2) \geq ov(s_3, s_1)$, then select s_3). Let $RR = \{g_1, g_2, \ldots, g_m\}$. If g_i is from a 2-cycle then let f_i be the second element of the cycle. If g_i is from a 3-cycle then let f_i be the superstring of two other strings in the cycle, ordered as on the cycle. Let $RR' = \{f_1, f_2, \ldots, f_m\}$.
4. Find an optimal assignment CCC on RR. Create two superstrings for $RR \cup RR'$ by splitting cycles in CCC and inserting strings from RR' (see Fig. 1). Cycles with even number of nodes are split exactly (i.e. every edge of a cycle is built in a superstring), in cycles with odd number of nodes the edge with a smallest overlap is left out. Let q_0 and q_1 be the result superstrings and let q be the shorter of them.

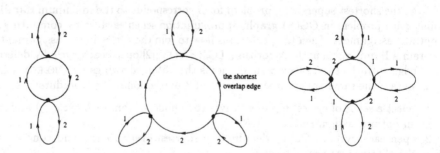

Fig. 1. Splitting CCC cycles together with 2,3-cycles from CC into two superstrings. Edges labeled with 1(2) are in the superstring $g_1(g_2)$.

5. Open each i-cycle, $i \geq 4$, in CC by deleting the edge with the smallest overlap; concatenate the resulting strings together with q to obtain α. Note that α is a superstring for R. Split each cycle in C in a node from R to obtain superstrings that begin and end with the strings from R. Let $\bar{\alpha}$ be the extended string of

α obtained by replacing each string of R with the superstring representing its cycle in C. Return $\bar{\alpha}$.

Analysis

Observe first that the algorithm runs in polynomial time, because an optimal assignment can be constructed in time $O(n^3)$ (see e.g. [10]), and for given a minimum assignment one can transform it into a canonical minimum one in $O(n|S|)$ time [15].

Let d_2, d_3, d_4 be, respectively, the total weight of the cycles in C that have representatives in 2-cycles, 3-cycles, and all i-cycles for $i \geq 4$, in assignment CC. Thus $d_2 + d_3 + d_4 = d(C)$. Let ov_2, ov_3 and ov_4 be, respectively, the total overlap present in the 2-cycles, 3-cycles, and all i-cycles for $i \geq 4$, in assignment CC. Then, we have: $ov(q_1) + ov(q_2) \geq ov_2 + ov_3 + ov_3/3 + \frac{2}{3}ov(CCC)$, since in each 3-cycle of CC we take twice the edge with the longest overlap and in CCC-cycles with the odd number of edges only the edge with the smallest overlap is deleted. Since $ov(CCC) \geq ov_{opt}(RR) = |RR| - opt(RR)$, we get

$$ov(q) \geq \frac{1}{2}ov_2 + \frac{2}{3}ov_3 + (|RR| - opt(RR))/3.$$

Let RR_i, $i = 2, 3$, denote the set $\{r \in RR : r$ is in a i cycle in $CC\}$. Then $ov(\alpha) \geq ov(q) + \frac{3}{4}ov_4 = \frac{1}{2}ov_2 + \frac{2}{3}ov_3 + \frac{3}{4}ov_4 + (|RR_2| + |RR_3| - opt(RR))/3$.

Lemma 16.

$$|RR_2| \geq ov_2 - d_2/2,$$
$$|RR_3| \geq ov_3/2 - d_3/2.$$

Proof. The first inequality follows from Lemma 14 and the second from Lemma 15.

Combining all observations above we get,

$$|\alpha| = |R| - ov(\alpha) = opt(R) + ov_{opt}(R) - ov(\alpha) \leq opt(R) + ov(CC) - ov(\alpha)$$

$$\leq opt(R) + opt(RR)/3 + \frac{1}{2}ov_2 + \frac{1}{3}ov_3 + \frac{1}{4}ov_4 - (|RR_2| + |RR_3|)/3$$

$$\leq opt(R) + opt(RR)/3 + \frac{1}{6}ov_2 + \frac{1}{6}ov_3 + \frac{1}{4}ov_4 + \frac{1}{6}(d_2 + d_3)$$

$$\leq opt(S) + opt(S)/3 + \frac{1}{3}d_2 + \frac{1}{3}d_3 + \frac{1}{2}d_4 + \frac{1}{6}(d_2 + d_3)$$

$$= \frac{4}{3}opt(S) + \frac{1}{2}(d_2 + d_3 + d_4) \leq opt(S)(\frac{4}{3} + \frac{1}{2}) = 1\frac{5}{6}opt(S)$$

Since $|\bar{\alpha}| = |\alpha| + d(C)$, we obtain $|\bar{\alpha}| \leq 2\frac{5}{6}opt(S)$.

5 Parallel Approximations of Superstring Length

In this section we present an $\mathcal{NC}$ algorithm with a logarithmic approximation ratio and an $\mathcal{RNC}$ algorithm with a constant approximation ratio.

5.1 The Weighted Set Cover Problem

Let $X = \{1, 2 \ldots, n\}$ and let $Y = \{Y_1, Y_2, \ldots, Y_m\} \subseteq 2^X$ be a family of its subsets. A *cover* of X is a subcollection $Y' \subseteq Y$ such that $\bigcup_{Y_i \in Y'} Y_i = X$. For each set $Y_i \in Y$ let $w(Y_i)$ denote its (positive) weight and for a subcollection $Y' \subseteq Y$ define its weight by $w(Y') = \sum_{Y_i \in Y'} w(Y_i)$. The *weighted set cover* problem is to find a cover Y^* of X of the minimum weight.

The weighted set cover problem is known to be $\mathcal{NP}$-hard [6]. A recent result of Lund and Yannakakis [9] shows that this problem cannot be approximated in $\mathcal{P}$ with ratio $c \log_2 n$ for any $c < 1/4$ unless $\mathcal{NP} = \mathrm{DTIME}(n^{O(1)})$. However there is known a polynomial-time algorithm that finds a logarithmic-factor approximation. The following lemma has been shown by Berger et al. [3].

Lemma 17. *For any $\varepsilon > 0$, there is an $\mathcal{NC}$ algorithm for the weighted set cover problem that runs in $O(\log^4 n \log m \log^2(nm)/\varepsilon^6)$ time, uses $O(n + \sum_{i=1}^{m} \mid Y_i \mid)$ processors, and produces a cover of weight at most $(1 + \varepsilon) \log n$ times the weight of a minimum cover.*

5.2 The $\mathcal{NC}$ Algorithm with Logarithmic Approximation Factor

For any s_i, s_j and $d, 0 \le d < \min\{\mid s_i \mid, \mid s_j \mid\}$, let u and v be strings of the length d such that $s_i = xu$ and $s_j = vy$ for some non-empty string x and y. If $u = v$ then we define $\mathrm{CON}(i, j, d) = xuy$; otherwise $\mathrm{CON}(i, j, d)$ is undefined. If $\mathrm{CON}(i, j, d)$ is defined, then let $\mathrm{CON}(i, j, d) = \{s_i, s_j\} \cup \{s_k : s_k$ is a substring of $\mathrm{CON}(i, j, d)\}$. In this way we have obtained a family FAM-CON of subsets of S. For each $\mathrm{CON}(i, j, d) \in$ FAM-CON define its weight $w(i, j, d) = \mid \mathrm{CON}(i, j, d) \mid = \mid s_i \mid + \mid s_j \mid - d$.

Let $C = \{\mathrm{CON}(i_1, j_1, d_1), \mathrm{CON}(i_2, j_2, d_2), \ldots, \mathrm{CON}(i_c, j_c, d_c)\}$ be a cover of S defined by a collection of sets from FAM-CON . We can obtain the corresponding superstring $S_C = \mathrm{CON}(i_1, j_1, d_1) \circ \mathrm{CON}(i_2, j_2, d_2) \circ \cdots \circ \mathrm{CON}(i_c, j_c, d_c)$. Since C is a cover of S, every string from S must be a substring of S_C. Note also that $w(C) = \sum_{1 \le k \le c} \mathrm{CON}(i_k, j_k, d_k) = \sum_{1 \le k \le c} \mid \mathrm{CON}(i_k, j_k, d_k) \mid = \mid S_C \mid$. The following fact can be easily derived (see eg., [8]).

Fact 18. *Let C^* be a minimum weighted set cover of S. Then $w(C^*) = \mid S_{C^*} \mid \le 2 \cdot opt(S)$.*

Now, suppose that we have found a set cover C such that $w(C) \le t \cdot w(C^*)$, for some t. Then clearly $\mid S_C \mid \le t \cdot \mid S_{C^*} \mid$. Thus, the superstring S_C has length at most $2 \cdot t \cdot opt(S)$. Hence using Lemma 17 we obtain the following theorem.

Theorem 19. *There is an $\mathcal{NC}$ algorithm that for any $\varepsilon > 0$, finds a superstring whose length is at most $(2 + \varepsilon) \log n$ times the length of a shortest superstring.*

A similar construction was used implicitly by Li [8] for a sequential algorithm.

5.3 The $\mathcal{RNC}$ Algorithm with Constant Approximation Factor

Blum et al. [2] presented the following sequential algorithm for the approximation of the shortest superstring. Let G_S be the overlap graph for the set of strings S. Find a maximum weight cycle-cover C on G_S, where $C = \{c_1, \ldots, c_p\}$ is the collection of cycles. For each cycle $c_i = i_1 \to \cdots \to i_r \to i_1$, let $\tilde{s}_i = s_{i_1} \circ \cdots \circ s_{i_r}$ where i_1 is arbitrary chosen. Then the final superstring is obtained by concatenating all together the strings $\tilde{s}_i$. Blum et al. [2] proved that this algorithm always finds a superstring of length at most $4 \cdot opt(S)$. It is well known that the problem of finding a maximum weight cycle-cover is equivalent to the problem of finding a maximum weight matching in bipartite graph. In general it is not known whether it can be done either $\mathcal{NC}$ or in $\mathcal{RNC}$. However, when the weights of the graph are given in unary one can find a maximum weight matching in $\mathcal{RNC}$ [17]. Since in our case the weights of G_S are given in unary, the construction given by Blum et al. [2] can be parallelized to get an $\mathcal{RNC}$ algorithm.

Theorem 20. *There exists an $\mathcal{RNC}$ algorithm that finds a superstring of length at most $4 \cdot opt(S)$.*

References

1. A. Arora, C. Lund, R. Motwani, M. Sudan, and M. Szegedy. Proof verification and hardness of approximation problems. 33rd FOCS, pp. 14–23, 1992.
2. A. Blum, T. Jiang, M. Li, J. Tromp, and M. Yannakakis. Linear approximation of shortest superstrings. 23rd STOC, pp. 328–336, 1991.
3. B. Berger, J. Rompel, and P. W. Shor. Efficient NC algorithms for set cover with applications to learning and geometry. 30th FOCS, pp. 54–59, 1989. The full version is to appear in *J. Algorithms*.
4. V. Chvatal. A greedy heuristic for the set-covering problem. *Mathematics of Operations Research* 4(1979), pp. 233–235.
5. J. Gallant, D. Maier, and J. Storer. On finding minimal length superstrings. *Journal of Computer and System Sciences* 20(1980), pp. 50–58.
6. M. R. Garey and D. S. Johnson. *Computers and Intractability: A Guide to the Theory of NP-completeness.* Freeman, New York, 1979.
7. R. Greenlaw, H. J. Hoover, and W. L. Ruzzo. A compendium of problems complete for $\mathcal{P}$. Technical Report, University of Washington, 1991.
8. M. Li. Towards a DNA sequencing theory (Learning a string). 31st FOCS, 1990.
9. C. Lund and M. Yannakakis. On the hardness of approximating minimization problems. 25th STOC, 1993.
10. C. Papadimitriou and K. Steiglitz. *Combinatorial Optimization: Algorithms and Complexity.* Prentice-Hall, 1982.
11. C. Papadimitriou and M. Yannakakis. Optimization, approximation, and complexity classes. 20th STOC, pp. 229–234, 1988.
12. H. Peltola, H. Soderlund, J. Tarhio, and E. Ukkonen. Algorithms for some string matching problems arising in molecular genetics. IFIP, pp. 53–64, 1983.
13. J. Storer. *Data Compression: Methods and theory.* Computer Science Press, 1988.
14. J. Tarhio and E. Ukkonen. A Greedy approximation algorithm for constructing shortest common superstrings. *Theoretical Computer Science* 57(1988), pp. 131–145.
15. S-H. Teng and F. Yao. Approximating shortest superstrings. 34th FOCS, 1993.
16. J-S. Turner. Approximation algorithms for the shortest common superstring problem. *Information and Computation* 83(1989), pp. 1–20.
17. V. V. Vazirani. Parallel graph matching. In J. H. Reif, editor, *Synthesis of Parallel Algorithms*, chapter 18, pp. 783–811. Morgan Kaufmann, 1993.

Separating Translates in the Plane: Combinatorial Bounds and an Algorithm *

Jurek Czyzowicz[1], Hazel Everett[2] and Jean-Marc Robert[3]

[1] Dép. d'informatique, Université du Québec à Hull
C.P. 1250, Succ. B, Hull, Canada, J8X 2X7
[2] Dép. de mathématiques et d'informatique, Université du Québec à Montréal
C.P. 8888, Succ. Centre-Ville, Montréal, Canada, H3C 3P8
[3] Dép. d'informatique et de mathématique, Université du Québec à Chicoutimi
555 boul. de l'Université, Chicoutimi, Canada, G7H 2B1

Abstract. In this paper, we establish two combinatorial bounds related to the separation problem for sets of n pairwise disjoint translates of convex objects: 1) there exists a line which separates one translate from at least $n - c_0\sqrt{n}$ translates, for some constant c that depends on the "shape" of the translates and 2) there is a function f such that there exists a line with orientation Θ or $f(\Theta)$ which separates one translate from at least $\lceil \frac{3n}{4} \rceil - 4$ translates, for any orientation Θ (f is defined only by the "shape" of the translate). We also present an $O(n \log(n + k) + k)$ time algorithm for finding a translate which can be separated from the maximum number of translates amongst sets of n pairwise disjoint translates of convex k-gons.

1 Introduction

Given a set of pairwise disjoint convex objects S in the plane, an object $\mathcal{O} \in S$ is *separated* from a subset $X \subseteq S$ if there exists a line l such that $\mathcal{O}$ lies on one side of l and the objects in X lie on the other side.

The combinatorial aspect of the separation problem involves finding bounds on the maximum number of objects which can be separated from one object. Many bounds have been presented for different classes of sets of convex objects. For a set of n convex objects, Hope and Katchalski [9] showed there exists an object which can be separated from at least $\lceil \frac{n+1}{12} \rceil$ objects. For a set of n line segments, Alon, Katchalski and Pulleyblank [1] claimed there exists a line segment which can be separated from at least $\lceil \frac{n-1}{4} \rceil$ line segments. In [6], Czyzowicz, Rivera-Campo, Urrutia and Zaks increased this number to $\lceil \frac{n+1}{3} \rceil$ if the line segments can be extended to form a collection of pairwise disjoint half-lines. They also proved that this bound is tight. For a set of n circles, Czyzowicz, Rivera-Campo and Urrutia [4] showed there exists a circle which can be separated from at least $\lfloor \frac{n-c}{2} \rfloor$ circles, for some constant c. This bound is almost tight since there is a set of $2m$ circles such that none of them can be separated from more

* This research was supported by FCAR, FODAR and NSERC.

than m circles [6]. For a set of n pairwise disjoint translates of a convex object, Alon, Katchalski and Pulleyblank [1] showed there is an orientation such that any line having that orientation intersects at most $c\sqrt{n \log n}$ translates, for some constant c that depends on the convex object. Hence, there exists a translate which can be separated from at least $n - c\sqrt{n \log n}$ translates.

The algorithmic aspect of the separation problem involves finding a line separating an object from the *maximum* number of objects. Note that the proof found in [4], for example, is constructive and yields an efficient algorithm for finding a line separating one circle from at least $\lfloor \frac{n-c}{2} \rfloor$ circles. This line, however, is not necessarily optimal. Aside from brute-force, the only algorithm we know of for finding an optimal separating line is for the case of n disjoint line segments in the plane: Everett, Robert and van Kreveld gave an $O(n^{3/2} \log^3 n)$ time algorithm to solve this problem [8].

In Section 2, we establish two combinatorial bounds related to the separation problem for sets of n pairwise disjoint translates of a convex object in the plane. Our first result shows there exists a line with orientation in any given interval which separates one translate from at least $n - c\sqrt{n}$ translates, for some constant c that depends on the "shape" of the translates and on the length of the interval. This bound is almost tight and only the constant can be improved. The second result shows there is a function $f_\mathcal{O}$ such that there exists a line with orientation Θ or $f_\mathcal{O}(\Theta)$ which separates one translate from at least $\lceil \frac{3n}{4} \rceil - 4$ translates, for any orientation Θ. The function $f_\mathcal{O}$ is defined only by the "shape" of the translates. In this case, the bound is optimal.

In Section 3, we present an $O(n \log(n + k) + k)$ time algorithm solving the separation problem for sets of n pairwise disjoint translates of convex k-gons in the plane.

2   Combinatorial bounds

In order to establish our first combinatorial result, we show that any set of n pairwise disjoint translates of a convex object has an *extreme supporting line* with orientation in any given interval which intersects only "few" translates. An upper (lower) supporting line of an object is a line intersecting the boundary of the object such that the interior of the object lies entirely below (above) the line. An upper (lower) supporting line of a translate is extreme with respect to S if it lies below (above) or on the upper (lower) supporting line with the same orientation of any other translate in S.

Theorem 1. *Let S be a set of n pairwise disjoint translates of a convex object $\mathcal{O}$. Let Θ and δ be such that $-\frac{\pi}{2} \le \Theta < \Theta + \delta \le \frac{\pi}{2}$. Then, there exists an extreme upper supporting line with orientation between Θ and $\Theta + \delta$ which intersects at most $c_{\mathcal{O},\delta}\sqrt{n}$ translates, for some constant $c_{\mathcal{O},\delta}$ that depends on $\mathcal{O}$ and δ.*

Proof. Suppose that $\mathcal{O}$ has diameter 1 and area α. Let l_i be the extreme upper supporting line with respect to S with orientation $\Theta + \frac{i\delta}{\sqrt{n}}$, and let σ_i be the strip delimited by the lines parallel to l_i and at distance 1 from it, for $0 \le i \le \lfloor\sqrt{n}\rfloor$. A translate intersects l_i and l_{i+1} if it lies completely in the rhombus R defined by $\sigma_i \cap \sigma_{i+1}$. Since these strips have width 2 and intersect at an angle of $\frac{\delta}{\sqrt{n}}$, the area of R is $4/\sin\frac{\delta}{\sqrt{n}}$ which can be approximated by $\frac{4}{\delta}\sqrt{n} + o(1)$ by using Taylor's series. Thus, the number of pairwise disjoint translates of $\mathcal{O}$ intersecting l_i and l_{i+1} is at most $\frac{4}{\alpha\delta}\sqrt{n} + o(1)$.

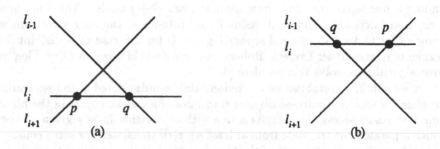

Fig. 1. The number of translates intersecting l_i.

Suppose there is a line l_i such that $l_{i-1} \cap l_{i+1}$ lies above it (see Fig. 1(a)). Since l_{i-1} is an extreme upper tangent, any translate intersecting l_i to the left of q has to intersect l_{i-1}, and any translate intersecting l_i to the right of p has to intersect l_{i+1}. Thus, any translate intersecting l_i must either intersect l_{i-1} or l_{i+1}. Hence, l_i intersects at most $2\frac{4}{\alpha\delta}\sqrt{n} + o(1)$ translates.

Now suppose that $l_{i-1} \cap l_{i+1}$ lies below l_i, for $0 < i < \lfloor\sqrt{n}\rfloor$ (see Fig. 1(b)). Let E be the polygonal convex chain $e_0 e_1 ... e_{\lfloor\sqrt{n}\rfloor}$ corresponding to the upper envelope of the l_i's. Let n_i and n_i^* denote the number of translates intersecting l_i and intersecting the interior of e_i, respectively. Since l_{i-1} is an extreme upper tangent, any translate intersecting l_i to the left of q has to intersect l_{i-1}. Similarly, any translate intersecting l_i to the right of p has to intersect l_{i+1}. Hence, $n_i \le n_i^* + 2\frac{4}{\alpha\delta}\sqrt{n} + o(1)$. Now, construct the incidence graph G as follows. For each edge e_i, put a vertex v_i just below it. For each translate $\mathcal{O}_i$, put a vertex u_i in $\mathcal{O}_i$ lying above or on E. If $\mathcal{O}_i$ intersects the interior of e_j, join u_i and v_j through a point in $\mathcal{O}_i \cap e_j$. Thus, u_i is connected to points of E and v_i is connected to points of e_i. Since the translates are convex and pairwise disjoint, G forms a planar $(n, \lfloor\sqrt{n}\rfloor)$-bipartite graph and contains at most $2(n+\lfloor\sqrt{n}\rfloor)-4$ edges [11]. Since $\sum n_i^*$ corresponds to the number of edges of G, there is an integer i_0 such that $n_{i_0}^* \le \frac{2(n+\lfloor\sqrt{n}\rfloor)-4}{\lfloor\sqrt{n}\rfloor}$. Therefore, l_{i_0} intersects at most $(2\frac{4}{\alpha\delta}+2)\sqrt{n}+2+o(1)$ translates. □

This theorem gives readily our first combinatorial result for the separation problem for sets of pairwise disjoint translates of a convex object in the plane.

Corollary 2. *Let S be a set of n pairwise disjoint translates of a convex object $\mathcal{O}$. Let Θ and δ be such that $-\frac{\pi}{2} < \Theta < \Theta + \delta < \frac{\pi}{2}$. Then, there exists a line with orientation between Θ and $\Theta + \delta$ such that one translate lies below it and at least $n - c_{\mathcal{O},\delta}\sqrt{n}$ translates lie above it, for some constant $c_{\mathcal{O},\delta}$ that depends on $\mathcal{O}$ and δ.*

This result holds also for sets of n pairwise disjoint convex objects of diameter at most 1 and area at least α. Nowhere in the proof of Theorem 1 we use the fact that the objects are translates. We simply use the fact that they are convex and pairwise disjoint and they all have diameter 1 and area α. Finally, we obtain the same result if the object separated from the maximum number of objects has to lie above the separating line instead of below it.

This combinatorial bound is almost optimal. In the next lemma, we give a construction showing that only the constant can be improved.

Lemma 3. *There is a set of $m\lfloor\frac{m}{\pi}\rfloor$ pairwise disjoint translates of a circle such that no translate can be separated from more than $m\lfloor\frac{m}{\pi}\rfloor - \lfloor\frac{m}{\pi}\rfloor - 1$ translates.*

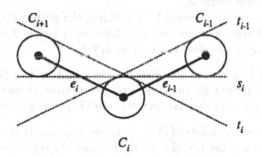

Fig. 2. The construction of Lemma 2.3.

Proof. Let P be a regular polygon with m sides of unit length. Let R be defined as the half of the height of the isosceles triangle determined by three consecutive vertices of P. By a simple trigonometric argument, we obtain that R is equal to $\frac{1}{2}\sin\frac{\pi}{m}$. Put a circle C_i of radius R centered on each vertex v_i of P and put $\lfloor\frac{m}{\pi}\rfloor - 1$ pairwise disjoint circles of radius R centered on each edge e_i of P. This is possible since the diameter of the circle is less than $\frac{\pi}{m}$. Thus, $m\lfloor\frac{m}{\pi}\rfloor$ pairwise disjoint circles are put around P. Let t_i be the common tangent of C_i and C_{i+1} which intersects the interior of P and let s_i be the separating tangent of C_i and C_{i+1} which is tangent to C_i inside P (see Fig. 2).

Consider any circle C_i centered on a vertex of the lower hull of P. Any upper tangent of C_i with slope between $slope(t_i)$ and $slope(s_i)$ is extreme and intersects all the circles centered on e_i. Similarly, any upper tangent of C_i with slope between $slope(s_i)$ and $slope(t_{i-1})$ is extreme and intersects all the circles centered on e_{i-1}. Now, consider any circle C centered on the interior of an edge e_i of the lower hull of P. The line t_i represents the unique extreme upper tangent of C and it intersects all the circles centered on e_i. Thus, any extreme upper tangent of a translate centered on the lower hull of P must intersect at least $\lfloor \frac{m}{\pi} \rfloor + 1$ circles. We obtain the same result for the extreme lower tangents of the translates centered on the upper hull of P. Hence, no circle can be separated from more than $m \lfloor \frac{m}{\pi} \rfloor - \lfloor \frac{m}{\pi} \rfloor - 1$ other circles. $\qquad\square$

In [5], Czyzowicz, Rivera-Campo, Urrutia and Zaks asserted that for a set of n isothetic rectangles, there exists a rectangle which can be separated from at least $\lfloor \frac{2n}{3} \rfloor - c$ rectangles by either a vertical or a horizontal line. This result can be generalized for sets of n pairwise disjoint translates of a convex object. Thus, we can define a function $f_{\mathcal{O}}$ which depends only on the "shape" of the translates such that there exists a translate which can be separated from at least $\lceil \frac{3n}{4} \rceil - 4$ translates by a line with orientation Θ or $f_{\mathcal{O}}(\Theta)$, for any orientation Θ. This bound is optimal. For example, the function $f_{\mathcal{O}}$ for a circle can be defined as simply as $f_{\mathcal{O}}(\Theta) = \Theta + \frac{\pi}{2}$. Hence, we can separate one unit-circle from at least $\lceil \frac{3n}{4} \rceil - 4$ unit-circles by a line parallel or orthogonal to any given line. We first present some technical lemmas.

Let l_1 and l_2 be two intersecting lines dividing the plane into four quadrants. A convex object $\mathcal{O}$ *crosses* a quadrant Q defined by l_1 and l_2 if it intersects both half-lines delimiting Q and does not contain $l_1 \cap l_2$.

Lemma 4. *Let $\mathcal{O}$ be a convex object in the plane. There is a function $f_{\mathcal{O}}$ such that the quadrants defined by any lines with orientation Θ and $f_{\mathcal{O}}(\Theta)$ can not be crossed by two pairwise disjoint translates of $\mathcal{O}$, for any orientation Θ.*

Proof. Let $\mathcal{O}_1$ be some translate of $\mathcal{O}$ in the plane. Suppose that Θ corresponds to the vertical orientation. Let l_1 and l'_1 be the left and the right vertical supporting lines of $\mathcal{O}_1$, respectively, and let p and q be two tangent points on l_1 and l'_1, respectively. Then, $f_{\mathcal{O}}(\Theta)$ is given by the orientation of the line pq. Now, let l_2 be a lower supporting line of $\mathcal{O}_1$ with orientation $f_{\mathcal{O}}(\Theta)$ and let r be a tangent point on l_2. Finally, let v be a point on the line segment pq such that the line segment rv is parallel to l_1. The points p, r and v form with the intersection point o of l_1 and l_2 two congruent triangles $\triangle pvr$ and $\triangle por$ (see Fig. 3).

Now, let l_a and l_b be two lines with orientation Θ and $f_{\mathcal{O}}(\Theta)$, respectively. Any quadrant defined by these two lines can not be crossed by two pairwise disjoint translates of $\mathcal{O}$. Consider the first quadrant. Copy the triangle $\triangle por$ on l_a and l_b. For any translate $\mathcal{O}'$ crossing the first quadrant, the triangle $\triangle p'v'r'$ corresponding to the triangle $\triangle pvr$ must also cross the quadrant. Thus, $\triangle p'v'r'$ must intersect op and or. Furthermore, since $\triangle p'v'r'$ does not contain o and since $\triangle pvr$ and $\triangle p'v'r'$ are congruent, $\triangle p'v'r'$ must also intersect pr. Hence, $\triangle p'v'r'$ intersects all the sides of $\triangle por$. Therefore, if there were two translates of $\mathcal{O}$ crossing the first quadrant, there would be two pairwise disjoint translates of $\triangle pvr$ intersecting all the sides of $\triangle por$ which is impossible. Similarly, the other quadrants can not be crossed by more than one translate. $\qquad\square$

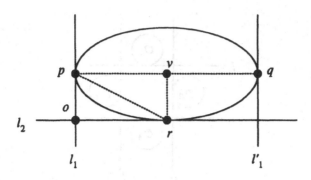

Fig. 3. The definition of the function f_O.

It is simple to prove that f_O is bijective and its range covers all the possible orientations if O is smooth. A convex object O is smooth if, for any point p on its boundary, there is only one line supporting O and containing p.

Lemma 5. *[10] Let l_1 and l_2 be two lines dividing the plane into four quadrants. Let O_1, O_2, O_3 and O_4 be four pairwise disjoint translates of a convex object O. If O_i crosses the ith quadrant, then any translate of O containing $l_1 \cap l_2$ must intersect at least one of the four translates O_i.*

We are now ready to establish our second combinatorial result for the separation problem for sets of pairwise disjoint translates of a convex object in the plane.

Theorem 6. *Let S be a set of n pairwise disjoint translates of a convex object O. There is a function f_O such that there exists a line with orientation Θ or $f_O(\Theta)$ which separates one translate from at least $\lceil \frac{3n}{4} \rceil - 4$ translates, for any orientation Θ.*

Proof. Let f_O be the function given by Lemma 4. Notice this function depends only on the shape of O and not on the set S. Suppose that Θ corresponds to the vertical orientation. Let l_0 be the rightmost vertical line without any translate lying strictly to its left and let l_2 be the leftmost vertical line without any translate lying strictly to its right. If l_0 lies on or to the right of l_2, S must have a vertical line transversal. In this case, it is easy to show there is a line with orientation $f_O(\Theta)$ separating one translate from at least $n - 4$ translates. The lines l_1 and l_3 with orientation $f_O(\Theta)$ are defined similarly. From now on, suppose that S does not have any line transversal with orientation Θ or $f_O(\Theta)$. Finally, let O_i be the extreme translates supported by l_i (see Fig. 4).

Let S_i be the set $\{O \in S | O$ intersects $l_i\}$. Then, l_i separates O_i from the translates in $S \backslash S_i$. By Lemmas 4 and 5, $S_i \cap S_{i+1}$ contains at most four translates.

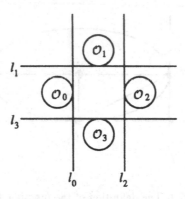

Fig. 4. Definition of the lines l_i.

Suppose that $S_0 \cap S_2 \neq \emptyset$. Then, the strip defined by l_0 and l_2 is too narrow to strictly contain a translate. Any translate intersecting l_1 must intersect either l_0 or l_2. Hence, S_1 contains at most 8 translates and l_1 separates $\mathcal{O}_1$ from at least $n - 8$ translates. We obtain a similar result if $S_1 \cap S_3 \neq \emptyset$.

Finally suppose that $S_0 \cap S_2 = \emptyset$ and $S_1 \cap S_3 = \emptyset$. Since only the translates belonging to $S_i \cap S_{i+1}$ are counted twice, $\sum_i |S_i| \leq n + 16$. Hence, there is a set S_{i_0} containing at most $\lfloor \frac{n+16}{4} \rfloor$ translates implying that l_{i_0} separates $\mathcal{O}_{i_0}$ from at least $\lceil \frac{3n}{4} \rceil - 4$ translates. $\qquad\qquad\square$

Once the function $f_\mathcal{O}$ is defined, it is simple to construct a set of $4m$ pairwise disjoint translates of a convex object for which no translate can be separated from more than $3m - 4$ translates by a line with orientation Θ or $f_\mathcal{O}(\Theta)$, for some fixed orientation Θ. We simply have to put the translates such that any extreme line l_i with orientation Θ or $f_\mathcal{O}(\Theta)$ supports exactly $m - 4$ translates and any pair of lines l_i and l_{i+1} intersect exactly four translates. Therefore, the bound presented in the above theorem is optimal.

3 The Algorithm

The algorithmic aspect of the separation problem consists in finding an object which can be separated from the maximum number of objects amongst a set of pairwise disjoint convex objects. In this section, we present an algorithm to solve the separation problem for sets of pairwise disjoint translates of convex polygons.

Let S be a set of n pairwise disjoint translates of a convex k-gon in the plane. In this section, we give an $O(n \log(n + k) + k)$ time algorithm for finding a line separating a translate from the maximum number of translates in S. Such a line is said to be an optimal separating line. It follows from Corollary 2 that X should contain at least $n - c_\mathcal{O} \sqrt{n}$ translates. Unfortunately, the proof of the corollary just gives a way to find good separating lines without giving any clue on how to find an optimal one. We address here the problem of finding an optimal solution.

We begin with some definitions. A line is a *common tangent* for a pair of objects if it is a lower or an upper supporting line for both objects. A line is a *separating tangent* for a pair of objects if it is a lower supporting line for one object and an upper supporting line for the other object. Finally, a separating tangent is *extreme* with respect to S if it is also an extreme lower supporting line or an extreme upper supporting line.

The algorithm is based on the existence of some special optimal solutions.

Lemma 7. *Let S be a set of pairwise disjoint convex objects in the plane. There is an optimal separating line for S corresponding to an extreme separating tangent.*

It is simple to transform any optimal separating line into an extreme supporting line simply by translation. Then, by rotating this extreme line around the object that it supports, we obtain an optimal separating line which is an extreme separating tangent.

The lemma suggests an approach to solve the separation problem. We have to find a way to enumerate efficiently the extreme separating tangents and a way to determine efficiently the number of translates lying below (or above) such a candidate line. For simplicity, we consider only the case where the translate separated from the other translates lies above the separating line. Therefore, we are interested only in the extreme lower supporting lines which are also separating tangents.

The main idea used by the algorithm is to reformulate the problem in the dual space. Let $\mathcal{D}$ be a dual transformation mapping a point $p = (x_p, y_p)$ in the primal space to the non-vertical line $\mathcal{D}(p)$ defined by the equation $b = -x_p \cdot a + y_p$ and mapping a non-vertical line l defined by the equation $y = a_l \cdot x + b_l$ to the point $\mathcal{D}(l) = (a_l, b_l)$. Since $\mathcal{D}$ is a bijection, its inverse $\mathcal{D}^{-1}$ is properly defined. This dual transformation can be extended for any compact convex object $\mathcal{O}$. In this case, $\mathcal{D}(\mathcal{O})$ corresponds to the dual points representing the lines intersecting $\mathcal{O}$ in the primal space. More precisely, $\mathcal{D}(\mathcal{O})$ is given by $\bigcup_{l \cap \mathcal{O} \neq \emptyset} \mathcal{D}(l)$. The boundary of $\mathcal{D}(\mathcal{O})$ consists of an upper part $up\_bd(\mathcal{D}(\mathcal{O}))$ and a lower part $low\_bd(\mathcal{D}(\mathcal{O}))$ corresponding to the dual of the upper and lower supporting lines of $\mathcal{O}$, respectively (see [7] for more details). A lower supporting line l is extreme with respect to S if and only if $\mathcal{D}(l)$ is above or on $low\_bd(\mathcal{D}(\mathcal{O}_i))$, for any $\mathcal{O}_i \in S$. Thus, it is sufficient to construct the upper envelope $UE_{\mathcal{D}(S)}$ of $\{low\_bd(\mathcal{D}(\mathcal{O}_i)) | \mathcal{O}_i \in S\}$ to describe all the extreme lower supporting lines. The upper envelope of a set of functions $F = \{f_i : \mathbb{R} \to \mathbb{R} | 1 \leq i \leq n\}$ is the function $UE_F(x) = \max_i f_i(x)$. Once $UE_{\mathcal{D}(S)}$ has been computed, the extreme separating tangents can be obtained easily by intersecting $UE_{\mathcal{D}(S)}$ and $up\_bd(\mathcal{D}(\mathcal{O}_i))$, for any $1 \leq i \leq n$. Each intersection point corresponds to an extreme lower supporting line which is also an upper supporting line of some translate. Hence, each intersection point corresponds to an extreme separating tangent. Obviously, we obtain all the extreme separating tangents in this way.

We are now ready to sketch our algorithm for solving the separation problem for sets of pairwise disjoint translates of convex polygons.

Algorithm Separation (sketch)

Input: A set S of n pairwise disjoint translates of a convex k-gon.
Output: An optimal separating line l for S.

1. Compute the upper envelope $UE_{\mathcal{D}(S)}$ of $\{low\_bd(\mathcal{D}(\mathcal{O}_i))|\mathcal{O}_i \in S\}$.
2. Compute the intersection of $UE_{\mathcal{D}(S)}$ and $up\_bd(\mathcal{D}(\mathcal{O}_i))$, for $1 \leq i \leq n$.
3. Traverse $UE_{\mathcal{D}(S)}$ and find an optimal separating line.

End of the Algorithm

The time complexity of the algorithm depends mainly on how fast the envelope can be constructed in Step 1 and how fast the intersection points can be found in Step 2.

Since any pair of disjoint translates of a convex object admit exactly one lower common tangent[1], any pair of polygonal chains in $\{low\_bd(\mathcal{D}(\mathcal{O}_i))|\mathcal{O}_i \in S\}$ intersect in exactly one point. This implies that the total size complexity of $UE_{\mathcal{D}(S)}$ is in $O(n+k)$. Hence, the envelope can be computed by a simple divide-and-conquer algorithm in $O(nk + n \log n)$ time. This algorithm needs a complete description of the polygonal chains which takes $O(nk)$ space.

To obtain a better algorithm, we take advantage of the fact that S contains translates of a convex k-gon. A translate can be represented by a description of the object and by the position in the plane of some reference point of the object. Thus, S is described by giving a description of the k-gon and the position of the n reference points which takes only $O(n + k)$ space.

The envelope $UE_{\mathcal{D}(S)}$ can be computed in $O(n \log n + k)$ time as follows. Dualize the set P of the n reference points and compute in $O(n \log n)$ time [2] the upper envelope $UE_{\mathcal{D}(P)}$ of these n lines. The envelope $UE_{\mathcal{D}(P)}$ corresponds to the dual of the upper hull $UH(P)$ of P. Let p be any point on $UE_{\mathcal{D}(P)}$. The line $\mathcal{D}^{-1}(p)$ must go through a vertex of $UH(P)$. This vertex gives the extreme translate $\mathcal{O}'$ with respect to the orientation of $\mathcal{D}^{-1}(p)$. Thus, the extreme lower supporting line parallel to $\mathcal{D}^{-1}(p)$ must support $\mathcal{O}'$ and can be obtained simply by translating $\mathcal{D}^{-1}(p)$ downward. This translation depends only on the shape of the translate.

Let K be a translate of the k-gon for which the reference point coincides with the origin. Construct the polygonal chain $low\_bd(\mathcal{D}(K))$ lying completely below the x-axis. Let $q = (x_q, y_q)$ be any point on $low\_bd(\mathcal{D}(K))$ and let l be the line passing through the origin and parallel to the lower supporting line $\mathcal{D}^{-1}(q)$ of K. The vertical distance between l and $\mathcal{D}^{-1}(q)$ is $|y_q|$. Therefore, if a line with slope x_q goes through the reference point of some translate, the lower supporting line with slope x_q of that translate can be obtained by translating downward the line by a vertical distance of $|y_q|$. This is the key idea on how to transform $UE_{\mathcal{D}(P)}$ into $UE_{\mathcal{D}(S)}$. Let A and B be two x-monotone curves and let $A \oplus B$

---

[1] For simplicity, we suppose that no pair of translates in S admit a vertical common tangent.

denote the set $\{(x, y' + y'')|(x, y') \in A, (x, y'') \in B\}$. Then, $UE_{\mathcal{D}(S)}$ is given by $UE_{\mathcal{D}(P)} \oplus low\_bd(\mathcal{D}(K))$. Since $UE_{\mathcal{D}(P)}$ and $low\_bd(\mathcal{D}(K))$ are both x-monotone polygonal chains, $UE_{\mathcal{D}(S)}$ can be computed in $O(n + k)$ time. Divide the plane into $O(n + k)$ slabs in which $UE_{\mathcal{D}(P)}$ and $low\_bd(\mathcal{D}(K))$ are composed only of one line segment. For each slab, the line segment of $UE_{\mathcal{D}(S)}$ can be computed in $O(1)$ time by considering only the endpoints of the line segments of $UE_{\mathcal{D}(P)}$ and $low\_bd(\mathcal{D}(K))$ lying in that slab.

The intersection points of $UE_{\mathcal{D}(S)}$ and $up\_bd(\mathcal{D}(\mathcal{O}_i))$ corresponding to the candidate lines can be computed easily once the envelope has been preprocessed. We first transform $UE_{\mathcal{D}(S)}$ to reduce polygonal chain intersection queries into line intersection queries.

Construct the polygonal chain $up\_bd(\mathcal{D}(K))$ lying completely above the x-axis. Let $q = (x_q, y_q)$ be any point on $up\_bd(\mathcal{D}(K))$. Now, let l be the line passing through the origin and parallel to the upper supporting line $\mathcal{D}^{-1}(q)$ of K. The vertical distance between l and $\mathcal{D}^{-1}(q)$ is y_q. Therefore, if a line with slope x_q goes through the reference point of some translate, the upper supporting line with slope x_q of that translate can be obtained by translating upward the line by a vertical distance of y_q. Let A and B be two x-monotone curves and let $A \ominus B$ denote the set $\{(x, y' - y'')|(x, y') \in A, (x, y'') \in B\}$. Then, compute the x-monotone polygonal chain $UE_{\mathcal{D}(S)} \ominus up\_bd(\mathcal{D}(K))$. This chain is denoted $\mathcal{C}$ and can be constructed in $O(n + k)$ time. Let $p = (x_p, y_p)$ be any point on $UE_{\mathcal{D}(S)}$ and let $p' = (x_p, y_p')$ be the corresponding point on $up\_bd(\mathcal{D}(K))$. Suppose that $UE_{\mathcal{D}(S)}$ and $up\_bd(\mathcal{D}(\mathcal{O}_i))$ intersect at p. Then, $\mathcal{D}^{-1}(p)$ represents an upper supporting line of $\mathcal{O}_i$ and $\mathcal{D}^{-1}((x_p, y_p - y_p'))$ represents the line passing through the reference point p_i and parallel to $\mathcal{D}^{-1}(p)$. By construction, $(x_p, y_p - y_p')$ must lie on $\mathcal{C}$. On the other hand, $(x_p, y_p - y_p')$ must lie also on $\mathcal{D}(p_i)$ since the dual transformation preserves the incidence relationship. Therefore, $\mathcal{C}$ and $\mathcal{D}(p_i)$ intersect at $(x_p, y_p - y_p')$. Similarly, if $\mathcal{C}$ and $\mathcal{D}(p_i)$ intersect at $(x_p, y_p - y_p')$, $UE_{\mathcal{D}(S)}$ and $up\_bd(\mathcal{D}(\mathcal{O}_i))$ must intersect at p.

Now, preprocess $\mathcal{C}$ for ray-shooting. This can be done in $O(s)$ time and each query can be answered in $O(\log s)$ time [3] where s is the size complexity of $\mathcal{C}$. Hence, $UE_{\mathcal{D}(S)}$ can be transformed and preprocessed for ray shooting in $O(n+k)$ time, and the nb_i intersection points between $UE_{\mathcal{D}(S)}$ and $up\_bd(\mathcal{D}(\mathcal{O}_i))$ can be found in $O(nb_i \cdot \log(n + k))$ time by succesive ray shooting queries. In fact, we first compute the intersection points between $\mathcal{C}$ and $\mathcal{D}(p_i)$, for any $p_i \in P$, and then, transform these points into the corresponding points of $UE_{\mathcal{D}(S)}$. The constant upper bound on nb_i given below implies that the $O(n)$ intersection points between $UE_{\mathcal{D}(S)}$ and the chains $up\_bd(\mathcal{D}(\mathcal{O}_i))$ can be found in sorted order in $O(n \log(n + k) + k)$ time.

Lemma 8. *Let S be a set of pairwise disjoint translates of a convex object. Let $\mathcal{O}$ be a translate in S. The number of upper supporting lines of $\mathcal{O}$ which are also extreme lower supporting lines of translates in S is constant.*

Proof. Let l_1 and l_2 be the vertical supporting lines of $\mathcal{O}$, and let l_3 be the upper

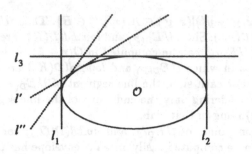

Fig. 5. The number of extreme tangents.

supporting line of $\mathcal{O}$ with orientation $f_{\mathcal{O}}(\frac{\pi}{2})$ where $f_{\mathcal{O}}$ is given in Lemma 4 (see Fig. 5).

Let $\mathcal{O}'$ and $\mathcal{O}''$ be two translates lying to the left of l_1 and let l' and l'' be the corresponding separating tangents which are upper supporting lines of $\mathcal{O}$. Suppose that the slope of l' is smaller than the slope of l''. Then, l' lies completely above l'' to the left of l_1. Since l' is a lower supporting line of $\mathcal{O}'$, $\mathcal{O}'$ lies completely above l''. Thus, at most one extreme separating tangent is defined by a translate to the left of l_1 (or, similarly, to the right of l_2). The same argument can be used to prove there are at most two extreme separating tangents defined by translates lying above l_3. One tangent must have a slope greater than $f_{\mathcal{O}}(\frac{\pi}{2})$ and the other must have a slope smaller than $f_{\mathcal{O}}(\frac{\pi}{2})$. The remaining extreme separating tangents are defined by translates intersecting either l_1 or l_2 above $\mathcal{O}$, and not lying strictly above l_3. It is simple to show that these translates must in fact intersect l_3. By Lemmas 4 and 5, we have at most six such translates, and each of them can define two extreme separating tangents supporting $\mathcal{O}$. Therefore, at most 16 extreme separating tangents can be upper supporting lines of $\mathcal{O}$. $\square$

Once the intersection points between the envelope $UE_{\mathcal{D}(S)}$ and the chains $up\_bd(\mathcal{D}(\mathcal{O}_i))$ have been computed, an optimal separating line can be determined by sweeping $UE_{\mathcal{D}(S)}$ from left to right. First, take the leftmost intersection point and determine the chains $up\_bd(\mathcal{D}(\mathcal{O}_i))$ lying below $UE_{\mathcal{D}(S)}$ to the left of the leftmost intersection point. These chains correspond to the translates lying below the candidate line. Then, traverse the envelope. Each intersection point corresponds to a translate which goes from the set of translates lying below the candidate line into the set of the translates intersecting the candidate line or vice versa. Hence, an optimal separating line can be found in $O(n + k)$ time once the intersection points are known.

We can now summarize our algorithm for solving the separation problem for sets of pairwise disjoint translates of convex polygons as follows.

Algorithm Separation

Input: A description of the convex k-gon K whose the reference point coincides with the origin and the set P containing the n reference points.

Output: An optimal separating line l for S.

1. Compute the envelope $UE_{\mathcal{D}(S)}$ as follows:
   (a) Compute the upper envelope $UE_{\mathcal{D}(P)}$.
   (b) Compute $UE_{\mathcal{D}(P)} \oplus low\_bd(\mathcal{D}(K))$ which corresponds to $UE_{\mathcal{D}(S)}$.
2. Compute the intersection of $UE_{\mathcal{D}(S)}$ and $up\_bd(\mathcal{D}(\mathcal{O}_i))$ as follows:
   (a) Compute the monotone chain $\mathcal{C}$ defined as $UE_{\mathcal{D}(S)} \ominus up\_bd(\mathcal{D}(K))$.
   (b) Preprocess $\mathcal{C}$ for ray shooting.
   (c) Find the intersection points of $\mathcal{D}(p_i)$ and $\mathcal{C}$ by using ray shooting queries, and transform these points into points of $UE_{\mathcal{D}(S)}$.
3. Traverse $UE_{\mathcal{D}(S)}$ and find an optimal separating line.

End of the Algorithm

This algorithm gives us the following result.

Theorem 9. *Let S be a set of n pairwise disjoint translates of a convex k-gon in the plane. An optimal separating line separating a translate from the maximum number of translates can be found in $O(n \log(n + k) + k)$ time.*

References

1. N. Alon, M. Katchalski, and W. Pulleyblank. Cutting disjoint disks by straight lines. *Disc. Comp. Geom.*, 4:239–243, 1989.
2. M. Atallah. Some dynamic computational geometry problems. *Comp. Math. Appl.*, 11:1171–1181, 1985.
3. B. Chazelle, H. Edelsbrunner, M. Gringi, L. J. Guibas, M. Sharir, and J. Snoeyink. Ray shooting in polygons using geodesic triangulations. In *Proc. of the 18th Int. Coll. on Automata, Languages and Programming*, pages 661–673, 1991.
4. J. Czyzowicz, E. Rivera-Campo, and J. Urrutia. A note on separation of convex sets. To appear in *Disc. Math.*
5. J. Czyzowicz, E. Rivera-Campo, J. Urrutia, and J. Zaks. Separating convex sets in the plane. In *Proc. of the Sec. Can. Conf. on Comp. Geom.*, pages 50–54, 1990.
6. J. Czyzowicz, E. Rivera-Campo, J. Urrutia, and J. Zaks. Separating convex sets in the plane. *Disc. Comp. Geom.*, 7:189–195, 1992.
7. H. Edelsbrunner, H.A. Maurer, F.P. Preparata, A.L. Rosenberg, E. Welzl, and D. Wood. Stabbing line segments. *BIT*, 22:274–281, 1982.
8. H. Everett, J.-M. Robert, and M. van Kreveld. An optimal algorithm for the ($\leq k$)-levels, with applications to separation and transversal. In *Proc. of the 9th Annual ACM Symp. on Comp. Geom.*, pages 38–46, 1993.
9. R. Hope and M. Katchalski. Separating plane convex sets. *Math. Scand.*, 66:44–46, 1990.
10. M. Katchalski, T. Lewis, and A. Liu. Geometric permutations of disjoint translates of convex sets. *Disc. Math.*, 65:249–259, 1987.
11. T. Nishizeki and N. Chiba. *Planar Graphs: Theory and Algorithms*. North-Holland, 1988.

Finding All Weakly-Visible Chords of a Polygon in Linear Time

(Extended Abstract)

Gautam Das * and Paul J. Heffernan and Giri Narasimhan **

Memphis State University, Memphis, TN 38152.

Abstract. A chord of a simple polygon P is *weakly-visible* if every point on P is visible from some point on the chord. We give an optimal linear-time algorithm which computes *all* weakly-visible chords of a simple polygon P with n vertices.

1 Introduction

In this paper we present an optimal-time algorithm which computes all *weakly-visible* chords of a simple polygon. For a simple polygon P with n vertices, our algorithm requires time $O(n)$. Previous results [3, 9] require $O(n \log n)$ time and compute only one weakly-visible chord. We also consider the case of rectilinear simple polygons, and show a much simpler linear-time algorithm to compute all weakly-visible chords of the polygon.

Two sets of points are said to be *weakly-visible* if *every* point in either set is visible from *some* point in the other set. A weakly-visible chord c of a polygon P is one such that c and P are weakly-visible. We state four versions of the weakly-visible chords problem for a polygon P: (1) determine whether a given chord c is weakly-visible; (2) determine whether there exists a weakly-visible chord; (3) return a weakly-visible chord c, if indeed such a chord exists; and (4) return *all* weakly-visible chords. Version 4 is the strongest, and an algorithm for it also solves the first three versions. In this paper we solve to optimality version 4 and prove the theorem given below. Although a polygon can have an infinite number of weakly-visible chords, the output can be described in a piece-wise manner using only $O(n)$ space as described later in the paper. In earlier papers [3, 9], version 3 has been solved in $O(n \log n)$ time.

Theorem 1 *Given a simple polygon P, there exists a linear-time algorithm that computes all weakly-visible chords of P.*

The question of weakly-visible chords falls in the larger area of weak-visibility in polygons, which has received much attention by researchers. A simple polygon P is weakly-visible from an edge e if e and $P \setminus e$ are weakly-visible. Any two points z and y of a polygon P partition P into two chains, which we call L and R, for left and right chains. A polygon is *LR-visible* for z and y if L and R are weakly-visible. A weakly-visible chord c of P is one such that c and P are weakly-visible. Weak-visibility of a

---

* *e-mail:* dasg@next1.msci.memst.edu; Supported in part by NSF Grant CCR-930-6822
** *e-mail:* giri@next1.msci.memst.edu; Supported in part by NSF Grant INT-911-5870

polygon from an edge was first studied in [1], and Sack and Suri [11] subsequently gave a linear-time algorithm which computes all weakly-visible edges of a simple polygon. Chen [2] gave a linear-time algorithm that finds the shortest weakly-visible edge, if one exists. An $O(n \log n)$-time algorithm that computes all LR-visible pairs x and y is given by Tseng and Lee [12], and Das, Heffernan and Narasimhan [4] subsequently gave a linear-time algorithm that computes all LR-visible pairs s and t. The weakly-visible chords problem was studied in [3, 9], and algorithms were developed which require $O(n \log n)$ time and compute only one weakly-visible chord. In this paper we present a linear-time algorithm which computes all weakly-visible chords.

This paper is of interest not only because we present an optimal result for an intriguing problem in polygonal visibility, but also on account of the techniques we employ, and because of the relationship between weakly-visible chords and other problems in polygonal visibility, such as LR-visibility. LR-visibility is a subproblem of weakly-visible chords, for it can be shown that two points x and y of P are the endpoints of a weakly-visible chord of P if and only if $\overline{xy}$ is a chord of P and P is LR-visible with respect to x and y. In the current paper, the linear-time algorithm for computing all LR-visible pairs in [4] is used as a subprocedure.

What is interesting about the techniques used here and in [4] is that both the linear-time algorithms output a mass of information, which when sifted appropriately can provide a wealth of visibility information for a simple polygon. Furthermore, the result and techniques reported here were used effectively by Das and Narasimhan [5] to solve to optimality the problem of finding the shortest weakly-visible segment (if one exists) in the interior of a simple polygon.

Another problem which is closely related to weak-visibility problems is the *two-guard* problem. While the two-guard problem has many formulations, we will state just one for the sake of illustration: a polygon P is walkable from point x to point y if one "guard" can traverse the left chain L and the other the right chain R from x to y while always remaining co-visible. Other formulations require the guards to move monotonically or that one guard traverses from y to x. For the two-guard problem, currently there exist optimal linear-time algorithms for various formulations for fixed x and y (version 1) [7], and $O(n \log n)$-time algorithms which find all pairs x and y (version 4) for various formulations [12]. The authors are currently working to develop optimal solutions for the all-pairs version (version 4) of various formulations of the two-guard problem, and we feel that our recent efforts are important steps towards this goal.

2 Preliminaries

In this section we define notation for this paper, and summarize the LR-visibility algorithm in [4], which will be used as a subprocedure. A *polygonal chain* in the plane is a concatenation of line segments or *edges* that connect *vertices*. If the segments intersect only at the endpoints of adjacent segments, then the chain is *simple*, and if a polygonal chain is closed we call it a *polygon*. In this paper, we deal with a simple polygon P, and its interior, $int(P)$. Two points $x, y \in P$ are *visible* if $\overline{xy} \subset P \cup int(P)$, i.e., $\overline{xy}$ is a *chord* of P. For $x, y \in P$, $P_{CW}(x, y)$ $(P_{CCW}(x, y))$ is the subchain obtained by traversing P clockwise (counterclockwise) from x to y.

The *ray shot* from a vertex v in direction d consists of "shooting" a "bullet" from v in direction d which travels until it hits a point of P. Formally, if r is the ray rooted at v in direction d, then the *hit point* of this ray shot is the point of $(P \setminus \{v\}) \cap r$ closest to v. Each reflex vertex defines two special ray shots as follows. Let v be a reflex vertex and v'' the vertex adjacent to v in the clockwise direction. Then the ray shot from v in the direction from v'' to v is called the *clockwise ray shot* of v. If v' is the hit point of the clockwise ray shot, then the subchain $P_{CW}(v, v')$ is the *clockwise component* of v. Counterclockwise ray shots and components are defined in the same way. A component is *redundant* if it is a superset of another component.

We assume that the input is in general position, which means that no three vertices are collinear, and no three lines defined by edges intersect at a common point. As noted in [8], the family of components completely determines LR-visibility of P, since a pair of points x and y admits LR-visibility if and only if each component of P contains either x or y. The definition of redundant gives the following.

Lemma 1 *A polygon P is LR-visible with respect to s and t if and only if each non-redundant component of P contains either s or t.*

If a polygon has more than two disjoint components, this lemma shows that it is not LR-visible. The LR-visibility algorithm in [4] outputs $O(n)$ pairs of subchains of the form (A_i, B_i) such that any point s on a subchain A_i is LR-visible to any point t on the corresponding subchain B_i. We now describe A_i and B_i more rigorously. The endpoints of non-redundant components partition P into a collection of intervals that we call *basic intervals*, and denote $A_1, \cdots, A_k$, ordered counterclockwise. (In the rest of the paper, we use the term *interval* to denote a subchain of the polygon's boundary). A basic interval may or may not contain either of its endpoints. It is possible for a degenerate basic interval consisting of a single point to exist. By lemma 1, all points of a basic interval form LR-visible pairs with the same collection of partners. Thus, we denote as B_i the set of points such that (x, y) is a LR-visible pair for all $x \in A_i$ and $y \in B_i$. The following two lemmas are proved in [4].

Lemma 2 *B_i is a connected set; that is, it is either the entire polygon P, or the empty set, or a non-empty subinterval of P composed of the union of adjacent basic intervals.*

Lemma 3 *Also, if $A_i \cap B_i \neq \emptyset$, then $B_i = P$.*

In [4], we gave a linear-time algorithm that constructs all LR-visible pairs of intervals $(A_1, B_1), \cdots, (A_k, B_k)$. The intervals $A_1, \cdots, A_k$ are disjoint and ordered counterclockwise on P. The intervals $B_1, \cdots, B_k$ are also ordered counterclockwise but are not necessarily disjoint. As one moves counterclockwise from A_i to A_{i+1}, one either leaves or enters a non-redundant component, which may result in either the starting or ending endpoint of B_i to move counterclockwise in order to form B_{i+1}. In the remaining sections we develop the algorithm for constructing weakly-visible chords. Additional notation is introduced where required.

3 Geometric Properties of the Problem

In this section we present the important geometric properties on which the algorithm depends. The actual algorithm is presented in the next section. The following lemma relates weakly-visible chords to LR-visibility and has a trivial proof.

Lemma 4 *For points x and y on P, the segment $\overline{xy}$ is a weakly-visible chord of P if and only if (1) $\overline{xy}$ is a chord of P and (2) P is LR-visible for x and y.*

The lemma suggests the algorithm: first compute LR-visibility using the algorithm in [4], then compute all chords $\overline{xy}$ such that $x \in A_i$ and $y \in B_i$.

We now develop further properties of basic intervals necessary for this task. The *kernel*, K, of a polygon P is defined as the collection of points in $P \cup int(P)$ which are visible from all points of P. The kernel is a convex set and a polygon with a non-empty kernel is called *star-shaped*. The points of P in the kernel are exactly those which intersect all components. It is clear that a point $x \in K \cap P$ forms a weakly-visible chord with every other point of P. This means that the set of weakly-visible chords containing a kernel point as an endpoint can be succinctly represented as the set $K \cap P$.

Lemma 5 *For a basic interval A_i, $B_i = P$ if and only if A_i is contained in K.*

Proof. Let $x \in A_i$. If $B_i = P$ then (x, x) is an LR-visible pair, so x intersects all components and thus is in the kernel. Suppose $A_i \subset K$. Let $x \in A_i$. Then x intersects all components, so (x, x) is an LR-visible pair; thus $x \in B_i$, and since $A_i \cap B_i \neq \emptyset$ we have $B_i = P$. □

We know that each basic interval consists entirely of points in K or points not in K, and we call a basic interval that consists of kernel points a *kernel interval*. A basic interval A_i which is not a kernel interval is disjoint from B_i; for such a case we define D_i as the interval of points encountered as one traverses counterclockwise from the ending point of A_i to the starting point of B_i, and E_i as the interval encountered counterclockwise from the ending point of B_i to the starting point of A_i. We call D_i and E_i the *side intervals* of A_i. The side intervals either contain or do not contain their endpoints in such a manner that the four intervals A_i, B_i, D_i and E_i partition P. It is possible for D_i and/or E_i to be empty.

As shown in [7], if w and v are points of P, and $SP(w, v)$ is the shortest path inside P directed from w to v, then any vertex of $SP(w, v)$ that lies on $P_{CW}(w, v)$ ($P_{CCW}(w, v)$) is a left (right) turn. We say that an interval F is *well-behaved* if the shortest path between its endpoints inside P only touches points of F and not the rest of the polygon. Thus if $P_{CCW}(w, v)$ is a well-behaved interval then $SP(w, v)$ contains no left turns. This is a stronger statement than simply saying that that $SP(w, v)$ is a convex chain, since it specifies the direction of any turns. The following lemmas prove that the A_is, D_is, and E_is are well-behaved, which is useful in their efficient computation as shown later.

Lemma 6 *Each non-kernel basic interval A_i is well-behaved.*

123

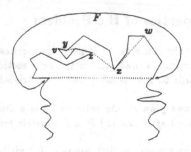

Fig. 1. Proof of lemma 6

Proof. Each basic interval is contained in some non-redundant component. Let $A_i = P_{CCW}(w, v)$, and let F be the non-redundant component containing it. Suppose $SP(w, v)$ contains a point z of $P_{CW}(w, v) \setminus \{w, v\}$. Let x (y) be the first point of $P_{CCW}(w, v)$ preceding (succeeding) z on $SP(w, v)$ (see Figure 1). The chord that forms F partitions P into two subpolygons, and since w and v are in the same subpolygon, any point on $SP(w, v)$ must also be in this subpolygon; thus $z \in F$. Since z is on $SP(w, v)$ it is a reflex vertex of P and therefore generates two components. The hit points of these components must both be on $P_{CCW}(x, y)$, as one can see by considering that the ray shots are contained in the subpolygon formed by $P_{CCW}(x, y)$ and $SP(x, y)$. Since z and its two hit points all lie on F, one of the two components generated by z is strictly contained in F, a contradiction of the fact that F is non-redundant. □

We next show that even the side intervals D_i and F_i are well-behaved. We first consider the case where P does not have two disjoint components.

Lemma 7 *Let P be a polygon with no two disjoint components. Then for each non-kernel A_i, the corresponding D_i and E_i are well-behaved.*

Proof. We first prove that any non-redundant chord is a weakly-visible chord. Suppose we have a non-redundant component F. Since no two components are disjoint, F intersects every other component. Furthermore, at least one endpoint of F intersects another component G unless G is nested inside F, which is not possible since F is non-redundant. Therefore the chord that forms F has endpoints which intersect every component of P, and thus by lemmas 1 and 4 this is a weakly-visible chord of P.

Thus there are (at least two) weakly-visible chords between each non-kernel A_i and its corresponding B_i. Any one of these weakly-visible chords has one endpoint on A_i and the other on B_i, and therefore separates D_i and E_i into different subpolygons. Thus, if D_i is not well-behaved it is because the shortest path between its endpoints contains a point of A_i or of B_i. Say it contains a point z of A_i that is not an endpoint of A_i. By an argument similar to that in the proof of lemma 6, z must be a reflex vertex whose hit points lie inside D_i. Thus z generates a component H that intersects both A_i and D_i yet contains neither. H does not intersect B_i, and does not contain the first point of A_i in counterclockwise order. This contradicts the fact that each point of A_i is LR-visible with each point of B_i. □

We now consider the case where P has at least two disjoint components. The following lemma is similar to the above, except that it somewhat less general.

Lemma 8 *Let P be a polygon with at least two disjoint components. If there exists a weakly-visible chord in P, or if there is more than one basic interval A_i with non-empty B_i, then each D_i and E_i is well-behaved.*

Proof. The proof of this lemma is more involved than lemma 7. In the proof we first show that if D_i (E_i) is not well-behaved, then it cannot be because of A_i or B_i. Then we show that if D_i (E_i) is not well-behaved due to obstruction by E_i (D_i), then the LR-visible pairs of points are not visible from each other, and consequently no weakly-visible chords exist. Details of the proof can be found in a full version of the paper. □

We introduce the following notation. Let $b(F)$ $(e(F))$ be the starting point (ending point) of an interval F encountered in the counterclockwise direction. The following lemma states that the visibility restrictions between points on A_i and points on B_i are imposed only by the side intervals.

Lemma 9 *Let x be a point on A_i and y a point on B_i. If the line $\overline{xy}$ intersects the polygon at a point $w \in P_{CCW}(x, e(A_i)) \cup P_{CCW}(b(B_i), y)$, then it also crosses D_i. Similarly, if $\overline{xy}$ intersects the polygon at a point $w \in P_{CCW}(y, e(B_i)) \cup P_{CCW}(b(A_i), x)$, then it also crosses E_i.*

By "crosses," we mean intersects such that there are points of D_i (E_i) on either side of the ray. The lemma essentially says that, if a point $x \in A_i$ cannot see a point $y \in B_i$, its visibility is blocked by D_i or E_i. The proof uses techniques similar to ones used in the previous lemmas, and we omit details in this version.

4 Computing Weakly-Visible Chords

4.1 Overview

We first give an overview of the algorithm. There are several preliminary steps. Our algorithm first constructs the kernel K using the linear time method of [10] and then constructs $K \cap P$. This latter step is easily accomplished since the algorithm of [10] can return the vertices of K which lie on P. If this is not empty, the algorithm outputs $(K \cap P, P)$ which describes all weakly-visible chords with one endpoint in the kernel.

Then the LR-visibility algorithm in [4] is run, which gives us the non-redundant components with endpoints in counterclockwise order, as well as the (A_i, B_i) pairs. If there is only one pair, we check if D_1 and E_1 are well-behaved, by running the shortest-path algorithm of [6] from their endpoints. If one of them is not well-behaved, the algorithm halts and reports that there are no weakly-visible chords.

We next determine in linear time whether there exist two disjoint components. Suppose we find that P does have a pair of disjoint components F and G. Any LR-visible pair of points must have one point on F and the other on G, so for any basic interval A_i in F we know that B_i is contained in G. It suffices to look only at A_i in F (and their corresponding B_i in G) and compute weakly-visible chords, if any. Suppose P does not have two disjoint components. In this case the algorithm examines each non-kernel A_i and its corresponding B_i in search of weakly-visible chords.

In either case, the basic step consists of determining the weakly-visible chords between a non-kernel A_i and its corresponding B_i. By the definition of these intervals we know that a point $x \in A_i$ forms an LR-visible pair with a point y if and only if $y \in B_i$. By lemma 4, therefore, we can construct the set of weakly-visible chords by determining for each point x of a basic interval A_i the points of B_i from which it is visible. Essentially, for each x we must determine the restrictions on its visibility with B_i. We also know that A_i, D_i and E_i are well-behaved, and visibility restrictions are only imposed by the side intervals. Below we describe how to compute chords between A_i and B_i and the transition from iteration i to $i+1$.

4.2 Implementation Details

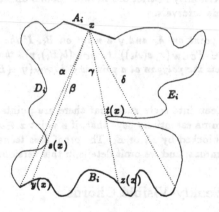

Fig. 2.

The visibility between a point $x \in A_i$ and B_i is determined by a pair of constructs we call the *pseudo-tangents*. Consider Figure 2. The pseudo-tangent from x to D_i (E_i) is the unique line directed from x through a point $s(x)$ of D_i ($t(x)$ of E_i) such that all of D_i (E_i) lies on or to the left (right) of the line (pseudo-tangents are different from *tangents*, which are also required to be chords). We denote the direction of the pseudo-tangent to D_i (E_i) by $\beta(x)$ ($\gamma(x)$). We give labels to two other special directions: the direction from x to $b(B_i)$ ($e(B_i)$) is denoted $\alpha(x)$ ($\delta(x)$). Sometimes we will use abbreviated notation, for example α instead of $\alpha(x)$, when the x under consideration is clear.

The point $x \in A_i$ is visible from some point of B_i if and only if the special directions satisfy the following relationship: $\alpha \leq_{ccw} \beta \leq_{ccw} \gamma \leq_{ccw} \delta$ (where $\leq_{ccw}$ means "precedes or equals in counterclockwise order, as viewed from z"). Let us consider the picture for a point $z \in A_i$ which is visible from B_i. In this event the pseudo-tangencies are actually tangencies, in the sense that they are chords of P. Let $y(z)$ ($z(x)$) be the other endpoint of the longest chord of P with one endpoint at z in direction β (γ). Since α, β, γ and δ are ordered counterclockwise, the points y and z lie on B_i. Points of B_i lying between $b(B_i)$ and y (z and $e(B_i)$) are not visible from z because visibility is blocked by D_i (E_i). However, all points of $P_{CCW}(y, z)$ are visible from z. Also, if $b(B_i)$ ($e(B_i)$) is the pseudo-tangent point of D_i (E_i), i.e. if $\alpha = \beta$ ($\gamma = \delta$), then $b(B_i)$ ($e(B_i)$) is also visible from z. Thus we see that if z is visible from B_i, its set of weakly-visible partners consists of a closed subinterval $P_{CCW}(y, z)$ of B_i, plus possibly one or both of the endpoints of B_i.

If z is not visible from B_i, then the four special directions are not ordered properly. If D_i (E_i) blocks all of B_i from z, then we have the subordering $\alpha \leq_{ccw} \delta <_{ccw} \beta$ ($\gamma <_{ccw} \alpha \leq_{ccw} \delta$). If the ordering is $\alpha \leq_{ccw} \gamma <_{ccw} \beta \leq_{ccw} \delta$, then neither D_i nor E_i totally block visibility individually, but together they do. In this case we can still define y and z as the extensions of the pseudo-tangencies until they hit B_i, where the opposite side interval is simply ignored. The fact that the ordering of β and γ is reversed means that z precedes y counterclockwise on B_i.

The algorithm uses the following strategy. We traverse P once counterclockwise with a point z, calculating for each z of a basic interval A_i the pseudo-tangent to the side interval D_i. We then traverse P once clockwise in order to compute for each z the pseudo-tangent to E_i. We determine for which z the ordering $\alpha \leq_{ccw} \beta \leq_{ccw} \gamma \leq_{ccw} \delta$ is obeyed, and for these z we compute the extensions y and z to obtain the partner interval $P_{CCW}(y, z)$. We also note whether $\alpha = \beta$ ($\gamma = \delta$) for any of these z, in which case $b(B_i)$ ($e(B_i)$) is also a partner. Since y and z are different for each z, and there are an infinitude of values of z, we must exhibit care in our manner of computing and storing the output; this issue will be addressed during the discussion below.

We describe the counterclockwise traversal of P with a point z, calculating for each z in A_i the pseudo-tangents to D_i; the procedure for E_i pseudo-tangents is symmetrical. If, for some z in A_i, β lies between α and δ, then we wish to compute the extension y. If $\alpha = \beta$ then the extension y equals $b(B_i)$. If $\delta <_{ccw} \alpha$ then z sees no point of B_i and the extension y is undefined.

We now show that the functions $\beta(z)$ and $y(z)$ are monotonic, i.e as z moves counterclockwise along the entire polygon, the direction $\beta(z)$ moves counterclockwise and $y(z)$ also moves counterclockwise along P. First, consider the motion of z within a basic interval A_i. As z moves counterclockwise the pseudo-tangent rotates counterclockwise around the *pseudo-tangent point* (the point on D_i which the pseudo-tangent touches), with the consequence that y (if defined) moves counterclockwise on B_i. The pseudo-tangent point may change to a point of D_i closer to $b(D_i)$, but this does not affect the monotone motion. Next consider the transition of z to the next basic interval. When z reaches $e(A_i) = b(D_i)$, the basic interval is updated to A_{i+1} and the side interval to D_{i+1}. The new side interval is obtained from D_i by subtracting A_{i+1} and possibly adding some additional basic intervals at the far end. There are two cases. In the first case (Figure 3(a)), the pseudo-tangents from z to both D_i and D_{i+1} are the same, hence $y(e(A_i)) = y(b(A_{i+1}))$. In the second case (Figure 3(b)), the pseudo-tangent from z to

D_{i+1} is counterclockwise to the the pseudo-tangent from z to D_i, hence $y(b(A_{i+1}))$ is counterclockwise of $y(e(A_i))$. Thus we see that $\beta(z)$ and $y(z)$ moves monotonically counterclockwise.

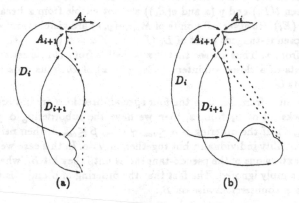

(a)   (b)

Fig. 3.

Another important observation regards the monotone motion of the pseudo-tangent point, $s(z)$. As z progresses counterclockwise within a basic interval A_i, s moves clockwise on D_i. Furthermore, when the side interval and the pseudo-tangent point are updated upon z reaching $e(A_i)$, the updating occurs in such a way that no point which has previously been the pseudo-tangent point can again become the pseudo-tangent point at a later time. This observation is clear from Figures 3(a) and (b), and is crucial in computing all pseudo-tangent points in overall linear time.

Since the side interval D_i is always well-behaved, the shortest path $SP(b(D_i), e(D_i))$ (which we will simply denote $SP(D_i)$) is a convex chain consisting of only right turns. This makes it easy to find and update tangencies. For $z = b(A_i)$, traverse $SP(D_i)$ from $e(D_i)$ until reaching the pseudo-tangent point (determining if a point of $SP(D_i)$ is the pseudo-tangent point is accomplished in constant time by comparing directions of the line from z with those of the adjacent edges). As z progresses counterclockwise we can update the pseudo-tangent point by continuing to traverse $SP(D_i)$ towards $b(D_i)$. No point of $SP(D_i)$ which is traversed will ever be a pseudo-tangent point in the future, and the total time of this operation is proportional to the number of vertices of $SP(D_i)$ traversed. The additional required ingredient, then, is an efficient manner of maintaining $SP(D_i)$ for the current D_i. We discuss this next.

Consider the first side interval D_1, which corresponds to the first basic interval A_1. It is composed of a collection of basic intervals $A_2, \cdots, A_j$. The side interval D_j is composed of the basic intervals $A_{j+1}, \cdots, A_l$, and is the first side interval that does not overlap with D_1. We will essentially deal with D_1 through D_j as a group. We will preprocess the entire group in a manner that allows efficient updating.

The preprocessing is the construction of the shortest path tree from $e(D_1)$ to all vertices of D_1. Since D_1 is well-behaved we perform this step in time proportional to the size of D_1, by a modification of the algorithm of [6]. At any time we will store $SP(D_i)$ in three pieces, namely as two shortest paths, $SP(b(D_i), e(D_1))$ and $SP(e(D_1), e(D_i))$, plus the bridge between them (see Figure 4). The chain $SP(b(D_i), e(D_1))$ is obtained from $SP(b(D_{i-1}), e(D_1))$ by using the shortest path tree. A depth-first search of this tree allows one to visit all vertices of D_1 according to the counterclockwise order on P, and thus allows us to maintain the current shortest path $SP(b(D_i), e(D_1))$. To maintain the other shortest path, $SP(e(D_1), e(D_i))$, we observe that updating from $SP(e(D_1), e(D_{i-1}))$ consists of adding zero or more basic intervals. Since each basic interval is well-behaved, the shortest path between its endpoints is convex. The bridge between two convex chains with a common endpoint can be found in time proportional to the number of vertices below the bridge on both chains. Since vertices below the bridge cannot be pseudo-tangent points, this time-complexity is acceptable to us.

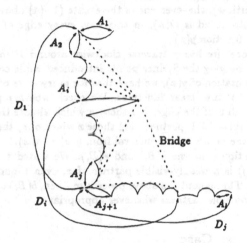

Fig. 4.

Thus, as z moves from A_{i-1} to A_i, we first update to obtain $SP(b(D_i), e(D_1))$ and then update to obtain $SP(e(D_1), e(D_i))$. Our true goal is $SP(D_i)$, and to obtain it we need the bridge between the two smaller shortest paths. Efficient computation and maintenance of the bridge is possible because the bridge endpoints display a property similar to that of the pseudo-tangents from z: the bridge endpoints progress monotonically clockwise, and no point which has previously been a bridge endpoint can again become a bridge at a later time.

In this manner we maintain the side interval while z traverses from A_1 to A_i, in total time proportional to the size of A_1 through A_l. If P has two disjoint components, then all side intervals intersect D, and a single iteration of this procedure suffices. If P does not have two disjoint components, then it is necessary to repeat the procedure, with the next step beginning with D_j and ending at D_l. This method "overlaps" portions of P, but no basic interval is used in more than two iterations, so maintaining $SP(D_i)$ while x traverses all of P requires only $O(n)$ time.

We discuss how $y(x)$ is efficiently computed and stored. Given x and the pseudo-tangent to D_i, we find y by extending the pseudo-tangent until it hits B_i. As x progresses counterclockwise, several events can occur. For example, (1) x can reach a vertex of A_i, (2) y can reach a vertex of B_i, or (3) the pseudo-tangent point s can pivot about an edge of $SP(D_i)$. Also, (4) when x reaches $e(A_i)$, both s and therefore y may need to be updated. Between events, however, we have an edge containing x and another containing y, and a point s which lies on $\overline{xy}$. Thus, even though we have an infinitude of values of x, each with a unique y, the function $y(x)$ can be described in constant time. A particular value $y(x)$ is found in constant time by computing the intersection of two lines.

In order to easily store $y(x)$, we introduce Steiner points at x and y (if x and/or y are not already vertices) whenever one of the events (1)-(4) above occurs. The total number of points introduced is $O(n)$. In this way, every edge of a non-kernel basic interval has a linear function $y(x)$.

A symmetric procedure has x traverse clockwise around P in order to compute $z(x)$ for each x. By merging the Steiner points introduced while computing $z(x)$ with those from the computation of $y(x)$, we have that for every edge of a non-kernel basic interval, $y(x)$ and $z(x)$ are linear functions. We check whether $y(x)$ precedes $z(x)$ counterclockwise for all x of the edge. For those x which violate this order, we return that they have no weakly-visible partners. For those x which obey the order, the weakly-visible partners are the points on the interval from $y(x)$ to $z(x)$.

A final consideration concerns $b(B_i)$ and $e(B_i)$. We stated that if $\alpha = \beta$ ($\gamma = \delta$) then $b(B_i)$ ($e(B_i)$) is a weakly-visible partner of x, even though it is outside the interval $P_{CCW}(y, z)$. Throughout the above procedure, then, $b(B_i)$ and $e(B_i)$ are stored separately as weakly-visible partners whenever appropriate.

5   The Rectilinear Case

In this section, we consider the simpler case where the given polygon P is a simple rectilinear polygon. Simple geometric observations about rectilinear polygons are used to construct a simpler algorithm for computing all non-redundant components and consequently all weakly-visible chords in the rectilinear case. It may be noted that the chords need not be rectilinear.

Before we proceed we need some notation. Components in rectilinear polygons are produced by two kinds of ray shots – vertical and horizontal. We call these components *vertical* (resp. *horizontal*) components. Furthermore, there are two kinds of horizontal (vertical) components – components that lie *above* (to the em left of) or *below* (to the right of) the ray shot. We call the four type of components as *left-vertical, right-vertical, above-horizontal,* and *below-horizontal* components respectively. The left-vertical and right-vertical types are called *complementary* types, as are the above-horizontal and the below-horizontal types. It is easy to see that two components of the same type are either disjoint or one contains the other (i.e., one is made redundant by the other). For example, there cannot be two partially overlapping left-vertical components. If P has more than 3 disjoint components, then by lemma 1 P has no weakly-visible chords. Hence for weakly-visible chords to exist there cannot be more than 12 non-redundant components. In fact, we can make a stronger statement. It can be proved that for a weakly-visible rectilinear polygon, if there are two disjoint non-redundant components of a certain type, then there are no non-redundant of the complementary type and there cannot be another non-redundant component that intersects both of them. Consequently for a weakly visible rectilinear polygon, if there are two disjoint left-vertical non-redundant components, then there are no right-vertical non-redundant components. We state the following lemma without proof.

Lemma 10 *There are at most 2 vertical non-redundant components, and at most 2 horizontal components in a LR-visible rectilinear polygon. Also every non-redundant vertical (resp. horizontal) component is y-monotone (resp. x-monotone).*

Without getting into the details, we mention here that in 4 sweeps of the entire polygon, all the non-redundant components can be identified.

References

1. Avis, D., Toussaint, G.T.: An optimal algorithm for determining the visibility of a polygon from an edge. IEEE Transactions on Computers 30 (1981) 910–914
2. Chen, D.Z.: Optimally computing the shortest weakly-visible edge of a simple polygon. Proc. Fourth ISAAC, LNCS 762 (1993) 323–332
3. Doh, J., Chwa, K.: An algorithm for determining visibility of a simple polygon from an Internal Line Segment. J. of Algorithms 14(1) (1993) 139–168
4. Das, D., Heffernan, P.J., Narasimhan, G.: LR-visibility in polygons. Proc. 5th Canadian Conference on Computational Geometry (1993) 303–308. Submitted to special issue of Computational Geometry - Theory and Appln.
5. Das, G., Narasimhan, G.: Optimal Linear-Time Algorithm for the Shortest Illuminating Line Segment in a Polygon. Proc. 10th Annual ACM Symp. on Computational Geometry (1994)
6. Guibas, L., Hershberger, J., Leven, D., Sharir, M., Tarjan, R.: Linear time algorithms for visibility and shortest path problems inside triangulated simple polygons. Algorithmica 2 (1987) 209–233
7. Heffernan, P.J.: An optimal algorithm for the two-guard problem. Proc. 9th Annual ACM Symp. on Computational Geometry (1993) 348–358
8. Icking, C., Klein, R.: The two guards problem. Proc. 7th Annual ACM Symp. on Computational Geometry (1991) 166–175
9. Ke, Y.: Detecting the weak visibility of a simple polygon and related problems. Tech. Report, The Johns Hopkins University (1987)
10. Lee, D.T., Preparata, F.P.: An optimal algorithm for finding the kernel of a polygon. Journal of the ACM, 26(3) (1979) 415–421
11. Sack, J.-R., Suri, S.: An optimal algorithm for detecting weak visibility. IEEE Transactions on Computers 39(10) (1990) 1213–1219
12. Tseng, L.H., Lee, D.T.: Two-guard walkability of simple polygons. manuscript (1993)

A Tight Lower Bound for
On-line Monotonic List Labeling

Paul F. Dietz[1], Joel I. Seiferas[1], and Ju Zhang[2]

[1] Computer Science Department, University of Rochester
Rochester, New York, USA 14627-0226
[2] Consumer Asset Management, Chemical Bank
380 Madison Avenue, 13th Floor, New York, New York, USA 10017

Abstract. Maintaining a monotonic labeling of an ordered list during the insertion of n items requires $\Omega(n \log n)$ individual relabelings, in the worst case, if the number of usable labels is only polynomial in n. This follows from a lower bound for a new problem, *prefix bucketing*.

1 Introduction

The *on-line list-labeling problem* can be viewed as one of linear density control. A sequence of n distinct items from some dense, linearly ordered set, such as the real numbers, is received one at a time, in no predictable order. Using "labels" from some discrete linearly ordered set of adequate but limited cardinality, the problem is to maintain an assignment of labels to the items received so far, so that the labels are ordered in the same way as the items they label. In order to make room for the next item received, it might be necessary to change the labels assigned to some of the items previously received. The *cost* is the total number of labelings and relabelings performed.

There are practical applications of on-line list labeling to the design of efficient data structures and algorithms. List labeling has been an especially fruitful approach to the *order maintenance problem* [Di82, Ts84, DS87, DZ90]. This problem involves the insertion and deletion of items into a linear list, and response to on-line queries on the relative order of items currently in the list. A low-cost on-line list-labeling algorithm provides an efficient solution (or sometimes a component of an even more efficient solution) to the order maintenance problem, provided its computational overhead is also low. For further discussion of this and other specific applications, see the earlier papers by Dietz and his collaborators [Di82, DS87, DZ90]. In addition, it seems likely that our problem and related problems of dynamic density control will prove fundamental to the spatially structured maintenance in bounded media of changing data, such as text and pictures on a computer screen [Zh93].

This paper is based on a portion of the third author's doctoral dissertation [Zh93]. The first author was supported in part by the National Science Foundation under grant CCR-8909667. We thank Jun Tarui and Ioan Macarie for their corrections and suggestions.

When the number of labels is at least $n^{1+\epsilon}$ for some $\epsilon > 0$, it is possible to limit the worst-case cost for on-line labeling of n items to $\mathcal{O}(n \log n)$ [Di82, Ts84, DS87]. The analyses are subtle; but the best of the algorithms are both simple and fast, and hence practically useful. In this paper we show that the upper bound is tight, and in fact that $\Omega(n \log n)$ relabelings are required even for an algorithm that is complicated and slow.

Our proof is a surprising adaptation of a lower-bound approach sketched by Dietz and Zhang [DZ90]. That approach seemed to be a dead end that addressed only strategies that satisfy the following "smoothness" property: The list items relabeled before each insertion form a contiguous neighborhood of the list position specified for the new item, and the new labels are as widely and equally spaced as possible. Although no good nonsmooth algorithms have been proposed or analyzed, it has seemed difficult to rule them out. (This is the usual sort of lower-bound predicament.) The key to our adaptation is to imagine appropriate dynamic "recalibrations" of the label space, in terms of which the arbitrary strategy does look fairly smooth.

To facilitate the elaboration and adaptation of the earlier argument, we formulate and separately attack variants of a previously unstated combinatorial "bucketing problem" that really lies at the heart of the argument. In the *unordered bucketing problem*, the challenge is to cheaply insert n items, one at a time, into k buckets. The cost of each insertion is the number of items (including the new one) in the bucket chosen for that insertion. The optimum total cost for the task as described so far is clearly $\Omega(n^2/k)$, but we allow an additional operation: Between insertions, we can redistribute the contents within any subset of the buckets, at a cost equal to the total number of items in those buckets. Now $\mathcal{O}(n \log n)$ is an upper bound on the required cost, by the well-known Hennie-Stearns strategy [HS66, Zh93], provided k is $\Omega(\log n)$. On the other hand, we prove in Sect. 3 that $\Omega(n \log n / \log k)$ is always a lower bound on the required cost, and we conjecture that $\Omega(n \log n)$ is a lower bound when k is $\mathcal{O}(\log n)$. We show in Sect. 2 that either lower bound leads to a similar lower bound for the problem of primary interest, the n-item, polynomial-label labeling problem.

The *prefix bucketing problem* is like the unordered bucketing problem, except with the constraint that, in terms of some fixed linear order of the buckets, the subset for each redistribution must be a *prefix* of the bucket list. (The Hennie-Stearns strategy still applies.) Under this constraint, each redistribution may as well move all items involved into the very last bucket of the chosen prefix. Section 2 actually shows that even a lower bound on this bucketing problem leads to a labeling lower bound. In Sect. 4, we prove the needed lower bound on prefix bucketing.

2 Relation to Bucketing

A relabeling algorithm is *normalized* if the items it relabels on each insertion, along with the newly inserted item, form a contiguous sublist of the list resulting from the insertion. Since *non*contiguous relabelings can safely be deferred until later, each labeling algorithm can be replaced at no additional cost by a normalized one.

To prove a lower bound for the labeling problem, we show that each normalized algorithm (and hence each algorithm of any kind) performs many relabelings when confronted with some bad-case sequence of insertion requests. That sequence, which will depend on the particular algorithm, will be determined by an "adversary" strategy that *interacts* with the algorithm – each next insertion will be into a gap that the adversary chooses based on the labeling decisions made by the algorithm in response to earlier insertion requests.

Intuitively, the most promising strategy for the adversary is to insert the next item into a gap between items in a part of the label space that is currently "relatively crowded". It seems difficult, however, to formulate an appropriate notion of crowdedness. Ordinary *density*, for instance, can vary depending on the size and choice of neighborhood. A more robust notion of a "dense point" is a gap *all* of whose neighborhoods are currently about as dense as the entire label space. The following lemma shows that such a dense point always does exist.

Dense-point Lemma. *Consider any nonnegative, integrable function f on the interval* $[0,1]$. *For each (nontrivial) subinterval I, define*

$$\rho(I) = \frac{1}{|I|} \int_I f(x)\,dx.$$

Then there is some point $x_0 \in [0,1]$ *such that* $\rho(I) \geq \frac{1}{2}\rho([0,1])$ *holds whenever I includes* x_0.

Proof. For the sake of argument, suppose not. Then, for each point x, select a spoiling interval that includes x and that is open in $[0,1]$. (An interval is "open in $[0,1]$" if it is the intersection of $[0,1]$ itself and an ordinary open interval of real numbers.) The selected open intervals cover the topologically compact set $[0,1]$, so some finite subfamily must do so. If any point lies in three or more intervals of the subfamily, then keep only the one that reaches farthest left and the one that reaches farthest right. This leaves a finite family $\mathcal{I}$ that covers each point in $[0,1]$ either once or twice, but each of whose members I satisfies $\rho(I) < \frac{1}{2}\rho([0,1])$. Therefore,

$$\int_{[0,1]} f(x)\,dx \leq \sum_{I \in \mathcal{I}} \int_I f(x)\,dx = \sum_{I \in \mathcal{I}} \rho(I)|I|$$

$$< \sum_{I \in \mathcal{I}} \frac{1}{2}\rho([0,1])|I| = \rho([0,1])\sum_{I \in \mathcal{I}} \frac{1}{2}|I| \leq \rho([0,1]) = \int_{[0,1]} f(x)\,dx,$$

a contradiction.
□

Corollary 1. *In each labeling, there is a label such that every label-space subinterval containing that label is at least half as dense as the entire label space. (The same applies to either of the two gaps that include the distinguished label, since they themselves are qualifying subintervals.)*

Proof. If the total number of labels is m, then consider a function f that is constantly 1 or 0 on each subinterval $((i-1)/m, i/m)$, depending on whether the i-th label is or is not in use, respectively. □

Corollary 2. *In each labeling, there is a label in the "middle population third" (where rounding is in favor of that middle third) such that every label-space subinterval containing that label is at least one-sixth as dense as the entire label space.*

Proof. Ignore the rounded-down left third and the rounded-down right third of the items, and cite Corollary 1. □

Although such a dense point always exists, it does not quite suffice always to insert into just any such gap. For example, the algorithm that always inserts at the midpoint of the requested gap will be able to maintain an essentially perfect spread without ever relabeling even a single item, if the adversary inserts into the sequence of (dense-point) gaps numbered

$$1, \ 2, 1, \ 4, 3, 2, 1, \ 8, 7, 6, 5, 4, 3, 2, 1, \ \ldots.$$

The problem is that the adversary is forfeiting an opportunity to use its insertions to selectively increase congestion in a particular locality.

To take advantage of its opportunity to create congestion, our adversary will try to keep the relocation of its insertion point "commensurate with" the relabeling response by the algorithm. I.e., unless it has forced the algorithm to move a lot of items away from the insertion point, it will continue to insert into and add congestion to the same neighborhood. To this end, it will actually maintain an entire *nest* of $k = \mathcal{O}(\log n)$ distinct *intervals* that converge down to the insertion point. The population of the smallest enclosing one of the intervals will be proportional to the number of relabelings performed. This will "justify" relocation of the insertion point to any appropriate gap in that interval, since the relabeling could have reconcentrated the population arbitrarily within the interval.

The nest of intervals $I_1 \supset I_2 \supset \cdots \supset I_k$ that our adversary maintains will satisfy the five essential conditions listed below. For each label interval I, we denote the number of currently assigned labels ("population") and the total number of labels by $\text{pop}(I)$ and $\text{area}(I)$, respectively. If $I \supset I'$, then the difference $I - I'$ consists of (at most) a left interval and a right interval; we denote their respective populations by $\text{leftpop}(I - I')$ and $\text{rightpop}(I - I')$.

(1) I_1 is the whole label space.

(2) $\text{pop}(I_k) = \mathcal{O}(1)$.

(3) For every i, $\text{area}(I_{i+1}) \leq \text{area}(I_i)/2$.

(4) For every i, $\text{pop}(I_{i+1}) = \Omega(\text{pop}(I_i))$.

(5) For every i, $\text{leftpop}(I_i - I_{i+1}) = \Theta(\text{rightpop}(I_i - I_{i+1}))$.

As we promised above, it follows from these conditions that $k = \mathcal{O}(\log n)$, and that the population of the smallest interval enclosing a batch of relabelings is at most some constant times the number of relabelings in the batch. Therefore, if we consider the successive differences, $I_i - I_{i+1}$, and the innermost interval, I_k, to be the buckets, then the algorithm solves the resulting (prefix) bucketing problem at a total cost that is at most some constant times the number of relabelings it performs. Since the former has to be $\Omega(n \log n / \log k) = \Omega(n \log n / \log \log n)$, for example, so does the latter.

Finally, the following lemma insures that we can appropriately restore the invariant after each batch of relabelings by the algorithm.

Restoration Lemma. *Each sufficiently long and populous interval I has a subinterval I' such that*

$$\text{area}(I') \leq \text{area}(I)/2,$$
$$\text{pop}(I') = \Omega(\text{pop}(I)), \text{ and}$$
$$\text{leftpop}(I - I') = \Theta(\text{rightpop}(I - I')).$$

Proof. From a dense point in the middle population third of I (provided by Corollary 2), expand through population leftward and rightward in proportion to the total populations in those directions (which can differ by at most a factor of 2), until half the area is covered. (If this requires a *fraction* of an item in either direction, then just stop one label short of that item's label.) □

3 Lower Bound for Unordered Bucketing

This section is devoted to the relatively easy proof of the following lower bound, which we conjecture can be tightened to $\Omega(n \log n)$ when k is $\mathcal{O}(\log n)$.

Theorem. *The cost for unordered bucketing of n items into k buckets is $\Omega(n \log n / \log k)$.*

Consider the following measure of a configuration's complexity:

$$C = \sum_{i=1}^{k} n_i \log n_i,$$

where n_i is the number of items in bucket i. (Since $\lim_{x \downarrow 0} x \log x = 0$, it works well to define $0 \log 0$ to *be* 0.) C starts out at 0; and, by the Complexity-range Lemma below, it finally reaches a value no smaller than

$$F = n \log n - n \log k.$$

We show below, however, that no operation increases C by more than $\mathcal{O}(\log k)$ times the cost of the operation. Therefore, the total cost will have to be at least $F/\log k = \Omega(n \log n / \log k)$.[1]

The main operation to consider is the reorganization of $k' \leq k$ buckets containing a total of $n' \leq n$ items. By definition, the cost of the operation is n'. And, by the Complexity-range Lemma below again, the increase in C is indeed at most

$$n' \log k' \leq n' \log k = \mathcal{O}(n' \log k).$$

The only other operation is insertion into an n'-item bucket. If $n' = 0$, then there is *no* change in C; so assume $n' \geq 1$. Then the cost is exactly $n' + 1$, and the increase in C is exactly

$$
\begin{aligned}
(n' + 1) &\log(n' + 1) - n' \log n' \\
&= \log(n' + 1) + n'(\log(n' + 1) - \log n') \\
&= \log(n' + 1) + n' \mathcal{O}(1/n'),
\end{aligned}
$$

which is certainly $\mathcal{O}((n' + 1) \log k)$.

Complexity-range Lemma. *If $n_1 + \cdots + n_k = n$, where each n_i is nonnegative, then $\sum n_i \log n_i$ lies between $n \log n$ and $n \log n - n \log k$.*

Proof. It is easy to argue that the sum is maximized when some n_i equals n, and minimized when every n_i equals n/k. □

4   Tight Lower Bound for Prefix Bucketing

This section is devoted to a proof of the following tight lower bound:

Theorem. *The cost for prefix bucketing of n items into $k = \mathcal{O}(\log n)$ buckets is $\Omega(n \log n)$.*

It will be convenient to have terminology for the current configuration of (a prefix of) a bucket list. While we are at it, to make Lemmas 2 and 3 possible below, we generalize to allow fractional numbers of items in a bucket. If k is a positive integer and n is a positive real number, then an (n, k)-*configuration* is a list $L = (n_1, \ldots, n_k)$ of k nonnegative real numbers such that $\sum n_i = n$, viewing n_i as the (possibly fractional) number of items in bucket i. L is *nondecreasing* if $n_i \leq n_{i+1}$ holds for every $i < k$; and it is *exponential, with ratio a and with k' initial 0's*, if $n_i = 0$ for every $i \leq k'$ and $n_i = a n_{i+1}$ for every $i \in \{k' + 1, \ldots, k - 1\}$.

---

[1] For bucketing to be nontrivial, n and k have to be at least 2. In that case, $\log n$ and $\log k$ are safely positive if we use some logarithmic base strictly between 1 and 2.

For each (n, k)-configuration $L = (n_1, \ldots, n_k)$, we define *two* measures of complexity:

$$C(L) = \sum_{i=1}^{k} n_i \log n_i,$$

$$M(L) = \sum_{i=1}^{k} i n_i.$$

Because the redistributions we consider move all items to the last bucket involved, we make the following anticipatory definitions:

$$\Delta C(L) = n \log n - C(L),$$
$$\Delta M(L) = kn - M(L).$$

Let d be a constant so large that $k \leq d \log n$. Assuming n is large enough so that $d \log n \leq n^{1/2}$, the complexity C starts out at 0 and finally reaches a value

$$C_{\text{final}} \geq n \log n - n \log k \geq \frac{1}{2} n \log n.$$

The measure M starts out at 0, grows monotonically, and finally reaches a value

$$M_{\text{final}} \leq kn \leq dn \log n.$$

Over all, therefore, the increase in C is at least $\frac{1}{2d}$ times the increase in M. Consider the steps on which we have

$$\Delta C < \tfrac{1}{4d} \Delta M,$$

where ΔC and ΔM are the respective increases in C and M. Such steps can account for at most half of the overall change in C. Therefore, we can restrict attention to the other steps, on each of which we must have

$$\Delta M \leq 4d \Delta C.$$

We show that, on each such step, regardless of its context, ΔC is at most some constant times the number of items involved, which is the bucketing cost, and hence that the total bucketing cost for such steps has to be $\Omega(n \log n)$. We saw in Sect. 3 that this fact holds for every insertion step; so we restrict further attention to the analysis of redistribution steps, and Corollary 5.1 below will complete the proof.

Actually we *directly* analyze such a redistribution step only when the (n, k)-configuration before the step is of a special form. The first sequence of lemmas below, culminating in Lemma 3, shows that it is no loss of generality to restrict attention to this form; and the final lemmas provide the needed estimates involving $\Delta C(L)$ and $\Delta M(L)$ when L is of this form.

Lemma 1. *For each (n, k)-configuration L that is not nondecreasing, there is a nondecreasing (n, k)-configuration L' with $C(L') = C(L)$ and $M(L') > M(L)$.*

Proof. Just reorder the configuration so that it *is* nondecreasing. □

Lemma 2. *For each nondecreasing (n, k)-configuration L that is not exponential, there is a nondecreasing (n, k)-configuration L' with $C(L') < C(L)$ and $M(L') = M(L)$.*

Proof. First, note that we lose no generality if we assume $k = 3$: If $L = (n_1, \ldots, n_k)$ is nondecreasing but not exponential, then there has to be some $i \leq k - 2$ such that (n_i, n_{i+1}, n_{i+2}) is an $(n_i + n_{i+1} + n_{i+2}, 3)$-configuration with these same properties. It is clear from the definitions of C and M that the desired conclusion for (n_i, n_{i+1}, n_{i+2}) will yield the conclusion for L, too.

For $k = 3$, the idea is to take

$$L' = (n_1 - x, n_2 + 2x, n_3 - x)$$

for some nonzero x. Since L is not exponential and k is only 3, there can be no initial 0's. In the case that $n_1 > (n_2/n_3)n_2$, x must satisfy $0 < x < n_1$; and, in the case that $n_1 < (n_2/n_3)n_2$, it must satisfy $-n_2/2 < x < 0$. *Whatever x is,* we will have $M(L') = M(L)$. It remains only to show that some eligible x will yield $C(L') < C(L)$.

For each prospective x, let $C(x)$ denote the resulting value $C(L')$. It is enough to show that

$$\lim_{x \downarrow 0} C'(x) < 0 \quad \text{if } n_1 > (n_2/n_3)n_2,$$

and that

$$\lim_{x \uparrow 0} C'(x) > 0 \quad \text{if } n_1 < (n_2/n_3)n_2.$$

Expressed more explicitly,

$$C(x) = f(n_1 - x) + f(n_2 + 2x) + f(n_3 - x),$$

where $f(x) = x \log x$. It is straightforward to check that the derivative $C'(x)$ does satisfy both requirements. □

Lemma 3. *For each (n, k)-configuration L, there is a nondecreasing, exponential (n, k)-configuration L' with $C(L') \leq C(L)$ and $M(L') \geq M(L)$.*

Proof. If the given configuration is not nondecreasing, then apply Lemma 1 one time. Then, calling the result L, consider the set $\mathcal{L}$ of nondecreasing (n, k)-configurations L' that satisfy $C(L') \leq C(L)$ and $M(L') = M(L)$. Since C is continuous on the topologically compact set $\mathcal{L}$, there is some L' in $\mathcal{L}$ that minimizes C. By Lemma 2, that (n, k)-configuration must be exponential. □

Let $L_{n,k+k',a,k'}$ denote the nondecreasing, exponential $(n, k + k')$-configuration with ratio a and with k' initial 0's. Note that, for every k', $\Delta C(L_{n,k+k',a,k'})$ equals $\Delta C(L_{n,k,a,0})$, and $\Delta M(L_{n,k+k',a,k'})$ equals $\Delta M(L_{n,k,a,0})$.

Lemma 4. *If* $a = 1$, *then*

$$\Delta C(L_{n,k,a,0}) = (\log k)n, \text{ and}$$

$$\Delta M(L_{n,k,a,0}) = \left(\frac{k-1}{2}\right)n.$$

If $a < 1$, *then*

$$\Delta C(L_{n,k,a,0}) = \left(\log A - \frac{B}{A}\log a\right)n, \text{ and}$$

$$\Delta M(L_{n,k,a,0}) = \left(\frac{B}{A} - 1\right)n,$$

where

$$A = \sum_{i=1}^{k} a^i \quad and \quad B = \sum_{i=1}^{k} i a^i.$$

Proof. The calculations are exact and easy. □

Corollary 4.1. *If* $a < 1$, *then*

$$\Delta C(L_{n,k,a,0}) < \left(\log \frac{1}{1-a}\right)n.$$

Proof. Just use the estimates

$$A < a/(1-a), B > a, \text{ and } a\log a < 0.□$$

Corollary 4.2. *For each fixed* k,

$$\lim_{a\uparrow 1} \Delta C(L_{n,k,a,0})/n = \Delta C(L_{n,k,1,0})/n = \log k, \text{ and}$$

$$\lim_{a\uparrow 1} \frac{\Delta M(L_{n,k,a,0})}{\Delta C(L_{n,k,a,0})} = \frac{\Delta M(L_{n,k,1,0})}{\Delta C(L_{n,k,1,0})} = \frac{k-1}{2\log k}.$$

Proof. If k is fixed, then $\lim_{a\uparrow 1} A = k$, $\lim_{a\uparrow 1} B = k(k+1)/2$, and $\lim_{a\uparrow 1} \log a = 0$, so that the results follow immediately from Lemma 4. □

Lemma 5. *There exists a pair of thresholds,* $a_0 < 1$ *and* k_0, *such that, whenever* $a_0 < a < 1$ *and* $k > k_0$,

$$\frac{\Delta M(L_{n,k,a,0})}{\Delta C(L_{n,k,a,0})} > 4d.$$

Proof. Choose k_0 large, and then choose $a_0 < 1$ large in terms of that k_0. The proof is by induction on $k \geq k_0$.

The base case, that

$$\frac{\Delta M(L_{n,k_0,a,0})}{\Delta C(L_{n,k_0,a,0})}$$

exceeds $4d$ whenever $a_0 < a < 1$, follows from Corollary 4.2. For the induction step, it is enough to show that

$$\frac{\Delta M(L_{n,k+1,a,0}) - \Delta M(L_{n,k,a,0})}{\Delta C(L_{n,k+1,a,0}) - \Delta C(L_{n,k,a,0})}$$

exceeds $4d$ whenever $a_0 < a < 1$ and $k \geq k_0$.

In terms of A and B, the goal is for the following to exceed $4d$:

$$E = \frac{[(B + (k+1)a^{k+1}) - (A + a^{k+1})] - [B - A]}{[(A + a^{k+1})\log(A + a^{k+1}) - (B + (k+1)a^{k+1})\log a] - [A \log A - B \log a]}$$

$$= \frac{ka^{k+1}}{A \log(1 + a^{k+1}/A) + a^{k+1}\log(A + a^{k+1}) - (k+1)a^{k+1}\log a}.$$

Since $0 < a < 1$ and k is large, $a^{k+1}/A < 1/k$ is small enough that

$$\log(1 + a^{k+1}/A) < a^{k+1}/A.$$

Since $A < k$ and $a^{k+1} < 1$, we certainly have

$$\log(A + a^{k+1}) < \log(k + 1).$$

Substituting these estimates, and cancelling a^{k+1}, we get

$$E > \frac{k}{1 + \log(k+1) - (k+1)\log a}.$$

Since k and a are large, this estimate finally does clearly exceed $4d$. □

The following corollary is just what we need to complete the proof.

Corollary 5.1. *If* (n,k)-*configuration* L *satisfies*

$$\frac{\Delta M(L)}{\Delta C(L)} \leq 4d,$$

then it also satisfies $\Delta C(L) = \mathcal{O}(n)$, *where the implicit constant depends on neither* n *nor* k.

Proof. By Lemma 3, since the conditions

$$C(L') \leq C(L) \quad \text{and} \quad M(L') \geq M(L)$$

respectively imply

$$\Delta C(L') \geq \Delta C(L) \quad \text{and} \quad \Delta M(L') \leq \Delta M(L),$$

it suffices to prove this when L is nondecreasing and exponential, say with ratio a, and with no initial 0's.

If $a = 1$, then the hypothesis gives us an upper bound on

$$\frac{\Delta M(L)}{\Delta C(L)} = \frac{k-1}{2 \log k} = f(k).$$

Since $\lim_{k \to \infty} f(k) = \infty$, this imposes some upper bound $b < \infty$ on k, so that we do get

$$\Delta C(L) = (\log k)n \leq (\log b)n = \mathcal{O}(n).$$

If $a < 1$, then we deal with three separate subcases: a "small" and k arbitrary, a "large" but k "small", and a and k both "large". First, however, we have to formulate appropriate size notions. Recall the thresholds a_0 and k_0 from Lemma 5. For each $k \leq k_0$, cite the first part of Corollary 4.2 to select a threshold $a_k < 1$ such that

$$\Delta C(L_{n,k,a,0})/n \leq 1 + \log k$$

holds whenever $a_k < a < 1$. Take $a_{max} = \max_{0 \leq k \leq k_0} a_k$.

Whenever $a \leq a_{max}$, Corollary 4.1 yields

$$\Delta C(L) \leq \left(\log \frac{1}{1-a} \right) n \leq \left(\log \frac{1}{1-a_{max}} \right) n.$$

Whenever $k \leq k_0$ and $a > a_{max} \geq a_k$, we have made sure that

$$\Delta C(L) \leq (1 + \log k)n \leq (1 + \log k_0)n.$$

And, whenever $k > k_0$ and $a > a_{max} \geq a_0$, Lemma 5 insures that $\Delta M(L)/\Delta C(L)$ exceeds $4d$. $\qquad\square$

5 Further Discussion

When the number of usable labels is *not* at least $n^{1+\epsilon}$ for any $\epsilon > 0$, the known upper bounds are not as low. With $\mathcal{O}(n)$ labels and exactly n labels, the respective bounds are $\mathcal{O}(n \log^2 n)$ [IKR81] and $\mathcal{O}(n \log^3 n)$ [AL90]. These bounds seem tight, and it can be shown that they are tight for smooth relabeling strategies [DZ90, Zh93, DSZ94]; but we do not yet see how to extend these results to non-smooth strategies, for which our $\Omega(n \log n)$ is still the only known lower bound. More generally, we would like a tight bound that is some nice *function*, say F, of the number n of usable labels, with $F(n) = \Theta(n \log^3 n)$, $F(cn) = \Theta(n \log^2 n)$ for each particular $c > 1$, and $F(n^{1+\epsilon}) = \Theta(n \log n)$ for each particular $\epsilon > 0$.

When the density of the labels in use grows large, the cost of further labeling becomes more closely related to an alternative natural cost measure: the number of *labels* spanned (rather than the number of *items*). Many of the same questions can be asked of this cost measure, and the answers and arguments might be independently interesting and enlightening. It turns out that the strongest version of the $\Omega(n \log^2 n)$ lower bound mentioned above (for smooth insertion of n items into a linearly bounded label space) is most natural in this setting, because then it turns out to hold regardless of the size of the label space; the lower bound on the standard cost follows as a corollary [DSZ94].

Even if it turns out that bucketing problems are not as closely related to on-line labeling for smaller numbers of labels, we would like to see tighter and more general analyses of their complexity as well. For $k = o(\log n)$ buckets, Jingzhong Zhang has proposed, via personal communication, a prefix-bucketing algorithm of cost $\mathcal{O}(n^{1+1/k}(k!)^{1/k})$. More careful analysis of his algorithm yields an expression that may be the exact optimum.

References

[AL90]   Andersson, A., Lai, T. W.: Fast updating of well-balanced trees. In: SWAT 90: 2nd Scandinavian Workshop on Algorithm Theory, Proceedings. (Lecture Notes in Computer Science, vol. 447.) Springer, New York Berlin Heidelberg, 1990, pp. 111–121

[Di82]   Dietz, P. F.: Maintaining order in a linked list. In: Proceedings of the Fourteenth Annual ACM Symposium on Theory of Computing. Association for Computing Machinery, New York, 1982, pp. 122–127

[DS87]   Dietz, P. F., Sleator, D. D.: Two algorithms for maintaining order in a list. In: Proceedings of the Nineteenth Annual ACM Symposium on Theory of Computing. Association for Computing Machinery, New York, 1987, pp. 365–372 (revision to appear in Journal of Computer and System Sciences)

[DSZ94] Dietz, P. F., Seiferas, J. I., Zhang, J.: Lower bounds for smooth list labeling. (in preparation)

[DZ90]   Dietz, P. F., Zhang, J.: Lower bounds for monotonic list labeling. In: SWAT 90: 2nd Scandinavian Workshop on Algorithm Theory, Proceedings. (Lecture Notes in Computer Science, vol. 447.) Springer, New York Berlin Heidelberg, 1990, pp. 173–180

[HS66]   Hennie, F. C., Stearns, R. E.: Two-tape simulation of multitape Turing machines. Journal of the Association for Computing Machinery **13** (1966) 533–546

[IKR81] Itai, A., Konheim, A. G., Rodeh, M.: A sparse table implementation of sorted sets. IBM Thomas J. Watson Research Center, Yorktown Heights, New York, Research Report RC 9146, 1981

[Ts84]   Tsakalidis, A. K.: Maintaining order in a generalized linked list. Acta Informatica **21** (1984) 101–112

[Zh93]   Zhang, J.: Density control and on-line labeling problems. Computer Science Department, University of Rochester, Rochester, New York, Technical Report 481 and Ph. D. Thesis, 1993

Trapezoid Graphs and Generalizations, Geometry and Algorithms*

STEFAN FELSNER

Bell Communications Research, 445 South Street, Morristown, NJ 07962, U.S.A., and Freie Universität Berlin, Institut für Informatik, Takustr. 9, 14195 Berlin, Germany.

E-mail address: felsner@inf.fu-berlin.de

RUDOLF MÜLLER

Technische Universität Berlin, Fachbereich Mathematik, Straße des 17. Juni 135, 10623 Berlin, Germany.

E-mail address: mueller@math.tu-berlin.de

LORENZ WERNISCH

Freie Universität Berlin, Inst. für Informatik, Takustr. 9, 14195 Berlin, Germany.

E-mail address: wernisch@inf.fu-berlin.de

Abstract. Trapezoid graphs are a class of cocomparability graphs containing interval graphs and permutation graphs as subclasses. They were introduced by Dagan, Golumbic and Pinter [DGP]. They propose an $O(n^2)$ algorithm for chromatic number and a less efficient algorithm for maximum clique on trapezoid graphs. Based on a geometric representation of trapezoid graphs by boxes in the plane we design optimal, i.e., $O(n \log n)$, algorithms for chromatic number, weighted independent set, clique cover and maximum weighted clique on such graphs. We generalize trapezoid graphs to so called k-trapezoidal graphs. The ideas behind the clique cover and weighted independent set algorithms for trapezoid graphs carry over to higher dimensions. This leads to $O(n \log^{k-1} n)$ algorithms for k-trapezoidal graphs. We also propose a new class of graphs called *circle trapezoid graphs*. This class contains trapezoid graphs, circle graphs and circular-arc graphs as subclasses. We show that clique and independent set problems for circle trapezoid graphs are still polynomially solvable. The algorithms solving these two problems require algorithms for trapezoid graphs as subroutines.

Mathematics Subject Classification (1991). 06A07, 05C85, 68R10 .

Key words. Algorithms, partially ordered sets, order dimension, trapezoid graphs, circle graphs, circular-arc graphs .

1. Introduction

Trapezoid graphs were introduced by Dagan, Golumbic, and Pinter [DGP]. Consider a channel, i.e., a pair of two horizontal lines. A *trapezoid* between these lines is defined by two points on the top and two points on the bottom line. A graph is a *trapezoid graph* if there exists a set of trapezoids corresponding to the vertices of the graph such that two vertices are joined by an edge iff the corresponding trapezoids intersect (see Figure 1). Motivated by a layer assignment problem in VLSI design the authors of [DGP] propose an algorithm

The research of the first author was supported by DFG under grant FE-340/2-1. The research of the third author was supported by ALCOM project II.

* This is an extended abstract of a full length article that will appear elsewhere.

An expanded version is avilable as preprint or ps-file at the authors.

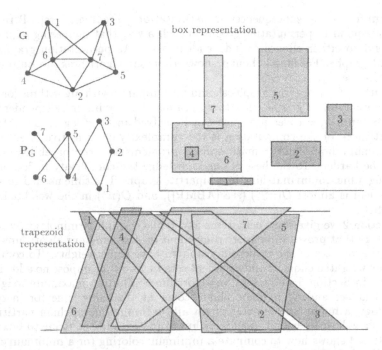

Figure 1. A trapezoid graph **G**, the order **P** and two representations.

computing the minimum number of colors in a proper coloring of such a graph in time $O(n^2)$ and less efficient backtracking algorithm finding a maximum clique in such graph (throughout the paper we assume that n is the number of vertices of the graph or order in question).

For our algorithms we will make use of another equivalent characterization of trapezoid graphs. To give this alternative characterization it is convenient to fix some terminology. If $x = (x_1, \ldots, x_k)$ and $y = (y_1, \ldots, y_k)$ are points in $\mathbb{R}^k$, then x is said to be *dominated* by y, denoted $x < y$, if x_i is less than y_i for all $i = 1, \ldots, k$. The order thus given between points in $\mathbb{R}^k$ is also called *dominance order*. This order can be extended to *boxes*, i.e., sets of the form $\{(x_1, \ldots, x_k) \in \mathbb{R}^k : l_i \leq x_i \leq u_i, 1 \leq i \leq k\}$ where $(l_1, \ldots, l_k)$ is the *lower corner* and $(u_1, \ldots, u_k)$ is the *upper corner* of the box. A box b dominates a box b' if the lower corner of b dominates the upper corner of b'. Note that points may be understood as boxes where the lower and upper corner coincides. If one of the two boxes dominates the other we say that they are *comparable*. Otherwise they are *incomparable*. Now the vertices of trapezoid graph may be represented by boxes with two boxes incomparable iff the corresponding vertices are joined by an edge (see Figure 1).

What makes the box representation useful is the additional dominance order on boxes that may be exploited by sweep line algorithms. All computation is done in a single sweep leading to $O(n \log n)$ algorithms for clique, independent set and cover problems on trapezoid graphs. Hence, these graphs are another class of graphs where very efficient algorithms for such problems can be given. There exists a lower bound for the number of comparisons needed to compute maximum increasing subsequences in permutations, Fredman [Fre]. Permutations correspond to permutation graphs in such a way that increasing sequences correspond to either cliques or independent sets. As permutation graphs are trapezoid graphs, Fredman's bound shows that our algorithms are optimal in the same sense.

Algorithms for trapezoid graphs should be compared with algorithms for general cocomparability graphs. For these graphs the maximum independent set and minimum clique cover problems can be solved in $O(n^2 \log n)$, see [McSp]. The bottleneck of the computation is the complexity of transitive orientation. The maximum clique and chromatic number problems on cocomparability graphs seem to be harder. To the best of our knowledge the complexity is dominated by finding a maximum matching in a bipartite graph. The time needed to solve this problem is almost $O(n^{2.5})$ (see [ABMP]), and $O(n^3)$ in the weighted case (see [PaSt]).

In Section 2 we give some definitions and replace graph terminology by order terminology that proves to be more convenient in designing our algorithms. We assume the vertices of the trapezoid graph to have some weights. To compute maximum weighted cliques or independent sets turns out to impose no additional difficulty. In Section 3 we present an algorithm computing maximum weighted independent set and a minimum clique cover at the same time (or in order terminology, a maximum weighted chain and a minimum antichain partition). We also show how to extend this algorithm from boxes in the plane to boxes in $\mathbb{R}^k$. Section 4 shows how to compute a minimum coloring (or a minimum chain partition). Unfortunately, this algorithm cannot be turned into an efficient one finding a maximum weighted clique (maximum weighted antichain). Hence, a different approach is proposed in Section 5 giving an efficient algorithm for the last problem.

In Section 6 we discuss a new class of graphs, called circle trapezoid graphs. A *circle trapezoid* is the region between two non-crossing chords of a circle. Alternatively, it is the convex hull of two disjoint arcs on the circle. *Circle trapezoid graphs*, *CT-graphs* for short, are the intersection graphs of families of circle trapezoids on a fixed circle. It is easily seen, that CT-graphs are a common generalization of trapezoid graphs, circle graphs and circular-arc graphs. We show, that in this large class of graphs the maximum clique and maximum independent set problems can still be solved in polynomial time.

2. Trapezoid graphs and trapezoid orders

The k-*dimensional box representation* (V, l, u) of a graph $\mathbf{G} = (V, E)$ consists of mappings $l: V \rightarrow \mathbb{R}^k$ and $u: V \rightarrow \mathbb{R}^k$ such that $l(v)$ is the lower and $u(v)$ the upper corner of a box box(v) where two vertices of the graph are joined by an edge iff their corresponding boxes are incomparable. If a graph has such a representation it is a k-*trapezoid graph*. If we additionally have a *weight* $w: V \rightarrow \mathbb{R}$ on the vertices of $\mathbf{G}$ then the k-trapezoid graph is *weighted*. The weight of a clique, i.e., a set of mutually joined vertices in the graph, is the sum of the weights of its elements. Similarly, the weight of an independent set, i.e., a set of vertices with no two of them joined by an edge, is the sum of the weights of its elements. We are mainly interested in the case $k = 2$ where we simply deal with *trapezoid graphs*.

As already mentioned in Section 1, we switch to the richer structure given by the dominance order on the boxes of a box representation. Let the boxes of a box representation of a trapezoid graph together with the dominance order be the corresponding *trapezoid order*. Note that chains in the trapezoid order correspond to independent sets in the trapezoid graph and antichains in the trapezoid order correspond to cliques in the trapezoid graph. A *maximal element* of a trapezoid order is one with no element dominating it. Each chain has exactly one maximal element. In contrast to the weight $w(v)$ of a box v in a trapezoid order we will often attribute a *chain weight* $W(v)$ to v which is the maximum weight of a chain with v as its maximal element.

Note that in the limiting case the box representation (V, l, u) of a trapezoid graph (V, E) may consist of points, i.e., $l(v) = u(v)$, for all $v \in V$. Such graphs are known as *permutation graphs* and the points with the dominance order in the plane as *2-dimensional order* (see, e.g., [Gol]). We denote such an order by (V, p) with $p(v) = l(v) = u(v)$. Before giving the actual algorithms for the trapezoid orders we will sometimes recall algorithms for 2-dimensional orders since they are easier to grasp while showing important features extendible to the general case.

We will often have to maintain a finite set of real numbers such that values may be inserted or deleted from it and the predecessor or successor of a given query value can be found. Using balanced trees (e.g., red-black trees described in Cormen, Leiserson, Rivest [CLR]) all these operations can be done in $O(\log n)$ time and linear space. If we further assume the benefits of a random access machine and assume that the values are taken from a finite range U then the above operations take only $O(\log \log n)$ time and linear space when implemented on a data structure of van Emde Boas [vEB]. Hence, under these assumptions, the $\log n$ factor in the running time of the algorithms for 2-dimensional trapezoid orders may be replaced by a $\log \log n$ factor.

Throughout the paper we assume that the points $l(v)$ and $u(v)$ of a box representation have mutually different x- and y-coordinates. Otherwise, we may obtain a box representation of the same order fulfilling this requirement by perturbing the corner points slightly. Appropriate perturbations can be computed in $O(n \log n)$ time.

3. Minimum antichain partition and maximum chain for k-trapezoidal orders

We first give a brief description of an algorithm solving the maximum chain problem for a 2-dimensional order (V, p) in the weighted case. Let the weights be given by $w: V \to \mathbb{R}^+$. First, the points are sorted, so that we can access them by increasing x-coordinate, i.e., from left to right. Secondly, we compute a function $W: V \to \mathbb{R}$, where $W(v)$ is the chain weight of v, i.e., the weight of a maximum weighted chain having v as its maximal element.

$W(v)$ is computed with the aid of a sweep line L moving from left to right and halting at every point $p(v)$. We maintain a set M of weighted markers on L so that the weight $W(m)$ for $m \in M$ is just the weight of a maximum weighted chain on the set of points dominated by m, i.e., on $\{v \in V : p(v) < m\}$. For each $m \in M$ origin(m) is the maximal element of the maximum weighted chain dominated by m. When reaching a point $p(v)$ we find the first marker m below $p(v)$ on L, set $W(v) = W(m) + w(v)$ and establish a link from v to origin(m). To update L we position a new marker m' with $W(m') = W(v)$ and origin(m') = v

at the y-coordinate of $p(v)$. Then we remove those markers above m' that have smaller weight. Note that although the number of markers removed in one step may be large, the overall number of insertions and removals of markers on L cannot exceed $2n$. Finally, starting from a point v with maximum chain weight $W(v)$ we go along the links to construct a heaviest chain.

In the case where the box representation (V, l, u) of a trapezoid order $\mathbf{P}$ is given we divide the action taken by the algorithms for 2-dimensional orders whenever the sweep line reaches a new element into two parts. The first part of the action, located at $l(v)$, is to compute the chain weight W of the new element v. The second part of the action, located at $u(v)$, is to make the chain weight of v available for further elements. The main difference to the permutation graph algorithm is that before releasing the information corresponding to v we have to check whether the information is still relevant when released.

The algorithm is given next. For convenience, we initialize the sweep line with a dummy point d with $W(d) = 0$ and $\text{origin}(d) = \text{nil}$, such that d is below all points $l(v)$ for $v \in V$.

> MAXCHAIN(V, l, u, w)
> **for each** p from left to right **do**
> $\quad m \leftarrow$ first marker below p on L
> $\quad$ **if** $p = l(v)$ for some $v \in V$ **then**
> $\quad\quad W(v) \leftarrow W(m) + w(v)$
> $\quad\quad \text{link}(v) \leftarrow \text{origin}(m)$
> $\quad$ **if** $p = u(v)$ for some $v \in V$ **then**
> $\quad\quad$ **if** $W(v) > W(m)$ **then**
> $\quad\quad\quad$ insert a new m_v at p_y in L
> $\quad\quad\quad W(m_v) \leftarrow W(v)$
> $\quad\quad\quad \text{origin}(m_v) \leftarrow v$
> $\quad\quad\quad$ remove all m' that are higher and lighter than m_v from L
> $v \leftarrow \text{origin}(\text{uppermost}(L))$
> $C \leftarrow \{v\}$
> **while** $\text{link}(v) \neq \text{nil}$ **do**
> $\quad v \leftarrow \text{link}(v)$
> $\quad C \leftarrow C \cup \{v\}$
> **return** C

LEMMA 3.1. *At the end of the main loop in MAXCHAIN the following invariant holds true. If y is an arbitrary point on L and m is the next marker below y, then a maximum weighted chain dominated by point y has weight $W(m)$.* $\square$

Of course, Lemma 3.1 implies that Algorithm MAXCHAIN computes a maximum weighted chain, since all boxes are dominated by the uppermost point on L after the sweep has completed.

As already noted the sweep line can be implemented so that find, insert and delete operations require $O(\log n)$ time. It is easily seen, that $3n$ is an upper bound for the number of these operations. This proves a $O(n \log n)$ time bound.

The unweighted case can be simulated by unit weights. As the weights of all markers are different the number of markers on L in the unweighted case cannot exceed the length ω of a maximum chain in $\mathbf{P}$. If each element of $\mathbf{P}$ has unit weight, then no two elements with the same chain weight are comparable.

Hence, collecting the elements of chain weight i in a set A_i yields a partition $A_1, \ldots, A_\omega$ of $\mathbf{P}$ into antichains. It is easily seen that the maximum weighted chain must contain one element of A_i, $i = 1, \ldots, \omega$. This proves this antichain partition to be minimal since a partition into fewer antichains would force at least two elements of the chain into one antichain, which is impossible. Hence, a minimal antichain partition is a byproduct of algorithm MAXCHAIN. We summarize these remarks in

THEOREM 3.2. *A maximum weighted chain and a minimum antichain partition of a trapezoid order on n points, given its box representation, can be computed in time $O(n \log n)$ and linear space.*

Given a box representation of $\mathbf{P}$ in some higher dimension $k > 2$ we use a $k-1$ dimensional range tree to compute a maximum chain and a minimum antichain partition of $\mathbf{P}$.

THEOREM 3.3. *If an order $\mathbf{P} = (V, P)$ is given by a box representation in $\mathbb{R}^k$, then a minimum antichain partition and a maximum chain of $\mathbf{P}$ can both be obtained in $O(n \log^{k-1} n)$ time and $O(n \log^{k-2} n)$ space.* □

4. Chain partitions of trapezoid orders

In this section we show how to partition a trapezoid order $\mathbf{P}$ into chains such that the number of chains used is minimal. Of course, this only makes sense if we assume unit weights on the elements of $\mathbf{P}$. Again, we begin with a short description of a similar algorithm for 2-dimensional orders which we then adapt for the case of a given box representation of $\mathbf{P}$.

An optimal chain partition for a point set can be obtained by a sweep of a line L from left to right in the following way. Assume the set of points to the left of the current position of L to be already optimally partitioned into chains. On L the maximal elements of the chains of this partition are maintained ordered by y-coordinates. When reaching a new point p we search for the point q on L which has maximal y-coordinate among all points on L that are below p. If q exists then p is appended as new maximal element to the chain of q, otherwise, p does not dominate any chain of the actual partition and we initialize a new chain consisting of p only. Finally, L is updated by inserting p and removing q. Note that p is assigned to a chain by a *first-fit* strategy.

Now suppose, that $\mathbf{P}$ is given by a box representation (V, l, u). We have to separate the action that has to be taken when the sweep line reaches a new element into two parts. The first part of the action, located at $l(v)$, is to find the chain of the already existing partition that will be extended by v. The second part, located at $u(v)$, is to make the chain with maximum v available for further elements. A chain C with maximum element v will be called *closed* as long as $u(v)$ has not been visited by L, otherwise C is *open*. The assignment of an element to a chain is again done by first-fit, however, the set of admissible chains is restricted to the open ones.

The algorithm for computing a minimum chain partition in a box representation is given as follows. We initialize the sweep line with a dummy point d such that d is below all points that will ever be inserted into L.

MINCHAINPARTITION(V, l, u)
  for each p from left to right **do**
    $q \leftarrow$ first element below p on L
    if $p = l(v)$ for some $v \in V$ **then**
      if $q = u(w)$ for some $w \in V$ **then**
        chain$(v) \leftarrow$ chain$(w) \cup \{v\}$
        remove q from L
      else $(q = d)$
        chain$(v) \leftarrow \{v\}$
    if $p = u(v)$ for some $v \in V$ **then**
      insert p at p_y in L
  return $\{$ chain$(v) : u(v) \in L\}$

The time consuming operations in this algorithm are the search, insert and remove operations for points on the sweep line L. With the use of a balanced search tree the running time of the algorithm is in $O(n \log n)$. If we assume the points to be presorted, the running time is in $O(n \log \alpha)$ where α is the number of chains in the partition.

THEOREM 4.1. *A minimum chain partition of a trapezoid order on n points, given its box representation can be computed in time $O(n \log n)$ and linear space.*

5. Maximum antichain for trapezoid orders

We first describe the geometry of antichains in a box representation. Our algorithm for maximum weighted antichains of trapezoid orders will be based on this geometric structure rather than on duality as the algorithms presented so far. First, we need some definitions.

Define the *shadow of a point* p as the set of points in the plain dominating p, i.e., shadow$(p) = \{q : q > p\}$. The *shadow of a set of points* is the union of the shadows of the elements. A downwards *staircase* is a sequence of horizontal and vertical line segments that may be obtained as the topological boundary of the shadow of a set of points. Note that any two different points on a staircase are incomparable. If S is a staircase then $A(S)$ denotes the set of elements whose box intersect S, i.e., $A(S) = \{v \in V : \text{box}(v) \cap S \neq \emptyset\}$.

LEMMA 5.1. *Given a trapezoid order by a box representation. If S is a staircase then $A(S)$ is an antichain. Moreover, if A is an antichain of* **P** *then there exists a staircase S such that $A \subseteq A(S)$.*   □

Given a weighted order (V, l, u) we define the *weight of a staircase* S as the sum of weights of all boxes intersecting S. If S is a staircase and $p \in S$, then we refer to the part of S that is above and to the left of p as *staircase ending in p* and again its weight is the sum of weights of intersecting boxes.

The following algorithm computes an antichain of maximum weight. It uses two different data structures. The sweep line L halts at every point $l(v)$ and $u(v)$, for $v \in V$. Roughly, it contains a list of weighted markers, so that the weight of marker m is the weight of a heaviest staircase ending in m. Moreover, a heaviest staircase ending in an arbitrary point y on L can be composed by joining the vertical line segment from y to the next marker m above y with a heaviest staircase ending in m. The second structure Δ contains a list of all *open* boxes, i.e., boxes which have their left sides already swept but not their right ones. The total weight of all open boxes the upper sides of which lie between points y_1 and y_2 on L with $y_1 \geq y_2$ is denoted by $\Delta(y_1, y_2)$.

> MAXANTICHAIN(V, l, u, w)
> **for each** p from left to right **do**
> $m \leftarrow$ first marker above p on L
>   **if** $p = l(v)$ for some $v \in V$ **then**
>   add $w(v)$ to all markers in interval $[l_y(v), u_y(v)]$
>   insert a new item in Δ at height $u_y(v)$ with weight $w(v)$
> $m^* \leftarrow$ next marker below p on L
>   **while** $W(m) + \Delta(m, m^*) > W(m^*)$ **do**
>   remove m^* from L
> $m^* \leftarrow$ next marker below p on L
>   **if** $p = u(v)$ for some $v \in V$ **then**
>   insert a new marker m_v at p_y in L
> $W(m_v) \leftarrow W(m) + \Delta(m, m_v)$
>   list$(m_v) \leftarrow$ list$(m) \cup \{p\}$
>   remove item at $u_y(v)$ from Δ
> $T \leftarrow$ staircase of points in list(lowest(L))
> **for each** $v \in V$ **do**
>   **if** v intersects T **then** $A \leftarrow A \cup \{v\}$
> **return** A

LEMMA 5.2. *At the end of the main loop in Algorithm MAXANTICHAIN we have the following invariant. If y is an arbitrary point on L and m the next marker above y, then a maximum weighted staircase that ends in y on L has weight* $W(m) + \Delta(m, y)$. $\qquad\Box$

After all boxes have been swept, structure Δ is empty (i.e., there is no box left open). Therefore, the lemma immediately implies that MAXANTICHAIN computes a maximally weighted antichain of a trapezoid order.

The sweep line L may be implemented by a balanced binary tree. One has to be careful only about adding some weight w to a whole interval $[a, b]$. Let each node of the tree have some extra field holding the increment in the weight for all nodes in its subtree. During a rebalancing rotation this fields must be corrected accordingly. But it is easily seen that only a constant number of such fields is affected. Consequently, the addition of some weight to an interval as well as insertion, deletion, predecessor and successor queries, and the computation of the weight of some element can all still be done in time $O(\log n)$.

Δ may be implemented by any one dimensional range tree where insertion, deletion, and query again takes $O(\log n)$ time. The main loop is executed n times and each step therein takes logarithmic time save the while loop. But in total the while loop is executed at most n times since each removed point must have been inserted before. Of course, the test for intersection of a box v with staircase T can be done in time $O(\log n)$. In summary, we obtain

THEOREM 5.3. *A maximum weighted antichain of a trapezoid order on n points, given its box representation, can be computed in time $O(n \log n)$ and linear space.*

6. Algorithms for circle trapezoid graphs

A *circle trapezoid* is the region in a circle that lies between two non-crossing chords and CT-*graphs* are the intersection graphs of families of circle trapezoids on a common circle. Figure 2 gives an example. In this section we develop polynomial algorithms for the maximum weighted clique and a maximum weighted independent set problems on CT-graphs.

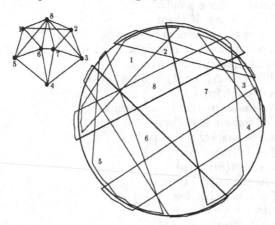

Figure 2. A circle trapezoid graph **G** with a representation

6.1. Crossing graphs and independent sets of CT-graphs

Let $\mathbf{G} = (V, E)$ be a CT-graph, of course, we will assume that a representation of **G** is given. Let p be an arbitrary point on the circle and let C_p be the set of vertices of **G** whose circle trapezoid contains p. Note that C_p induces a clique of **G**, therefore, an independent set of **G** can contain at most one element from C_p. Using p as the 'origin' of the circle and fixing an orientation (clockwise) of the circle we can define a unique representation for circle trapezoids. The representation consists of 5-tuple $(t_1, t_2, t_3, t_4, \sigma)$. The first four components are the corners of the circle trapezoid in clockwise order starting from p. The fifth component σ is a sign, $+$ or $-$, where $+$ indicates that p is contained in one of the arcs of the circle trapezoid.

Define a *double interval* as a pair (I_1, I_2) of intervals on the real line, where I_2 is a subinterval of I_1, i.e., $I_2 \subset I_1$. Let $I = (I_1, I_2)$ and $J = (J_1, J_2)$ be double intervals. We say I *contains* J if $J_1 \subset I_2$ and call them *disjoint* if $I_1 \cap J_1 = \emptyset$. Two double intervals are called *crossing* if they are not disjoint and non of them is contained in the other. Call a graph $\mathbf{G} = (V, E)$ a *crossing graph* if its vertices can be put in one to one correspondence to a collection of double intervals such that two vertices of **G** are adjacent if and only if their corresponding double intervals cross. It is not hard to see that the class of crossing graphs contains both, trapezoid graphs and overlap graphs (recall that a graph is an overlap graph if and only if it is a circle graph). Our next lemma relates CT-graphs and crossing graphs.

LEMMA 6.1. *Let* $\mathbf{G} = (V, E)$ *be a CT-graph given by a representation and* C_p *be the set of all vertices of* **G** *whose circular trapezoid share a specified point* p *on the circle. The subgraph of* **G** *induced by* $V \setminus C_p$ *is a crossing graph.* $\square$

For $v \in V$ let $N[v]$ denote the set of neighbors of v together with v itself and let $G(v)$ be the subgraph of G induced by $V \setminus N[v]$. Also, let G_p denote the subgraph induced by $V \setminus C_p$. A maximum independent set I of G is either a maximum independent set in G_p or there is a $v \in C_p$, such that $I = I' \cup \{v\}$ where I' is a maximum independent set of $G(v)$. Since $C_p \subseteq N[v]$ for all $v \in C_p$ the lemma shows that each of the above graphs $G(v)$, as well as G_p are crossing graphs. This reduces the detection of a maximum independent set of a CT-graph to at most n maximum independent set problems on crossing graphs. We therefore turn the attention to the maximum independent set problem for crossing graphs. Our algorithm for this problem is very much alike the algorithm given by Gavril [Gav] (see also Golumbic [Gol]) for the case of overlap graphs.

For a pair of double intervals we have defined the relations containment, disjointness and crossing and by definition every pair of double intervals is in exactly one of these relations. The containment is a antisymmetric and transitive relation, i.e., an order relation. For the disjointness we only need the first interval of each double interval, therefore, we can transitively orient disjoint pairs by the relation 'lies entirely to the left', this gives an interval order.

To compute the maximum independent set of a crossing graph $G = (V, E)$ given by a family $\mathcal{I}$ of double intervals we proceed as follows. First, the containment order $P = (V, P)$ and the interval order $Q = (V, Q)$ corresponding to $\mathcal{I}$ are extracted and a linear extension $L = v_1, \ldots, v_n$ of P is computed. We artificially extend P and L by an element v_{n+1} of weight 0, such that $v_{n+1} > v_i$ for all $i = 1, \ldots, n$. This preprocessing can be accomplished in time $O(n^2)$. Next, the following algorithm is called.

MAXINDEPENDENTSET(P, Q, L)
 for $i = 1$ to $n + 1$ **do**
  $U_i \leftarrow \{v_j : v_j < v_i \text{ in } P\}$
  $C \leftarrow$ maximum W-weighted Q-chain of elements of U_i
  $W(v_i) \leftarrow w(v_i) + \sum_{v \in C} W(v)$
  $I(v_i) \leftarrow \{v_i\} \cup \bigcup_{v \in C} I(v)$
 return $I(v_{n+1})$

It is important to note, that U_i only contains elements v_j with $j < i$, hence, the weights $W(v_j)$ of all elements in U_i have already been computed before the ith round. The following invariance of the algorithm is easily proved. *At the end of the ith round for all $j \leq i$ the weight $W(v_j)$ is the weight of a maximum independent set $I(v_j)$ containing only elements $v \in U_j \cup \{v_j\}$, i.e., elements with $v \leq v_j$ in P.* From this invariant $I(v_{n+1})$ is a maximum independent set for G.

Clearly, every but the second instruction in the loop can be executed in $O(n)$ time. The second instruction itself is a maximum chain computation in an interval order. This problem is known to be solvable in linear time. Hence, algorithm MAXINDEPENDENTSET solves the maximum weighted independent set problem for crossing graphs in $O(n^2)$.

Recall that the solution for the maximum independent set problem for CT-graphs is a either a maximum independent set in G_p or one of the sets $I = I' \cup \{v\}$ where $v \in C_p$ and I' is a maximum independent set in $G(v)$. It can be shown that having applied algorithm MAXINDEPENDENTSET to G_p the problem for a each of the graphs $G(v)$, $v \in C_p$, can be solved by a single maximum chain computation in an interval order, i.e., in $O(n)$ time.

THEOREM 6.2. *Given a CT-graph* **G** *with a circular trapezoid representation then the maximum weighted independent set problem for* **G** *can be solved in* $O(n^2)$.

6.2. Cliques of CT-graphs

Let $\mathbf{G} = (V, E)$ be a CT-graph, given by a circular trapezoid representation. A clique C of **G** is called a *Helly-clique* with respect to the representation, if there is a point p in the interior of the circle, such that the circle trapezoid of every vertex $v \in C$ contains p. Our algorithm for the maximum clique problem on CT-graphs proceeds in two stages. In the first stage it determines a maximum Helly-clique of **G**, in the second stage a maximum non-Helly-clique, NH-*clique* for short, is computed. The larger of the two cliques is a maximum clique for **G**.

The determination of a maximum Helly-clique is a purely geometrical problem. It is not hard to design an algorithm that finds a maximum Helly-clique in $O(n^2 \log n)$ time.

We turn to the case of NH-cliques. An NH-*triangle* T in **G** is a three element clique, such that the circular trapezoids of the three vertices have no point in the circle in common. Let $T = \{x, y, z\}$ be an NH-triangle of **G** and for $v \in V$ let $B(v)$ be the circular trapezoid of v. Let c be one of the chords bounding $B(x)$, we call c the *inner chord* of x if $B(x)$ is on one side of c and $B(y) \cap B(z)$ is on the other side of c. The side of c not containing $B(x)$ is the *inner side* of x. The intersection of the inner sides of the elements of an NH-triangle T is the *enclosed area* of T, of course, the enclosed area is a triangle in the interior of the circle. The definitions are illustrated in Figure 2, the set $\{6, 7, 8\}$ is an NH-triangle, the inner chords of their circular trapezoids are bold.

A family $\mathcal{F}$ of subsets of a set X has *Helly number* k if for every subfamily $\mathcal{S}$ of $\mathcal{F}$ the property that any k members of $\mathcal{S}$ have non-empty intersection implies that the intersection over all members of $\mathcal{S}$ is non-empty. It is well known, that the Helly number of convex objects in the plane is 3, see [Hel]. As a consequence we obtain.

LEMMA 6.3. *Every NH-clique in a CT-graph contains an NH-triangle.* ☐

Based on this lemma an algorithm to compute a maximum NH-clique in a CT-graph **G** can proceed as follows. Enumerate all NH-triangles in **G** and for each NH-triangle T compute a maximum clique among the cliques containing T. Our approach is only slightly more sophisticated. An NH-triangle T is called a *maximal triangle* for a clique C, if the enclosed area of T is maximal among the enclosed areas of NH-subtriangles of C, i.e., there is no NH-subtriangle T' of C, whose enclosed area contains the enclosed area of T.

LEMMA 6.4. *If C is a clique in* **G** *containing an NH-triangle T as a maximal triangle, then every $v \in C$ has the property that $B(v)$ intersects each of the inner chords c_x for $x \in T$.* ☐

For an NH-triangle T let $V(T)$ be the set of vertices satisfying the condition of the lemma, i.e., $v \in V(T)$ if the intersection of $B(v) \cap c_x \neq \emptyset$ for all $x \in T$.

The six endpoints of the inner chords c_x of the elements of T partition the circle into six disjoint arcs $A_1, \ldots, A_6$. The numbering of these arcs is assumed to be consecutive. We call the arcs A_i and A_{i+3} for $i = 1, 2, 3$ a pair of *opposite* arcs.

LEMMA 6.5. *If $v \in V(T)$, then there is at least one $i \in \{1,2,3\}$, such that $B(v)$ intersects A_i and A_{i+3}.* □

Let $f: V(T) \to \{1,2,3\}$ be a function such that $B(v) \cap A_{f(v)} \neq \emptyset$ and $B(v) \cap A_{f(v)+3} \neq \emptyset$ for every $v \in V(T)$. Also let $V_i(T) = \{v \in V(T) : f(v) = i\}$ for $i = 1,2,3$.

LEMMA 6.6. *If $i,j \in \{1,2,3\}$ with $i \neq j$ then for all $v_i \in V_i(T)$ and $v_j \in V_j(T)$ the circular trapezoids $B(v_i)$ and $B(v_j)$ intersect, i.e., v_i and v_j are adjacent in G.* □

As a consequence we obtain that, if C_i is a clique in the subgraph $\mathbf{G}_i(T)$ of $\mathbf{G}$ induced by $V_i(T)$ for $i = 1,2,3$, then $T \cup C_1 \cup C_2 \cup C_3$ is a clique in $\mathbf{G}$. Moreover, and more important.

LEMMA 6.7. *C is a maximum clique containing T as a maximal triangle, exactly if $C_i = C \cap V_i(T)$ is a maximum clique in $\mathbf{G}_i(T)$ for each $i = 1,2,3$.* □

It can be shown that the problem of finding a maximum clique in $\mathbf{G}_i(T)$ for $i = 1,2,3$ is a maximum clique problem in a trapezoid graph. Moreover, the circular trapezoid representation of $\mathbf{G}_i(T)$ induces a box representation. Therefore, a largest clique containing T as NH-triangle can be found in $O(n \log n)$. Unfortunately there can be as much as $O(n^3)$ NH-triangles in a representation of $\mathbf{G}$. We summarize the results of this subsection.

THEOREM 6.8. *The maximum weighted clique problem for CT-graphs can be solved in $O(n^4 \log n)$.*

REFERENCES

[ABMP] H. Alt, N. Blum, K. Mehlhorn and M. Paul, Computing a maximum cardinality matching in a bipartite graph in time $O(n^{1.5}(m/\log n)^{.5})$, *Inf. Proc. Letters* **37** (1991), 237–240.

[CLR] T.H. Cormen, C.E. Leiserson and R.L. Rivest, *Introduction to Algorithms*, The MIT Press, 1989.

[DGP] I. Dagan, M.C. Golumbic and R.Y. Pinter, Trapezoid Graphs and their Coloring, *Discr. Appl. Math.* **21** (1988), 35–46.

[Fre] M.L. Fredman, On Computing the Length of Longest Increasing Subsequences, *Discr. Math.* **11** (1975), 29–35.

[Gav] F. Gavril, Algorithms for a Maximum Clique and a Maximum Independent Set of a Circle Graph, *Networks* **3** (1973), 361–273.

[Gol] M.C. Golumbic, *Algorithmic Graph Theory and Perfect Graphs*, Academic Press, New York, 1980.

[Hel] E. Helly, Über die Menge konvexer Körper mit gemeinsamen Punkten, *Jahresb. d. Dt. Mathem. Ver.* **32** (1923).

[McSp] R. McConnel and J. Spinrad, Linear-Time Modular Decomposition and Efficient Transitive Orientation of Undirected Graphs., *Proc. 5. Annual Symp. on Discr. Alg.* (1994).

[PaSt] C.H. Papadimitriou and K. Steiglitz, *Combinatorial Optimization: Algorithms and Complexity*, Prentice-Hall, 1983.

[vEB] P. van Embde Boas, Preserving order in a forest in less than logarithmic time and linear space,, *Inf. Proc. Letters* **6** (1977), 80–82.

Optimal Parametric Search on Graphs of Bounded Tree-width

David Fernández-Baca* and Giora Slutzki

Department of Computer Science, Iowa State University, Ames, IA 50011

Abstract. We give linear-time algorithms for a class of parametric search problems on weighted graphs of bounded tree-width. We also discuss the implications of our results to approximate parametric search on planar graphs.

1 Introduction

Parametric search has been the center of a considerable amount of research in recent times due to its numerous applications to optimization and computational geometry [CoMe93, Tol93, CEGS92]. Much of this work stems from two fundamental papers by Megiddo [Meg79, Meg83], where he introduced a powerful tool for parametric search. In the context of optimization problems, the application of Megiddo's technique tends to follow a common pattern. Suppose we have an algorithm A that allows us to determine the value of a certain function f for any λ within a certain range, and that we wish to locate a critical value λ^* for f — the nature of this value depends on the application. Megiddo argued that we can proceed by simulating the execution of A to determine its computation path at λ^*. To do so, the operations of A must be executed symbolically, manipulating functions of λ instead of numbers. The difficult part is handling the branching points of A, where decisions must be made about the path to follow without explicit knowledge of λ^*. These questions are resolved by invoking an oracle, which is often closely related to algorithm A. Since oracle calls are expensive, they must be used sparingly. Megiddo showed that if these operations can be *batched* (i.e., grouped and ordered in such a way as to permit many of them to be resolved by a single oracle call), the total amount of work to solve the parametric problem can often be made at most a polylogarithmic factor slower than that of algorithm A. The polylogarithmic slowdown in going from non-parametric to parametric algorithms remains even when using Cole's clever technique [Cole87].

In this paper we shall show that any parametric search problem whose underlying non-parametric version is *regular* [BLW87, BPT92] can be solved in linear time on graphs of bounded tree-width. This matches the complexity of the algorithms for the non-parametric versions of these problems, and hence our algorithms are optimal. Regular optimum subgraph problems include maximum cut,

---

* Supported in part by the National Science Foundation under grant No. CCR-8909626 and grant No. CCR-211262.

minimum-weight dominating set, minimum-weight vertex cover, traveling salesman problem, and many others [BLW87, BPT92]. In their non-parametric versions one is given vertex- and/or edge-weighted graph and one must find an optimum (minimum or maximum cost) subgraph satisfying a specified property. In the parametric versions, weights are linear functions of a parameter λ and $Z_G(\lambda)$ denotes the cost of the optimum solution at λ. The λ^* to be sought will depend on the specific problem. The applications of parametric search in optimization include sensitivity analysis, Lagrangian relaxation (which requires finding the maximizer of Z_G), and minimum ratio optimization [Meg79, CoMe93, FeSl94].

Our work was motivated by that of Frederickson [Fre90], who showed that a number of location problems on trees can be solved in (optimal) linear time by a novel application of Megiddo's method. We borrow several of his ideas here, including repeated graph contraction, accompanied by a rapid narrowing of the search interval, and the use of not one, but a sequence of increasingly faster oracles. We also introduce some new ideas which we have shown to have applications outside of the context of bounded tree-width graphs. The most useful of these is a method that combines exhaustive enumeration with repeated pruning. The technique is based on two observations: First, each one of our search problems can be solved easily once Z_G has been constructed. Second, Z_G can be constructed by simulating the behavior of A for *all* possible values of λ. Of course, doing the latter is impractical, but the idea becomes more reasonable if we settle for obtaining a complete description of Z_G within a small interval which is known to contain λ^*. To do this, we divide the computation of Z_G into a sequence of phases, each of which is faster than the previous one and yields ever smaller intervals. By the time the algorithm is complete, Z_G will have been constructed in its entirety within a suitable interval; at that point the search can proceed by exhaustive enumeration. The decomposition of the computation into phases relies heavily on a linear-time algorithm for constructing tree-decompositions satisfying a strong balance condition.

The non-parametric versions of the problems we study here can all be solved in linear time using a standard dynamic programming approach which takes the parse tree of the input graph and traverses it from the bottom up, building solutions for ever larger subgraphs through a combination of solutions for smaller subgraphs (see section 3 and [ALS91, BLW87, BPT92]). This is an inherently local approach, which does not yield readily to fast parametric algorithms. For example, narrowing down the search interval for subregions of the graph does not directly translate into a corresponding narrowing of the search for the global maximizer. On the other hand, during the bottom up traversal of the parse tree, some information that will be of global use *can* be gathered. We achieve linear time algorithms by devising a way in which this data can be collected and used efficiently.

Our results lead to $O(n)$-time approximation algorithms for several NP-hard parametric search problems on planar graphs and to some observations on the structure of functions describing the cost of approximate solutions to parametric problems on planar graphs.

2 Constructing Balanced Parse Trees in Linear Time

Our parametric search algorithm requires the construction of a parse tree for the input graph satisfying a strong balance condition. This tree will be used in the next section to obtain a subdivision of the computation into increasingly faster phases. In this section, we define this notion more precisely and show how the parse tree can be constructed in linear time. We first review the closely-related notions of tree-decompositions and parse trees.

A *tree decomposition* of an undirected graph G is a labeled tree (T, χ), where χ is the labeling function for T, such that for all $i \in V(T)$, $\chi(i) = \chi_i \subseteq V(G)$, and such that: (1) $\bigcup_{i \in V(T)} \chi_i = V(G)$; (2) for every $(u, v) \in E(G)$, $\{v, u\} \subseteq \chi_i$ for some $i \in V(T)$; (3) if j lies on the path of T from i to k, then $\chi_i \cap \chi_k \subseteq \chi_j$. The *width* of a (T, χ) is $\max_{i \in V(T)}(|\chi_i| - 1)$. The *tree-width* of a graph G is the minimum over all tree-decompositions (T, χ) of G of the width of (T, χ).

We write Γ_w to denote the set of all graphs of tree-width at most w. Classes of graphs of bounded tree-width have been surveyed by Bodlaender [Bod88] and van Leeuwen [vLe90]. Bodlaender [Bod93] has shown that the problem of determining whether a graph has tree-width at most w and, if so, constructing a tree decomposition of width at most w can be solved in linear time for every fixed w.

A graph G is a *k-terminal graph* if it is given together with a list $\mathsf{terms}(G) = \langle t_1, \ldots, t_s \rangle$, $1 \leq s \leq k$, of distinct vertices of G called *terminals*. A *k-terminal composition operator* φ of *arity* $r \geq 0$ joins r k-terminal graphs $G_1, \ldots, G_r$ to obtain a new k-terminal graph $G = \varphi(G_1, \ldots, G_r)$ by identifying the terminals of the composing graphs in some precisely prescribed way. The terminals of G are obtained from the terminals of the composing graphs [Wim87, BPT92]. We shall write $\mathcal{U}[k, r]$ to denote the set of all k-terminal graph composition operators φ of arity at most r, and where, for each φ of arity zero, $|V(\varphi)| \leq k$. Note that for every fixed k and r, $|\mathcal{U}[k, r]|$ is bounded by a constant.

Let $\mathcal{R} \subseteq \mathcal{U}[k, r]$. Denote by $\mathcal{R}_0$ the subset of $\mathcal{R}$ consisting of all operators of arity 0. $\mathcal{R}_0$ is a set of k-terminal graphs whose elements are called *primitive* graphs. We write $\mathcal{R}^*$ to denote the set of all k-terminal graphs G such that either $G \in \mathcal{R}_0$ or G can be expressed as $G = \varphi(G_1, \ldots, G_r)$ for some operator $\varphi \in \mathcal{R}$ and some set of graphs $G_1, \ldots, G_r \in \mathcal{R}^*$. The equality $G = \varphi(G_1, \ldots, G_r)$ is called a *decomposition of G with respect to $\mathcal{R}$*.

A *parse tree* T for $G \in \mathcal{R}^*$ is a tree constructed as follows. If $G \in \mathcal{R}_0$, T consists of a single node v with label $\delta(v) = G$. Otherwise, let $G = \varphi(G_1, \ldots, G_r)$ be decomposition of G with respect to $\mathcal{R}$. Then T consists of a root v, with label $\delta(v) = \varphi$ and whose children are the parse trees of $G_1, \ldots, G_r$. For any node $v \in T$, the subtree of T rooted at v represents a graph $G_v \in \mathcal{R}^*$, such that (1) $G_v = \delta(v)$, if v is a leaf, or (2) $G_v = \delta(v)(G_{v_1}, \ldots, G_{v_r})$, if the children of v are $v_1, \ldots, v_r$.

The decomposition algorithm. It is known (see, e.g., [Wim87]) that graphs of constant-bounded tree-width can be expressed as the composition of graphs with bounded number of terminals according to some finite set of rules. Moreover, there exists a subset $\mathcal{R}' \subseteq \mathcal{U}[4w + 4, 5]$ such that every $G \in \Gamma_w$ has a $O(\log n)$-height parse tree over $\mathcal{R}'$ (see [Lag90, FeSl94]; a similar result was independently proved by Frederickson [Fre93]). For our current purposes, we shall require a somewhat stronger property than logarithmic height. We shall say that a decomposition $G = \varphi(G_1, \ldots, G_r)$ with respect to $\mathcal{R}$ is ϵ-balanced if $|V(G_i)| \geq \epsilon |V(G)|$, for some constant ϵ that depends only on $\mathcal{R}$. A parse tree (T, δ) of an n-vertex graph G is said to be ϵ-balanced if for every internal node v of T with children $v_1, \ldots, v_r$, $G_v = \delta(v)(G_{v_1}, \ldots, G_{v_r})$ is ϵ-balanced. The height of an ϵ-balanced parse tree T is clearly $O(\log n)$.

Suppose we have a procedure DECOMPOSE, whose details will be supplied later, which obtains a $1/16$-balanced decomposition with respect to $\mathcal{U}[9w + 6, 3]$ of any $(9w+6)$-terminal graph of tree-width w. To obtain a $1/16$-balanced parse tree T of G, we do as follows: If $|V(G)| \leq 9w + 6$, G is a primitive graph whose parse tree consists of a single node; otherwise, apply DECOMPOSE to G to obtain a balanced decomposition $G = \varphi(G_1, \ldots, G_r)$, label the root of T with φ, and recurse on the G_i's. We will argue that, at the expense of a one-time-only linear-time preprocessing step, DECOMPOSE can be implemented to run in $O(\log n)$ time. Thus, the total work required to construct the decomposition, aside from preprocessing, will be described by the recurrence $T(n) \leq \sum_{i=1}^{q} T(n_i) + b \log n$, for some constant b, where $1 \leq q \leq 3$, $\sum_{i=1}^{q} n_i = n$ and $n_i \leq 15n/16$. It can readily be shown that $T(n) = O(n)$.

Implementation of DECOMPOSE. DECOMPOSE is closely patterned after a procedure by Lagergren [Lag90]. Like Lagergren's method (and like a similar algorithm by Reed [Ree92]), it relies on the familiar notion of a separator. Given a subset $Q \subseteq V(G)$ and a real number α, $S \subseteq V(G)$ is said to be an α-Q-separator if there exists a partition (A_1, A_2, S) of $V(G)$ such that every path from A_1 to A_2 goes through S and $|A_i \cap Q| \leq \alpha|Q - S|$, $i = 1, 2$. If $Q = V(G)$, S will be referred to simply as an α-separator. Robertson and Seymour showed that, for every $G \in \Gamma_w$ and every $Q \subseteq V(G)$, G has a $2/3$-Q-separator S with $|S| \leq w+1$ [RoSe86].

In describing DECOMPOSE, we write terms(G) to denote the set of vertices in terms(G).

DECOMPOSE

Step 1. Find a $7/8$-separator S_1 of G and the associated partition (A_1, A_2, S_1) of $V(G)$. For $i = 1, 2$, let $G_i = G[A_i \cup S_1]$ and let terms(G_i) consist of $S_1 \cup (\text{terms}(G) \cap A_i)$, in any order. Assume w.l.o.g. that $|\text{terms}(G_1)| \geq |\text{terms}(G_2)|$.

Step 2. If $|\text{terms}(G_1))| \leq 9w + 6$, return the decomposition $G = \varphi(G_1, G_2)$, for the appropriate $\varphi \in \mathcal{U}[9w + 6, 3]$.

Step 3. If $|\text{terms}(G_1)| > 9w + 6$, find a $2/3$-Q-separator S_2 of G_1, where $Q = \text{terms}(G_1)$, and the associated partition (B_1, B_2, S_2) of $V(G_1)$. For $i = 1, 2$, let $H_i = G[B_i \cup S_2]$ and let terms(H_i) consist of $S_2 \cup (\text{terms}(G_1) \cap B_i)$, in any order. Assume w.l.o.g. that $|V(H_1)| \leq |V(H_2)|$.

Step 4. If $|V(H_1)| \geq n/16$, return the decomposition $G = \varphi(H_1, H_2, G_2)$, for the appropriate $\varphi \in \mathcal{U}[9w + 6, 3]$.

Step 5. If $|V(H_1)| < n/16$ (and, hence, $|V(H_2)| > n/16$), let $G_2' = G_2 \cup H_1$ and let $\mathbf{terms}(G_2')$ consist of the vertices in $\mathrm{terms}(G_2) \cup \mathrm{terms}(H_1)$, in any order. Return the decomposition $G = \varphi(H_2, G_2')$, for the appropriate $\varphi \in \mathcal{U}[9w + 6, 3]$.

For reasons of space, we omit the proof of correctness of this procedure and only state that it is not hard to verify that each of the at most three graphs entering the decomposition returned by DECOMPOSE has no more than $15n/16$ vertices and no fewer than $n/16$; it is only slightly more difficult to show that each of them has at most $9w+6$ terminals. A correctness proof for a similar procedure can be found in [FeSl94]. The most time-consuming steps of DECOMPOSE are the separator computations in Steps 1 and 3. We shall sketch how to implement these computations in $O(\log n)$ time, at the expense of a one-time-only linear-time preprocessing step.

Lagergren's original version of DECOMPOSE was used to construct tree decompositions of graphs [Lag90]. For our problem, we can assume that Bodlaender's algorithm has provided us, in linear time, with a tree-decomposition of width w [Bod93]. From a width-w tree-decomposition of $G \in \Gamma_w$, we can easily obtain in linear time a *good* tree-decomposition, i.e., a decomposition (T, χ) such that: (i) T is a rooted binary tree, (ii) $|\chi_i| = w + 1$ for every $i \in V(T)$, (iii) $|\chi_i \cap \chi_j|$ is either w or $w + 1$ for every $(i, j) \in E(T)$, and (iv) for every internal node $i \in V(T)$, i has at least one child j such that $\chi_i \neq \chi_j$. We will use good tree decompositions to compute separators efficiently.

In what follows, when no confusion can arise, a tree decomposition (T, χ) of G, shall be referred to by mentioning only T, leaving χ implicit. Also, given any subtree T' of a tree-decomposition T of a graph G, we shall write $G[T']$ to denote the subgraph of G induced by $\bigcup \{\chi_j : j \in V(T')\}$. If χ' is the restriction of χ to $V(T')$, then, clearly, (T', χ') is a tree-decomposition of $G[T']$. We have the following lemma.

Lemma 2.1 *Let (T, χ) be a good tree decomposition of an n-vertex graph G. Suppose $e = (u, v)$ is a centroid edge in T (i.e., each subtree in $T - e$ has at most $3|V(T)|/4$ vertices), and let T_1 and T_2 be the subtrees of $T - e$, with $G_i = G[T_i]$, $i = 1, 2$. Then, for $i = 1, 2$, $|V(G_i)| \geq n/8$. Furthermore, $S = \chi_u \cap \chi_v$ separates $V(G_1)$ from $V(G_2)$.*

For each fragment H of G manipulated during the construction of the parse tree, we will maintain (1) a good tree decomposition T_H of H represented using Sleator and Tarjan's dynamic trees data structure with partitioning by size [SlTa83], and (2) a *skeleton* for $\mathrm{terms}(H)$, to be used in computing Q-separators. Since T is a good tree decomposition, we can use Lemma 2.1 and compute a vertex separator in H by finding a centroid edge in T. As Goodrich and Tamassia have shown [GoTa91], a dynamic tree representation of the tree allows one to locate a centroid edge in $O(\log n)$ time. The same data structure allows one to link trees [SlTa83], as required in Step 5, where the tree-decompositions of G_2 and H_1 must be combined into a decomposition for G_2'.

A skeleton for $Q = \text{terms}(H)$ is a tree T_Q constructed as follows. Choose any mapping f of the vertices $v \in Q$ to nodes $i \in V(T_H)$ such that $v \in \chi_i$, with possibly more than one v mapped to the same i. Let L be the set of nodes in T_H to which at least one node of Q has been mapped. A tree T_L for L is defined recursively as follows. If $L = \emptyset$, T_L is empty. Otherwise, let r be the least common ancestor (lca) in T_H of the nodes in L. Let T_1 and T_2 be the subtrees of T_H rooted at the children of r, and let $L_i = L \cap V(T_i)$, $i = 1, 2$. Then T_L has r as its root and as its subtrees it has the trees T_{L_1} and T_{L_2} for L_1 and L_2, respectively. Finally, make $T_Q = T_L$. Note that, since $|Q| = O(w)$, $|V(T_Q)| = O(w)$. Since dynamic trees support logarithmic-time lca queries [SlTa83], the skeleton can be computed in $O(w \log n)$ time.

To each $i \in V(T_Q)$ we assign a weight w_i equal to the number of vertices of Q mapped to it by f. A *centroid vertex* c in T_Q is a vertex whose removal leaves no subtree having total weight exceeding $(\sum_{i \in V(T_Q)} w_i)/2$. Since T_Q is of size $O(w)$, such a vertex can be found in $O(w)$ time. It is not hard to show that χ_c is a $2/3$-Q-separator in H. Dynamic trees allow one to construct, in $O(\log n)$ time, skeletons for the fragments of H resulting from the removal of the separator. They also allow one to combine in $O(\log n)$ time the skeletons for tree decompositions that are joined in Step 5.

The required preprocessing consists of building a good tree decomposition and constructing the initial dynamic tree representation of T. The former takes linear time, starting from a decomposition of width w, while the latter takes linear time, as discussed by Goodrich and Tamassia [GoTa91]. All subsequent updates take only $O(\log n)$ time.

3 Parametric Search Problems

As before, let $Z_G(\lambda)$ describe the cost of the optimum solution as a function of λ. Since the cost of every feasible solution is a linear function of λ, Z_G is a piecewise-linear function which will be concave or convex, depending on whether the problem involves minimization or maximization. The *breakpoints* of Z_G are the points at which the slope of Z_G changes. These points partition the λ-axis into a sequence of intervals, each of which is the maximal connected set of λ-values for which a given subgraph H is optimal.

For concreteness, we shall deal with problems where Z_G is concave (i.e., the objective is to find a minimum cost subgraph), and we are asked to find λ^* such that $Z_G(\lambda^*) = \max_\lambda Z_G(\lambda)$. All our results are easily transferable to cases where Z_G is convex, or to other variants of parametric search, including sensitivity analysis or locating the roots of Z_G, which is useful in minimum-ratio optimization [Meg83].

We shall say that a value λ_0 has been *resolved* if it has been determined whether or not $\lambda_0 \leq \lambda^*$. To resolve a value λ_0 we invoke an *oracle* $\mathcal{B}$. It is known (see, for example, [CoMe93]) that an oracle can be obtained from an algorithm $\mathcal{A}$ that evaluates $Z_G(\lambda)$ for any λ. To do this, it suffices to execute $\mathcal{A}$, with λ fixed at λ_0 and then do additional problem-specific work that is $O(1)$ [FeSl94]. For example, in the problem of maximizing Z_G, the $O(1)$ work is to determine the slope of Z_G at λ_0. Thus, the running time of this oracle will be asymptotically equal to that of $\mathcal{A}$, which, for the problems we are considering here, is linear. In most applications of Megiddo's method, the oracle is applied a manner reminiscent of binary search to a set of candidate parameter values, leading to a search procedure that is at least a logarithmic factor slower than their non-parametric counterpart, algorithm $\mathcal{A}$. To improve on this, we divide the computation into a sequence of phases, such that each phase uses an oracle that is faster than the one for the previous phase. We first need to give some more details on the non-parametric algorithm $\mathcal{A}$.

The non-parametric algorithm. Here we summarize the key ideas behind the linear-time dynamic programming algorithm for finding optimum subgraphs satisfying a regular property. More rigorous treatments of the subject can be found elsewhere [ALS91, BPT92, BLW87].

A graph-subgraph pair is a pair (G, H) where G is a graph and H is a subgraph of G; the cost of (G, H) is the sum of the weights of the vertices and edges in H. Problems that are regular on graphs of bounded tree-width share two properties: (1) the set of graph-subgraph pairs (G, H) where $G \in \Gamma_w$ can be partitioned into a finite number of equivalence classes and (2) there is a finite set of rules, expressed in tables that are fixed for each problem, whereby graph-subgraph pairs can be combined to obtain new pairs. Let $\mathcal{C} = \{C_1, \ldots, C_N\}$ be the set of equivalence classes with respect to some regular property P. Not all these classes will contain pairs (G, H) where H satisfies P; those classes that do are called *accepting*. For $i = 1, \ldots, N$, let $z_G^{(i)}$ be the value of the least cost pair $(G, H) \in C_i$. Let T be a bounded-degree linear-size parse tree of G. To compute z_G, the cost of the optimum solution, process T from the bottom up, computing $z_{G_v}^{(i)}$ for every $v \in V(T)$ and every class C_i. If v is a leaf, and hence G_v is a primitive graph, then all the $z_{G_v}^{(i)}$'s are computed by exhaustive enumeration in constant time, since there are only a constant number of pairs (G_v, H) in each class. Otherwise, for each equivalence class C_i, we will have tables giving a finite number of ways in which a pair $(G_v, H) \in C_i$ can be expressed as a combination of pairs of the form $(G_{v_j}, H_j) \in C_{i_j}$ where v_j is a child of v. If we know the $z_{G_{v_j}}^{(i_j)}$'s, we can combine this information via additions and comparisons in constant time to obtain $z_{G_v}^{(i)}$ [BLW87, BPT92]. After every node in the parse tree has been processed, we can compute the cost z_G of the optimum solution by using

$$z_G = \min\Big\{\{+\infty\} \cup \{z_G^{(i)} : C_i \text{ is an accepting class}\}\Big\}. \tag{1}$$

The algorithm just described can easily be used to evaluate $Z_G(\lambda_0)$ for any given λ_0 in $O(n)$ time. It can also be "lifted" so that, instead of operating on graphs with numerical weights, it operates on graphs whose weights are linear functions of λ [FeSl94, Meg83]. For every node $v \in T$ let $Z_{G_v}^{(i)}(\lambda)$ describe the cost of the optimum subgraph of G_v for C_i. Since the operations conducted by the algorithm are additions and comparisons, it can easily be shown that the $Z_{G_v}^{(i)}$'s are concave piecewise linear functions of λ. It is known that every $Z_{G_v}^{(i)}$ has a polynomial number of breakpoints in $|V(G_v)|$ [FeSl94].

Dividing the parse tree into levels. From this point on, we concentrate on the problem of finding λ^*. Assume that we are given a 1/16-balanced parse tree T of $G \in \Gamma_w$. The first step of our algorithm is a linear-time preprocessing of the input to identify certain sets of nodes in the tree; these sets will be called *levels*. For this, we use a procedure MARK-TREE, which takes as inputs the root v of T and a number r. Assuming $n = |V(G)| \geq r$, it marks a set of nodes $L = \{u_1, \ldots, u_k\} \subseteq V(T)$ with the property that, if T_i is the subtree of T rooted at u_i, for $1 \leq i \leq k$, then: (1) the T_i's are disjoint, (2) every leaf of T is a leaf in some T_i, and (3) $|V(G_{u_i})| = \Theta(r)$. MARK-TREE is straightforward: If $|V(G_v)| \leq 16r$, it marks v. Otherwise, MARK-TREE is applied recursively to each child u of v. It can be readily verified that every marked vertex u satisfies $r \leq |V(G_u)| \leq 16r$. Assuming we have precomputed $|V(G_u)|$ for every node u in the tree (which can be done in $O(n)$ time), MARK-TREE will take $O(n/r)$ time.

Suppose now that we are given a sequence of numbers $r_1 < r_2 < \cdots < r_k$ and a balanced parse tree T, with root v, of the input graph G. Apply MARK-TREE(v, r_k) and declare all marked nodes to be *level k* nodes. Now, for $j = k, k-1, \ldots, 2$, do the following. For each level j node u, apply MARK-TREE(u, r_{j-1}) and declare the newly marked nodes to be level $j-1$ nodes. Finally, declare the leaves of T to be level 0 nodes. Identifying levels k through 0 takes $O(n)$ time.

For $j = 1, \ldots, k$, let L_j denote the set of level j nodes. By construction, $|L_j| = O(n/r_j)$. Suppose $v \in L_j$. Then the subtree of T rooted at v shall be called a *level j subtree*. Similarly, $Z_v^{(i)}(\lambda)$, the function describing the cost of the optimum solution for the ith equivalence class for the graph G_v, shall be called a *level j function*. In an earlier paper [FeSl94], we proved that for every regular problem there exists a constant $\beta > 0$, independent of G, with the following property: For every $v \in T$ and every equivalence class C_i, $Z_v^{(i)}$ has $O(r^\beta)$ breakpoints, where $r = |V(G_v)|$. Thus, any level j function has $O(r_j^\beta)$ breakpoints. From now on, we shall assume that we are considering some specific (but arbitrarily chosen) regular problem, so that β is fixed.

Suppose that, for some interval $\mathcal{I} = [\lambda_L, \lambda_R]$ such that $\lambda^* \in \mathcal{I}$, the breakpoints of each level j function $Z_v^{(i)}$ were stored in a balanced binary search tree. Together with each breakpoint we store the interval of $Z_v^{(i)}$ to its right, as well as the equation of the line segment corresponding to that interval (note that, to do this properly, we may have to introduce an artificial breakpoint at λ_L). The height of the binary search tree is $O(\log r_j)$, since the number of elements in it is polynomial in r_j. Evaluating any level j function at λ_0 thus takes only $O(\log r_j)$ time, as it reduces to a search over the tree. Let us call what we have just described an *efficient representation* of $Z_v^{(i)}$ within interval $\mathcal{I}$.

Armed with efficient representations of the level j functions, we can use only these functions to evaluate $Z_G(\lambda_0)$ for any $\lambda_0 \in \mathcal{I}$ as follows. First, evaluate every level j function at λ_0. Since there are $O(n/r_j)$ level j functions to evaluate (a constant number per node in that level), the total work to do so will be $O(n \log r_j/r_j)$. We then evaluate $Z_G(\lambda_0)$ in the usual bottom-up fashion, but instead of starting at the leaves we start at the vertices of L_j. Since the number of vertices of T above L_j is $O(n/r_j)$, the remainder of the computation takes only $O(n/r_j)$ time, which means that the total time is $O(n \log r_j/r_j)$. This evaluator can be used to implement the oracle in $O(n \log r_j/r_j)$ time as follows. Suppose the value to be tested is λ_0. If $\lambda_0 \notin \mathcal{I}$, we can in $O(1)$ time determine the relative position of λ_0 with respect to λ^*; otherwise, evaluate $Z_G(\lambda_0)$ and use the result to resolve λ_0. Unfortunately, unless we have chosen a very special interval $\mathcal{I}$, just constructing each level j function will take $O(r_j^{\beta+1})$ time [FeSl94], for a total time of $O(nr_j^\beta)$ time. We now describe how to get around this problem.

The search. Our algorithm conducts k *phases*, constructing a sequence of increasingly faster procedures $\mathcal{A}_0, \mathcal{A}_1, \ldots, \mathcal{A}_k$, for evaluating Z_G. Each procedure $\mathcal{A}_i$ will have an associated interval $\mathcal{I}_i$ such that $\lambda^* \in \mathcal{I}_i$. The intervals will be nested — i.e., $\mathcal{I}_0 \supseteq \mathcal{I}_1 \supseteq \cdots \supseteq \mathcal{I}_k$ — and will contain a rapidly decreasing number of breakpoints of Z_G. The initial evaluator, $\mathcal{A}_0$, will have $\mathcal{I}_0 = [-\infty, +\infty]$ and will work by running the $O(n)$-time dynamic programming algorithm described earlier. For $j = 1, \ldots, k$, $\mathcal{A}_j$ will rely on efficient representations of the level $j-1$ functions to evaluate Z_G within $\mathcal{I}_j$ in $O(n \log r_j/r_j)$ time. Each algorithm $\mathcal{A}_j$ leads to a corresponding oracle $\mathcal{B}_j$ with the same asymptotic complexity. In order to construct the successive evaluators efficiently, we rely on the following lemma, where, as before, β is a constant such that the number of breakpoints of $Z_G^{(i)}$ for any n-vertex graph G is $O(n^\beta)$.

Lemma 3.1 *Let $\mathcal{I}$ be an interval within which the total number of breakpoints of the level $j-1$ functions is $O(n/r_j)$. Then the total number of breakpoints of the level j functions within $\mathcal{I}$ is $O(nr_j^{\beta-1}/r_{j-1}^\beta)$. Furthermore, given efficient representations of the level $j-1$ functions within $\mathcal{I}$, we can construct efficient representations of all level j functions within $\mathcal{I}$ in $O(nr_j^\beta \log r_j/r_{j-1}^{\beta+1})$ time.*

We sketch some of the ideas behind the proof. Consider any level j tree T'; call its root v. As argued earlier G_v has $O(r_j)$ nodes and, hence, so does T'. Tree T' contains $O(r_j/r_{j-1})$ level $j-1$ nodes; thus, if we delete all nodes in T' lying below level $j-1$, we obtain a tree T'' with $O(r_j/r_{j-1})$ nodes. Now consider any interval $\mathcal{I}' \subseteq \mathcal{I}$ between consecutive breakpoints of level $j-1$ functions associated with leaves of T''. Within $\mathcal{I}'$, any level $j-1$ function is a straight line. We can construct $Z_v^{(i)}$ within $\mathcal{I}'$ by considering only T'', and the level $j-1$ functions at the leaves of T''. Thus, within $\mathcal{I}'$, $Z_v^{(i)}$ will behave like the function describing the cost of the optimum solution from C_i for a graph of size $O(r_j/r_{j-1})$ with weights that are linear in λ. The upper bound in [FeSl94] translates into an upper bound of $O((r_j/r_{j-1})^\beta)$ on the number of breakpoints of $Z_v^{(i)}$ within $\mathcal{I}'$. Adding up the contributions of all the level j functions over all $O(n/r_j)$ possible intervals, the total number of breakpoints is $O(nr_j^{\beta-1}/r_j^\beta)$.

We next compute efficient representations of all level 0 functions within $\mathcal{I}_0$, using exhaustive enumeration for each function. This takes $O(n)$ time, since there are $O(n)$ such functions, each of them corresponding to a primitive graph (i.e., a graph of constant-bounded size), and having $O(1)$ breakpoints. Let P_0 be the set of all breakpoints of the level 0 functions. We can find an interval $\mathcal{I}$ such that $P_0 \cap \mathcal{I} = O(n/r_1)$ using $O(\log r_1)$ invocations of the oracle $\mathcal{B}_0$ in the usual way (see, e.g., [Meg83]): Choose a median element of P_0 and apply the oracle to it; depending on the outcome of the call, either resolve all values larger than the median or all values smaller than the median. In either case, at least one half of the elements in P_0 will be resolved; these values can therefore be removed from further consideration and the interval $\mathcal{I}$ is updated accordingly. After q invocations of the oracle, the number of elements of P_0 that remain unresolved will be $O(n/2^q)$; thus, with $q = O(\log r_1)$ calls, we can find the desired interval. The total overhead incurred in median computations is $O(n)$, while the total time spent in oracle calls is $O(n \log r_1)$.

For $j \geq 1$, suppose we have an interval $\mathcal{I}$ containing λ^* and $O(n/r_j)$ breakpoints of level $j-1$ functions. Suppose also that we have constructed an appropriate $\mathcal{A}_{j-1}$. Using Lemma 3.1, we can construct efficient representations of all level j functions within $\mathcal{I}$ in $O(nr_j^\beta \log r_j/r_{j-1}^{\beta+1})$ time. The total number of breakpoints of these functions is $O(nr_j^{\beta-1}/r_{j-1}^\beta)$. Using $O(\log r_{j+1})$ calls to $\mathcal{B}_{j-1}$, we can narrow the search down to a new interval $\mathcal{I}$ containing λ^* and only $O(n/r_{j+1})$ breakpoints of level j functions. The total work required to do the necessary median computations is $O(n/r_{j+1})$, while the total work required for the oracle calls is $O(n \log r_j/r_{j+1})$. Hence, the total work at level j, $1 \leq j \leq k$, is $O(nr_j^\beta \log r_j/r_{j-1}^{\beta+1} + n \log r_j/r_{j+1})$.

The time taken by building the decomposition and identifying the levels is only $O(n)$. The work at level 0 is $O(n \log r_1)$. Now, suppose $r_1 = c$ and that, for $j = 1, \ldots, k-1$, $r_{j+1}/r_j = c$, for some constant $c > 1$. Then, $r_j = c^j$ and the total work is $O(n(c^{\beta-1} \sum_{j=1}^k j/c^{j-1} + \sum_{j=1}^k j/c^{j+1}))$, which is clearly $O(n)$, independent of the number of terms, because c and β are constants. If $r_k = \Theta(n)$, and if v denotes the root of the parse tree, at the end of the kth phase, we have an interval $\mathcal{I}$ within which for every C_i, the number of breakpoints of $Z_G^{(i)}$ within $\mathcal{I}$ is $O(1)$. This implies that Z_G has only $O(1)$ breakpoints within this interval, as it is the lower envelope of the $Z_G^{(i)}$'s (this follows from equation (1)). Moreover, we have an evaluator $\mathcal{A}_k$ that computes Z_G in $O(\log n)$ time for each $\lambda_0 \in \mathcal{I}$. Thus, in $O(\log n)$ additional time, we can construct Z_G within this interval and search directly for λ^*. Therefore, the entire search takes linear time.

4 Further Results

Our results can be extended to multiparameter problems using the techniques of [CoMe93, AgFe92] and to problems with nonlinear concave weights using the techniques of Toledo [Tol93]. For lack of space, we will omit the discussion of these results. Instead, we devote this section to sketching some results on approximate parametric search problems on planar graphs (see also [Tol93]).

Baker [Bak83] has developed a scheme whereby several NP-hard optimization problems can be solved approximately on planar graphs. In the case of maximization problems, for each fixed k, her approach computes in linear time a solution that is at least $k/(k+1)$ times optimal (or, in the case of minimization problems, at most $(k-1)/k$ times optimal). Baker's idea relies on a decomposition of the planar input graph into k-outerplanar graphs. Like her, we use the maximum independent set problem to illustrate the technique. Suppose G is a planar graph with some given planar embedding. A node is *on layer 1* if it lies on the exterior face. A cycle of layer i nodes is a *layer i face* if it is an interior face in the subgraph induced by the layer i nodes. Given a layer i face f, G_f is the subgraph induced by all nodes inside f in the embedding. The nodes on the exterior face of G_f are the *layer $i+1$* nodes. A planar embedding is *k-outerplanar* if it has no nodes with layer number greater than k and a graph is *k-outerplanar* if it has a k-outerplanar embedding. It was shown by Bodlaender [Bod88] that k-outerplanar graphs have treewidth at most $3k-1$. For any given k, we can obtain $k+1$ k-outerplanar graphs from G; the r-th of these, $0 \le r \le k$, is denoted G_r and is constructed by removing from G every vertex with layer number congruent to r (mod $k+1$). Solve the problem optimally on each of the $k+1$ graphs and then, from amongst the answers obtained, we return the largest solution. Baker's algorithm is easily lifted to the parametric setting. The functions Z_{G_r} describing the optimum solutions to each subproblem have polynomially-many breakpoints, since the G_r's have bounded tree-width and independent set is a regular property. The function F_G describing the solution returned by the algorithm as a function of λ is the upper envelope of the $k+1$ Z_{G_r}'s. Thus, for each fixed k, F_G has polynomially-many breakpoints. Searching for the maximum of F_G will require k searches in the Z_{G_r}'s. Since k is fixed, our earlier results imply that the search will take $O(n)$ time. Furthermore, $\max_\lambda F_G$ will be at least $k/(k+1)$ times $Z_G(\lambda^*)$.

References

[AgFe92] R. Agarwala and D. Fernández-Baca. Weighted multidimensional search and its application to convex optimization. DIMACS Technical Report 92-51, November, 1992. To appear in *SIAM J. Comput.*.

[ALS91] S. Arnborg, J. Lagergren, and D. Seese. Easy problems for tree-decomposable graphs. *J. Algorithms*, 12:308–340.

[Bak83] B.S. Baker. Approximation algorithms for NP-complete problems on planar graphs. In *Proceedings of 24th Annual Symposium on Foundations of Computer Science*, pp. 265–273, 1983.

[BLW87] M.W. Bern, E.L. Lawler, and A.L. Wong. Linear time computation of optimal subgraphs of decomposable graphs. *J. Algorithms*, 8:216–235, 1987.

[BPT92] R.B. Borie, R.G. Parker, and C.A. Tovey. Automatic generation of linear-time algorithms from predicate-calculus descriptions of problems on recursively-constructed graph families. *Algorithmica*, 7:555–582, 1992.

[Bod88] H.L. Bodlaender. Some classes of graphs with bounded tree-width. *Bulletin of the EATCS*, **36** (1988), 116-126.

[Bod93] H.L. Bodlaender. A linear time algorithm for finding tree-decompositions of small tree-width. In *Proceedings of the 25th Annual ACM Symposium on Theory of Computing*, pp. 226–233, 1993.

[CEGS92] B. Chazelle, H. Edelsbrunner, L. Guibas, and M. Sharir. Diameter, width, closest line pair, and parametric searching. In *Proceedings of the 8th Annual ACM Symposium on Computational Geometry*, pp. 120–129 (1992).

[CoMe93] E. Cohen and N. Megiddo. Maximizing concave functions in fixed dimension. In *Complexity in Numerical Computations*, P.M. Pardalos, ed., pp. 74–87, World Scientific Press 1993.

[Cole87] R. Cole. Slowing down sorting networks to obtain faster sorting algorithms. *J. Assoc. Comput. Mach.*, 34(1):200–208, 1987.

[FeSl94] D. Fernández-Baca and G. Slutzki. Parametric problems on graphs of bounded tree-width. *J. Algorithms*, 16:408–430 (1994).

[Fre90] G.N. Frederickson. Optimal algorithms for partitioning trees and locating p-centers in trees. Technical Report CSD-TR 1029, Department of Computer Science, Purdue University, October 1990.

[Fre93] G.N. Frederickson. Maintaining regular properties in k-terminal graphs. Manuscript, 1993.

[GoTa91] M.T. Goodrich and R. Tamassia. Dynamic trees and dynamic point location. In *Proceedings of the 23rd Annual Symposium on Theory of Computing*, 1991, pp. 523–533.

[Lag90] J. Lagergren. Efficient parallel algorithms for tree-decomposition and related problems. In *Proceedings of 31st Annual Symposium on Foundations of Computer Science*, pp. 173–182 (1990).

[Meg79] N. Megiddo. Combinatorial optimization with rational objective functions. *Math. Oper. Res.*, 4:414–424 (1979).

[Meg83] N. Megiddo. Applying parallel computation algorithms in the design of serial algorithms. *J. Assoc. Comput. Mach.*, 30(4):852–865, 1983.

[Ree92] B. Reed. Finding approximate separators and computing tree-width quickly. Proceedings of *24th Annual Symposium on Theory of Computing*, pp. 221–228, 1992.

[RoSe86] N. Robertson and P.D. Seymour. Graph minors II: Algorithmic aspects of tree-width. *J. Algorithms*, 7:309–322, 1986.

[vLe90] J. van Leeuwen. Graph Algorithms. In J. van Leeuwen (ed.) *Handbook of Theoretical Computer Science*, MIT Press, Cambridge, Mass., 1990.

[SlTa83] D.D. Sleator and R.E. Tarjan. A data structure for dynamic trees. *Journal of Computer and System Sciences*, 26:362–391 (1983).

[Tol93a] S. Toledo. Maximizing non-linear convex functions in fixed dimension. In *Complexity in Numerical Computations*, P.M. Pardalos, ed., pp. 429–446. World Scientific Press 1993.

[Tol93] S. Toledo. Approximate parametric searching. *Manuscript*, 1993.

[Wim87] T.V. Wimer. Linear algorithms on k-terminal graphs. Ph.D. Thesis, Report No. URI-030, Clemson University (1987).

Lower Bounds for Dynamic Algorithms

Michael L. Fredman

Rutgers University, New Brunswick NJ 08903, USA

1 The Cell Probe Model of Computation

The focus of this paper concerns the cell probe model of computation and its utilization for assessing the inherent complexity of dynamic algorithms. The following section discusses some concrete examples. The concluding section contains discussion concerning the strengths and weaknesses of the method, and mentions an open problem.

The cell probe model of computation was introduced by Yao [16]. This fascinating paper addresses the question of how many memory accesses are required to perform membership queries using a space-restricted dictionary. The paper also suggests that this approach might be of general use for analyzing other data structures in a random access type model of computation.

In a nutshell the cell probe model defines the time complexity of a sequential computation to be the number of memory accesses to a random access memory of fixed word size. The number of bits b in a single word is a parameter of the model. All other computations are free. The model allows for unusual representations of data, indirect addressing etc. As a consequence, lower bounds proved in this model are very general as contrasted with structured lower bounds which are proved in models (e.g. algebraic, comparison, pointer manipulation) that require restrictions on the manner in which data is represented and manipulated. Given this framework, there are two problem categories in which non-trivial lower bounds might be forthcoming: executing individual queries concerning a static object represented by a space-restricted data structure; and executing intermixed sequences of update and query tasks in an on-line manner, where updates potentially change the object being queried. Algorithms for the second problem category are referred to as *dynamic algorithms*. As we shall see immediately, lower bounds for the static problems of the first category can sometimes be used to obtain lower bounds for dynamic algorithms.

2 Representative Results

We describe seven topics for which lower bounds have been obtained.

The Predecessor Problem. The first complete success story for lower bounds in the cell probe model is found in Ajtai's remarkable paper [1]. The problem is to represent a static subset S of the universe $U = \{1, 2, \cdots, n\}$ to accommodate the query, $pred(x)$, which returns the greatest element in S that does not exceed x. It had been shown in the well-known paper of Willard [14] that these queries can be done in $\log \log n$ time using a data structure with space linear in $|S|$. Ajtai's result shows that constant time for the $pred(x)$ query is not possible for any data structure of space which is polynomial in $|S|$, when the word size is $\log n$. Xiao [15] strengthened the result, demonstrating a $\log \log n / \log \log \log n$ lower bound for the $pred(x)$ query when the word size is $polylog(n)$. As a consequence of having extended beyond $\log n$ the word size under which the lower bound holds, Xiao [15] is able to infer a lower bound for dynamic predecessor queries. Without any space restrictions, Xiao demonstrates that the amortized costs of $pred(x)$, $insert(x)$, and $delete(x)$ are bounded below by $\log \log n / \log \log \log n$. This nearly matches the upper bound of $\log \log n$ from the well-known paper of Van Emde Boas, Kaas, and Zijlstra [5].

Partial Sums. Given an array, $A[j]$, $1 \leq j \leq n$, and an integer k, $k \leq n+1$, we require a data structure for the operations, $update(j, \delta)$, which implements $A[j] \leftarrow A[j] + \delta$, and $sum(j)$, which returns $\sum_{i<j} A[i] \pmod{k}$. Tree based data structures perform these operations in time $\log n$ per operation. Fredman and Saks [8] establish an amortized $\log n / \log \log n$ lower bound in the cell probe model with $polylog(n)$ word size. Motivated by this lower bound, Dietz [3] constructs a new data structure with matching, hence optimal, complexity, provided that the quantity δ in the operation $update(j, \delta)$ is constrained to be at most $polylog(n)$ in magnitude. This data structure only requires word size $\log n$.

List Representation. This problem requires that we represent a list in a manner that permits the operations, $access(j)$, which reports the item occupying the jth position in the list, $insert(j, x)$, which places x in position j of the list, and $delete(j)$, which deletes the jth item from the list. Balanced tree schemes can be used to accomplish these tasks in time $\log n$ per operation when n items are present in the list. Fredman and Saks [8] establish an amortized $\log n / \log \log n$ lower bound in the cell probe model with $polylog(n)$ word size. Dietz [3] constructs a new data structure with complexity that matches the lower bound and only requires word size $\log n$.

Disjoint Set Union. The inverse Ackermann lower bound holds for the disjoint set union problem in the cell probe model with $\log n$ word size [8].

Dynamic Traveling Salesman Tours. Certain heuristic algorithms for finding good traveling salesman tours require a data structure that performs the following operations on tours: $Next(a)$, which returns the city that follows a in the current tour (tours are assigned an orientation); $Prev(a)$, which returns the city that precedes a in the current tour; $Between(a, b, c)$, which returns "true" provided that when traversing the tour starting at a, b is encountered before c; and $Flip(a, b, c, d)$, which replaces edges (a, b) and (c, d) with edges (b, c) and (a, d). The Flip operation assumes that $a = Next(b)$ and $d = Next(d)$. Data structures exist [7] that perform these operations in $\log n$ amortized time, when n cities are present. Moreover, a $\log n / \log \log n$ lower bound holds for the amortized complexities of these tasks in the cell probe model of computation with $polylog(n)$ word size [7].

Connectivity Testing in Planar Graphs. Given a planar embedded graph G, we consider the following operations: *insert*(u, v) inserts the edge (u, v) into G provided that this insertion does not destroy the planarity of the embedding; *delete*(u, v) deletes the edge (u, v) from G, and *Query*(u, v) returns "Yes" provided that u and v are connected in G. As shown be Eppstein et.al. [6] these operations can be performed in time $\log n$ (for n node graphs), and a $\log n/\log\log n$ lower bound holds in the cell probe model with polylog(n) word size, as shown by Rauch [12].

Dictionary Problem. Dietzfelbinger et.al. [4] establish constant amortized bounds for insertions, deletions and search queries in a dictionary, provided that randomization is available. For deterministic computations, Sundar [13] establishes a lower bound of $\log\log n/\log\log\log n$ for the amortized cost per operation for n intermixed insertions and search operations in the cell probe model with polylog(n) word size, provided that the size of the data structure is at all times bounded by a polynomial in the size of the dictionary.

3 Discussion

The main strength of the cell probe model of computation is that it provides a single coherent framework for assessing the inherent complexity of dynamic algorithms. The examples in the previous section range from purely combinatorial to numerical in character and we find it unnecessary to introduce an ad-hoc model of computation to fit the problem, whose generality is then subject to question. Results established in the cell probe model are not subject to possible reconsideration as, for example, the necessity of having to reconsider the $n\log n$ lower bounds for sorting in light of fusion trees [9].

A second example is worthy of detailed consideration. The lower bound established by La Poutré [11] for the disjoint set union problem in the pointer machine model holds for problem instances in which the sequence of union operations are known in advance, a situation for which RAM algorithms provide constant amortized time solutions [11]. Thus, in isolation the La Poutré result leaves open the possibility that RAM algorithms, which fall outside of the pointer machine framework, may avoid the inverse Ackermann limits. Moreover, the cell probe model with $\log n$ bit word size is adequate for simulating pointer machines with a bounded number of pointers per record. On the other hand, the pointer machine lower bound does circumvent a limitation of the cell probe approach. As the word size grows beyond $\log n$ the cell probe lower bound breaks down, as shown by Ben-Amram and Galil [2]; the precise point at which that happens is not known, however. In contrast, the La Poutré result allows for arbitrarily large amounts of directional information to be stored in nodes, which can be utilized to direct a search during the execution of a find operation. The immediate cost of a find operation then becomes the length of the shortest path leading to the node containing the set name that answers the query. That the inverse Ackermann bound still holds under these circumstances is an insight that the cell probe result does not provide.

The above discussion concerning disjoint set union points out a weakness of the cell probe method; the model is unrealistically powerful, particularly as the word size increases. This problem is mitigated in part by the fascinating work of Ben-Amram and Galil [2]. By restricting the operations on words to the standard RAM arithmetics, Ben-Amram and Galil show that most cell probe bounds go through without *any* restrictions on word size. The results break down, however, with the inclusion of bitwise Boolean operations.

A second weakness of the cell probe method is that lower bound proofs are not easily constructed. To date we have not seen any lower bounds that exceed $\log n/\log\log n$ complexity, and we pose as our main open problem the challenge of deriving for a natural problem a lower bound exceeding $\log n/\log\log n$ in the cell probe model with word size at least $\log n$. The partial sum problem is a possible candidate as suggested by recent lower bound improvements for the arithmetic complexity of the problem [10], and most certainly its multi-dimensional generalization is a good candidate.

References

1. M. Ajtai: A lower bound for finding predecessors in Yao's cell probe model. Combinatorica 8,3 (1988) 235-247
2. Ben-Amram, A., Galil, Z: Lower bounds for data structure problems on RAMs. Proceedings of the 32nd Symposium on Foundations of Computer Science (1991), 622-631
3. Dietz, P.: Optimal algorithms for list indexing and subset rank. Algorithms and data structures : Workshop WADS '89, Ottawa, Canada (1989), 39-46
4. Dietzfelbinger, M., Karlin, A., Mehlhorn, K., Meyer auf der Heide, F., Rohnhert, H. and Tarjan, R.: Dynamic perfect hashing: upper and lower bounds. Proceedings of the 29th Symposium on Foundations of Computer Science (1988), 524-531
5. Van Emde Boas, P., Kaas, R., and Zijlstra, E.: Design and implementation of an efficient priority queue. Math. Systems Theory 10 (1977), 99-127
6. Eppstein, D., Italiano, G., Tamassia, R., Tarjan, R., Westbrook, J., and Yung, M.: Maintenance of a minimum spanning forest in a dynamic planar graph. J. Algorithms 13 (1992), 33-54
7. Fredman, M., Johnson, D. S., McGeoch, L. A., and Ostheimer, G.: Data structures for traveling salesmen. Proceedings of the Fourth Annual ACM-SIAM Symposium on Discrete Algorithms (1993), 145-154
8. Fredman, M., Saks, M.: The cell probe complexity of dynamic data structures. Proceedings of the 21st Annual ACM Symposium on Theory of Computing (1989), 345-354
9. Fredman, M., Willard, D.: Surpassing the information theoretic bound with fusion trees. J. Computer and System Sciences 47,3 (1993), 424-436
10. Hampapuram, H., Fredman, M.: Optimal bi-weighted binary trees and the complexity of maintaining partial sums. Proceedings of the 34th Symposium on Foundations of Computer Science (1993), 480-485
11. LaPoutre, J.: Lower bounds for the union-find and the split-find problem on pointer machines. Proceedings of the 22nd Annual ACM Symposium on Theory of Computing (1990), 34-44.

12. Rauch, M.: Improved data structures for fully dynamic biconnectivity. Proceedings of the 26th Annual ACM Symposium on Theory of Computing (1994), to appear

13. Sundar, R.: A lower bound for the dictionary problem under a hashing model, Proceedings of the 32nd Symposium on Foundations of Computer Science (1991), 612-621; and *personal communication*

14. Willard, D.: Log-logarithmic worst case range queries are possible in space $O(N)$. Information Processing Letters 17 (1983), 81-89

15. Xiao, B.: New bounds in cell probe model. Doctoral Dissertation, University of California, San Diego, 1992

16. Yao, A.: Should tables be sorted? J. Assoc. Comput. Mach. **28**,3 (1981), 615-628

Sequential and Parallel Algorithms for Embedding Problems on Classes of Partial k-Trees

Arvind Gupta*   Naomi Nishimura†

April 22, 1994

Abstract

We present sequential and parallel algorithms for various embedding problems on bounded degree partial k-trees and k-connected partial k-trees; these include subgraph isomorphism and topological embedding, known to be NP-complete for general partial k-trees. As well as contributing to our understanding of the types of graphs for which these problems are tractable, this paper introduces methods for solving problems on graphs. In particular, we make use of the tree-decomposition of the graph to apply techniques used to solve problems on trees to solve problems on partial k-trees.

1 Introduction

In devising sequential and parallel algorithms for subgraph isomorphism and topological embedding of bounded degree partial k-trees and k-connected partial k-trees, we make advances in two streams of research. One stream of research is the identification of problems for which a bound on the tree-width of a graph allows a more efficient solution than in the general case; a partial k-tree is also known as a graph of bounded tree-width. The other stream is the identification of classes of graphs for which these embedding problems are tractable: the subgraph isomorphism problem, and consequently the more general problem of topological embedding, is known to be NP-complete for general graphs [GJ79], and remains so even for graphs of bounded tree-width [Sys82]. The algorithms presented in this paper make use of techniques developed in each of these streams, and in addition introduce general methods for enabling the techniques in one stream to be applied to problems in the other.

Graphs of bounded tree-width include many natural classes of graphs, such as trees, outerplanar graphs, series-parallel graphs, and Halin graphs. Of great interest for our algorithms is the fact that any graph in this class can be represented as a special type of tree, namely a *tree-decomposition*. Such graphs have the property that there are small separators which break the graph into a tree-like structure. This property has led to efficient algorithms for a number of problems which are difficult to solve on general graphs. The basic idea of our algorithms is to attempt to process a graph with respect to its tree-decomposition. We use some techniques developed for handling trees and introduce some new techniques applicable to more general types of graphs.

Previous sequential algorithms for embedding problems on various classes of graphs include algorithms for subgraph isomorphism on trees [Mat78], two-connected outerplanar graphs [Lin89], two-connected series-parallel graphs [LS88], bounded degree partial k-trees [MT92], and k-connected partial k-trees [MT92], and algorithms for topological embedding on bounded degree partial k-trees [MT92] and k-connected partial k-trees [MT92]. Earlier work in a parallel setting includes algorithms for subgraph isomorphism [LK89, GKMS90, GN92] and topological embedding on trees [GN92] and a number of problems on bounded degree partial k-trees including the subgraph isomorphism problem [Bod88b], but not for topological embedding or for subgraph isomorphism on k-connected partial k-trees.

The contributions of our work to the understanding of embedding problems are two-fold. We present the first parallel algorithms for subgraph isomorphism of k-connected partial k-trees and for topological embedding of bounded degree partial k-trees and k-connected partial k-trees. Although there are different algorithms known for the other problems considered in this paper, one of our contributions is to provide a general framework that allows us to extract the similarities between trees and

---

*School of Computing Science, Simon Fraser University, Burnaby, Canada, V5A 1S6. email: arvind@cs.sfu.ca, FAX (604) 291-3045. Research supported by the Natural Sciences and Engineering Research Council of Canada, the Center for System Sciences and the Advanced Systems Institute.

†Department of Computer Science, University of Waterloo, Waterloo, Ontario, Canada, N2L 3G1. email: nishi@plg.uwaterloo.ca, FAX (519) 885-1208. Research supported by the Natural Sciences and Engineering Research Council of Canada.

partial k-trees, subgraph isomorphism and topological embedding, and sequential and parallel settings. Our algorithms are conceptually simple, exploiting the relatively simple structure of a subtree isomorphism algorithm [Mat78], and thereby making clear the connection between trees and partial k-trees. At the same time, our methods are sufficiently flexible to allow us to introduce modifications that address the differences between the problems under consideration; we retain the same structure for sequential topological embedding algorithms, parallel subgraph isomorphism algorithms, and ultimately parallel topological embedding algorithms.

The outline of the remainder of the paper is as follows. In the next section we give definitions. Section 3 contains a description of some the difficulties encountered in attempting a straightforward adaptation of the subtree isomorphism algorithm to form algorithms for our problems, as well as techniques which can be applied to overcome these difficulties. The section ends with a general outline of all eight algorithms. In Section 4, we present the sequential algorithm for subgraph isomorphism when one input graph is k-connected, along with the alterations needed for the case when one input graph is of bounded degree. Then, in Sections 5 and 6 we present the main ideas behind the algorithms for sequential topological embedding, parallel subgraph isomorphism, and parallel topological embedding. Finally, in Section 7 open problems are presented. For details of results omitted from this paper, the reader is referred to the expanded technical report [GN93].

2 Preliminaries

For the problems discussed in this paper, each input graph will be a *partial k-tree*, which we will structure as a special type of tree called a *tree-decomposition*. Both of these notions are discussed in greater detail in Section 2.2.

As our algorithms entail identifying a tree-decomposition, we introduce a graph that includes all tree-decompositions, called a *tree-decomposition graph*; the tree-decomposition graph will be discussed in greater detail in Section 3.2. In order to avoid confusion, we will refer to *nodes* in the original graphs, *vertices* in a tree-decomposition, and *tdg-vertices* in the tree-decomposition graph. Terminology relevant to many of these structures is discussed in Section 2.1.

2.1 Graphs

We assume a basic familiarity with graphs and trees; the reader is referred to a standard reference [BM76] for the appropriate background. We denote the vertex and edge sets of a graph G by $V(G)$ and $E(G)$. Unless otherwise specified, all our graphs will be connected.

In the course of our algorithms, we will divide a graph G into pieces using a subset A of $V(G)$. We will denote the set of connected components of $G \backslash A$ by $C_G(A)$. For G_1 and G_2 subgraphs of G, A *separates* G_1 and G_2 if the only nodes that G_1 and G_2 have in common are in A and there is no edge in G from a node in $G_1 \backslash A$ to one in $G_2 \backslash A$. In this case, we will refer to A as a *separator*.

All trees in this paper will be rooted (and therefore directed). We will say that a vertex x is a *descendant* of a vertex y if there is a (directed) path from x to y of nonnegative length. For a directed acyclic graph G and a node a of G, we will say that node b is a *child* of node a if there is an edge from b to a in the graph.

For T a tree we will distinguish between an arbitrary connected *subgraph* of T, and a *subtree* of T consisting of a vertex and all of its descendants. We use T_v to denote the subtree of T rooted at v. In addition, we will be concerned with pieces of the tree that arise from removing a subtree from another subtree. For $v \in V(T)$ and $w \in V(T_v)$, the subgraph $T_v \backslash T_w$ denotes the subgraph obtained by removing from T_v all proper descendants of w (notice that $w \in T_v \backslash T_w$).

In addition, we will refer to (*induced*) subgraphs of a graph G. By a subgraph of G we will mean a graph G' such that $V(G') \subseteq V(G)$ and $E(G') \subseteq E(G)$. For S a subset of nodes of G, the *node-induced subgraph* (or just *induced subgraph*) of G with respect to S, or $G_{[S]}$, is the subgraph of G consisting of all nodes in S and all edges in G between two vertices in S.

2.2 Tree-decompositions and partial k-trees

In order to apply tree techniques to graphs, we represent graphs using structures first introduced by Robertson and Seymour [RS86].

Definition: Let G be a graph. A *tree-decomposition* for G is a pair (T^G, χ^G) where T^G is a tree and $\chi^G : V(T^G) \rightarrow \{$subsets of $V(G)\}$ satisfying:

1. for every $e = (u, v) \in E(G)$, there is an $x \in V(T^G)$ such that $u, v \in \chi^G(x)$; and

2. for $x, y, z \in V(T^G)$, if y is on the path from x to z in T^G then $\chi^G(x) \cap \chi^G(z) \subseteq \chi^G(y)$.

The *width* of a tree-decomposition (T^G, χ^G), $tw(T^G, \chi^G)$, is $\max\{|\chi^G(x)| - 1 : x \in V(T^G)\}$ and the tree-width of a graph G, $tw(G)$, is the minimum width over all its tree-decompositions.

When clear from context, we may drop the superscript and write (T, χ). In addition, we may refer to $\chi^H(x)$ as the *label* of x, for $x \in V(T^G)$. A tree-decomposition of a graph can be created in linear time [Bod93].

Although in the above definition, labels of vertices of T^G consist of sets of nodes in G, we will assume that for every x, $\chi^G(x)$ is an (ordered) sequence instead of a set (although we will use set notation where convenient). For $x, y \in V(T^G)$, y a child of x, suppose $a, b \in \chi^G(x) \cap \chi^G(y)$. All our tree-decompositions will have the property that a and b occur in the same relative order in both $\chi^G(x)$ and $\chi^G(y)$. Moreover, we will say that an edge is in $\chi^G(x)$ if it is an edge in the graph induced by $\chi^G(x)$.

We can extend the notion of a node-induced subgraph to a set of nodes appearing in $\chi^G(x)$ for nodes x in $V(T^G)$. Namely, for a subgraph S of T^G, we set $G_{\{S\}}$ to be the subgraph of G induced by the nodes in $\bigcup\{\chi^G(x) : x \in V(S)\}$. Then for S a connected subgraph of T^G and χ_S^G the restriction of χ^G to S, (S, χ_S^G) is a tree-decomposition of $G_{\{S\}}$.

Intuitively, a tree-decomposition gives insight into the separators of the graph. In particular, for (T^G, χ^G) a tree-decomposition of G and $x \in V(T^G)$, $\chi^G(x)$ separates the subgraphs of G induced by $(T_{c_1}^G, \chi^G), \ldots, (T_{c_\ell}^G, \chi^G), (T^G \backslash T_x^G, \chi^G)$ where $c_1, \ldots, c_\ell$ are the children of x. The following lemma is a consequence of the definition of a tree-decomposition and the discussion above. The separator is crucial to the development of our algorithms.

Lemma 2.1. *Let (T^G, χ^G) be a tree-decomposition of a graph G. Let y be a non-root vertex of T^G with parent x. Then $\chi^G(x) \cap \chi^G(y)$ separates $G_{\{T_y\}}$ from $G_{\{T \backslash T_x\}}$.*

Clearly the tree-decomposition of a graph G is not unique and in general, if G has tree-width k it has tree-decompositions of width between k and $|V(G)|$. For fixed k, there has been considerable effort devoted to finding algorithms which determine if a graph has tree-width at most k, and if so, find a tree-decomposition of bounded width.

A graph of bounded tree-width is also known as a partial k-tree [Ros74]. A k-tree is a graph from which it is possible to obtain a k-clique (K_k) by a sequence of eliminations of degree k vertices whose neighbours form a k-clique; a partial k-tree is any subgraph of a k-tree. The sequence of vertex eliminations described above is called a *perfect elimination ordering* for G.

Every graph G is a partial k-tree for some k, since G is trivially a partial $|V(G)|$-tree. The problem of determining the minimum k for which a graph is a partial k-tree is NP-complete [ACP89]. This problem is in fact equivalent to finding the minimum k for which G has tree-width k, since it can be shown that G is a partial k-tree if and only if $tw(G) = k$.

2.3 Embeddings

The problem of topological embedding is a generalization of the problem of subgraph isomorphism. Intuitively, a subgraph isomorphism algorithm determines a mapping from nodes in G to nodes in H and from edges in G to edges in H. A topological embedding algorithm instead finds a mapping from nodes in G to nodes in H and from edges in G to paths in H, such that the images of the edges in G form node-disjoint paths in H.

Alternatively, if there is a subgraph of H that is isomorphic to G, we can transform H into G by removing nodes and edges not found in that subgraph. Each node of H is either preserved by the transformation, if it is in the image of $V(G)$ under the mapping, or it is deleted. If there is a topological embedding of G in H, we can transform H into G by removing nodes and edges, and then by shrinking paths into edges. The process of shrinking a path consists of contracting edges in the path. We can view the edges in the path as being contracted from one path endpoint or the other into the middle of the path, where each time an edge is contracted, the intermediate node on the path is *collapsed* into a path endpoint. In the transformation of H into G, there are nodes which are preserved (the image of the mapping of the $V(G)$), nodes which are collapsed (interior nodes in the paths), and nodes which are deleted.

3 Adapting tree techniques

Our algorithms make use of methods developed to solve problems on trees by making use of the tree-like structure of the tree-decomposition of a graph. To provide a framework in which to discuss techniques needed to create working algorithms, we briefly describe one method of solving subtree isomorphism sequentially and then delineate the problems encountered in attempting to adapt this algorithm to more general classes of graphs. We demonstrate our techniques with respect to the subgraph isomorphism problem on k-connected partial k-trees and then show how they can be applied to other embedding problems. In Section 6 we discuss further developments used in forming parallel algorithms.

Matula [Mat78] was the first to show a polynomial time procedure for the subtree isomorphism problem. He used a dynamic programming approach to work level by level up the tree, at each step combining information using bipartite matching; this approach has been used to develop sequential and parallel algorithms for embedding problems on trees and bounded degree graphs of bounded tree-width [LK89, GKMS90, GN92, Bod88a, Bod88b]. Suppose we are given two trees T and T' and we wish to determine whether T is isomorphic to a subtree of T'. We proceed by working from leaves to root in T, finding in turn for each node v all possible mappings of T_v into T'. For a node v with children c_1 through c_k, the mappings of T_v will be determined using the previously computed mappings of T_{c_1} through T_{c_k}. Because each vertex is a separator of the tree, by identifying a node v we are immediately identifying a set of previously processed connected components, namely T_{c_1} through T_{c_k}, and a connected component yet to be processed, namely $T \backslash T_v$.

In the subtree algorithm, the structure of T can be used to determine the order in which nodes are processed. In particular, the rooting of a tree provides an orientation, making it meaningful to talk about processing children before a parent. This orientation is particularly important as information gained in processing children is used in processing a parent.

Viewed as a problem on graphs of tree width one, we can view the processing of nodes as taking place simultaneously in the tree-decompositions of the source tree T and the target tree T'. Since the trees themselves can act as tree-decompositions, we have the tree-decompositions of the source and target graphs as inputs.

When our inputs are graphs of tree-width greater than one, however, we no longer have as input canonical tree-decompositions of the two graphs. Although we can easily obtain tree-decompositions of the graphs (as discussed in Section 2), tree decompositions are not necessarily unique and isomorphic graphs may not be characterized by isomorphic tree-decompositions. We can fix a tree decomposition for one of the two graphs and root it to give an orientation to that graph, but it is not clear how this orientation can be used in the other graph. It is in fact this problem that seems to induce "NP-completeness" in the general case for subgraph isomorphism of partial k-trees. We will see in Section 4, that we can use the fact that either the source graph is of bounded degree or k-connected to solve this problem. In the remainder of this section, we introduce tools created to address these problems, namely normalized tree-decompositions and tree-decomposition graphs.

3.1 Normalized tree-decompositions

In this section, we define a special type of tree-decomposition needed for the subgraph isomorphism algorithm. We will say that (T^H, χ^H) is a *normalized* tree-decomposition if the vertices of T^H alternate between *separator vertices* and *clique vertices* such that the root of T^H is a separator vertex, the leaves of T^H are clique vertices, and the following properties hold. Each child y of a particular separator vertex x (with label of size at most k) will have a label consisting of all the nodes in the label of x, in order, with a node, say a, appended to the end. Any child z of y will have as its label a subset of the nodes in $\chi^H(y)$; this subset must include a. To ensure that $|V(T^H)|$ is linear in the size of H, we also ensure that no siblings have the same label.

A normalized tree-decomposition roughly corresponds to a perfect vertex elimination ordering for the underlying graph completed to a k-tree, where the new node a corresponds to the node being eliminated at a particular point in the sequence. It is not difficult to transform any tree-decomposition into a normalized one.

Lemma 3.1. *Let H be a partial k-tree with n nodes. Then, there is an $O(n^2)$ algorithm which, on input H returns a normalized tree-decomposition (T^H, χ^H) of H of width k.*

3.2 Tree-decomposition graphs

As outlined above, our algorithm constructs a normalized tree-decomposition of H and then proceeds up this tree level by level attempting to construct a similar tree-decomposition of G. Here we will construct a representation graph whose vertices correspond to vertices in such a tree-decomposition of G, if one exists. We give a high-level explanation of the construction. In this section we focus on the subgraph isomorphism problem for k-connected partial k-trees; modifications needed for the other algorithms will be presented in later sections.

We construct a graph called a *tree-decomposition graph* of G, $TDG(G)$; we refer to the vertices of this graph as *tdg-vertices*. Intuitively, a tdg-vertex corresponds to a potential vertex in a tree-decomposition of G; it contains information about the structure of that vertex as well as its placement in the tree-decomposition. A tree-decomposition graph consists of two types of tdg-vertices, *separator vertices* and *clique vertices*, intuitively corresponding to the k-clique to which a new node is attached and the resulting $(k+1)$-clique, respectively, in a node-elimination ordering of a k-tree.

A separator vertex α of $TDG(G)$ is a triple $(Sep^\alpha, \sigma^\alpha, Part^\alpha)$, where Sep^α is a set of nodes of G called the *separator of* α, σ^α is a permutation on Sep^α called the *ordering on* Sep^α, and $Part^\alpha = (P_0^\alpha, P_1^\alpha, \ldots, P_r^\alpha)$ is the *partition induced by* α where P_i^α is a single connected component of $G \backslash Sep^\alpha$. We call P_0^α the *parent partition* and for $i > 0$, P_i^α a *child partition*.

The separator is a set of nodes forming the label of a vertex in a tree-decomposition of G. The ordering on Sep^α indicates the ordering on the nodes, as necessitated by the definition of a label of a tree-decomposition vertex as a sequence rather than a set of nodes. The tdg-vertices represent all possible separators and all possible orderings.

In any tree-decomposition of G with a vertex x labelled by the nodes in Sep^α (with any ordering), the subgraph of non-descendants of x and the subtrees rooted at the children of x induce subgraphs of G, each consisting of a single connected component from $C(Sep^\alpha)$ (this is a consequence of the fact that G is k-connected). For convenience, each connected component can be represented by the smallest numbered node appearing in that component. $Part^\alpha$ specifies the assignment of connected components to subgraphs of the tree-decomposition. In particular, P_0^α is the set of nodes in the connected component assigned to the subgraph of non-descendants of x and the P_i^α's for $i \neq 0$ to the subtrees rooted at the r children of x. The tdg-vertices represent all possible ways of making such assignments.

A clique vertex is defined similarly; the main difference is that in a clique vertex there must exist a node which did not exist in its parent. A clique vertex β is a triple $(Clique^\beta, \gamma^\beta, Region^\beta)$, where $Clique^\beta$ is a set of nodes in G, γ^β is a permutation on $Clique^\beta$, and $Region^\beta$ is a set of regions defined below. The set of partitions, $Region^\beta$, consists of the set R_0^β of nodes forming the parent partition of $Clique^\beta$, plus pairs (R_i^β, S_i^β), where S_i^β is $Clique^\beta$ less one node x_i (x_i not the new node) and R_i^β is G less the component of $C(S_i^\beta)$ which contains x_i.

Just as the tdg-vertices correspond to possible vertices in a tree-decomposition of G, the edges in the tree-decomposition graph correspond to potential edges in a tree-decomposition. To capture this notion, we group children of a node into clusters, such that the edge between the parent and only one of the children in a cluster can exist. We can think of the clusters as a level of OR gates between the parents and the children. Then, a clique vertex is in the jth cluster of a separator vertex (the parent) if the clique vertex contains all the nodes in the parent plus a new node in the jth partition, and if the sets of connected components match up in the appropriate way.

The requirements for a separator vertex to be in the cluster of a clique vertex are similar in general. In the case of a k-connected partial k-tree, there is only one possible choice for a separator and thus the clusters are trivial.

In order to make use of tree-decomposition graphs in our algorithm, we first establish a number of properties concerning the size and structure of the graphs.

Lemma 3.2. *Let $TDG(G)$ be the tree-decomposition graph of a k-connected partial k-tree G, and let $n = |V(G)|$. Then, $TDG(G)$ has size $O(n^{k+1})$ and can be created in time $O(n^{k+2})$.*

One can show that $TDG(G)$ is acyclic, and that for any tdg-vertex α, there is a one-to-one correspondence between the children of α and the subgraphs of G induced by the partition $Part^\alpha$.

3.3 General algorithm sketch

Since the same techniques are used in all eight algorithms discussed in this paper, the algorithms share a general structure. In this section we mention the points common to all the algorithms; in later sections we will highlight the differences between them.

Each algorithm first creates a normalized tree-decomposition of H. In the parallel algorithms, the tree-decomposition of H undergoes further processing to ensure sublinear running time.

Next, the tree decomposition graph of G is formed. Each algorithm entails searching in $TDG(G)$ for a tree-decomposition of G which mimics that of H. Since the transformation of H into G is dependent on whether the algorithm is for subgraph isomorphism or for topological embedding, the definition of $TDG(G)$ will also depend on which problem is being solved. As well, $TDG(G)$ will depend on whether we are solving the problem for G being bounded degree or k-connected.

To talk about the similarity between a tdg-vertex in $TDG(G)$ and a vertex in T^H, we define a special type of tdg-mapping between the two structures. Ultimately, we wish to show that if there exists a tdg-mapping between any tdg-vertex and the root of T^H, then we can conclude that there exists a subgraph of H isomorphic to G, in the case of the subgraph isomorphism problem, or that G can be topologically embedded in H, in the case of the topological embedding problem.

Finally, each algorithm proceeds by processing the nodes of T^H from bottom to top, at each point determining for each tdg-vertex in $TDG(G)$ whether there is a tdg-mapping from the tdg-vertex to the node of T^H. The results of this determination for a particular node of T^H are stored in a bit vector or array for that node. The size of the array and the amount of information to be stored in an entry depend on the particular problem being solved. In each case, the values in the array for a particular vertex of T^H will depend on values in the arrays at children of that vertex.

The running time of the algorithm is then the time to create a tree-decomposition of H, added to the time to normalize that tree-decomposition (and add further structure, if needed), added to the time to create $TDG(G)$, and finally added to the time to fill in each array in T^H. This last quantity will consist of the number of vertices in T^H multiplied by the time to fill in all entries in one array, in the case of a sequential algorithm, and in the case of a parallel algorithm, the depth of the newly structured T^H multiplied by the time to fill in all entries in one array.

4 Sequential subgraph isomorphism

In this section we present an $O(n^{k+4.5})$ time algorithm to determine, given a k-connected partial k-tree G and a partial k-tree H, whether H contains a subgraph isomorphic to G. We then show how this algorithm can be modified to yield an $O(n^{k+2})$ time algorithm for the subgraph isomorphism problem for G a bounded degree partial k-tree, and algorithms for topological embedding. Although there are other algorithms known for these problems [MT92], we present ours as natural generalizations of the approaches taken in solving subgraph isomorphism on trees which, in turn, generalize naturally to sequential topological embedding algorithms and parallel algorithms for all the problems under consideration.

Here we list only those details not previously discussed in the general algorithm sketch: in particular, the nature of the tdg-mapping and the bit-vector used. In order to be able to discuss the mapping of clique (respectively, separator) tdg-vertices in $TDG(G)$ to clique (respectively, separator) vertices of T^H, we need to define an isomorphism between the subgraphs of G induced by the two sets of nodes.

Definition: Given a $TDG(G)$ clique vertex β and a vertex x in T^H, we will say that β is *order preserving isomorphic to x* if $k + 1 = |Clique^\beta| = |\chi^H(x)|$ and $G_{[Clique^\beta]}$ is isomorphic to a subgraph of $H_{[\chi(x)]}$, such that the sequence of nodes of G in $Clique^\beta$ ordered by the permutation γ^β map to the nodes in $\chi^H(x)$, in order, in that isomorphism.

The definition with respect to separator vertices is analogous, and hence omitted. More generally, we require isomorphisms from subgraphs of $TDG(G)$ to subtrees of a tree-decomposition of H; the definition with respect to separator vertices is analogous.

Definition: Suppose there is a clique tdg-vertex β of $TDG(G)$ and a vertex x of T^H such that $G_{(\beta)}$ is isomorphic to a subgraph of $H_{(T_x)}$. Let f be an isomorphism between these two graphs. Then, f is a *tdg-isomorphism* from β to x if:

1. f is an order preserving isomorphism of β into x; and

2. for each child α of β there is a unique child y of x such that (the appropriate restriction of) f is a tdg-isomorphism from α to y.

Theorem 4.1. *Let G and H be partial k-trees for k greater than zero, and let G be k-connected. Let $n = |V(G)| + |V(H)|$. Then there is an $O(n^{k+4.5})$ time algorithm to determine whether or not G is isomorphic to a subgraph of H.*

Proof of Theorem:

If there is a tdg-isomorphism from any sink in $TDG(G)$ to the root of T^H, we will then conclude that there exists a subgraph of H which is isomorphic to G. With every vertex x of T^H we associate a bit vector, B_x, of length $|TDG(G)|$ with one bit for each tdg-vertex of $TDG(G)$. The bit corresponding to tdg-vertex α of $TDG(G)$ is denoted by $B_x[\alpha]$. We will set $B_x[\alpha]$ to 1 if and only if there is a tdg-isomorphism from α to x, proceeding up from the leaves of T^H. For a vertex x of T^H and tdg-vertex α of $TDG(G)$, $B_x[\alpha]$ is determined after all $B_y[\beta]$ have been determined where y is a child of x and β is any tdg-vertex of $TDG(G)$.

We now show how to determine whether or not $B_x[\alpha] = 1$. If x has no children, then $B_x[\alpha] = 1$ if and only if α is order preserving isomorphic to x, and $Part^\alpha = (P_0)$ (which implies that α has no children). This can be determined in constant time.

Now suppose that x has at least 1 child. Let $y_1, \ldots, y_m$ be the children of x and $\beta_1, \ldots, \beta_r$ be the children of α. When we process x and α, vectors for all children of x have been completed; in particular, we have obtained all tdg-isomorphism information concerning children of α and children of x. More formally, for $1 \leq i \leq m$, $1 \leq j \leq r$, $B_{y_i}[\beta_j]$ has been determined.

We construct a bipartite graph (X, Y) where there is a vertex in X for each β_i and a vertex in Y for each y_j. There is an edge from β_i to y_j if $B_{y_j}[\beta_i] = 1$. Then we set $B_x[\alpha]$ to 1 if there is a matching in (X, Y) in which all of X is used and there is an order preserving isomorphism from α to x which is consistent with the matching.

By finding tdg-isomorphisms from tdg-vertices in $TDG(G)$ to vertices in T^H, the process of filling in the entries in the bit vectors can be seen as selecting tdg-vertices that represent a tree-decomposition of G corresponding to that of H. Since there are nodes in H which do not appear in G, it is possible that there is not a one-to-one correspondence between selected tdg-vertices and vertices of T^H. This case consists of passing information from grandchildren to grandparents, which is easily accomplished within the stated time bounds.

To prove that the algorithm is correct, we can show by induction that for any α and x, there is a subgraph of $H_{\{T_x\}}$ isomorphic to $G_{(\alpha)}$ if and only if there is a tdg-isomorphism from α to x. To verify the running time, we note that determining $B_x[u]$ can be accomplished in time $O(n^{2.5})$ (due to the cost of bipartite matching). $\blacksquare$

We give a brief outline of the changes to the above algorithm for the case where G is of bounded degree but not necessarily k-connected. Recall that in the k-connected case when a separator was removed from G, each remaining piece was connected. When we created the tree-decomposition graph of G, to form a partition into parent and child partitions, it was sufficient to identify which one connected component belonged to the parent partition. Now, we have to consider the cost of assigning multiple connected components to a particular partition. Moreover, in the previous case we could assume that the size of a clique was always $k + 1$ and that of a separator was always k; in this more general case, we can no longer make such an assumption.

Aside from technical changes required in the definitions of $TDG(G)$ and tdg-isomorphism, the biggest changes occur in the analysis of the algorithm. Although there is no longer a single component per partition, the bound on the number of partitions makes the number of choices constant. Furthermore, the matching problem is now a problem that can be solved in constant time, yielding the following result:

Theorem 4.2. *Let $k, d > 0$ and let G and H be partial k-trees, G with degree at most d. Let $n = |V(G)| + |V(H)|$. Then there is an $O(n^{k+2})$ time algorithm to determine whether or not G is isomorphic to a subgraph of H.*

5 Sequential topological embedding

In an algorithm with a similar structure, the tree-decomposition graph and the array used in processing H can be modified to allow us to solve the topological embedding problem. We present only a few high-level concepts here.

As mentioned in Section 2.3, we can view the topological embedding of G in H as the determination of a subgraph of H such that each node in the subgraph is either in the image of the mapping of the nodes of G to the nodes of H, or is an intermediate node in a path to which one of the edges of G is mapped. A node of H which is the intermediate node in such a path is said to have been *collapsed* into one of the endpoints of the path; special care has to be taken to ensure that all collapsing can be detected by the algorithm.

To account for collapsed nodes of H, we modify $TDG(G)$ to include tdg-vertices which mimic the structure of the section of T^H involving the addition of a node which is collapsed. In particular, a collapsed node may be the new node added in a clique in T^H; to account for this possibility, we allow a separator node α to have as a child a clique node β such that $Sep^\alpha = Clique^\beta$.

To determine whether or not a tdg-vertex α can tdg-map to a vertex x of T, we must be able to map each node in Sep^α (or $Clique^\alpha$) into a node in $\chi^H(x)$ and each edge in $G_{[Sep^\alpha]}$ (or $G_{[Clique^\alpha]}$) into a path in H. Due to the presence of collapsed nodes in H, particular care must be taken in the combining of mappings obtained for children of x. The arrays at each x hold information sufficient to allow the correct combination of children mappings, with several entries for each α.

For each x and α, the array contains entries for each partition of the edges in $G_{[Sep^\alpha]}$ (or $G_{[Clique^\alpha]}$) into the sets $Above$, $Below$, and $Inside$. An edge is in the set $Below$ if the entire path to which that edge is mapped is contained in $H_{\{T_y\}}$ for some child y of x. If an edge is in the set $Inside$, the edge can be mapped to a path in $H_{\{T_x\}}$, of which at least one edge is contained in $H_{[\chi(x)]}$. $Above$ contains all edges that require the use of edges in $H_{\{T \setminus T_x\}}$. When filling in the array for x using the arrays for the children of x, it is necessary to be sure that the partitions are consistent, namely that each edge appears somewhere in H.

Since the edges in G must be mapped to node-disjoint paths in H, we must keep track of the collapsed nodes to ensure that each is used in no more than one path. The detection of a path in H can be seen as occurring in several states. Suppose the edge (u, v) in G is mapped to the path $p_1, \ldots, p_d, \ldots p_a, q_1, \ldots, q_b, r_1, \ldots, r_e, \ldots, r_c$ in H. First u is mapped to a node p_i, say p_d, when tdg-vertex α_1 is tdg-mapped to vertex x_1 in T^H. Then, in a series of ancestors of α_1 tdg-mapping into a series of ancestors of x_1, p_d is collapsed into p_{d+1}, p_{d+1} is collapsed into p_{d+2}, and finally p_{a-1} is collapsed into p_a. Similarly, v is tdg-mapped to some r_j, say r_e, when tdg-vertex α_2 is tdg-mapped to vertex x_2 in T^H, and subsequently r_e is collapsed into r_{e-1} and so on until r_2 is collapsed into r_1. There is an ancestor β of α_1 and α_2 that maps into an ancestor y of x_1 and x_2, such that y contains both p_a and r_1. In the mapping of β to y, q_1 through q_f are collapsed into p_a and q_{f+1} through q_b are collapsed into r_1 to complete the path. It is not difficult to see that the segments p_1 through p_{d-1} and r_{e+1} through r_c are determined through a similar series of collapses.

The above example contains the two ways in which paths are determined: the collapsing of a series of nodes occurs when the edge is in the set $Above$ and the mapping of β to y when the edge is in the set $Inside$. When an edge is in the set $Inside$, it suffices to map that edge into a node-disjoint path. In terms of the definition below, f is the mapping of u to p_a and v to r_1, and g is the mapping which collapses the q_i's into p_a and r_1.

Definition: Given a $TDG(G)$ vertex α and a vertex x in T^H, we will say that α *embeds in x subject to Inside, f, and g*, for $Inside$ a subset of the edges in $G_{[Clique^\alpha]}$ (or $G_{[Sep^\alpha]}$), if and only if the graph with vertex set $Clique^\alpha$ (or Sep^α) and edge set $Inside$ is topologically embeddable in $H_{[\chi(x)]}$ with respect to (f, g).

Handling the situation in which an edge (u, v) is in the set $Above$ requires more book-keeping. As we saw in the previous example, the node-disjoint path to which (u, v) is mapped may be partially determined by a series of collapses. An intermediate node on the path can be collapsed into another intermediate node or into an endpoint; moreover, it is also possible for an endpoint to be collapsed into an intermediate node. If there is a function f governing the mapping from α to x and another function f' governing the mapping from a child of α to a child of x, then either $f(u)$ is collapsed into $f'(u)$ or $f'(u)$ is collapsed into $f(u)$. It is important to be able to distinguish which is the endpoint and which an intermediate node. Moreover, in ensuring that the mappings chosen at a particular α and x are consistent with mappings chosen at children of α and x, we must be sure that the image of f and the nodes collapsed by g are disjoint from each other as well as from the sets of nodes collapsed in children of α and x. As well, the collapsed nodes in this mapping must have degree two with respect to the embedding thus far, so that the collapsing constitutes the mapping of an edge to a path, rather than to a more complicated structure.

To make the consistency checks possible, we keep track of which nodes in $\chi^H(x)$ were collapsed in descendants. In determining whether or not these sets of nodes are disjoint for various children, it is only necessary to consider nodes in $\chi^H(x)$. That is, since T^H is a tree decomposition, we know that the only nodes that could possibly be contained in both $H_{\{T_y\}}$ and $H_{\{T_z\}}$ are those in $\chi^H(y) \cap \chi^H(z)$, which must be contained in $\chi^H(x)$, for y and z children of x.

In addition, at a particular point in the mapping we record the *path degrees* of the nodes in $\chi^H(x)$ with respect to the mapping thus far. The path degree of a node is the number of partially-created paths for which the node is a current endpoint. For our purposes, it will suffice to know whether the number is 0, 1, or more than 1. It is important to note that although in the course of collapsing nodes a particular node u in G may be mapped to many different nodes in H, only one of the nodes in H will be the image of the node in G with respect to the topological embedding of G in H. That single node will have degree at least as great as the degree of u in G; any other node to which u is mapped must be an intermediate node in a path, and hence must have degree two in $H\backslash\{$nodes not in the image of f or in the domain of $g\}$. If the current endpoint is known not to be the image of u (in particular, we have collapsed a high-degree node into it), we mark it with a path degree of -1. We can then use the stored path degrees to ensure that only legal collapsings take place.

By appropriate use of the above concepts and those developed for the subgraph isomorphism problem, we obtain the following results:

Theorem 5.1. *Let $k, d > 0$ and let G and H be partial k-trees, where G is k-connected. Let $n = |V(G)| + |V(H)|$. Then there is an $O(n^{k^2+3k+6.5})$ time algorithm to determine whether or not G can be topologically embedded in H.*

Theorem 5.2. *Let $k, d > 0$ and let G and H be partial k-trees, G with degree at most d. Let $n = |V(G)| + |V(H)|$. Then there is an $O(n^{k+2})$ time algorithm to determine whether or not G can be topologically embedded in H.*

6 Parallelizing the algorithms

The overall complexity of a parallel algorithm depends in various ways on the choice of tree-decomposition algorithm used. Among the best known tree-decomposition algorithms, there are tradeoffs between running time, processor count, depth of the resulting tree-decomposition, and width of the resulting tree-decomposition. Since our algorithms for the k-connected case depend on a resulting tree-decomposition of width k and since the number of processors needed to form a tree decomposition graph is a function of the width of the tree decomposition, we cannot make use of either Lagergren's algorithm, which using $O(n)$ processors and $O(\log^3 n)$ time to produce a tree-decomposition of width $6k + 5$ [Lag90], or Reed's algorithm, which uses $n/\log n$ processors and $O(\log^2 n)$ time to obtain a tree-decomposition of width $3k$ [Ree92]. Instead we would opt to use either the fastest known parallel tree-decomposition algorithm, running in time $O(\log n)$ and using $O(n^{3k+4})$ processors [Bod88b], or an earlier algorithm using $O(n^{2k+5})$ processors and $O(\log^2 n)$ time [CH88] each producing a width k tree decomposition.

The choice of tree-decomposition algorithm is further complicated by the importance of the depth of the resulting tree-decomposition on the running time of the subgraph isomorphism algorithm. As the tree-decomposition is processed level by level, the depth is a multiplicative factor in the running time. To obtain a sublinear running time, we could use a result of Bodlaender, who obtains a logarithmic depth tree decomposition of width $3k+2$ in $O(\log n)$ time using $O(n^{3k+4})$ processors [Bod88b], resulting again in the problems associated with a decomposition of width greater than k.

We instead consider a type of algorithm slightly more complicated than the straightforward one described above. Instead of reducing the depth of the tree-decomposition of H, we instead apply a technique developed by Jordan [Jor69] and Brent [Bre74] which, in a logarithmic number of iterations, breaks T^H into constant size pieces. Full details of this technique can be found in an earlier paper on the subject [GN92]. The key idea is that we can recursively divide a tree into smaller subgraphs using Brent restructuring; information about such restructuring can be captured by a *Brent tree of T*. In particular, we transform a normalized tree-decomposition of H into a Brent tree of logarithmic depth, allowing us to process in parallel all Brent-vertices at a particular level (representing subgraphs of T^H). Details in the expanded technical report include parallel algorithms for the normalization of a tree decomposition, formation of a Brent tree, and creation of a tree-decomposition graph, as well as methods for combining results obtained for subgraphs of T^H into a result for all of T^H. The following theorems are obtained; we assume the existence of a tree-decomposition so that we can demonstrate the complexities of our algorithms independent of their being dominated by the current best parallel tree-decomposition algorithms.

Theorem 6.1. *For $k > 0$, let H be a partial k-tree and let G be a k-connected partial k-tree. Let $n = |V(G)| + |V(H)|$. Then there is an $O(\log^3 n)$ time, $O(n^{max\{3k+4, 2k+7.5\}})$ processor randomized CRCW PRAM algorithm to determine whether or not G is isomorphic to a subgraph of H.*

Theorem 6.2. *For $k > 0$, let H be a partial k-tree and let G be a bounded degree partial k-tree. Let $n = |V(G)| + |V(H)|$. Then there is an $O(\log n)$ time, $O(n^{3k+4})$ processor CRCW PRAM algorithm to determine whether or not G is isomorphic to a subgraph of H.*

Theorem 6.3. *For $k > 0$, let H be a partial k-tree and let G be a k-connected partial k-tree. Let $n = |V(G)| + |V(H)|$. Then there is an $O(\log^3 n)$ time, $O(n^{k^2+4k+9.5})$ processor randomized CRCW PRAM algorithm to determine whether or not G can be topologically embedded in H.*

Theorem 6.4. *For $k > 0$, let H be a partial k-tree and let G be a bounded degree partial k-tree. Let $n = |V(G)| + |V(H)|$. Then there is an $O(\log n)$ time, $O(n^{3k+4})$ processor CRCW PRAM algorithm to determine whether or not G can be topologically embedded in H.*

We note that since the tree decomposition algorithms are written for CRCW PRAMs, our running times assume concurrent writing. Since the algorithms depend on matching, for which in the general case RNC but not NC algorithms are known, the algorithms for the k-connected case are randomized and those for the bounded degree case are deterministic.

7 Conclusions and open problems

We have presented algorithms for subgraph isomorphism and topological embedding on two classes of partial k-trees. Since these problems are NP-complete for general partial k-trees, it is evident that the result cannot be fully generalized. However, it is possible that some degree of generalization might be possible; this is left as an open question. Moreover, although the results in this paper delineate the boundary between the tractable and the intractable with respect to these embedding problems for classes of partial k-trees, there is still the need for the delineation of such a boundary for other classes of graphs with respect to these problems.

Subgraph isomorphism is only one of many problems known to have a polynomial time solution for trees but to be NP-complete for general graphs. It might be possible to attempt to apply to other problems of this type the methods developed in this paper for adapting tree techniques to techniques for general graphs.

The search for polynomial time algorithms for classes of partial k-trees is itself a topic of interest. It would be instructive to determine for what other problems a bound on degree or a restriction to k connectedness would yield a more efficient algorithm than in the general case.

References

[ACP89] S. Arnborg, D. Corneil, and A. Proskurowski, "Complexity of finding embeddings in a k-tree," *SIAM Journal of Algebraic and Discrete Methods* **8**, pp. 277-284, 1987.

[ALS91] S. Arnborg, J. Lagergren, and D. Seese, "Problems easy for tree-decomposable graphs," *Journal of Algorithms* **12**, 2, pp. 308ff, 1991.

[Bod88a] H. Bodlaender, "Dynamic programming on graphs with bounded treewidth," *Proceedings of the 15th International Colloquium on Automata, Languages and Programming*, pp. 105-118.

[Bod93] H. Bodlaender, "A linear time algorithm for finding tree-decompositions of small treewidth," *Proceedings of the 25th Annual ACM Symposium on the Theory of Computing*, pp. 226-234, 1993.

[Bod88b] H. Bodlaender, "NC-algorithms for graphs with bounded tree-width," Technical Report RUU-CS-88-4, University of Utrecht, 1988.

[Bod90] H. Bodlaender, "Polynomial Algorithms for Graph Isomorphism and Chromatic Index on Partial k-trees," *Journal of Algorithms* **11**, pp. 631-643, 1990.

[BM76] J. Bondy, and U.S.R. Murty,*Graph Theory with Applications*, North-Holland, 1976.

[Bre74] R. Brent, "The parallel evaluation of general arithmetic expressions," *Journal of the ACM* **21**, 2, pp. 201-206, 1974.

[CH88] N. Chandrasehkaran and S.T. Hedetniemi, "Fast parallel algorithms for tree decomposition and parsing partial k-trees," *Proceedings 26th Annual Allerton Conference on Communication, Control, and Computing*, 1988.

[GJ79] M. R. Garey and D. S. Johnson, *Computers and Intractability: A Guide to the Theory of NP-completeness*, Freeman, San Francisco, 1979.

[GKMS90] P. Gibbons, R. Karp, G. Miller, and D. Soroker, "Subtree isomorphism is in random NC," *Discrete Applied Mathematics* **29**, pp. 35-62, 1990.

[GN92] A. Gupta and N. Nishimura, "The parallel complexity of tree embedding problems," *to appear in Journal of Algorithms*. A preliminary version has appeared in *Proceedings of the Ninth Annual Symposium on Theoretical Aspects of Computer Science*, pp. 21-32, 1992.

[GN93] A. Gupta and N. Nishimura, "Sequential and parallel algorithms for embedding problems on classes of partial k-trees," Technical Report CS-93-55, University of Waterloo, December 1993.

[Jor69] C. Jordan, "Sur les assemblages de lignes," *Journal Reine Angew. Math.* **70**, pp.185-190, 1869.

[Lag90] J. Lagergren, "Efficient parallel algorithms for tree-decompositions and related problems," *Proceedings of the 31st Annual IEEE Symposium on the Foundations of Computer Science*, pp. 173-181, 1990.

[Lin89] A. Lingas, "Subgraph isomorphism for biconnected outerplanar graphs in cubic time," *Theoretical Computer Science* **63**, pp. 295-302, 1989.

[LK89] A. Lingas and M. Karpinski, "Subtree isomorphism is NC reducible to bipartite perfect matching," *Information Processing Letters* **30** pp. 27–32, 1989.

[LS88] A. Lingas and M. M. Syslo, "A polynomial-time algorithm for subgraph isomorphism of two-connected series parallel graphs," *Proceedings of the 15th International Colloquium on Automata, Languages, and Programming* pp. 394–409, 1988.

[MT92] J. Matoušek and R. Thomas, "On the complexity of finding iso- and other morphisms for partial *k*-trees," *Discrete Mathematics* **108**, pp. 343–364, 1992.

[Mat78] D. Matula, "Subtree isomorphism in $O(n^{5/2})$," *Annals of Discrete Mathematics* **2**, pp. 91–106, North-Holland, 1978.

[Ree92] B. Reed, "Finding approximate separators and computing tree width quickly," *Proceedings of the 24th ACM Symposium on the Theory of Computing*, pp. 221–228, 1992.

[Ros74] D.J. Rose, "On simple characterization of k-trees," *Discrete Mathematics* **7**, pp. 317–322, 1974.

[RS84] N. Robertson and P. Seymour, "Graph Minors III. Planar tree-width," *Journal of Combinatorial Theory (Ser. B)* **36**, pp. 49–64, 1984.

[RS86] N. Robertson and P. Seymour, "Graph Minors II. Algorithm aspects of tree-width," *Journal of Algorithms* **7**, pp. 309–322, 1986.

[Sys82] M. M. Syslo, "The subgraph isomorphism problem for outerplanar graphs," *Theoretical Computer Science* **17**, pp. 91–97, 1982.

On Intersection Searching Problems Involving Curved Objects

Prosenjit Gupta[1]   Ravi Janardan[1]   Michiel Smid[2]

Abstract

Efficient solutions are given for the problem of preprocessing a set of linear or curved geometric objects (e.g. lines, line segments, circles, circular arcs, d-balls, d-spheres) such that the ones that are intersected by a curved query object can be reported (or counted) quickly. The problem is considered both in the standard setting (where one is interested in all the objects intersected) and in a generalized setting (where the input objects come aggregated in disjoint groups and one is interested in the disjoint groups that are intersected). The solutions are based on geometric transformations, simplex compositions, persistence, and, for the generalized problem, on a method to progressively eliminate groups that cannot possibly be intersected.

1 Introduction

In a generic instance of an *intersection searching problem* a set, S, of n geometric objects must be preprocessed, so that the k objects of S that are intersected by a query object q can be reported (or counted) efficiently. Examples of S and q are points, lines, line segments, rays, hyperplanes, and simplices. These problems are rich in applications and space and query-time-efficient solutions are known for many of them [Ede87].

Most previous work on these problems assumes that the input objects and the query object are linear or piecewise-linear. To our knowledge, the case where the input and/or the query are curved has been investigated systematically only in [AvKO93, KOA90, Sha91, AM92]. In [AvKO93], searching on curved objects (e.g., circles, disks, circular arcs, Jordan arcs) with linear query objects (e.g., lines, line segments, halfspaces, rays) is considered. In [KOA90] (resp. [Sha91]), the input is a set of disks and the query is a line or line segment (resp. a point).

---

[1]Department of Computer Science, University of Minnesota, Minneapolis, MN 55455, U.S.A. Email: {pgupta, janardan}@cs.umn.edu. The research of these authors was supported in part by NSF grant CCR–92–00270.

[2]Max-Planck-Institut für Informatik, D-66123 Saarbrücken, Germany.   Email: michiel@mpi-sb.mpg.de. This author was supported by the ESPRIT Basic Research Actions Program, under contract No. 7141 (project ALCOM II).

In [AM92] range searching on point sets in $\mathcal{R}^d$ with ranges defined by a constant number of bounded-degree polynomials is considered.

1.1 Overview of results

In this paper, we make further contributions to intersection searching in the curved setting by presenting efficient solutions to three broad classes of problems[3]. For the first two classes below, we will discuss only the reporting version since the counting problems can be solved similarly.

The first thrust of the paper is the design of efficient solutions to the following general problem: *"Preprocess a set S of n linear objects so that the k objects that are intersected by a curved query object can be reported efficiently."* Thus, this part of our work complements the results in [AvKO93, KOA90, Sha91]. Table 1 summarizes our results.[4]

The second thrust of the paper is the following problem: *"Preprocess a set S of n curved objects so that the ones that are intersected by a curved query object can be reported efficiently."* This problem was left open in [AvKO93]. Table 2 summarizes our results.

Finally, we consider a generalization of the preceding two classes: Here S consists of n linear or curved objects and the objects come aggregated in disjoint groups. If we assign each object a color, according to the group it belongs to, then our goal is to report the distinct colors of the objects intersected by a query (rather than reporting all the intersected objects as in the case of the standard problem). Such *generalized intersection searching* problems are rich in applications and have been considered recently in [JL93, GJS93b, AvK93, GJS94] in the context of linear input and query objects. The challenge in these problems is to obtain solutions whose query times are sensitive to the output size, namely the number, i, of distinct colors intersected (not the number, k, of intersected objects, which can be much larger). Typically, we seek query times of the form $O(f(n) + i \cdot g(n))$, where $f(n)$ and $g(n)$ are polylogarithmic. We also consider the counting version, where we seek polylogarithmic query time. Tables 3 and 4 summarize our results.

1.2 Overview of techniques and contributions

Our results are based on three main approaches. The first approach transforms the curved problem at hand to a simplex range searching problem and solves the latter by suitably composing together known techniques such as partition trees, cutting trees, and spanning paths of low stabbing number. This general approach has been used in [AvKO93]. What makes this part of our work interesting is that the characterization of intersection and the appropriate transform(s) to use

---

[3]In this paper, the term "curved" means circular or circle-like objects such as circles, disks, circular arcs, annuli, d-balls, and d-spheres. A d-sphere is the boundary of a closed d-ball.

[4]Throughout, $\epsilon > 0$ is an arbitrarily small constant. Whenever ϵ appears in a query time (resp. space) bound, the corresponding space (resp. query time) bound contains a multiplicative factor which goes to ∞ as $\epsilon \to 0$.

Input objects	Query object	Space	Query time
Lines	Circular arc	$n^{2+\epsilon}$	$\log n + k$
		$n \log n$	$n^{1/2+\epsilon} + k$
Line segments	Disk	$n^{2+\epsilon}$	$\log n + k$
		$n \log n$	$n^{1/2+\epsilon} + k$
	Circle	$n^{2+\epsilon}$	$\log n + k$
		$n \log^2 n$	$n^{1/2+\epsilon} + k$
Points	Annulus	$n^{3+\epsilon}$	$\log n + k$
		$n \log n$	$\sqrt{n} \log^2 n + k$

Table 1: Summary of results for intersection reporting on linear objects with variable radius curved query objects. k denotes the output size.

Input objects	Query object	Space	Query time
d-balls	d-ball	$n \log \log n$	$n^{1-1/\lfloor(d+2)/2\rfloor}(\log n)^{O(1)} + k$
		$n^{d+2+\epsilon}$	$\log n + k$
d-spheres	d-sphere	n	$n^{1-1/(d+2)}(\log \log n)^{O(1)} + k$
		$n^{d+2+\epsilon}$	$\log n + k$
Circular arcs	Circle	n	$n^{3/4+\epsilon} + k$
		$n^{3+\epsilon}$	$\log n + k$
Circles	Circular arc	n	$n^{3/4+\epsilon} + k$
		$n^{3+\epsilon}$	$\log n + k$

Table 2: Summary of results for intersection reporting on variable radii curved objects using a variable radius curved query object.

Input objects	Query object	Space	Query time
Disks	Point	$n \log n$	
	Line	$n^{2+\epsilon}$	
			$\log n + i \log^2 n$
Line Segments	Disk	$n^2 \log n$	
Lines			
Points	Annulus	$n^3 / \log n$	

Table 3: Summary of results for generalized intersection reporting with variable-radius curved objects. i denotes the output size.

Input objects	Query object	Space	Query time
Points	Annulus	$n^4 \log^2 n$	$\log^2 n$
Lines	Circle/Disk		
Line segments	Disk	$n^4 \log n$	$\log n$
Disks			

Table 4: Summary of results for generalized intersection counting with variable-radius curved objects.

are not always apparent; indeed, in some cases, we need to apply successively more than one transform (moreover, in the correct order). In addition to using several known transforms we also introduce some new ones.

Our second approach, which we apply to the generalized reporting problems, is as follows: We start out by assuming that all colors in S are intersected and then progressively refine this estimate by identifying those colors that cannot possibly be intersected. To do this efficiently, we store the distinct colors of S in a suitable data structure and perform *standard intersection detection* tests on appropriately-defined subsets of S.

Our third approach, which we use for the generalized counting problems is as follows: We partition the plane into suitable regions within which the answer to a query is invariant, order the regions suitably, and apply persistence. However the querying strategy is quite subtle because the answer to the query is not available explicitly but rather is embedded implicitly in a region-specific total order on the input objects. We show how the search within this total order can be reduced to a generalized 1-dimensional range counting problem, which can be solved efficiently.

Thus the contribution of the paper is a uniform framework to solve efficiently a wide variety of intersection searching problems involving curved objects.

2 Preliminaries

2.1 Geometric transformations

We use the following transforms [Ede87, AvKO93]. (Some new transforms are introduced in Section 5.)

1. Let C be a circle in the plane, with center (a, b) and radius r and let $p = (c, d)$ be a point in $\mathcal{R}^2$. Define $\varphi(C)$ to be the plane $z = a(2x - a) + b(2y - b) + r^2$ in $\mathcal{R}^3$ and define $\psi(p)$ to be the point $(c, d, c^2 + d^2)$ in $\mathcal{R}^3$. $\psi(p)$ is the vertical projection of p onto the paraboloid $U : z = x^2 + y^2$. $\varphi(C)$ is the unique plane in $\mathcal{R}^3$ which intersects the paraboloid U in the vertical projection of C onto U. Note that $\psi(p)$ lies below (resp. on, above) $\varphi(C)$ iff p lies inside (resp. on, outside) C.

2. Let C be as in **1.** above and let ℓ be the non-vertical line $y = mx + c$. Let ℓ^+ (resp. ℓ^-) be the closed halfplane lying above (resp. below) ℓ.[5] Define $\mu(C)$ to be the point (a, b, r) in $\mathcal{R}^3$ and $\omega(\ell^+)$ (resp. $\omega(\ell^-)$) to be the halfspace $z \geq (mx - y + c)/\sqrt{1 + m^2}$ (resp. $z \geq (-mx + y - c)/\sqrt{1 + m^2}$) in $\mathcal{R}^3$. Then ℓ^+ (resp. ℓ^-) intersects C iff $\mu(C) \in \omega(\ell^+)$ (resp. $\mu(C) \in \omega(\ell^-)$). Moreover, ℓ intersects C iff $\mu(C) \in \omega(\ell^+) \cap \omega(\ell^-)$.

3. Let $\mathcal{F}$ be the well-known duality transform: In $\mathcal{R}^2$, $\mathcal{F}$ maps the point $p = (a, b)$ to the non-vertical line $\ell : y = ax - b$ and vice versa. In $\mathcal{R}^3$, $\mathcal{F}$ maps

---

[5]In general, if h is a non-vertical hyperplane in $\mathcal{R}^d$, $d \geq 2$, then we use h^+ (resp. h^-) to denote the closed halfspace lying above (resp. below) h.

$p = (a, b, c)$ to the non-vertical plane $\ell : z = ax + by - c$ and vice versa. Clearly, parallel lines (resp. planes) are mapped to points with the same y-coordinate (resp. x- and y-coordinate) and, moreover, p lies above (resp. on, below) ℓ iff $\mathcal{F}(p)$ lies below (resp. on, above) $\mathcal{F}(\ell)$.

2.2 Simplex composition

Let S be a set of n geometric objects in $\mathcal{R}^d$. Let D be a data structure for some query problem on S, with space and query time bounds $O(f(n))$ and $O(g(n))$, respectively. Suppose that we wish to now solve our query problem w.r.t. a subset S' of S satisfying some condition. Assume that S' can be specified by putting S in 1-1-correspondence with a set $\mathcal{P}$ of points in $\mathcal{R}^d$ and letting S' correspond to the subset $\mathcal{P}'$ of $\mathcal{P}$ lying in a query simplex. How (and how fast) can we solve the query problem on S'? In [vK92], van Kreveld investigates this problem, which he calls *a simplex composition on $\mathcal{P}$ to D*, and proves the following result, which we shall use extensively:

Theorem 2.1 [vK92] *Let S, D, and $\mathcal{P}$ be as above. For an arbitrarily small constant $\epsilon > 0$, simplex composition on $\mathcal{P}$ to D yields a data structure (i) of size $O(n^\epsilon(n^d + f(n)))$ and query time $O(\log n + g(n))$, or (ii) of size $O(n + f(n))$ and query time $O(n^\epsilon(n^{1-1/d} + g(n)))$, or (iii) of size $O(m^\epsilon(m + f(n)))$ and query time $O(n^\epsilon(g(n) + n/m^{1/d}))$, for any $n \le m \le n^d$, assuming $f(n)/n$ is nondecreasing and $g(n)/n$ is nonincreasing.* $\square$

3 Intersection searching on lines

We show how to preprocess a set S of n lines in the plane so that the ones that are intersected by a variable-radius circular arc can be or reported efficiently. We first give a solution for the case where the query arc is a circle $C = C(q, r)$, with center $q = (a, b)$ and radius r. Let $\ell : y = mx + c$ be any line in S.

Lemma 3.1 *C intersects ℓ iff (i) $(a, b) \in \ell^+$ and $(a, b, r) \in \omega(\ell^-)$ or (ii) $(a, b) \in \ell^-$ and $(a, b, r) \in \omega(\ell^+)$.* $\square$

Consider lines ℓ satisfying condition (i) of the Lemma (the discussion is symmetric for those satisfying condition (ii)). We need to (a) find those lines ℓ which lie below (a, b) and (b) among these lines find those lines ℓ' for which (a, b, r) lies in the halfspace $\omega(\ell'^-)$. By duality, condition (a) clearly reduces to a halfplane range search. Condition (b) reduces to a halfspace range search in $\mathcal{R}^3$ if we use $\mathcal{F}$ to map $\omega(\ell^-)$ to a point in $\mathcal{R}^3$ for each line ℓ satisfying (a) and map (a, b, r) to a plane in $\mathcal{R}^3$. We preprocess the points $\mathcal{F}(\omega(\ell^-))$ into a data structure D for halfspace range searching given in [AHL90]. Using the notation of Section 2.2, we have $f(n) = O(n \log n)$ and $g(n) = O(\log n)$. Applying a halfplane composition based on Theorem 2.1(ii) (resp. Theorem 2.1(i)) to D, (with $\mathcal{P} = \{\mathcal{F}(\ell) : \ell \in S\}$ and $d = 2$), we get:

Lemma 3.2 *A set S of n lines in the plane can be preprocessed into a data structure of size $O(n^{2+\epsilon})$ (resp. $O(n \log n)$) such that the k lines that are intersected by any variable-radius query circle, can be reported in time $O(\log n + k)$ (resp. $O(n^{1/2+\epsilon} + k)$).* □

Now we consider the general case. Let γ be a variable-radius query arc. Let $circ(\gamma)$ denote the circle that γ is a part of, let γ' denote the closure of $circ(\gamma) - \gamma$, let $chord(\gamma)$ denote the line segment joining the endpoints a and b of γ, and let $center(\gamma)$ and $radius(\gamma)$ denote, respectively, the center and radius of $circ(\gamma)$.

Lemma 3.3 [AvKO93] *A line ℓ intersects a circular arc γ iff (i) ℓ separates the endpoints of γ, or (ii) both endpoints of γ and the circular arc γ' lie on the same side of ℓ, and ℓ intersects $circ(\gamma)$.* □

Lines ℓ satisfying condition (i) of Lemma 3.3 must intersect $chord(\gamma)$ and so can be handled as follows: By duality, ℓ intersects $chord(\gamma)$ iff $\mathcal{F}(\ell)$ lies in the doublewedge, W, formed by $\mathcal{F}(a)$ and $\mathcal{F}(b)$. In [CW89] it is shown how to report the points in a fixed-size convex polygon in $O(n)$ space and $O(\sqrt{n} \log n + k)$ time. We simply apply this query twice, once for each wedge of W. Alternatively, we can use two halfplane compositions based on Theorem 2.1(i) to solve the problem in $O(n^{2+\epsilon})$ space and $O(\log n + k)$ query time.

For condition (ii), we need to (a) find those lines ℓ such that $\overline{ab}$ and γ' are on the same side of ℓ and (b) among these lines find those that intersect $circ(\gamma)$. Wlog assume that γ lies above $\overline{ab}$. Clearly, a and b must both lie below any line ℓ satisfying condition (a) above. That is, by duality, $\mathcal{F}(\ell)$ must lie in $\mathcal{F}(a)^- \cap \mathcal{F}(b)^-$. Thus condition (a) reduces to two halfplane compositions.

Since condition (b) can be tested as in Lemma 3.2, we can find lines satisfying condition (ii) by applying two halfplane compositions to the structure of Lemma 3.2. Applying Theorem 2.1(i) (resp. Theorem 2.1(ii)), we get:

Theorem 3.1 *A set S of n lines in the plane can be preprocessed into a data structure of size $O(n^{2+\epsilon})$ (resp. $O(n \log n)$), so that given any variable-radius query circular arc γ, we can report the k lines intersected by γ in $O(\log n + k)$ (resp. $O(n^{1/2+\epsilon} + k)$) time.* □

4 Intersection searching on line segments

Here S is a set of line segments in the plane and we wish to report the ones intersected by a variable-radius query circle C. For a segment s, we denote by $strip_s$ the region between the two lines that pass through the endpoints of s and are perpendicular to s. To simplify the discussion, we ignore the degenerate case where a segment endpoint lies on C.

Lemma 4.1 *Let $int(C)$ be the interior of the closed disk bounded by a circle C. Then C intersects a segment $s = \overline{ab}$ iff at least one of the following is true:*

(i) int(C) *contains exactly one endpoint of s, or (ii)* center$(C) \in$ strip$_s$ *and C intersects the supporting line, ℓ_s, of s.* $\square$

We determine the segments satisfying condition (i) of the Lemma as follows: We map C to the plane $H = \varphi(C)$ and map each segment $s = \overline{ab}$ to a segment $s' = \overline{\psi(a)\psi(b)}$ in 3-space. Now if $s = \overline{ab}$ is such that $a \in int(C)$ and $b \in ext(C)$, then we have $\psi(a) \in H^-$ and $\psi(b) \in H^+$. To report such segments, we build a spanning path Π of stabbing number $O(\sqrt{n})$ (w.r.t. disks) on the endpoints of the segments in S and then build an s-tree T on Π. (For background see [CW89].) For any node $v \in T$, let $S'(v) = \{\psi(p) \mid p \in S(v)\}$. At v, we store an instance, $I(v)$, of the halfspace reporting structure given in [AHL90]. $I(v)$ is constructed on a set $S''(v)$ of points that is obtained by including for each point $\psi(p) \in S'(v)$, the other endpoint of the segment of which $\psi(p)$ is an endpoint. To answer a query, we first find the set V_{H^+} of so-called canonical nodes w.r.t. H^+. At each node $v \in V_{H^+}$, we apply a halfspace reporting query using H^- and for each point $\psi(p)$ found, we report the segment $s \in S$ having p as an endpoint.

For condition (ii), let us first consider how to report the segments $s \in S$ such that center$(C) \in$ strip$_s$. By applying the transform $\mathcal{F}$, we dualize each strip to a vertical line segment and dualize center(C) to a line. Thus the problem is equivalent to reporting the vertical line segments that are intersected by $\mathcal{F}(\text{center}(C))$. This problem is solvable via two halfplane compositions. In Section 3, we already saw how to report the segments $s \in S$ such that ℓ_s is intersected by C. Thus condition (ii) of Lemma 4.1 can be handled by applying two halfplane compositions to the structure of Lemma 3.2.

Theorem 4.1 *A set S of n line segments in the plane can be preprocessed into a data structure of size $O(n\log^2 n)$ (resp. $O(n^{2+\epsilon})$) such that the k segments intersecting a variable radius query circle C, can be reported in time $O(n^{1/2+\epsilon} + k)$ (resp. $O(\log n + k)$).* $\square$

5 Querying curved objects with curved objects

5.1 Querying d-spheres with a d-sphere

By a *d-sphere*, S_i, we mean the boundary of the closed d-ball B_i, with radius r_i and center $C_i = (b_{i1}, b_{i2}, \ldots, b_{id})$, $1 \le i \le n$. (For example, a 2-sphere is a circle.) Let $C_Q = (q_1, q_2, \ldots, q_d)$ be the center of Q and r_Q its radius. Q intersects S_i iff the Euclidean distance between C_i and C_Q is at most $r_i + r_Q$ and at least $|r_i - r_Q|$, i.e., Q intersects S_i iff $(r_i - r_Q)^2 \le \sum_{j=1}^{d}(b_{ij} - q_j)^2 \le (r_i + r_Q)^2$. Consider the following transformations, α and β, which map Q to a hyperplane in $\mathcal{R}^{d+2}$ and the transformation τ which maps S_i to a point in $\mathcal{R}^{d+2}$.

$$\alpha(Q): x_{d+2} = 2q_1 x_1 + \cdots + 2q_d x_d + 2r_Q x_{d+1} - q_1^2 - \cdots - q_d^2 + r_Q^2.$$

$$\beta(Q): x_{d+2} = 2q_1 x_1 + \cdots + 2q_d x_d - 2r_Q x_{d+1} - q_1^2 - \cdots - q_d^2 + r_Q^2.$$

$$\tau(S_i) = (b_{i1}, \ldots, b_{id}, r_i, b_{i1}^2 + \cdots + b_{id}^2 - r_i^2).$$

It is easily verified that Q intersects S_i iff $\tau(S_i) \in \alpha(Q)^- \cap \beta(Q)^+$. The points $\tau(S_i)$ satisfying this condition can be found via two halfspace compositions. From Theorem 2.1 we conclude:

Theorem 5.1 *Let* $S = \{S_1, S_2, \ldots, S_n\}$ *be a collection of d-spheres in* $\mathcal{R}^d$, $d \geq 2$, *of possibly different radii.* S *can be preprocessed into a data structure of size* $O(n)$ *(resp.* $O(n^{d+2+\epsilon})$*) such that the d-spheres that are intersected by a variable-radius d-sphere,* Q, *can be counted or reported in time* $O(n^{1-1/(d+2)}(\log\log n)^{O(1)} + k)$ *(resp.* $O(\log n + k)$*). Here* $\epsilon > 0$ *is an arbitrarily small constant and* k *is the number of intersected d-spheres.* □

We remark that, for $d = 2$, the $O(n^{4+\epsilon})$ space bound given by Theorem 5.1 can be improved to $O(n^{3+\epsilon})$, without affecting the query time, by extending a result in [AS91].

5.2 Querying circular arcs with a circle

We show the following:

Theorem 5.2 *A set* S *of* n *circular arcs in the plane, of possibly different radii, can be preprocessed into a data structure of size* $O(n)$ *(resp.* $O(n^{3+\epsilon})$*) such that the* k *arcs that are intersected by a variable-radius query circle* C *can be reported in time* $O(n^{3/4+\epsilon} + k)$ *(resp.* $O(\log n + k)$*).* □

We use the notation of Section 3. Wlog assume that the arcs in S are x-monotone since any arc can be decomposed into at most three x-monotone pieces. Let $l(\gamma)$ and $r(\gamma)$ be γ's left and right endpoints, respectively. Let $disk(C)$ be the closed disk bounded by C. Let ℓ_l^{in} (resp. ℓ_l^{out}) be the halfplane bounded by the line joining $center(\gamma)$ and $l(\gamma)$ and containing (resp. not containing) γ. Define ℓ_r^{in} and ℓ_r^{out} similarly. The following lemma follows from [Pel92]:

Lemma 5.1 *A circle* C *and an* x-monotone *arc* γ *intersect iff one of the following is true: (i)* C *separates the endpoints of* γ *or (ii)* C *intersects* $circ(\gamma)$ *and either (a)* $l(\gamma) \notin disk(C)$ *and* $r(\gamma) \notin disk(C)$, *and* $center(C) \in \ell_l^{in} \cap \ell_r^{in}$ *or (b)* $l(\gamma) \in disk(C)$ *and* $r(\gamma) \in disk(C)$, *and* $center(C) \in \ell_l^{out} \cap \ell_r^{out}$. □

To handle condition (i), we apply the transforms ψ and φ to the endpoints of each γ and to C respectively and convert the problem to one of reporting those line segments $\overline{\psi(l(\gamma))\psi(r(\gamma))}$ in $\mathcal{R}^3$ whose endpoints are in opposite halfspaces of $\varphi(C)$. This can be done via two halfspace compositions in $O(n)$ (resp. $O(n^{3+\epsilon})$) space and $O(n^{2/3+\epsilon} + k)$ (resp. $O(\log n + k)$) query time (Theorem 2.1).

Consider condition (ii)(a). Using ψ and φ, the condition "$l(\gamma) \notin disk(C)$ and $r(\gamma) \notin disk(C)$" becomes "$\psi(l(\gamma))\psi(r(\gamma))$ lies above $\varphi(C)$ in $\mathcal{R}^3$". This can be expressed as two halfspace compositions in $\mathcal{R}^3$. Next, note that the wedge $\ell_l^{in} \cap \ell_l^{out}$ is of one of the three forms $t_1^+ \cap t_2^+$, $t_1^- \cap t_2^-$, or $t_1^+ \cap t_2^-$ for some lines t_1 and t_2. Thus, by applying the transform $\mathcal{F}$ to C, t_1, and t_2, the condition "$center(C) \in \ell_l^{in} \cap \ell_r^{in}$" becomes "$\mathcal{F}(center(C))$ is below or above or intersects $\overline{\mathcal{F}(t_1)\mathcal{F}(t_2)}$". Each of these three cases can be expressed as two halfplane compositions in $\mathcal{R}^2$. Similarly for condition (ii)(b).

We can now apply Theorem 2.1. As the structure D we take the $O(n)$-space structure of Theorem 5.1 (for $d = 2$) or the $O(n^{3+\epsilon})$-space structure mentioned at the end of Section 5.1 to report the circles $circ(\gamma)$ intersected by C. We then build a 4-level tree structure, where the nodes at the innermost level are augmented with instances of D and the two outermost levels apply halfspace compositions and the next two levels apply halfplane compositions. This yields Theorem 5.2. (We note that in [AS91], the counting problem is solved in $O(n^{3+\epsilon})$ space and $O(\log n)$ query time.)

6 Generalized intersection reporting

6.1 A general technique

Let S be a set of n colored geometric objects and let q be any query object. In preprocessing, we store the distinct colors in S at the leaves of a balanced binary tree CT (in no particular order). For any node v of CT, let $C(v)$ be the colors stored in the leaves of v's subtree and let $S(v)$ be the objects of S colored with the colors in $C(v)$. At v, we store a data structure $DET(v)$ to solve the following *standard detection* problem on $S(v)$: Decide whether or not q intersects any object of $S(v)$. $DET(v)$ returns "true" iff there is an intersection.

To answer a generalized reporting query on S, we do a depth-first search of CT and query $DET(v)$ with q at each node v visited. If v is a non-leaf node then we continue searching below v iff the query returns "true"; if v is a leaf, then we output the color stored there iff the query returns "true".

Theorem 6.1 *Assume that a set of n geometric objects can be stored in a data structure of size $O(M(n))$ such that it can be detected in $O(f(n))$ time whether or not a query object intersects any of the n objects. Then a set S of n colored geometric objects can be preprocessed into a data structure of size $O(M(n) \log n)$ such that the i distinct colors of the objects in S that are intersected by a query object q can be reported in time $O(f(n) + i \cdot f(n) \log n)$.* $\square$

Theorem 6.1 generalizes a method used in [GJS94] for the generalized halfspace range searching problem in $d \geq 4$ dimensions. For lack of space, we give just one application of Theorem 6.1.

6.2 An application: generalized disk stabbing

We show the following:

Theorem 6.2 *A set S of n colored disks in $\mathcal{R}^2$ can be stored in a data structure of size $O(n \log n)$ so that the i distinct colors of the disks stabbed by any query point can be reported in time $O(\log n + i \log^2 n)$.* □

We need to solve the following (standard) problem: Preprocess a set of n disks such that it can be detected if a query point lies in the union of the disks. Using the transformations φ and ψ, we map each disk D to the halfspace $\varphi(D)^-$ and the query q to the point $\psi(q)$ in $\mathcal{R}^3$. The complement of the union of the disks maps to an upper convex polytope. It is then sufficient to check if $\psi(q)$ lies in this polytope. The dual problem is to check if a query plane cuts a convex polytope. The latter problem can be solved using a data structure due to Dobkin and Kirkpatrick [DK83], which occupies $O(n)$ space and can be queried in $O(\log n)$ time. We now apply Theorem 6.1 to get Theorem 6.2.

7 Generalized intersection counting

The following problem is illustrative of the techniques we use: Let S be a set of n colored sites (points) in the plane. Let $Ann(q, r_1, r_2)$ be the variable-radius query annulus, with radius r_1 and r_2 ($r_1 \le r_2$) and center q. We wish to count the distinct colors of the sites contained in $Ann(q, r_1, r_2)$. We begin by partitioning the plane into regions that satisfy a certain distance invariance property, as specified in the following lemma:

Lemma 7.1 *Let S be a set of n sites in the plane. Let A be the arrangement of the perpendicular bisectors of the line segments joining pairs of points in S. Let f be any face of A and let p_1 and p_2 be any two points in f. Then the ordering of the sites by non-decreasing distance from p_1 is the same as the ordering of the sites by non-decreasing distance from p_2.* □

We preprocess A for fast planar point location [Ede87]. Given $Ann(q, r_1, r_2)$ we locate q in a face f of A. Let $\mathcal{E}(f) = s_1, s_2, \ldots, s_n$ be the ordering of the sites w.r.t. f, as given by Lemma 7.1. By definition of $\mathcal{E}(f)$, the sites at distance at least r_1 and at most r_2 from q (i.e., the ones in $Ann(q, r_1, r_2)$) are contiguous in $\mathcal{E}(f)$, say $s_j, \ldots, s_k$, for some $j \ge 1$ and $k \le n$. Thus, our problem becomes: "Given a sequence of colored integers $1, 2, \ldots, n$ on the real line (namely, the indices of the sites in $\mathcal{E}(f)$), count the i distinct colors of the points lying in the interval $[j, k]$." This is just an instance of the *generalized 1-dimensional range counting* problem for which a dynamic solution was given in [GJS93b], which uses $O(n \log n)$ space, has a query time of $O(\log^2 n)$ and undergoes $O(\log^2 n)$ memory modifications per update. Let D_f be the structure for face f. Thus, given j and k, we can answer our problem in $O(\log^2 n)$ time using D_f.

But how do we find j and k efficiently? We find j (and symmetrically k) in $O(\log n)$ time using binary search as follows: Note that s_j is the leftmost site in $\mathcal{E}(f)$ such that $d(q, s_j) \ge r_1$. Let T_f be a red-black tree storing the sites according to the order given by $\mathcal{E}(f)$. Let s be the site at T_f's root. If

$d(q,s) \geq r_1$ (resp. $d(q,s) < r_1$) then we visit the left (resp. right) subtree of the root recursively. Let l be the leaf where the search runs off T and let s' be the site stored at l. If the search at l branched left then j is simply the index of s' in $\mathcal{E}(f)$; otherwise, j is the index of the site stored in the inorder successor of l.

Clearly, the total space is $O(n^5 \log n)$. We can reduce this to $O(n^4 \log^2 n)$ by applying persistence to the faces of $\mathcal{A}$. The following lemma provides a suitable ordering of the faces for this purpose:

Lemma 7.2 *Let P be any planar subdivision with m vertices. There exists an ordering, T_P, of the faces of P such that T_P has length $O(m)$, consecutive faces in T_P share an edge, and T_P visits each face of P at least once.* □

For faces f and f' of $\mathcal{A}$, let $\Delta(f,f')$ denote the number of positions in which $\mathcal{E}(f)$ and $\mathcal{E}(f')$ differ. Using Lemma 7.2, we can prove:

Lemma 7.3 *Let S be a set of n sites in the plane. Let $\mathcal{A}$ be the arrangement of the perpendicular bisectors of the line segments joining pairs of sites in S. There is an ordering, $f'_1, f'_2, f'_3, \ldots, f'_t$, of the t faces of $\mathcal{A}$ such that $\sum_{i=1}^{t-1} \Delta(f'_i, f'_{i+1}) = O(n^4)$.* □

We can now use the ordering $f'_1, \ldots, f'_t$ provided by Lemma 7.3 to store all the T_f's and D_f's persistently. We build $T_{f'_1}$ and $D_{f'_1}$ and then scan $f'_2, \ldots, f'_t$. For $i \geq 2$, we determine the elements in $\mathcal{E}(f'_{i-1})$ whose ranks change in $\mathcal{E}(f'_i)$, delete them from $T_{f'_{i-1}}$ and $D_{f'_{i-1}}$ and reinsert them with their new ranks. These updates are done in a persistent way and yield $T_{f'_i}$ and $D_{f'_i}$. By Lemma 7.3, there are $O(n^4)$ updates. Moreover each update causes $O(\log^2 n)$ memory modifications ($O(1)$ for T_f and $O(\log^2 n)$ for D_f).

Theorem 7.1 *A set S of n colored points in the plane can be preprocessed into a data structure of size $O(n^4 \log^2 n)$ such that the distinct colors of the points lying inside any variable-radius query annulus can be counted in $O(\log^2 n)$ time.* □

References

[AHL90] A. Aggarwal, M. Hansen, and T. Leighton. Solving query-retrieval problems by compacting Voronoi diagrams. In *Proc. 18th Annual ACM Symposium on Theory of Computing*, pages 331–340, 1990.

[AM92] P. K. Agarwal and J. Matoušek. On range searching with semialgebraic sets. In *Proc. 17th Internat. Sympos. Math. Found. Comput. Sci.*, LNCS 629, pages 1–13. Springer-Verlag, 1992.

[AS91] P.K. Agarwal and M. Sharir. Counting circular arc intersections. In *Proc. 7th Annual Symposium on Computational Geometry*, pages 10–20, 1991.

[AvK93]   P.K. Agarwal and Marc van Kreveld. Connected component and
          simple polygon intersection searching. In *Proc. 3rd WADS*, LNCS
          709, pages 36–47. Springer-Verlag, 1993.

[AvKO93] P.K. Agarwal, Marc van Kreveld, and Marc Overmars. Intersection
          queries for curved objects. *Journal of Algorithms*, 15:229–266, 1993.

[CW89]   B. Chazelle and E. Welzl. Quasi-optimal range searching in spaces
          with finite VC-dimension. *Discrete and Computational Geometry*,
          4:467–489, 1989.

[DK83]   D.P. Dobkin and D.G. Kirkpatrick. Fast detection of polyhedral in-
          tersection. *Theoretical Computer Science*, 27:241–253, 1983.

[Ede87]   H. Edelsbrunner. *Algorithms in Combinatorial Geometry*. Springer–
          Verlag, 1987.

[GJS94]   P. Gupta, R. Janardan, and M. Smid. Efficient algorithms for gener-
          alized intersection searching on non-iso-oriented objects. To appear
          in *Proc. 10th Annual ACM Symposium on Computational Geometry*,
          1994.

[GJS93b] P. Gupta, R. Janardan, and M. Smid. Further results on generalized
          intersection searching problems: counting, reporting and dynamiza-
          tion. In *Proc. WADS 1993, LNCS 709*, pages 361–372. Springer
          Verlag, 1993. (To appear in *Journal of Algorithms*.)

[JL93]   R. Janardan and M. Lopez. Generalized intersection searching prob-
          lems. *International Journal on Computational Geometry & Applica-
          tions*, 3(1):39–69, 1993.

[KOA90]   M. Van Kreveld, M. Overmars, and P. K. Agarwal. Intersection
          queries in sets of disks. In *Proc., SWAT 1990, LNCS 447*, pages
          393–403. Springer Verlag, 1990.

[Pel92]   M. Pellegrini. A new algorithm for counting circular arc intersections.
          TR-92-010, International Computer Science Institute, Berkeley, 1992.

[Sha91]   M. Sharir. The k-set problem for arrangement of curves and surfaces.
          Discrete and Computational Geometry, 6:593–613, 1991.

[vK92]   M. van Kreveld. *New results on data structures in computational
          geometry*. PhD thesis, Department of Computer Science, University
          of Utrecht, Utrecht, the Netherlands, 1992.

Improved Approximations of Independent Sets in Bounded-Degree Graphs

Magnús M. Halldórsson   Jaikumar Radhakrishnan

School of Information Science   Theoretical Computer Science Group
Japan Adv. Inst. of Science and Tech.   Tata Institute of Fundamental Research
Ishikawa 923-12, JAPAN   Bombay, India
magnus@jaist.ac.jp   jaikumar@tifrvax.bitnet

Abstract. Finding maximum independent sets in graphs with bounded maximum degree is a well-studied NP-complete problem. We study two approaches for finding approximate solutions, and obtain several improved performance ratios.

The first is a subgraph removal schema introduced in our previous paper. Using better component algorithms, we obtain an efficient method with a $\Delta/6(1 + o(1))$ performance ratio. We then produce an implementation of a theorem of Ajtai et al. on the independence number of clique-free graphs, and use it to obtain a $O(\Delta/\log\log\Delta)$ performance ratio with our schema. This is the first $o(\Delta)$ ratio.

The second is a local search method of Berman and Fürer for which they proved a fine performance ratio but by using extreme amounts of time. We show how to substantially decrease the computing requirements while maintaining the same performance ratios of roughly $(\Delta+3)/5$ for graphs with maximum degree Δ. We then show that a scaled-down version of their algorithm yields a $(\Delta + 3)/4$ performance, improving on previous bounds for reasonably efficient methods.

1   Introduction

An *independent set* in a graph is a set of vertices in which no two are adjacent. The problem of finding an independent set of maximum cardinality is of central importance in graph theory and combinatorial optimization.

Given that the problem is NP-hard, the approach that holds the greatest promise is in developing heuristics that find high-quality approximate solutions. The performance ratio of such an algorithm is defined to be the worst-case ratio of the size of the optimal solution to the size of the algorithm's solution. In spite of considerable effort, no algorithm is known for the independent set problem with a performance ratio less than $O(n/\log^2 n)$ [6], where n is the number of vertices in the input graph. Results in recent years on interactive proof systems, culminating in the celebrated paper of Arora et al [3], show that no constant factor approximation can be expected, and in fact, a $n^{1/4}$ ratio appears out of reach [4].

Given this apparent hardness of the general problem, it is natural to ask what restrictions make the problem easier to approximate. Perhaps the most natural and frequently occurring case is when the maximum vertex degree is bounded above by a constant. Just as the independent set (or clique) problem occurs in various context when modeling pairwise conflicts among elements, the bounded-degree variant occurs naturally when key parameters of the problems are fixed.

For the bounded-degree version (B-IS), the exact problem remains NP-complete, but the approximation problem becomes considerably easier. In fact, any algorithm that finds a *maximal* independent set has a performance ratio of Δ on graphs with maximum degree Δ. This problem is also among the original MAX SNP-complete problems [11], so the results of [3] imply that there is a constant $c > 1$ for which c-approximation becomes NP-hard, thus no polynomial-time approximation schema can exists (unless $P = NP$). This naturally leads us to ask for the best possible constant within which B-IS can be approximated.

We address this question by studying two recent approaches. The former is an algorithm schema that involves removing small cliques from the graph. The idea – which originates in [6], with traces back to Erdős[7] – comes from the observation that graphs without small cliques contain provably larger independent sets than general graphs do. Moreover, these larger solutions can be found effectively. For graphs with few disjoint cliques, we can manually remove all the cliques and find the promised improved solution on the remainder. On the other hand, graphs with many disjoint cliques cannot contain a very large independent set, providing an upper bound on the optimal solution. Hence, in either case, our performance ratio will be improved.

We previously used this schema to improve the approximation of the minimum degree greedy algorithm from $(\Delta+2)/3$ to $\Delta/3.81(1+o(1))$ [8]. This time, using new analysis of a simple local search algorithm of Khanna et al. [10], we obtain a surprisingly strong $\Delta/6(1 + o(1))$ ratio.

We also use this schema to answer a tantalizing question: Given that all B-IS approximation results so far have merely improved the coefficient in front of Δ, is a $o(\Delta)$ performance ratio possible? The answer is affirmative; we present an algorithm with a $O(\Delta/\log\log\Delta)$ performance ratio. As a crucial step, we give a deterministic implementation of an existential theorem of Ajtai, Erdős, Komlós, and Szemerédi [1] on the independence number of sparse graphs containing no small cliques.

The latter approach that we consider is a method due to Berman and Fürer [5], that can be characterized as a local search algorithm that additionally searches in the complement of the current solution. Their algorithm yields excellent performance ratios on graphs of low maximum degree: $(\Delta+3)/5 + 1/h$, when Δ is even, and $(\Delta + 3.25)/5 + 1/h$ when Δ is odd, for any fixed constant h.

Unfortunately, the method is extremely time consuming. The local search neighborhood, the set of solutions searched in for an improvement, involves all solutions whose difference (or distance) from the current one is a connected graph with σ vertices. In order to obtain their result, we must search a neighborhood of size $\sigma = 32h\Delta^{4h}\log n$. That implies a search complexity of $n(4\Delta)^{\sigma} \approx n^{64h\Delta^{4h}\log\Delta}$. In particular, the dependence of the complexity on the quality of the approximation h is doubly exponential. For instance, in order to obtain a ratio of 1.6 when $\Delta = 4$, this approach requires $n^{2^{90}}$ time complexity.

We address this feasibility concern by tightening the analysis, and thereby reducing the neighborhood size requirements considerably. In particular, we obtain identical performance ratios while shrinking the neighborhood size to $4h^{2.6}\Delta \log n$, eliminating the exponential dependence on h. While still not exactly practical, this is close to the limitations of this approach.

We also observe that small neighborhoods yield surprisingly good performance ratios. Using only 2Δ size neighborhood, which can easily be implemented in $O(n^2)$ time, we obtain a $(\Delta + 3)/4$ ratio. This improves on the previous best $(\Delta + 2)/3$ ratio for practical bounded-degree independent set algorithms [8].

Notation We use fairly standard graph terminology. For the graph in question, usually denoted by G, n denotes the number of vertices, Δ maximum degree, $\overline{d}$ average degree, α independence number (or size of the maximum independent set), and τ independence fraction (or the ratio of the independence number to the number of vertices). For a vertex v, $d(v)$ denotes the degree of v, and $N(v)$ the set of neighbors of v.

For an independent set algorithm Alg, $Alg(G)$ is the size of the solution obtained by the algorithm on graph G. The performance ratio of the algorithm in question is defined by

$$\rho = \max_G \frac{\alpha(G)}{Alg(G)}.$$

2 Subgraph Removal Approach

We present a strategy for approximating independent sets, based on removing cliques. Using this idea, we first obtain an asymptotically improved $O(\Delta/\log \log \Delta)$ performance ratio, via a constructive proof of a graph theorem. We then use two practical algorithms from the literature to obtain improved bounds for graphs of intermediate maximum degree.

Generic Clique Removal Schema

We present an algorithm schema, indexed by a cardinal k and a collection of subordinate procedures. One is algorithm General-BDIS-Algorithm for finding independent sets in general (bounded-degree) graphs. The others are methods for finding independent sets in ℓ-clique free graphs, possibly one for each value of ℓ, $3 \leq \ell \leq k$.

CliqueRemoval$_k(G)$
    $A_0 \leftarrow$ General-BDIS-Algorithm(G)
    for $\ell = k$ downto 2 do
        $S \leftarrow$ CliqueCollection(G,ℓ)
        $G \leftarrow G - S$

$A_\ell \leftarrow K_\ell$-free-BDIS-Algorithm(G)
od
Output A_i of maximum cardinality
end

The algorithm CliqueCollection finds in G a maximal collection of disjoint cliques of size ℓ; in other words, S is a set of mutually non-intersecting cliques of size ℓ such that the graph $G - S$ contains no ℓ cliques. Such a collection can be found in $O(\Delta^{\ell-1}n)$ time by searching exhaustively for a $(\ell - 1)$-clique in the neighborhood of each vertex. That is polynomial whenever $\ell = O(\log_\Delta n)$.

We now present two instances of this schema.

AEKS

Ajtai, Erdős, Komlós and Szemerédi [1] proved the following result about K_ℓ free graphs.

Theorem 1 (AEKS) *There exists an absolute constant c_1 such that for any K_ℓ-free graph G,*

$$\alpha(G) \geq c_1 \frac{\log((\log \bar{d})/\ell)}{\bar{d}} n.$$

We have obtained an algorithm AEKS that constructs such an independent set in polynomial time, by derandomizing the parts of the proof of [1] where probabilistic existence arguments are used. For lack of space, its somewhat lengthy description is omitted here.

It suffices to use the following simplified algorithm to approximate independent sets. An independent set is *maximal* (MIS) if adding any further vertices to the set violates independence. An MIS is easy to find and provides a sufficient general upper bound of $n/(\Delta + 1)$.

AEKS-SR(G)
$G' \leftarrow G -$ CliqueCollection$(G, c_1 \log \log \Delta)$
return max(AEKS(G'), MIS(G))
end

Theorem 2. *The performance ratio of AEKS-SR is $O(\Delta/\log \log \Delta)$.*

Proof. Let k denote $c_1 \log \log \Delta$, and let n' denote the size of $V(G')$. The independence number collects at most one from each k-clique, for at most

$$\alpha \leq n/k + n' \leq 2 \max(n/k, n'),$$

while the size of the solution found by AEKS-SR is at least

$$\text{AEKS-SR}(G) \geq \max(\frac{1}{\Delta+1}n, \frac{k}{\Delta}n') \geq \frac{k}{\Delta+1} \max(n/k, n').$$

The ratio between the two clearly satisfies the claim. ∎

Observe that the combined method runs in polynomial time for Δ as large as $n^{1/\log \log n}$.

Effective method for moderately large maximum degree

While the clique removal method in combination with AEKS yields a good asymptotic performance ratio, Δ must be quite high for the gained $\log \log \Delta$ factor to overcome the large constants involved (implicit in the proof of [1]).

We now turn our attention to practical methods that can benefit from the clique removal schema. We present methods that combine to yield an asymptotic $\Delta/6(1 + o(1))$ performance ratio, improving on the best previous known ratios for moderate to large values of Δ. This involves an algorithm of Shearer [12] for 3-clique-free graphs, and a simple local search algorithm for other k-clique-free graphs as well as for use as the general BDIS algorithm.

2-opt. Khanna et al. [10] studied a simple local search algorithm that we have named 2-opt. Starting with an initial maximal independent set, it tries all possible ways of adding two vertices and removing only one while retaining the independence property. We say that a triple $\langle v_1, v_2, u \rangle$ is a *2-improvement* of an independent set I iff vertices v_1, v_2 are outside of I, u is in I, and adding the former two to I while removing the latter retains the independence property. Since I can be assumed to be a maximal independent set, it suffices to look at pairs adjacent to a common vertex in I.

The method is a simplification of the algorithm presented in the following section (it omits the recursive call). Using proper data structures, it can be implemented in $O(poly(\Delta)n)$ time.

The following was shown by Khanna et al [10].

Lemma 3. 2-opt $\geq \dfrac{1 + \tau}{\Delta + 2} n$.

They proved a $\Delta/2.44(1+o(1))$ performance ratio of this algorithm combined with another simple algorithm. A better bound can be obtained via a technique of Nemhauser and Trotter (see [9] for application) which effectively allows one to assume without loss of generality that the independence fraction τ is at most $1/2$. A ratio of $(\Delta + 2)/3$ then follows easily, for the combination of these two methods. But we digress.

We can get improved bounds for k-clique free graphs.

Lemma 4. *On a k-clique free graph G,* 2-opt$(G) \geq \dfrac{2}{\Delta + k} n$.

Proof. Since A is a maximal independent set, each vertex in $V - A$ must have at least one edge coming into the set A. If the graph has no k-clique, then for each $u \in A$, at most $k - 1$ vertices can be adjacent only to u and no other vertex in A. Thus, at most $(k - 1)|A|$ vertices can be adjacent to only one vertex in A. Hence, if we sum up the necessary degrees of vertices from $V - A$ into A, we find that $|A|\Delta \geq (n - |A|) + (n - (k - 1)|A|)$, which yields the lemma. ∎

Shearer. Shearer [12] proved the following theorem, improving a previous result of Ajtai, Komlós and Szemerédi [2].

Theorem 5 (Shearer [12]) *Let* $f_s(d) = (d \log_e d - d + 1)/(d-1)^2$, $f_s(0) = 1$, $f_s(1) = \frac{1}{2}$. *For a triangle-free graph* G, $\alpha(G) \geq f_s(\bar{d})n$.

Moreover, he gave a simple algorithm (which we name after him) attaining the bound, which repeatedly selects any vertex v of degree d_v satisfying

$$(d_v + 1)f_s(\bar{d}) \leq 1 + (\bar{d}d_v + \bar{d} - 2 \sum_{w \in N(v)} d(w))f_s'(\bar{d}),$$

removes it and its neighbors from the graph, and repeats until the graph is empty. Using an appropriate data structure to maintain the f-values of the vertices, the algorithm can be implemented in time $O(poly(\Delta)n)$. In fact, the claim is also obtained in fully linear time by a simple randomized greedy algorithm, that randomly selects a non-adjacent vertex in each step. We shall only need the obvious corollary that $\mathsf{Shearer}(G) \geq f_s(\Delta)n \approx n(\log \Delta)/\Delta$.

Analysis. To improve the approximation further, we apply the method of Nemhauser and Trotter on each incarnation of G. That will allow us to assume that nothing will be left after the edges (2-cliques) are removed.

We obtain the following explicit, if less than compact, bound on the performance ratio. H_k is the k-th Harmonic number.

Theorem 6. CliqueRemoval$_k$, *using* 2-opt *and* Shearer *attains a performance ratio of at most*

$$\left[\frac{\Delta}{2} + 2 + \frac{k}{2}\left(H_{k-1} + \frac{1}{3f_s(\Delta)} - \frac{3}{2} + \frac{\Delta}{3}\right)\right]/(k+1)$$

for graphs of maximum degree $\Delta \geq 5$.

Proof. Let n_t denote the number of vertices in the t-clique free graph. Thus, $n \geq n_k \geq \ldots \geq n_3 \geq n_2 \geq 0$. From applying Nemhauser-Trotter, we may assume $n_2 = 0$.

The size of the optimal solution is τn, which can be bounded by

$$\tau n \leq \frac{1}{2}(n_3 - n_2) + \cdots + \frac{1}{k}(n_{k+1} - n_k) = \sum_{i=3}^{k} \frac{1}{i(i-1)} n_i + \frac{1}{k} n. \qquad (1)$$

Our algorithm is guaranteed to output a solution of size at least

$$\max\left[\frac{1+\tau}{\Delta+2}n, \max_{4 \leq t \leq k} \frac{2}{\Delta+t} n_t, f_s(\Delta)n_3\right].$$

Thus, the performance ratio ρ attained by the algorithm is bounded by

$$\rho \leq \min\left[\frac{\tau n}{\frac{1+\tau}{\Delta+2}n}, \frac{\tau n}{\frac{2}{\Delta+t}n_t}, \frac{\tau n}{f_s(\Delta)n_3}\right].$$

From this we derive, respectively, that

$$\tau \geq \frac{\rho}{\Delta + 2 - \rho}, \tag{2}$$

$$n_t \leq \frac{\tau}{\rho} \frac{\Delta + t}{2} n, \quad t = 4, 5, \ldots, k \tag{3}$$

$$n_3 \leq \frac{\tau}{\rho} \frac{1}{f_s(\Delta)} n. \tag{4}$$

Combining (1), (4) and (3), we find that

$$\tau \leq \frac{\tau}{\rho} s_{\Delta,k} + \frac{1}{k}$$

where

$$s_{\Delta,k} = \frac{1}{6 f_s(\Delta)} + \sum_{i=4}^{k} \frac{\Delta + i}{2i(i-1)} = \frac{1}{2} \left[\frac{1}{3 f_s(\Delta)} + \left(H_{k-1} - \frac{3}{2} \right) + \Delta \left(\frac{1}{3} - \frac{1}{k} \right) \right].$$

Thus,

$$\tau \leq \frac{1}{k(1 - s_{\Delta,k}/\rho)}. \tag{5}$$

Hence, from (2) and (5)

$$\frac{\rho}{\Delta + 2 + \rho} \leq \frac{1}{k(1 - s_{\Delta,k}/\rho)}$$

which simplifies to

$$\rho \leq \frac{\Delta + 2 + k s_{\Delta,k}}{k+1}$$

and the claim follows. ∎

It is now easy to compute the ratio for particular values of Δ. Selected values are presented in section 4.

It is also easy to see that if Δ and k are assumed to be growing functions, then the $\Delta/3$ term will dominate for a $\Delta/6$ asymptotic ratio.

Corollary 7. CliqueRemoval, *using* 2-opt *and* Shearer, *attains a performance ratio of* $\Delta/6 \, (1 + o(1))$.

3 Local Search Approach

The Algorithm

The algorithm, due to Berman and Fürer [5], is a type of a local search algorithm – with a twist. The locality, or neighborhood, is the natural one: the set of solutions that differ (in terms of symmetric set difference) from the current one in only few vertices. Starting with an arbitrary solution (e.g. a maximal independent set), we search through all neighboring solutions for one that is strictly larger. This process is repeated until no further improvements can be found.

The twist to the tale is a recursive application of this method on what can be thought of as the *complement* of the solution. Once a non-improvable solution A is found, the method is applied to $\text{COMP}(A)$, defined to be the subgraph induced by nodes with at least two neighbors in A (when $\Delta = 3$, one neighbor suffices). Since the maximum degree of this subgraph is two less, the recursion ends on degree two graphs which we solve optimally. Given this new alternative solution, if larger, we make it our current solution and continue to try to improve it; otherwise, we exit and proclaim optimality under this type of search.

```
LS(G)
    if (Δ ≤ 2) return MaximumIndependentSet(G)
    A ← MaximalIndependentSet(G)
    repeat
        try all possible t-improvements
        A₂ ← LS(Comp(A))
        if (|A₂| ≥ |A|)
            A ← A₂
    until (no more improvements are found)
end
```

Paraphrasing, in each step we search for an *improvement* which is a vertex set whose symmetric difference with the current solution is larger and still independent. A em *t-improvement* adds t and removes $t-1$ vertices from the current solution. It is easy to see that it suffices to look for an improvement that induces a connected graph, and since the maximum degree is bounded by Δ, there are at most $\Delta^{2(2t-1)}n$ connected subgraphs of size $2t - 1$. The size of the locality searched, indicated by t, is therefore the crucial factor in the complexity. We say that a solution is *t-optimal* if it has no *t*-improvement nor an improvement in Comp, and call an algorithm finding such a solution *t-opt*.

Tools for analysis

We analyze the guaranteed size of any t-optimal solution. Let A refer to such a solution, and B refer to some hypothetical optimal solution. The *lace* of a vertex in B is defined to be the set of adjacent vertices in A. The gist of the analysis is to show that few vertices can have a small lace, thus limiting the size of B given the limited adjacency capacity of A that results from the degree bound.

Notation. Let C denote the intersection of A and B. Let B_1 denote the subset of B of of vertices whose laces are unit size, (i.e. adjacent to precisely one vertex of A), and let A_1 be the set of vertices of A adjacent to the vertices of B_1. Let $A_0 = A - A_1 - C$ and $B_0 = B - A_1 - C$. We further split B_0 into B_2 and B_3, namely vertices with lace of size two, and three or more, respectively.

Finally, denote the cardinalities of the above sets by their respective lower case letter. Denote by ρ^Δ the performance ratio of this algorithm on graphs of maximum degree Δ.

Simple bounds. The sum of the lace sizes can add up to at most the adjacency capacity of A. This gives us the first crucial bound.

$$b_1 + 2b_2 + 3b_3 \leq \Delta a. \tag{6}$$

The second bound is on the size of B_1. By definition, $b_1 \geq a_1$. On the other hand, a vertex in A adjacent to two vertices in B_1 forms a 2-improvement. Thus the equality is strict under 2-opt.

$$b_1 = a_1 \tag{7}$$

We shall also use it in the form of $b_1 + c = a_1 + c$.

Bounds on the number of pairs. From lemma 3.3 of [5], $4h \log n + 2$-opt guarantees

$$b_2 \leq (1 + \frac{1}{h})a_0 \tag{8}$$

This is shown by constructing a graph, whose vertex set is A_0 and whose edges are formed by the laces of nodes in B_2. At least one element of a lace must be inside A_0; those with only one endpoint in A_0 are modelled as a self-loop on that node. If the number of edges in the resulting multigraph exceeds the above bound, it can be shown that there exists an induced subgraph with at most $4h \log n - 1$ nodes containing strictly more edges. Since the graph may be assumed to be connected, it contains at most two self-loops. This then maps to a set of at most $4h \log n$ nodes in B_2, two nodes in B_1 and a corresponding set of nodes in A, that together constitute an improvement.

Bounds obtained by recursion. The recursive application provides us with a bound on b_0. If no larger solution can be found, the maximum independent set in $\text{Comp}(A)$ cannot be greater than a by a factor of more than the performance ratio for $(\Delta - 2)$-degree graphs. But, this complement – the set of vertices adjacent to at least two vertices in A – must contain B_0 (and B_1 when $\Delta = 3$). Thus we obtain:

$$b_0 \leq \rho^{\Delta-2}a \tag{9}$$

$$b_1 + b_0 \leq \rho^{\Delta-1}a. \tag{10}$$

Derivative. We now introduce the concept of a *derivative* of the partitions A_0 and B_0, similar to the ideas of [5, lemma 3.5]. Removing A_1 from the graph splits B_0 and A_0 again into two parts: $B' = B_0 = B_0' \cup B_1'$, $A' = A_0 = A_0' \cup A_1'$. This procedure can be continued, producing second and third derivatives A_0'', $A_0^{(3)}$ etc.

The important observation is that inequalities (7), (6), and (8) hold equally for derivatives, with an additional Δ factor in neighborhood size.

Lemma 8. *A t-improvement of A' implies a $t\Delta$-improvement of A.*

Proof. Consider the t vertices of B' in a t-improvement of A'. Each lace contains some vertex in $A' = A_0$ and thus at most $\Delta - 1$ vertices in A_1. The vertices in A_1 are in one-to-one correspondence with vertices in B_1, whose laces in turn do not contain any further vertices in A. The combination of the abovementioned vertices includes at most $t\Delta$ vertices in B, and one fewer in A, and thus forms a $t\Delta$-improvement of A. ∎

We need to relate the sizes of the 'unit' sets, A_1, A'_1 etc. Each vertex in B'_1 is adjacent to a particular vertex in A'_1. In addition, it must be adjacent to some vertex in A_1, since otherwise it belongs to B_1. Each vertex in A_1 is adjacent to a vertex in B_1, and thus at most $\Delta - 1$ vertices in B'_1. Hence, we obtain the useful bound:

$$b'_1 \le (\Delta - 1)a_1. \tag{11}$$

Applying the tools

Fast local search. We first give a simple proof of a $(\Delta + 3)/4$ performance ratio for 2Δ-opt, which is an improvement over previously known bounds for practical methods.

Add (7) and (6) and take the derivative:

$$b_0 \le \frac{\Delta + 1}{2} a_0. \tag{12}$$

For $\Delta = 3$, we add half (12), half (9), and once (7) to get

$$\rho^3 \le 3/2.$$

In general, we add $2/(\Delta + 1)$ times (12), $(\Delta - 1)/(\Delta + 1)$ times (9), and once (7):

$$\rho^\Delta \le 1 + \frac{\Delta - 1}{\Delta + 1} \rho^{\Delta - 2}.$$

This is bounded by $(\Delta + 3)/4$, and is slightly better for small even values of Δ.

Slow local search, faster. Let us now derive performance bounds identical to [5].

Adding (6), (8), and twice (7) yields

$$b \le \frac{\Delta + 1 + 1/h}{3} a + \left(\frac{1 - 1/h}{3}\right)a_1$$

and s-th derivative gives

$$b^{(s)} = b_1^{(s)} + b_0^{(s)} \le \frac{\Delta + 1 + 1/h}{3} a^{(s)} + \frac{1 - 1/h}{3} a_1^{(s)}.$$

That implies that by using derivatives of (7) and (11), we get

$$b_0 = b_1' + b_1'' + \ldots + b_1^{(s-1)} + b_1^{(s)} + b_0^{(s)}$$

$$\leq a_1' + a_1'' + \ldots + a_1^{(s-1)} + \frac{\Delta + 1 + 1/h}{3}a^{(s)} + \frac{1 - 1/h}{3}a_1^{(s)}$$

$$= \frac{\Delta + 1 + 1/h}{3}a_0 + \frac{1 - 1/h}{3}a_1^{(s)} - \frac{\Delta - 2 + 1/h}{3}(a_1' + a_1'' + \ldots + a_1^{(s-1)})$$

$$\leq \frac{\Delta + 1 + 1/h}{3}a_0 + \frac{1/(\Delta - 1)^{s-1} - 1/h}{3}a_1^{(s)}$$

Thus, as long as $h \leq (\Delta - 1)^{s-1}$, that is $s \geq 1 + \lceil \log h / \log(\Delta - 1) \rceil$, we obtain the essential inequality:

$$b_0 \leq \frac{\Delta + 1 + 1/h}{3}a_0. \tag{13}$$

We need s derivatives, resulting in an *improvement* requirement of

$$8h\Delta^s \log n \leq 8h\Delta h^{\log \Delta / \log \Delta - 1} \log n \leq 8h^{2.6}\Delta \log n.$$

The analysis of the performance guarantee now follows from the argument of [5, sec. 4]. Add $\frac{3}{\Delta+1}$ times (13), $\frac{\Delta-2}{\Delta+1}$ times (9) (for $\Delta = 3$, use (10)), and once (7) to obtain the recurrence:

$$\rho^\Delta \leq \left[1 + \frac{1}{h(\Delta + 1)} + \frac{\Delta - 2}{\Delta + 1}\right]\rho^{\Delta-2}$$

which yields the desired ratios of $(\Delta + 3)/5 + 1/h$ $((\Delta + 3.25)/5 + 1/h)$ for even (odd) Δ.

In conclusion:

Theorem 9. LS *attains a* $(\Delta + 3)/4$ *performance ratio in time* $O(\Delta^\Delta n^2)$, *and a* $(\Delta + 3)/5 + 1/h$ *(plus 0.25 if Δ is odd) ratio in time* $n^{O(h^{2.6}\Delta \log \Delta)}$.

4 Comparison of Results

The results presented excel at different ranges of values of Δ. The local search approach is best for small values, the $\Delta/6$-ratio from subgraph removal is best for intermediate values, while eventually for large enough values, the asymptotically superior $O(\Delta/\log\log\Delta)$ bound wins. The intermediate one beats the $(\Delta+3)/5 + 1/h$ ratio of [5] for $\Delta \geq 613$, and the $(\Delta+3)/4$ ratio for $\Delta \geq 31$. We are left with four nearly incomparable results ruling the approximability landscape. Bounds for selected values of Δ are given below:

Δ	CliqRem	[5]	2Δ-opt	[8]
10	3.54	2.60	3.25	4.00
33	8.92	7.25	9.00	11.66
100	23.01	20.60	25.75	34.00
1024	201.57	205.40	256.75	342.00
8192	1535.20	1639.00	2048.75	2731.33

An interesting future topic would be to strengthen the lower bounds; currently, no lower bound is known to us that increases as a function of maximum degree.

References

1. M. Ajtai, P. Erdős, J. Komlós, and E. Szemerédi. On Turán's theorem for sparse graphs. *Combinatorica*, 1(4):313–317, 1981.
2. M. Ajtai, J. Komlós, and E. Szemerédi. A note on Ramsey numbers. *J. Combin. Theory Ser. A*, 29:354–360, 1980.
3. S. Arora, C. Lund, R. Motwani, M. Sudan, and M. Szegedy. Proof verification and intractability of approximation problems. In *Proc. 33nd IEEE Symp. on Found. of Comp. Sci.*, pages 14–23, Oct. 1992.
4. M. Bellare and M. Sudan. Improved non-approximability results. To appear in STOC '94, May 1994.
5. P. Berman and M. Fürer. Approximating maximum independent set in bounded degree graphs. In *Proc. Fifth ACM-SIAM Symp. on Discrete Algorithms*, Jan. 1994.
6. R. B. Boppana and M. M. Halldórsson. Approximating maximum independent sets by excluding subgraphs. *BIT*, 32(2):180–196, June 1992.
7. P. Erdős. Some remarks on chromatic graphs. *Colloq. Math.*, 16:253–256, 1967.
8. M. M. Halldórsson and J. Radhakrishnan. Greed is good: Approximating independent sets in sparse and bounded-degree graphs. To appear in STOC '04, May 1994.
9. D. S. Hochbaum. Efficient bounds for the stable set, vertex cover, and set packing problems. *Disc. Applied Math.*, 6:243–254, 1983.
10. S. Khanna, R. Motwani, M. Sudan, and U. Vazirani. On syntactic versus computation views of approximability. Manuscript, Dec. 1993.
11. C. Papadimitriou and M. Yannakakis. Optimization, approximation, and complexity. *J. Comput. Syst. Sci.*, 43:425–440, 1991.
12. J. B. Shearer. A note on the independence number of triangle-free graphs. *Discrete Math.*, 46:83–87, 1983.

Asymptotically Optimal Election on Weighted Rings

Lisa Higham[1]
Computer Science Department,
University of Calgary
Calgary, Alberta, T2N 1N4,
Canada

Teresa Przytycka[2]
Dept. of Mathematics and Computer Science,
Odense University,
DK-5230 Odense M,
Denmark

Abstract. In general, in a network of asynchronous processors, the cost to send a message will differ from one communication link to another. In such a setting, it is desirable to factor the cost of links into the cost of distributed computation. Assume that associated with each link is a positive *weight* representing the cost of sending one message along the link and the cost of an algorithm executed on a *weighted* network is the sum of the costs of all messages sent during its execution. We determine the asymptotic complexity of distributed leader election on a weighted unidirectional asynchronous ring assuming this notion of cost, by exhibiting a simple algorithm and a matching lower bound for the problem.

1 Introduction

Consider a network of asynchronous processors that communicate via message passing. The standard measure of the cost of a distributed algorithm on such a network is the number of messages sent. This measure assumes that the cost of sending a message along any link is equal to one. In practice, the cost of sending a message may depend upon the link that the message traverses. This motivates the study of distributed algorithms where the cost of transmitting a message over a link is factored into the communication complexity of the algorithm. We consider a *weighted network* of processors where each link e of the network has associated with it a positive weight $w(e)$, which is the *cost* of sending a message along link e. The cost of a distributed algorithm for a given network and input is the maximum, over all message delay patterns, of the sum of the costs of all message traffic that occurs while executing the algorithm on that input.

The leader election problem is to design a distributed algorithm that distinguishes exactly one processor from among all the processors of the network as a unique processor called the *leader*. In this paper, we study the cost of leader election when the network topology is an asynchronous unidirectional weighted ring. Leader election on asynchronous unweighted rings has been very

[1] This research was supported in part by the Natural Sciences and Engineering Research Council of Canada. Email higham@cpsc.ucalgary.ca

[2] Part of this research was carried out while visiting the University of Calgary in August 1993. Email przytyck@imada.ou.dk

well studied. Several early paper appeared before Peterson [5] and Dolev *et.al.*[1] independently solved the unidirectional version of the problem using $O(n \log n)$ messages. By the results of Pachl *et.al.*[4] these algorithms are asymptotically optimal. Some effort has been made to reduce the constant [5, 1, 3] leading recently to the constant 1.271 [3]. Also there has been substantial work generalizing and strengthing the lower bound for election on rings and other networks under a variety of assumptions about the model. See [2] for a list of research addressing algorithms and lower bounds related to the leader election problem.

Running an algorithm designed for an unweighted network on a weighted network will, in general, not be efficient. Let W be the sum of the weights of all links of a weighted ring. Peterson's classical algorithm [5], when executed on the weighted ring will incur a cost of $\Omega(W \log n)$. The results of this paper show that this is not optimal. Let $\mathcal{R}$ be a ring with n_i edges having weight in $(2^{i-1}, 2^i]$. We present an algorithm for the leader election problem on unidirectional weighted rings that has cost $O(\sum_{n_i \geq 1} n_i 2^i \lg(n_i + 1))$. We show that this algorithm is optimal in the following sense: Given a multiset W of weights where n_i weight are in the interval $(2^{i-1}, 2^i]$ and a leader election algorithm A, we can design a ring $\mathcal{R}$ with edge weights equal to the set W such that the weighted message cost of A on $\mathcal{R}$ is $\Omega(\sum_{n_i \geq 1} n_i 2^i \lg(n_i + 1))$.

Our algorithm for weighted rings is in some sense a generalization of the basic algorithm for the unweighted case [3]. The perspective of this algorithm facilitates an extension to the weighted case. However, the analysis in the weighted case requires completely different techniques. The new contribution of our lower bound is that it explicitly incorporates the weights into the result. The basic idea to achieve this bound is derived from the work of Pachl, Korach and Rotem [4]. However, we need to adjust the technique to overcome the complications introduced by weighted links.

2 The leader election algorithm for weighted rings

2.1 Algorithm description

Although, in the literature, the leader election algorithms for unweighted rings are presented in a variety of ways, there is a high-level perspective that can be used to describe them all (see [2]). Initially each processor creates an *envelope* containing a *label* set to its own identifier, a *round number* (or sometimes a round parity bit), and possibly additional information and forwards the envelope to its neighbour. Upon receipt of an envelope, a processor applies a *casualty test*, which compares the contents of the envelope with the processor's stored information, to determine whether or not to destroy the envelope. If the receiving processor determines not to destroy the envelope, it applies a *promotion test*, to determine whether or not to increment the round number. It then updates the content of the envelope and its own information as required and forwards the envelope to its neighbour. Eventually only one envelope remains and a leader is elected. The various algorithms differ in four ways: the content of an envelope in addition to label and round number, the local information stored by each processor, the specification of the casualty test, and the specification of the promotion test.

In both the basic algorithm [3] which we refer to as the MIN-MAX algorithm, and our algorithm for weighted rings, called WEIGHTED ELECT, the label of an envelope is never changed. In MIN-MAX, each envelope contains only its label and a round number initialized to 1. Each processor stores the label and the round of the last envelope it sent. The casualty test is simply: the envelope and the receiving processor have the same round number and this round number is odd (respectively, even) and the label of the envelope is larger (respectively, smaller) than that stored by the processor. The promotion test is simply: the envelope and the receiving processor have the same round number.

One way to visualize MIN-MAX is to imagine that execution proceeds in rounds. In an odd round any envelope that directly follows an envelope with label smaller than it's own label is destroyed while in an even round any envelope that directly follows an envelope with a larger label is destroyed. Notice that in MIN-MAX, as well as in other election algorithms for unweighted rings, in every round (or sometimes in every second round) messages traffic covers every link of the ring. The central idea in WEIGHTED ELECT is to accelerate processing of envelopes that have travelled a large weighted distance by promoting them to a higher round as soon as they incur a sufficient weighted cost. Algorithm WEIGHTED ELECT can be thought of as combining MIN-MAX with this idea of "early promotion by weighted distance". Early promotion ensures that message traffic does not necessarily cover every link in each round, thus reducing the weighted distance an envelope travels before the algorithm terminates.

For algorithm WEIGHTED ELECT, in addition to the *label*, each envelope contains a *round* and a *credit*. Both are initialized as a function of the weight of the link adjacent to the processor that created the envelope. The initial credit is proportional to this weight and the initial round number is the logarithm of this weight. The label of an envelope remains unchanged as long as the envelope survives. The round and credit are adjusted during the course of the algorithm. Throughout the algorithm, each processor stores the label and the round of the last envelope that it sent. The casualty test for WEIGHTED ELECT is: the round number of the received envelope is less than that of the last envelope sent, or the casualty test of MIN-MAX holds. If an envelope is not destroyed then it may be promoted resulting in an increased credit and larger round. The promotion test for WEIGHTED ELECT is: the credit is less than the weight of the outgoing edge or the promotion test of MIN-MAX holds. For any surviving envelope (whether promoted or not) the processor reduces its credit by the weight of its adjacent edge before sending the envelope forward.

The complete protocol for WEIGHTED ELECT is given in Figure 1. The protocol for each processor is parameterized by its identifier (*proc-id*) and the weight of its outgoing edge (*adj_wt*). Four consecutive rounds of WEIGHTED ELECT are grouped together to form a *phase*; hence round r is represented by an ordered pair (p,t) where p is the phase number, $t \in \{0,1,2,3\}$ and $r = 4 * p + t$.

The pseudo code assumes the following three tests that are employed when an envelope containing label id, and round (p,t) and credit cnt arrives at a processor that has recorded a label fwd_id and a round (fwd_p, fwd_t) and has an outgoing edge with weight adj_wt.

Casualty-test
  $(p < fwd\_p)$ or
  $((p,t) = (fwd\_p, fwd\_t)$ and $t \in \{1,3\}$ and $id > fwd\_id)$ or
  $((p,t) = (fwd\_p, fwd\_t)$ and $t \in \{0,2\}$ and $id < fwd\_id)$
Promotion-test
  $((p,t) = (fwd\_p, fwd\_t)$ and $t \in \{1,3\}$ and $id < fwd\_id)$ or
  $((p,t) = (fwd\_p, fwd\_t)$ and $t \in \{0,2\}$ and $id > fwd\_id)$ or
  $(p > fwd\_p$ and $cnt < adj\_wt)$.
Leader-test $id = fwd\_id$ and $(p,t) = (fwd\_p, fwd\_t)$.

Processor($proc$-id, adj_wt):

```
id ←—proc-id ; p ←—⌈lg adj_wt⌉ ; t ←—0 ; cnt ←—2^{p+1} ;
fwd_id ←— −∞ ; fwd_p ←— −1 ; fwd_t ←— 0 ;

repeat
      if not Casualty-test then
          if Promotion-test then
                  t ←—(t+1) mod 4 ;
                  if t= 0 then p ←—p+1 ;
                  cnt ←—2^{p+t+1} ;
              fi
              fwd_id ←—id ;
              fwd_p ←—p ;
              fwd_t ←—t ;
              send(id, p, t, cnt − adj_wt) ;
      fi
          receive(id, p, t, cnt) ;
      until Leader-test .
```

Fig. 1. Algorithm WEIGHTED ELECT

2.2 Correctness of weighted elect

Correctness of WEIGHTED ELECT follows immediately after establishing:

safety the algorithm never deletes all message envelopes,
progress if there is more than one envelope then after a finite number of messages the number of envelopes is reduced, and
correct termination the algorithm elects a leader exactly when one envelope remains.

Because the ring is unidirectional and the algorithm is deterministic and message-driven with messages processed in first-in-first-out order, the messages received by each processor and the order in which each processor processes its messages is entirely determined by the initial configuration of identifiers and edge weights. Thus the scheduler is powerless to influence the outcome of the computation. We emphasize that for message-driven algorithms on *unidirectional rings*, correctness under any fixed scheduler implies correctness under all schedulers. Thus we can assume without loss of generality that envelopes are processed in such a way that envelopes with the least round are processed first. That is, we assume that an envelope with a given round number is not delivered until there does not exist an envelope of smaller round number.

Suppose, contrary to safety, that some run of WEIGHTED ELECT removes all envelopes. Then there is a maximum round (p, t) achieved by any envelope. Suppose t is odd, and let S be the set of identifiers in envelopes that achieve round (p, t). According to Casualty-test, an envelope in round (p, t) with identifier i can only be destroyed by meeting a processor that either (1) last forwarded an envelope with round larger than (p, t), or (2) last forwarded an envelope with round equal to (p, t) and identifier less than i. Since (p, t) is the maximum round, case (1) is impossible. Furthermore, in case (2), the envelope in S with minimum identifier cannot be destroyed. However, any envelope that survives the Casualty-test is forwarded. If it is forwarded without promotion then its credit is diminished. So eventually its credit will be insufficient to cross an edge and it will be promoted at which point its round increases contradicting that (p, t) was the maximum round. A similar argument applies if t is even.

Suppose contrary to progress, that after some point, $k \geq 2$ envelopes remain alive. Then eventually each of these envelopes will receive a credit at least as large as the weight of the ring. At this point each envelope has a large enough credit to allow it to travel to the processor that forwarded the successor envelope. Then, if all non-destroyed envelopes have the same round number and if t is odd (respectively, even) the envelope with maximum label (respectively, minimum label) must be destroyed. If all envelopes do not have the same round number then the envelope that follows an envelope of larger round number must be destroyed. Thus in both cases progress is assured.

The algorithm cannot prematurely elect a leader because a processor will receive an envelope with *id* equal to its *fwd_id* and with $(p, t) = (fwd\_p, fwd\_t)$ if and only if there are no other envelopes, thus passing the Leader-test and confirming correct termination.

2.3 Message complexity of weighted elect

We first introduce some definitions and notation. The pth *phase* consists of all the message traffic of envelopes with round (p, t) for $t \in \{0, 1, 2, 3\}$. Since algorithm

WEIGHTED ELECT never changes the label of an envelope for the duration of its existence, we use *envelope a* as an abbreviation for the envelope with label a. For an envelope a in phase p, let $host_p(a)$ denote the processor that promoted the envelope to phase p (that is, from round $(p-1, 3)$ to round $(p, 0)$). The *weighted distance* from processor x to processor y, denoted $\delta(x, y)$, is the sum of the weights of all links between processor x and processor y, travelling in the direction of the ring. Envelope b is the *immediate successor in phase p* of envelope a in phase p, if the first envelope with phase q, where $q \geq p$, encountered after envelope a travelling in the direction of the ring, is envelope b. Let envelope b in phase q be the immediate successor of envelope a in phase p. The *horizon* of envelope a in phase p is $\delta(host_p(a), host_q(b))$, that is, the horizon of a is the weighted distance from its host in phase p to the host of its immediate successor.

Let n_i be the number of links with weight in $(2^{i-1}, 2^i]$ in an asynchronous ring with distinct identifiers. Let d_p denote the number of envelopes that participate in phase p. An envelope that participates in phase p is *sparse in phase p* if its horizon is greater than 2^p, otherwise it is *dense*. Let s_p denote the number of envelopes that are sparse in phase p.

The next three lemmas allow us to bound the number of sparse envelopes and the total number of envelopes in each phase as a function of the weights on the ring.

Lemma 1. *The number s_p of sparse envelopes that participate in phase p satisfies $s_p \leq \sum_{i=0}^{p} n_i 2^{i-p}$.*

Proof. The result follows immediately from the definition of s_p.

We expect each pair of successive rounds in a phase to reduce the number of dense envelopes by at least one half because first any envelope meeting a processor with *fwd_id* smaller than its own *id* and in the same round is eliminated and in the following round any envelope meeting a processor with *fwd_id* larger than its own is eliminated. This is made precise in the next lemma.

Lemma 2. *The number d_p of envelopes that participate in phase p satisfies the recurrence:*

$$d_0 = n_0; \qquad d_{p+1} \leq \frac{d_p - s_p}{4} + n_{p+1} + s_p \qquad p \geq 0$$

Proof. The proof relies on the following observation:

Fact 3. Let x be an envelope that survives to phase $p + 1$. Assume that at the beginning of phase p, x is immediately followed by k dense envelopes. Then the $\min\{k, 3\}$ dense envelopes that immediately follow x do not survive to phase $p+1$. The fact holds because consecutive dense envelopes have enough credit to reach the host of the next envelope with the same round number. By applying the min-max comparison for four consecutive rounds it is easily checked that if x survives for four rounds, the $\min\{k, 3\}$ dense envelopes that follow x must be eliminated.

Consider a maximal chain of dense envelopes in round $(p, 0)$. Let s be the non-dense envelope (a sparse envelope or an envelope with higher phase) that immediately precedes the first dense envelope of this chain. If x is a dense envelope in this chain that survives to phase $p + 1$ and is followed by at least 3 dense envelopes then, by fact 3, we can attribute 3 eliminated envelopes to x. Suppose x is followed by fewer than 3 dense envelopes. If s survives than we attribute to x the 3 eliminated envelopes that, by fact 3, follow s at the beginning of phase p. If s did not survive to phase $p + 1$ then in our count of surviving envelopes we can count x as eliminated instead of s. So at most $(d_p - s_p)/4 + s_p$ phase p envelopes survive to phase $p + 1$. In addition, there are n_{p+1} new envelopes that begin in phase $p + 1$.

Lemma 4. *The number d_p of envelopes that participate in phase p satisfies $d_p < 4 \sum_{i=0}^{p} n_i 2^{i-p}$.*

Proof. By lemma 2, $d_p \leq (d_{p-1})/4 + n_p + s_{p-1}$. Thus, by lemma 1, $d_p < (d_{p-1})/4 + n_p + \sum_{i=0}^{p-1} n_i 2^{i-p+1}$. Solving this recurrence with $d_0 = n_0$ yields: $d_p < 4 \sum_{i=0}^{p} n_i 2^{i-p}$.

Theorem 5. *Let $\mathcal{R}$ be a ring with n_i edges having weight in $(2^{i-1}, 2^i]$. Then the weighted message cost of WEIGHTED ELECT on $\mathcal{R}$ is $O(\sum_{n_i \geq 1} n_i 2^i \lg(n_i + 1))$.*

Proof. Denote the worst case weighted message cost of WEIGHTED ELECT on ring $\mathcal{R}$ by $\text{cost}(\mathcal{R})$. There are at most d_p envelopes participating in round (p, t), each of which travels at most a weighted distance of 2^{p+i+1}. Since there are four rounds per phase, phase p costs less than $d_p 2^{p+5}$. Let $\hat{n}_i$ be the least integer that is a power of 2 and satisfies $\hat{n}_i \geq n_i + 1$ and let $I = \max\{i : n_i > 0\}$. Algorithm WEIGHTED ELECT terminates when one envelope remains, and, by lemma 4, the number of remaining envelopes in round $p \geq I$ is at most $4 \sum_{i=0}^{p} n_i 2^{i-p} = 4 \sum_{i=0}^{I} n_i 2^{i-p} \leq 4 \sum_{i=0}^{I} \hat{n}_i 2^{i-p}$. Thus the total number of rounds is at most the minimum p satisfying $(4 \sum_{i=0}^{I} \hat{n}_i 2^i)/2^p \leq 1$. Denote the smallest such integer p by $\hat{R}$. Thus

$$\text{cost}(\mathcal{R}) < \sum_{p=0, d_p \geq 1}^{\hat{R}} d_p \cdot 2^{p+5}$$

$$\in O\left(\sum_{p=0, d_p \geq 1}^{\hat{R}} \sum_{i=0}^{p} n_i 2^{i-p} \cdot 2^p \right) \qquad \text{by lemma 4}$$

$$\in O\left(\sum_{p=0, d_p \geq 1}^{\hat{R}} \sum_{i=0}^{p} \hat{n}_i 2^i \right).$$

We will now show that $S = \sum_{p=0, d_p \geq 1}^{\hat{R}} \sum_{i=0}^{p} \hat{n}_i 2^i \in O(\sum_{n_i \geq 1} \hat{n}_i 2^i \lg \hat{n}_i)$ which implies the theorem.

Let $A(p, i)$ be an array with the entry in row p and column i given by $A(p, i) = \hat{n}_i 2^{i-p}$ for $p \geq i$ and $A(p, i) = 0$ for $p < i$, as show in table 1. Let $D_p = \sum_{i=1}^{p} A(p, i)$. Then the elements of the summation S can be interpreted as the entries of this table multiplied by the corresponding multiplier. Specifically, $S \leq \sum_{0 \leq p \leq \hat{R}, D_p \leq 1, 0 \leq i \leq I} A(p, i) \cdot 2^p$. Note that entries from a row p such that $D_p < 1$

r/i	0	1	2		j		I	multiplier
0	$\hat{n}_0$	0	0				0	1
1	$\frac{\hat{n}_0}{2}$	$\hat{n}_1$	0					2
2	$\frac{\hat{n}_0}{2^2}$	$\frac{\hat{n}_1}{2}$	$\hat{n}_2$					2^2
$\vdots$	$\vdots$	$\vdots$	$\vdots$	$\ddots$				$\vdots$
$\lg \hat{n}_2 + 2$	$\vdots$	$\vdots$	1					$\hat{n}_2 2^2$
$\lg \hat{n}_0$	1	$\vdots$						$\hat{n}_0$
$\lg \hat{n}_1 + 1$		1						
					$\hat{n}_j$			
					$\frac{\hat{n}_j}{2}$			
					$\vdots$		0	
					$\vdots$		$\hat{n}_I$	
					$\vdots$		$\frac{\hat{n}_I}{2}$	
					$\vdots$		$\vdots$	
							1	
$\vdots$								
$\hat{R}$	$\frac{n_0}{2^{\hat{R}}}$				$\frac{n_j}{2^{\hat{R}-j}}$		$\frac{n_I}{2^{\hat{R}-I}}$	$2^{\hat{R}}$

Table 1. Summands in Summation S

do not contribute to the sum S. Among the remaining entries of array A we consider three types. Entries $A(i, p)$ such that $A(i, p) \geq 1$ are called *whole entries* (denoted $\mathcal{W}$). Other entries are called *fractional* entries. Among fractional entries we distinguish *heavy entries* (denoted $\mathcal{H}$) and *light entries* (denoted $\mathcal{L}$) to be defined later.

Claim 6. $\sum_{A(p, i) \in \mathcal{W}} A(p, i) 2^p \in O(\sum_{i=0}^{I} \hat{n}_i 2^i \lg \hat{n}_i)$.

Proof. Note that, for any fixed column i and $p \geq i$, $A(p, i) \cdot 2^p = \hat{n}_i 2^i$. Thus, for any column i, each of the elements in the column from entry $A(i, i) = \hat{n}_i$ down to $A(i + \lg \hat{n}_i, i) = 1$, multiplied by the corresponding value of the "multiplier" column, are by definition exactly those included in the summation $\sum_{i=0}^{I} \hat{n}_i 2^i \lg \hat{n}_i$.

Therefore, it remains to show that the contribution of the fractional elements in the summation is of the same order.

For each column i satisfying $\hat{n}_i > 0$, let level(i) be the row number k satisfying $A(k, i) = 1$. Let $T_k = \{i : \text{level}(i) = k\}$ and $t_k = |T_k|$. A fractional entry $A(p, i)$ is called *heavy* if $\text{level}(i) + t_{\text{level}(i)} \leq p$. A fractional entry which is not heavy is called *light*.

The claim below states that the contribution of all heavy fractional elements to the sum S is of the same order as the contribution of whole elements.

Claim 7. For any k such that $T_k \neq \emptyset$

$$\sum_{i \in T_k} \sum_{p=k+1}^{k+t_k} A(p, i) 2^p \leq 3 \sum_{i \in T_k} \hat{n}_i 2^i \lg \hat{n}_i$$

Proof. By definition of T_k, for all $i \in T_k$, $\hat{n}_i / 2^{k-i} = 1$. Thus $\hat{n}_i \cdot 2^i = 2^k$. So, $i, j \in T_k$ and $i \neq j$ implies $\hat{n}_i \neq \hat{n}_j$. Therefore, for all $i \in T_k$ the corresponding $\hat{n}_i$ are distinct and are powers of two. Hence:

$$\sum_{i \in T_k} \hat{n}_i \cdot 2^i \lg \hat{n}_i = 2^k \cdot \lg(\textstyle\prod_{i \in T_k} \hat{n}_i) \geq 2^k \cdot \lg(2^0 \cdot \ldots \cdot 2^{t_k - 1}) > 2^k \cdot (t_k)^2 / 3$$

On the other hand, for any $i \in T_k$ and for any row p, $A(p, i) \cdot 2^p = n_i 2^i = 2^k$. Hence:

$$\sum_{i \in T_k} \sum_{p=k+1}^{k+t_k} A(p, i) \cdot 2^p = (t_k)^2 2^k \leq 3 \sum_{i \in T_k} \hat{n}_i \cdot 2^i \lg \hat{n}_i$$

To complete the proof we estimate the contribution of light fractional elements. First we show the following claim:

Claim 8. For any p
$$\sum_{i, A(p,i) \in \mathcal{L}} A(p, i) 2^p \in O(2^p).$$

Proof.

$$\sum_{i, A(p,i) \in \mathcal{L}} A(p, i) 2^p = \sum_{k, k+t_k < p} \sum_{i \in T_k} A(p, i) 2^p = \sum_{k, k+t_k < p} t_k 2^k$$

Since $k < p$ we have $t_k < p - k$ and thus the last sum is bounded by $\sum_{k=1}^{p-1}(p-k)2^k \in O(2^p)$.

By claim 8, all light fractional entries that are in the same row as some whole entry, contribute to the sum S approximately the same amount as that whole entry. Thus we need to take care of light fractional entries that do not belong to the same row as a whole entry. Note that each row j that contains a whole entry can be directly followed by at most $\log D_j$ rows that do not contain whole entries. (Any further row p that does not contain a whole entry has $D_p < 1$ and thus is not counted in the summation). By claim 8, the contribution of light fractional entries in all these rows is bounded by $\sum_{p=j+1}^{j+\log D_j} 2^p \leq 2^{j+1+\log D_j} \in O(D_j 2^j)$. Thus the contribution of light fractional entries that belong to a row that does not contain a whole entry is dominated by the contribution of the closest row that contains a whole entry. This concludes the proof of theorem 5.

3 Lower bound for election on weighted rings

In this section we establish that the algorithm WEIGHTED ELECT is asymptotically optimal for asynchronous unidirectional weighted rings with distinct identifiers. We borrow and adapt the notation, techniques, and some terminology from Pachl et.al. [4] and, because of space constraints here, rely on some of their lemmas.

An asynchronous unidirectional weighted ring R with n processors is denoted by a sequence $R = ((id_0, w_0), \ldots, (id_{n-1}, w_{n-1}))$, called a *labelling sequence*, where id_i is the identifier of the ith processor and w_i is the weight of the link from the ith to the $(i + 1)^{\text{st}}$ processor. Let A be a leader election algorithm for weighted asynchronous unidirectional rings. Then the *cost of A on ring R*, denoted $\text{cost}_A(R)$, is the total weighted cost of all messages sent by A when executed on ring R. Let W be a multiset of n weights, and let I be a set of n distinct identifiers. (Elements of both W and I are assumed to be positive integers.) Let $\mathcal{R}(I, W)$ be the set of all rings $R = ((id_0, w_0), \ldots, (id_{n-1}, w_{n-1}))$ such that $\{id_0, \ldots, id_{n-1}\} = I$ and $\{w_0, \ldots, w_{n-1}\} = W$. Denote by $\text{cost}_A(I, W)$ the maximum over all rings $R \in \mathcal{R}(I, W)$ of $\text{cost}_A(R)$. Denote by $\text{cost}_A(W)$ the maximum over all sets I of n distinct identifiers of $\text{cost}_A(I, W)$. Finally, the *cost of leader election for weights in W* is the minimum over all leader election algorithms A of $\text{cost}_A(W)$. The theorem we will establish is:

Theorem 9. *Let W be a multiset of weights with n_i weights in the interval $(2^{i-1}, 2^i]$. Then the cost of leader election for weights in W is $\Omega(\sum_{n_i \geq 1} n_i 2^i \lg n_i)$.*

Proof. Call a ring R with edge weights taken from W *well-constructed* over W if, for each i, all n_i weights in $(2^{i-1}, 2^i]$ are on consecutive links. Such a sequence of links with weights in the same weight class form a *segment*. Let $\mathcal{R}_W$ denote the class of rings that are well-constructed over W. Assume A is any leader election algorithm for unidirectional weighted rings. To established the theorem, we show that the average weighted message cost of A on $\mathcal{R}_W$ is $\Omega(\sum_{n_i \geq 1} n_i 2^i \lg n_i)$. For each ring R in $\mathcal{R}_W$ insert barriers between the segments of R, and run algorithm A on R. That is, schedule the messages of A so that all message traffic between segments is delayed arbitrarily while message delay within each segment is just one time unit, and run A under this scheduler only until all messages are queued at the barriers. Clearly this can only decrease the total cost of the message traffic; we show that the total cost of the messages sent in only this part of the execution suffices to give the lower bound. Hence, to establish the lower bound we need only show that the average cost of a segment constructed from the n_i weights in weight class i is bounded below by $\Omega(n_i 2^i \lg n_i)$. To achieve this we examine the expected message traffic that ensues within a segment.

Define the *trace* of a message envelope created by the kth processor when it arrives at the pth processor to be the sequence $(id_k, w_k), (id_{k+1}, w_{k+1}) \ldots (id_p, w_p)$. Because the ring is unidirectional, the trace of a message captures the maximum possible information that a message may possess.

If s is a sequence, then $C(s)$ denotes the set of cyclic permutations of s, and $len(s)$ denotes its length. A sequence t is a *subsequence* of s if $s = utv$ for some sequences u and v. Denote by D the set of all finite nonempty labelling sequences $((id_1, w_1), (id_2, w_2) \ldots (id_n, w_n))$ where $\{id_1, \ldots id_n\}$ is a set of distinct integers, and $\{w_1, \ldots w_n\}$ is a multiset of weights in $(2^{i-1}, 2^i]$.

For $s \in D$ and $E \subseteq D$, define

$$B(s, E) = |\{t : t \in E \text{ and } t \text{ is a subsequence of } s\}| \qquad \text{and}$$

$$B_k(s, E) = |\{t : t \in E \text{ and } len(t) = k \text{ and } t \text{ is a subsequence of } s\}|.$$

A set $E \subseteq D$ is *exhaustive* if the following two properties hold:

1. *Prefix property*: if $tu \in E$ and $len(t) \geq 1$ then $t \in E$,

2. *Cyclic permutation property*: if $s \in D$ then $C(s) \cap E \neq \emptyset$.

For any algorithm A for unidirectional rings, define $m(s, A)$ to be the number of messages sent by A on a ring with labelling sequence s, when a barrier is placed between s_n and s_1.

Lemma 10. *For every leader election algorithm A, there exists an exhaustive set $E(A) \subseteq D$ such that $m(s, A) \geq B(s, E(A))$.*

Proof. Define $E(A)$ to be the set of those $s \in D$ for which a message with trace s is sent when executed on a ring labelled s. The lemma follows immediately from the following claims which are proved by Pachl et.al [1].

Claim 11. $E(A)$ is an exhaustive set.

Claim 12. If trace t is sent by A when executed on a ring labelled with t, then trace t is sent by A when executed on any sequence that contains t as a subsequence.

Now Lemma 10 is used to establish the expected weighted cost of messages traffic within a segment. Let W be a multiset of n weights in $(2^{i-1}, 2^i]$. Let I be a set of n distinct integer identifiers. Let $\mathcal{R}$ be the set of all labelling sequences $s = ((id_1, w_1), (id_2, w_2) \ldots (id_n, w_n))$ where $\{id_1, \ldots id_n\} = I$ and $\{w_1, \ldots w_n\} = W$.

Lemma 13. *For any algorithm A, the average number of messages $m(s, A)$ over all labelling sequences $s \in \mathcal{R}$ is bounded below by $H_n \cdot n - n$.*

Proof.

$$ave\{m(s, A)\}_{s\in\mathcal{R}} = \frac{1}{|\mathcal{R}|} \sum_{s\in\mathcal{R}} m(s, A) \geq \frac{1}{|\mathcal{R}|} \sum_{s\in\mathcal{R}} B(s, E(A))$$

$$\geq \frac{1}{|\mathcal{R}|} \sum_{k=1}^{n} \sum_{s\in\mathcal{R}} B_k(s, E(A))$$

For fixed k and a fixed $s \in \mathcal{R}$, there are $n - k$ subsequences of s with length k, so there are $|\mathcal{R}|(n - k)$ length k subsequences over all $s \in \mathcal{R}$. Partition these into $\frac{|\mathcal{R}|(n-k)}{k}$ sets where each set consists of all cyclic permutations of one sequence. By the cyclic permutation property, each set has at least one element in common with $E(A)$. Hence:

$$ave\{m(s, A)\}_{s\in\mathcal{R}} \geq \frac{1}{|\mathcal{R}|} \sum_{k=1}^{n} \frac{|\mathcal{R}|(n - k)}{k} = n\sum_{k=1}^{n} \frac{1}{k} - n \in \Omega(n \lg n)$$

Since on average $\Omega(n_i \lg n_i)$ messages are sent over the segment formed from the n_i elements of W that are in the weight class $(2^{i-1}, 2^i]$ and each message incurs a weighted cost of at least 2^{i-1}, the lower bound for each segment is $\Omega(n_i 2^i \lg n_i)$. Hence the lower bound for leader election on weighted rings follows from lemma 13 by summing for each segment.

References

1. D. Dolev, M. Klawe, and M. Rodeh. An $O(n \log n)$ unidirectional distributed algorithm for extrema finding in a circle. *J. Algorithms*, 3(3):245–260, 1982.
2. L. Higham and T. Przytycka. A simple, efficient algorithm for maximum finding on rings. Technical Report 92/494/32, University of Calgary, 1992.
3. L. Higham and T. Przytycka. A simple, efficient algorithm for maximum finding on rings. In *Lecture Notes in Computer Science #725*, pages 249–263. Springer Verlag, 1993. Proc. 7th International Workshop on Distributed Algorithms.
4. J. Pachl, E. Korach, and D. Rotem. Lower bounds for distributed maximum finding. *J. Assoc. Comput. Mach.*, 31(4):905–918, 1984.
5. G. Peterson. An $O(n \log n)$ algorithm for the circular extrema problem. *ACM Trans. on Prog. Lang. and Systems*, 4(4):758–752, 1982.

Optimal Algorithms for Broadcast and Gossip in the Edge-Disjoint Path Modes *

(Extended Abstract)

Juraj Hromkovič** , Ralf Klasing, Walter Unger, Hubert Wagener***

Department of Mathematics and Computer Science
University of Paderborn
33095 Paderborn, Germany

Keywords: communication algorithms, parallel computations.

Abstract. The communication power of the one-way and two-way edge-disjoint path modes for broadcast and gossip is investigated. The complexity of communication algorithms is measured by the number of communication steps (rounds). The main results achieved are the following:

1. For each connected graph G_n of n nodes, the complexity of broadcast in G_n, $B_{min}(G_n)$, satisfies $\lceil \log_2 n \rceil \leq B_{min}(G_n) \leq \lceil \log_2 n \rceil + 1$. The complete binary trees meet the upper bound, and all graphs containing a Hamiltonian path meet the lower bound.

2. For each connected graph G_n of n nodes, the one-way (two-way) gossip complexity $R(G_n)$ $(R^2(G_n))$ satisfies

$$\lceil \log_2 n \rceil \leq R^2(G_n) \leq 2 \cdot \lceil \log_2 n \rceil + 1,$$
$$1.44 \ldots \log_2 n \leq R(G_n) \leq 2 \cdot \lceil \log_2 n \rceil + 2.$$

All these lower and upper bounds are tight.

3. All planar graphs of n nodes and degree h have a two-way gossip complexity of at least $1.5 \log_2 n - \log_2 \log_2 n - 0.5 \log_2 h - 2$, and the two-dimensional grid of n nodes has the gossip complexity $1.5 \log_2 n - \log_2 \log_2 n \pm O(1)$, i.e. two-dimensional grids are optimal gossip structures among planar graphs. Similar results are obtained for one-way mode too.

Moreover, several further upper and lower bounds on the gossip complexity of fundamental networks are presented.

---

* This work was partially supported by grants Mo 285/4-1, Mo 285/9-1 and Me 872/6-1 (Leibniz Award) of the German Research Association (DFG), and by the ESPRIT Basic Research Action No. 7141 (ALCOM II).

** This author was partially supported by SAV Grant No. 88 and by EC Cooperation Action IC 1000 Algorithms for Future Technologies.

*** This author was supported by the Ministerium für Wissenschaft und Forschung des Landes Nordrhein-Westfalen.

1 Introduction and Definitions

This paper is devoted to the problem of information dissemination in interconnection networks. We investigate the three basic communication tasks, **broadcast**, **accumulation**, and **gossip** which can be described as follows. Assume that each vertex (processor) in a graph (network) has some piece of information. The **cumulative message** of G, $Cum(G)$, is the set of all pieces of information originally distributed in all vertices of G. To solve the **broadcast** [accumulation] problem for a given graph G and a vertex u of G, we have to find a communication strategy (using the edges of G as communication links) such that all vertices in G learn the piece of information residing in u [that u learns the cumulative message of G]. To solve the **gossip** problem for a given graph G, a communication strategy must be found such that all vertices in G learn the cumulative message of G. Since the above stated communication problems are solvable only in connected graphs, we note that from now on we use the notion "graph" for connected undirected graphs.

The meaning of a "communication strategy" depends on the communication mode. A communication strategy is realized by a **communication algorithm** consisting of a number of **communication steps (rounds)**. The rules describing what can happen in one communication step (round) are defined exactly by the communication mode. Here, we consider the following modes:

1. **One-way [Two-way] vertex-disjoint paths mode** (1VDP mode [2VDP mode])
   One round can be described as a set $\{P_1, \ldots, P_k\}$ for some $k \in \mathbb{N}$, where $P_i = x_{i,1}, \ldots, x_{i,\ell_i}$ is a simple path of length $\ell_i - 1$, $i = 1, \ldots, k$, and the paths are vertex-disjoint. $\{P_1, \ldots, P_k\}$ is called the **set of active paths** of this round. The executed communication of this round in one-way mode consists of the submission of the whole actual knowledge of $x_{i,1}$ to x_{i,ℓ_i} via path P_i for any $i = 1, \ldots, k$. $x_{i,1}$ is called the **sender** of P_i, and x_{i,ℓ_i} is called the **receiver** of P_i. [The executed communication of this round in two-way mode consists of the complete exchange of the actual knowledge between $x_{i,1}$ and x_{i,ℓ_i} for any $i = 1, \ldots, k$.] The inner nodes of path P_i (nodes different from the end-points $x_{i,1}$ and x_{i,ℓ_i}) do not learn the message submitted from $x_{i,1}$ to x_{i,ℓ_i} [exchanged between $x_{i,1}$ and x_{i,ℓ_i}], they are only used to realize the connection between $x_{i,1}$ and x_{i,ℓ_i}.
2. **Farley's edge-disjoint paths mode** (FEDP mode)
   One round can be again described as a set of active paths $\{P_1, \ldots, P_k\}$, $P_i = x_{i,1}, \ldots, x_{i,\ell_i}$ for $i = 1, \ldots, k$, satisfying the following conditions:
   (2.1) $\forall i,j \in \{1, \ldots, k\}, i \neq j$: P_i and P_j are edge-disjoint,
   (2.2) $\{x_{i,1} \mid i = 1, \ldots, k\} \cap \{x_{i,\ell_i} \mid i = 1, \ldots, k\} = \emptyset$, i.e. no node may simultaneously be sender and receiver in one round,
   (2.3) $|\{x_{i,1} \mid i = 1, \ldots, k\}| = |\{x_{i,\ell_i} \mid i = 1, \ldots k\}| = k$, i.e. no node may be the sender (receiver) for more than one path.

The executed communication is interpreted in the same way as in 1VDP mode.

3. **One-way [Two-way] edge-disjoint paths mode** (1EDP mode [2EDP mode])

One round is again described as a set of active paths $P = \{P_1, \ldots, P_k\}$ for some $k \in I\!N$, where $P_i = x_{i,1}, \ldots, x_{i,\ell_i}$ for $i = 1, \ldots, k$. P must satisfy the above conditions (2.1), (2.2), (2.3), and additionally the next one:

(2.4) $\{x_{i,1}, x_{i,\ell_i} \mid i = 1, \ldots, k\} \cap \{x_{r,s_r} \mid r \in \{1, \ldots, k\}, s_r \in \{2, \ldots, \ell_r - 1\}\} = \emptyset$, i.e. the nodes of the paths of P can be partitioned into three disjoint sets: the set of senders, the set of receivers, and the set of **connectors**, where a connector v is only used to transfer the pieces of information via the paths involving v.

The executed communication is interpreted in the same way as in the VDP modes.

The VDP modes were introduced and investigated in [FHMMM92, HKS93, HKSW93]. The FEDP mode was introduced by Farley in [Fa80], where the complexity of broadcast in this mode was investigated. The EDP modes are introduced here as a mode which communication power lies between VDP modes and FEDP mode. But the main reason to consider the EDP modes for information dissemination is not the fact that EDP modes are modes between some known communication modes. The main reason to consider them is that EDP modes are realistic for some models of parallel computers (some of the algorithms designed here were implemented on Transputer systems supporting exactly this kind of communication).

The main aim of this paper is to study the gossip and broadcast complexity of interconnection networks in EDP modes as well as to compare these modes with FEDP and VDP modes.

Now, let us fix the notation used in this paper. For any graph $G = (V, E)$, $V(G) = V$ denotes the set of vertices of G, and $E(G) = E$ denotes the set of edges of G. In what follows, we denote the complexity of broadcast, accumulation and gossip as B, A and R. For any given graph G and a vertex u of G, let $B_u(G)$ [$A_u(G)$] denote the number of rounds (complexity) of the optimal broadcast [accumulation] algorithm from [to] u in G in 1EDP mode. Because each accumulation algorithm for u and G can be "reversed" (the sequence of rounds is reversed and also the direction of information flow in each round) to obtain a broadcast algorithm for u and G (and vice versa), we have $B_u(G) = A_u(G)$ for any u and G, $u \in V(G)$. For any graph G, we define

$$\mathbf{B}(\mathbf{G}) = \max\{B_u(G) \mid u \in V(G)\},$$
$$\mathbf{B_{min}}(\mathbf{G}) = \min\{B_u(G) \mid u \in V(G)\}.$$

Furthermore, let $R(G)$ and $R^2(G)$ denote the number of rounds of the optimal gossip algorithm for G in the 1EDP and 2EDP mode respectively. Let also $b = (1 + \sqrt{5})/2$ throughout the paper.

The paper is organized as follows. Section 2 is devoted to the broadcast problem in 1EDP mode (note that the complexity for broadcast is the same in 1EDP and 2EDP mode). The only result known about the FEDP mode is the fact that one can broadcast in any graph G of n nodes from any $v \in V(G)$ in $\lceil \log_2 n \rceil$ rounds. The whole paper [Fa80] is devoted to the proof of this rather technical result. We note that the method used in [Fa80] does not work in our weaker 1EDP mode. So, we develop a new method designing a communication algorithm broadcasting in $\lceil \log_2 n \rceil + 1$ rounds in any graph G of n nodes from some v of G. Thus, we get

(1) $\lceil \log_2 n \rceil \leq B_{\min}(G_n) \leq \lceil \log_2 n \rceil + 1$

for any G_n of n nodes. Furthermore, we prove that (1) cannot be improved by showing that complete binary trees have broadcast complexity exactly $\lceil \log_2 n \rceil + 1$.

Section 3 is devoted to the gossip problem in EDP modes. (Note that the gossip complexity was not investigated in any edge-disjoint paths mode till now.) Using (1) and the known lower bound results about gossiping in complete graphs in standard modes [EM89, Kn75], we get

(2) $\lceil \log_2 n \rceil \leq R^2(G_n) \leq 2 \cdot \lceil \log_2 n \rceil + 1$,
(3) $1.44 \ldots \log_2 n = \log_{\phi}(\lfloor n/2 \rfloor) + 2 \leq R(G_n) \leq 2 \cdot \lceil \log_2 n \rceil + 2$

for any graph G_n of n nodes in Subsection 3.1. Here, we also show that the upper bounds of (2) and (3) are tight. The gossip in 2-dimensional grids Gr_n^2 of $n = m^2$ nodes and in planar graphs is investigated in Subsection 3.2. We establish

(4) $R^2(Gr_n^2) = 1.5 \cdot \log_2 n - \log_2 \log_2 n \pm O(1)$.

The most interesting observation connected with (4) is that the best known gossip algorithm in Gr_n^2 (even in any planar graph) in two-way vertex-disjoint mode takes $1.5 \log_2 n$ rounds. This supports the conjecture that for the 2-dimensional grids (planar graphs) the edge-disjoint mode may be more powerful than the vertex-disjoint mode. Furthermore, we prove

(5) $R^2(Pl(n, h)) \geq 1.5 \log_2 n - \log_2 \log_2 n - 0.5 \log_2 h - 2$

for every planar graph $Pl(n, h)$ of n nodes and degree h. Thus, the 2-dimensional grid belongs to the best gossip structures in the 2EDP mode among all planar graphs of bounded degree.

We also extend the technique used above in order to derive improved upper bounds on the gossip complexity in the 1EDP mode for the 2-dimensional grid and for all planar graphs. These bounds are again closed (up to $O(1)$), but only if we restrict the class of all one-way gossip algorithms to some special subclass of "regular" gossip algorithms [HKSW93].

Subsection 3.3 is devoted to gossiping in some fundamental interconnection networks. Extending the technique from Subsection 3.2, optimal (up to an additional constant) gossip algorithms for d-dimensional grids are designed. Applying methods developed in [HKSW93] for the vertex-disjoint paths modes, some fundamental bounded-degree interconnection networks are presented with the gossip complexity differing at most by $2 \log_2 \log_2 n$ from the gossip complexity of the complete graph of n nodes.

Concluding the introduction, we call attention to the fact that the main contributions of this paper are

(a) the general broadcast strategy for each graph (Section 2), and
(b) the gossip algorithm in 2-dimensional grids (Subsection 3.2).

The main reason for this claim is not only the fact that (a) and (b) lead to optimal communication algorithms, but that in order to get these results some essentially new ideas in design (proof) methods for the study of disjoint-path communication algorithms have been developed in this paper. Moreover, the proof method developed to get (b) provides a technique enabling to essentially improve some gossip algorithms for vertex-disjoint path modes [HKSW93], but we omit the formulation of these consequences in this extended abstract.

2   Accumulation and Broadcast in 1EDP Mode

In this section we investigate the communication complexity of accumulation and broadcast in 1EDP mode. Note that the more powerful FEDP mode enables to broadcast (accumulate) in $\lceil \log_2 n \rceil$ rounds in any graph of n nodes [Fa80], and that the weaker VDP modes require even $\Omega(n)$ broadcast (accumulation) complexity for some families of graphs of n nodes [HKS93]. Now, we prove the (somewhat surprising) upper bound $\lceil \log_2 n \rceil + 1$ on $B_{\min}(G)$ for any graph G of n nodes, which shows that the power of 1EDP mode is much closer to the power of FEDP mode than to the power of 1VDP mode.

When considering arbitrary graphs G, we can restrict our attention to trees, because accumulation as well as broadcast can be performed by applying an appropriate algorithm to some spanning tree of G. Note also that the problems of broadcasting and accumulation are dual to each other, i.e. reversing the communication pattern for accumulation results in a communication pattern for broadcasting. Thus we will concentrate on the accumulation problem for trees.

Recall that in the accumulation problem each vertex has some piece of information and a communication strategy has to be found for collecting all these pieces in some vertex v. The following notation is convenient for representing our communication strategy.

Definition 1. Let $G = (V, E)$ be any graph. A set of vertices $K \subset V$ is called *knowledge set*, if the pieces of information residing in the vertices of K form the cumulative message, i.e. all pieces of information are collected in K.

The size of minimal knowledge sets measures in a natural way the progress of an accumulation algorithm.

Let $T = (V, E)$ be some tree, we will refer to any vertex of degree > 2 in T as a critical vertex, while all other vertices are called non-critical.

Our communication strategy for a tree of n vertices consists of two principal phases:

PHASE 1: Collect all pieces of information in a subset S of non-critical vertices with $|S| \leq \lceil n/2 \rceil$. That is, $S \subseteq \{v \in V \mid degree(v) \leq 2\}$ becomes a knowledge set.

PHASE 2: In each of $\lceil \log_2 n \rceil - 1$ communication rounds, reduce the size of a given knowledge set K by a factor of two, more precisely, we make $K' \subset K$ with $|K'| = \lceil \frac{|K|}{2} \rceil$ a knowledge set. Clearly, at the end of this phase the knowledge set obtained consists of a single vertex. Thus, accumulation is completed.

Next, we will give the details of the two phases.

In the first phase, the algorithm sends all information from critical nodes to non-critical ones, and additionally produces a knowledge set of size $\lceil n/2 \rceil$. This initial phase takes at most two steps.

Lemma 2. *Two rounds of communication in* $1EDP$ *mode suffice for* PHASE 1 *of the accumulation algorithm.*

Outline of the Proof. Let T be a tree with n vertices. We show by induction on n that there exists a communication pattern, such that within two rounds

1. a knowledge set S with $|S| = \lceil n/2 \rceil$ is produced,

2. each critical vertex acts as sender in one of the rounds, and is a connector or idle in the other round,

3. each non-critical vertex is idle in at least one round, and eventually acts as either receiver or sender in the other round,

4. each edge is used in at most one communication path,

5. in case that n is odd, an arbitrary leaf can be chosen that is not involved in any communication. We call such a leaf *free vertex*.

These properties immediately imply that the communication can be implemented in 1EDP mode. Since all critical nodes are senders, and never receive any message, all these vertices can be excluded from any knowledge set. This shows that $S \subseteq \{v \mid v$ is a non-critical vertex of $T\}$ holds.

For $n = 1$ the statement trivially holds. Let us assume that it holds for $n - 1$.

If n is odd, we remove an arbitrary leaf ℓ. The communication pattern for $T \setminus \{\ell\}$ producing a knowledge set S' is used for T, and $S' \cup \{\ell\}$ clearly forms a knowledge set for T of size $\lceil n/2 \rceil$.

If n is even, we remove a leaf ℓ of maximal depth from T. The pattern for $T \setminus \{\ell\}$ is modified and extended. Actually we distinguish three cases: ℓ has no sibling in T, ℓ has one sibling in T, and ℓ has ≥ 2 siblings in T. To show some of the technical details we discuss here case 3.

In this case we remove ℓ from T and choose a sibling $s(\ell)$ in $T \setminus \{\ell\}$ as the free vertex. Note that $p(\ell)$ is critical in this case and thus acts as a sender in one round. For the other round, where $p(\ell)$ is idle or acts as connector, we schedule the additional path $P = (s(\ell), p(\ell), \ell)$. The new knowledge set is obtained by substituting $s(\ell)$ by ℓ. The invariants can easily be checked now.

Note that both initial rounds may be performed simultaneously, i.e. in one round, if only FEDP mode is requested, since any edge is used in at most one communication path. $\qquad\qquad\qquad\qquad\qquad\qquad\qquad\qquad\qquad\qquad\qquad\qquad\qquad\quad\square$

The crucial task in PHASE 2 is the reduction of a knowledge set by a factor of two. The following lemma shows that such a reduction is always possible with one round of communication.

Lemma 3. *Let K be any knowledge set for T consisting of non-critical vertices only. Then there exist a set $K' \subset K$ with $|K'| = \lceil \frac{|K|}{2} \rceil$ and a communication pattern that can be implemented in one round in $1EDP$ mode, such that after performing this communication, K' is a knowledge set for T.*

Outline of the Proof. W.l.o.g. we assume that $|K|$ is even and show that some K' with $|K'| = \frac{|K|}{2}$ exists satisfying the specification of the lemma.

To determine K' and the associated communication pattern, we first decompose the tree T into subtrees $T_1, \ldots, T_k$. All occurring communications will be local to these subtrees. We call an edge e even, if the removal of e partitions T into two subtrees containing an even number of members of K each. Otherwise an edge is called odd. The partition of T into $T_1, \ldots, T_m$ is obtained by removing all even edges. Note that after removing an even edge, the remaining edges keep their type with respect to the obtained subtrees. Let K_i be the set of members of K residing in T_i. Then each subtree T_i has the following properties (we omit the proofs of this):

1. T_i contains an even number of elements of K.

2. All edges in T_i are odd with respect to T_i.

3. Each vertex of K_i is a leaf in T_i.

4. Each inner vertex of T_i has even degree.

Note that it suffices to show that there is some set $K_i' \subset K_i$ with $|K_i'| = \frac{|K_i|}{2}$ collecting all pieces of information from K_i within one communication round, for all i.

The set K_i' together with the associated communication pattern is constructed incrementally. We choose any two members k and k' from K_i, and include k' in K_i'. The pieces of information resident in k will be transferred to k' by communicating along the unique path between k and k' in T_i. To avoid re-usage of communication links, we next remove all edges on the path between k and k'. This decomposes T_i into subtrees $T_i^1, \ldots, T_i^l$. For any of these subtrees (except the isolated leaves k and k') properties 1.), 2.), 3.) and 4.) given above also hold, since each inner vertex of T_i modified during this removal of edges loses exactly two edges. Now we proceed in the same way with any subtree T_i^j containing some member of $K_i \setminus \{k, k'\}$. An easy induction shows that this iterative pairing process leads to a knowledge set of size $\frac{|K_i|}{2}$. $\square$

Theorem 4. *For any graph G of n nodes*

$$\lceil \log_2 n \rceil \leq A(G) = B(G) \leq \lceil \log_2 n \rceil + 2,$$
$$\lceil \log_2 n \rceil \leq A_{min}(G) = B_{min}(G) \leq \lceil \log_2 n \rceil + 1.$$

Proof. $A_{min}(G) = B_{min}(G) \leq \lceil \log_2 n \rceil + 1$ follows directly from the above discussion. Obviously $A(G)$ exceeds $A_{min}(G)$ by at most one. The lower bound is obvious. $\square$

Remark: Recalling that the two initial rounds of our accumulation scheme can be implemented in one round in FEDP mode, we obtain an alternative, and less technical proof of Farley's result, namely that in FEDP mode $A_{min}(G) = B_{min}(G) \leq \lceil \log_2 n \rceil$ holds.

The next result shows that our accumulation (broadcast) strategy cannot be improved. Let $C2T_h$ denote the complete binary tree of depth h for any $h \in \mathbb{N}$.

Lemma 5. *For any positive integer $h \geq 4$ (and $n = 2^{h+1} - 1$),*

$$A_{min}(C2T_h) = A(C2T_h) = h + 2 = \lceil \log_2 n \rceil + 1.$$

Proof. $C2T_h$ is a graph of degree 3, which means that each 1EDP is a 1VDP algorithm too. For 1VDP mode, this assertion was established in [FHMMM92]. $\square$

3   Gossip

In this section we consider gossiping on planar graphs, BF_k, CCC_k and d-dimensional grids. We omit all proofs in this section.

3.1   General Bounds

Following the fact that one one-way gossip algorithm in a graph G can be constructed as the concatenation of an accumulation algorithm to some node v and a broadcast algorithm from v in G, we get

$$R(G) \leq \min_{u \in V(G)} \{A_u(G) + B_u(G)\} = 2 \cdot B_{\min}(G).$$

Note that the broadcast algorithm in this concatenation can be always taken as the "reverse" of the preceding accumulation algorithm. For the two-way mode, we get in this way that the last round of the accumulation algorithm is the same as the first round of the broadcast algorithm and therefore can be omitted from the communication scheme. Thus, we obtain

$$R^2(G) \leq 2 \cdot B_{\min}(G) - 1.$$

Applying the results of Section 2, we have the following result.

Theorem 6. *For any graph G_n of n nodes, $n \geq 2$,*

(i) $\lceil \log_2 n \rceil \leq R^2(G_n) \leq 2 \cdot \lceil \log_2 n \rceil + 1,$

(ii) $\log_b(\lfloor n/2 \rfloor) + 2 \leq R(G_n) \leq 2 \cdot \lceil \log_2 n \rceil + 2.$

The lower bound of (i) and (ii) are tight because one can gossip in the complete graph of n nodes in this number of rounds. The fact that the upper bounds are tight is provided by the following theorem.

Theorem 7. *For each complete binary tree $C2T_h$ of depth $h \geq 3$ (and $n = 2^{h+1} - 1$ nodes),*

(i) $2h + 3 = 2 \cdot \lceil \log_2 n \rceil + 1 \leq R(C2T_h) \leq 2h + 4,$

(ii) $2h + 2 = 2 \cdot \lceil \log_2 n \rceil \leq R^2(C2T_h) \leq 2h + 3.$

The aim of the next subsections is to search for the gossip complexity of concrete interconnection networks in the interval given by Theorem 6.

3.2 Gossiping in 2-Dimensional Grids and Planar Graphs

This subsection contains technically the most interesting result of this section. First, the following result for the 2-dimensional grid Gr_n^2 of size $m \times m = n$ is established.

Theorem 8. *For each* $n = m^2$, $m \in I\!N$,

$$R^2(Gr_n^2) = 1.5 \cdot \log_2 n - \log_2 \log_2 n \pm O(1).$$

The optimality of the above stated result is still underlined by the following theorem claiming that Gr_n^2 is an optimal gossip structure in the class of all planar graphs of bounded degree. The proof is based on the bisection result of Diks et al. [DDSV93] and on some generalization of the lower bound proof technique working for vertex-disjoint path communication [HKSW93].

Theorem 9. *For any planar graph* $Pl(n, h)$ *of* n *nodes and degree bounded by* h,

$$R^2(Pl(n, h)) \geq 1.5 \log_2 n - \log_2 \log_2 n - 0.5 \log_2 h - 2.$$

Note that the quickest gossip algorithm in 2VDP mode in planar graphs takes $1.5 \log_2 n$ rounds, and we conjecture that the real difference between the 2VDP and the 2EDP mode in planar graphs is of order $\log_2 \log_2 n$.

We are able to extend the above stated results for one-way mode.

Theorem 10. *For every* $n = m^2$, $m \in I\!N$,

$$R(Gr_n^2) \leq (1 + (\log_b 2)/2) \cdot \log_2 n - (2 - \log_b 2) \cdot \log_2 \log_2 n + O(1)$$
$$= 1.72... \log_2 n - 0.56... \log_2 \log_2 n + O(1).$$

Although we are not able to get a corresponding closed lower bound (up to $O(1)$) for general gossip strategies, we have obtained such matching lower bounds for some special subclass of "regular" gossip algorithms. This lower bound holds not only for grids but for all planar graphs. We omit the exact formulation of this lower bound in this Extended Abstract, because of the technicalities involved in the definition of "regularity" of gossip algorithms. Note that nontrivial lower bounds for unrestricted one-way gossip algorithms are known only for very simple structures as cycles and complete graphs [EM89, KCV92, HJM90].

3.3 Gossiping in Fundamental Interconnection Networks

In this subsection, we establish some estimations on the gossip complexity of some fundamental interconnection networks. Extending the design and proof technique used in the previous subsection, the optimal gossip algorithms for d-dimensional grids Gr_n^d of size $n = m^d$ can be constructed.

Theorem 11. *For $d \geq 3$, $n \in \mathbb{N}$,*

(i) $R^2(Gr_n^d) = (1 + 1/d) \cdot \log_2 n - \log_2 n \log_2 n \pm O(d)$,

(ii) $R(Gr_n^d) \leq (\log_b 2 + (2 - \log_b 2)/d) \cdot \log_2 n + O(d)$

$$= (1.44.. : + 0.56.../d) \cdot \log_2 n + O(d).$$

Note that we can show that our algorithms work also in VDP modes, which means that the 2EDP mode is not essentially more powerful than the 2VDP mode for d-dimensional grids, $d \geq 3$. Clearly, this contrasts to the two-dimensional case, where the difference seems to be of order $\log_2 \log_2 n$.

Since the next results are completely based on the proof techniques developed in [HKSW93], we omit the proofs. For $k \in \mathbb{N}$, let BF_k denote the butterfly network, CCC_k the cube-connected-cycles network, and Q_k the hypercube network of dimension k.

Theorem 12. *For every $X_k \in \{BF_k, CCC_k, Q_k\}$ of n nodes and dimension k,*

$$R(X_k) \leq R(K_n) + O(\log_2 \log_2 n).$$

Theorem 13. *For every $Y_k \in \{BF_k, CCC_k\}$ of n nodes and dimension k,*

$$R^2(Y_k) \leq R^2(K_n) + O(\log_2 \log_2 n).$$

The above theorems show that some of the fundamental networks of constant degree have a gossip complexity very close to the gossip complexity of complete graphs. Obviously, this is a very positive observation. An interesting consequence of Theorems 12 and 13 is that BF_k and CCC_k are much better constant-degree structures for gossiping in EDP modes than d-dimensional grids.

4 Conclusion

In this paper, we gave an optimal broadcast strategy in 1EDP mode and some first results (some of them optimal) about the gossip complexity in edge-disjoint path modes. Note that all our results proved for 2EDP mode can be simply transformed into FEDP mode of Farley, too. There are several problems left open here, and we formulate some of them which are of our main interest.

Problem 4.1. Prove or disprove our conjecture that the gossip complexity of planar graphs in 2VDP mode is at least $1.5 \log_2 n$. This would show that 2EDP mode is more powerful than 2VDP mode for planar graphs, and especially for two-dimensional grids.

Problem 4.2. Prove a lower bound on the gossip complexity in the 1EDP mode for the 2-dimensional grid and all planar graphs for more general classes of one-way algorithms than the class of all "regular" algorithms.

References

[DDSV93] K. Diks, H.N. Djidjev, O. Sýkora, I. Vrťo, "Edge separators of planar and outerplanar graphs with applications", *Journal of Algorithms* 14 (1993), pp. 258-279.

[EM89] S. Even, B. Monien, "On the number of rounds necessary to disseminate information", *Proc. 1st ACM Symp. on Parallel Algorithms and Architectures* (SPAA'89), 1989, pp. 318-327.

[Fa80] A.M. Farley, "Minimum-Time Line Broadcast Networks", *Networks* 10 (1980), pp. 59-70.

[FHMMM92] R. Feldmann, J. Hromkovič, S. Madhavapeddy, B. Monien, P. Mysliwietz, "Optimal algorithms for dissemination of information in generalized communication modes", *Proc. Parallel Architectures and Languages Europe* (PARLE'92), Lecture Notes in Computer Science 605, Springer Verlag 1992, pp. 115-130.

[HJM90] J. Hromkovič, C. D. Jeschke, B. Monien, "Optimal algorithms for dissemination of information in some interconnection networks (extended abstract)", *Proc. MFCS'90*, Lecture Notes in Computer Science 452, Springer Verlag 1990, pp. 337-346.

[HKMP93] J. Hromkovič, R. Klasing, B. Monien, R. Peine, "Dissemination of Information in Interconnection Networks (Broadcasting and Gossiping)", manuscript, University of Paderborn, Germany, Feb. 1993, to appear as a book chapter in: F. Hsu, D.-Z. Du (Eds.), *Combinatorial Network Theory*, Science Press & AMS, 1994.

[HKS93] J. Hromkovič, R. Klasing, E.A. Stöhr, "Gossiping in vertex-disjoint paths mode in interconnection networks", *Proc. 19th Int. Workshop on Graph-Theoretic Concepts in Computer Science* (WG '93), Lecture Notes in Computer Science, Springer Verlag 1993, to appear.

[HKSW93] J. Hromkovič, R. Klasing, E.A. Stöhr, H. Wagener, "Gossiping in Vertex-Disjoint Paths Mode in d-Dimensional Grids and Planar Graphs", *Proc. of the First Annual European Symposium on Algorithms* (ESA '93), Lecture Notes in Computer Science 726, Springer Verlag 1993, pp. 200-211.

[KCV92] D.W. Krumme, G. Cybenko, K.N. Venkataraman, "Gossiping in minimal time", *SIAM J. Comput.* 21 (1992), pp. 111-139.

[KMPS92] R. Klasing, B. Monien, R. Peine, E. Stöhr, "Broadcasting in Butterfly and DeBruijn networks", *Proc. STACS'92*, Lecture Notes in Computer Science 577, Springer Verlag 1992, pp. 351-362.

[Kn75] W. Knödel, "New gossips and telephones", *Discrete Math.* 13 (1975), p. 95.

[Le92] F.T. Leighton, "Introduction to Parallel Algorithms and Architectures: Array, Trees, Hypercubes", *Morgan Kaufmann Publishers* (1992).

Recent results in hardness of approximation

Johan Håstad

Department of Numerical Analysis & Computer Science
Royal Institute of Technology
S-100 44 Stockholm, SWEDEN
johanh@nada.kth.se

Abstract

We survey some of the recent results in proving that approximating some NP-hard optimization problems remains NP-hard. This is a survey paper and it contains no new results.

1   Introduction

One of the basic notions of complexity theory is NP-completeness [11]. It is a basic fact that all NP-complete problems are hard. At a closer look, however, it turns out that different NP-complete problems are very different in nature. Some turn out to be easy on random instances while some seem to remain difficult even for most instances. Although there is a developed theory of NP-completeness when a probability distribution is put on the input [22, 18], this difference is not well understood.

When it comes to NP-hard optimization problems[1] there is also another parameter of interest. Namely suppose that we ask not for the optimal solution, but only a solution whose value is within a reasonably small factor of the optimal value. Note that we are here asking a worst case question. We want the algorithm to perform reasonably well for all instances. Since finding the optimal solution is interreducible for NP-complete problems one could expect that the cost of finding approximate solutions would be similar for different NP-complete problems. Looking more closely at the situation there is no obvious reason that this should be the case. The problem is that while traditional NP-reductions transform optimal solutions into optimal solution, they do not transform almost optimal solutions into almost optimal solutions. It was also noticed early that while some problems, like bin-packing and traveling salesperson with triangle inequality[2], were very easy to approximate, some other problems like clique

---

[1] We will be interested in the set of optimization problems where the corresponding decision problem is NP-complete, i.e. find the largest clique, shortest traveling salesperson tour etc

[2] For the best approximation algorithms see [21] and [10] respectively

seemed inherently hard. A very extensive list of approximation results for many problems is contained in the thesis by Kann [20].

2 Early results

Already from early on in the theory of NP-completeness, approximation properties of the optimization versions of NP-complete problems were studied. In particular, the famous book by Garey and Johnsson [14] contains an extensive discussion on this topic. As an example of a simple early result let us take graph coloring. One simply observes that since it is NP-complete to three-color a planar graph it is NP-hard to approximate the chromatic number within a factor 4/3 (if we accept the four-color theorem). The method used is quite general. To prove that a certain optimization[3] problem O is hard to approximate one starts with a NP-complete language L and defines a reduction f from L to O such that if $x \in L$ then the optimal value of $f(x)$ is A while if $x \notin L$ then the optimal value is at most B. Then there cannot be a polynomial time approximation algorithm that achieves a ratio which is better than B/A. This approach remains today the basic tool for proving hardness results. The main change is new techniques for constructing very powerful reductions.

One of the first systematic treatments of comparing the approximability of various optimization problems was done by Papadimitriou and Yannakakis [26]. They defined new reductions which preserved approximability and also introduced the class MAX-SNP, a class which contains many problems that all can be approximated within a constant. Under the given reducibility that class has many complete problems (one example is MAX-3SAT, i.e. given a 3SAT formula satisfy as many clauses as possible) that has the property of being approximable within any constant iff all problems in the class has this property.

Another early paper that is worth mentioning is the result by Berman and Schnitger [8] which showed that if it is possible to approximate maximal clique within a factor n^ϵ for any $\epsilon > 0$ in polynomial time, then it is possible to approximate MAX-3SAT within any constant in probabilistic polynomial time.

3 The start of a new era

To describe later developments we first have to discuss interactive proof systems. In a proof system there is a verifier and one or more provers. The prover(s) try to convince the verifier of a certain fact. In complexity theory the verifier is usually restricted to be probabilistic polynomial time. There are sometimes restrictions on the prover(s), but for this note, let us assume that they are infinitely powerful. A proof system is a proof system for a certain language L with error probability ϵ if when $x \in L$ then the probability that the prover(s)

---

[3]We assume for this discussion that O is a maximization problem

can convince the verifier of this fact is 1, while if $x \notin L$ the probability that the prover(s) falsely can convince the verifier that $x \in L$ is $\leq \epsilon$. To have a robust notion one assumes that $\epsilon \leq 1/2$, but it is not difficult to see that by repeating a proof many time with independent random coins we can reduce the error probability to $2^{-Q(n)}$ for any polynomial Q.

The traditional framework of NP can easily be formalized in this way. In this case the verifier can be deterministic polynomial time and we need only one prover. Suppose for instance that we are interested in the case when L is satisfiability. Then, given a formula φ the prover sends an assignment of the variables to the verifier. If the assignment satisfies the formula the verifier accepts and otherwise it rejects. It is easy to see that we have a correct proof system with no error, i.e. with $\epsilon = 0$.

Interactive proofs were introduced by Goldwasser, Micali and Rackoff [16] and Babai and Moran [5] in slightly different ways. The two models were later proved equivalent from a language recognition point of view by Goldwasser and Sipser [17]. In these proof systems the verifier is probabilistic polynomial time and it exchanges messages with a single prover.

Later, Ben-Or, Goldwasser, Kilian and Wigderson [9] introduced multiprover interactive proofs. In order for more than one prover to be useful, it is important that the various provers cannot talk to each other. One can simply view this as interrogating two suspects[4] separately in different rooms. Everyday life experience shows that this is much more efficient than allowing the two suspects to communicate.

It was quite soon discovered by Fortnow, Rompel and Sipser [13] that if we were only interested in the recognition power of polynomial time verifiers then only two provers were needed. They also introduced another, very appealing model of interactive proofs. In this model we have only one prover, but it has to commit to its answers on all questions before any question is asked. It turns out that this model has the same power as multi-prover interactive proofs. A short reflection on the case of the two suspects in different rooms might convince the reader that this is reasonable. If the two suspects in advance can fix a successful strategy, i.e. agree on all answers, then there is no problem, while if there is no such strategy the suspects will be very lucky if they can avoid inconsistencies.

We call proofs when one prover fixes all answers in advance *oracle interactive proofs* because the single prover acts like an oracle. These kind of proofs have also been called probabilistically checkable proofs [2].

The power of interactive proofs remained unknown for quite some time. In fact, only few examples (most famous was probably the case of graph-nonisomorphism [15]) outside NP were known to have either single or multiple prover interactive proofs. The first chock was delivered in a sequence of steps started by Lund, Fortnow, Karloff and Nisan [24] and finished by Shamir [27] through showing that single prover interactive proofs can do all of PSPACE. The

---

[4] A suspect should be thought of as a prover

second powerful result which was no longer a chock was when Babai, Fortnow and Lund [3] established that multi-prover (or one oracle prover) interactive proofs could recognize all of nondeterministic exponential time. It would take us too far to describe these protocols in detail, let us just mention that using algebraic techniques related to low degree polynomials was a very important part.

These results have no obvious relation to approximating NP-hard approximation problems, but the first such connection was given by Feige, Goldwasser, Lovász, Safra and Szegedy [12]. Since this connection is of fundamental importance let us describe it in some detail. For concreteness let us consider an oracle interactive proof with a probabilistic polynomial time verifier of a language L which is complete for nondeterministic exponential time. On an input x of length n the verifier flips $p_r(n)$ many random coins and asks the oracle $p_q(n)$ questions for some polynomial p_r and p_q. We can assume that the oracle answers one bit on each question and that the error probability ϵ is 2^{-n}. Let us assume that the ith question is q_i, the answer is a_i and let us denote the random coins used by the verifier by r. Note that q_i is uniquely determined by r and a_j for $j \leq i - 1$ since the verifier is deterministic once the random bits are fixed. Now consider the following graph G_x:

The nodes of G_x are given by $(r, (q_i, a_i)_{i=1}^{p_q(n)})$ where r, q_i and a_i are as indicated above. The graph G_x only contains the nodes that constitute a possible verifier-oracle interaction where the verifier has random coins r and which would have caused the verifier to accept. Two nodes $(r^1, (q_i^1, a_i^1)_{i=1}^{p_q(n)})$ and $(r^2, (l_i^2, a_i^2)_{i=1}^{q(n)})$ are connected iff whenever $l_i^1 = l_j^2$ then $a_i^1 = a_j^2$.

Please note that G_x depends on x since x affects which questions are asked and whether the verifier accepts the input.

To see the importance of G_x observe that a node gives a transcript of a successful interaction. Two nodes are connected if the two conversations could take place with the same oracle. Note also that a node is not connected to any other node with the same value of r. This follows since if the two answer sequences are consistent and start with the same random coins r, they lead to the same future questions and hence the same total interaction.

We claim that G_x has a clique of size S iff there is an oracle which causes the the verifier to accept with probability $S2^{-p_r(n)}$. To see this assume first that there is such an oracle. Consider all vertices $(r, (q_i, a_i)_{i=1}^{p_q(n)})$ in the graph such that the given oracle answers q_i with a_i for all i. By the definition of the graph these form a clique. Each node in the clique corresponds to an accepting conversation that takes place with a probability $2^{-p_r(n)}$. This follows since once r takes the given value the rest of the conversation follows automatically. Hence, if the probability of acceptance is $S2^{-p_r(n)}$ the size of the corresponding clique is S.

To see the converse we just need to observe that a set of nodes determine a partial oracle by setting the answers according to the answers in the nodes. If the nodes form a clique, the oracle is well defined and we again have the same relation between the size of the clique and probability of acceptance.

This means that when $x \in L$ the size of the maximum clique in the above graph is $2^{p(n)}$ while if $x \notin L$ it is bounded by $2^{p(n)-n}$.

We claim that size of G_x is bounded by $2^{p_r(n)+p_q(n)}$. This follows since we have $2^{p_r(n)}$ different values of r and since the questions q_i are determined by r and previous answers, each possible answer sequence gives at most one node and there are $2^{p_q(n)}$ such sequences.

Now suppose we had an algorithm A that runs in polynomial time and approximates the size of the maximal clique within a factor of 2. Then consider the following algorithm

1. Construct G_x

2. Run A on G_x to get answer S.

3. If $S \geq 2^{p_r(n)-1}$ answer 1 and otherwise answer 0.

By the above properties of G_x this algorithm would correctly recognize L. Furthermore, by the bound on the size of G_x it runs in time $2^{O(p(n)+q(n))}$. Since L was assumed to be complete for nondeterministic exponential time, this would imply that nondeterministic exponential time is equal to exponential time. A statement most people would believe to be false!

In fact by running the original interactive protocol several times it is possible to decrease the probability of incorrect acceptance and improve the above result such that for any $\epsilon > 0$, the existence of a polynomial time algorithm that approximates clique within a factor $2^{(\log n)^{1-\epsilon}}$ implies that nondeterministic exponential time is equal to deterministic exponential time.

Most people are more familiar with complexity classes which are closer to P and NP. It turns out that one can indeed scale down the interactive proof results to give very efficient interactive proofs for NP. In particular, [12][5] using the techniques of [3] was able to obtain a proof system for NP that used $O(\log n \log \log n)$ random coins and answer bits from the oracle. This implied that if clique could be approximated within some constant factor then NP would be contained in deterministic time $n^{O(\log \log n)}$. A very unlikely event!

4   Many new results

This first very nice result opened up two obvious directions of research. The first direction was simply to strengthen the results technically. I.e. getting larger factor of inapproximability and to weaken assumptions to something like $P \neq NP$. The second direction was to apply the existing techniques to other optimization problems.

---

[5]For different reasons [4] had also scaled down the results of [3] to proof system for NP.

The key for the first approach was clearly to get even more efficient interactive proofs. As indicated above the two crucial parameters are the number of random coins used and the number of answer bits needed. In particular if one wanted to base the intractability of approximating clique on the assumption that $NP \neq P$, one would need to reduce both these measures to at most $O(\log n)$. This was first done by Arora and Safra [2]. This result was later strengthened by Arora, Lund, Motwani, Sudan and Szegedy by showing that it was possible to use $O(\log n)$ random bits and $O(1)$ bits from the oracle. The interactive proofs were by this time very complicated. A proof had a recursive structure and in everyday terms it was a proof that there existed a proof that there existed a proof. Forgetting the complications let us just see what it says more explicitly.

Suppose a verifier is interested in whether a given formula of length n is satisfiable. The prover writes down a polynomial length string and the verifier flips $O(\log n)$ coins and reads a constant (independent of n!) number of bits in the proof. After this the verifier is convinced that the formula is satisfiable. The probability that he has been fooled is 1% (over his own coin flips).

This might sound like science fiction but is none the less true. The result has the following beautiful consequences.

Theorem 1 *[1] There is a constant $\epsilon > 0$ such that unless $P = NP$ the size of the maximal clique cannot be approximated within a factor n^ϵ.*

Theorem 2 *[1] The is a constant $c > 0$ such that unless $P = NP$ MAX-3SAT cannot be approximated within a factor $1 - c$.*

After these result were obtained methods were developed to deal also with many other problems. Let us just mention the most famous problems for which new results were proved. Lund and Yannakakis [25] proved that for some constant $\epsilon > 0$, chromatic number could not be approximated with n^ϵ unless $P = NP$. For set cover (i.e. given a family of sets, find the smallest number of sets whose union contains all the elements) the old approximation algorithm of Johnson [19] and Lovàsz [23], giving approximation within a factor $O(\log n)$, was proved to probably be optimal. Lund and Yannakakis [25] and Bellare, Goldwasser, Lund and Russel[6] proved that set cover can not be approximated within a factor $o(\log n)$ in polynomial time. unless NP is contained in deterministic time $n^{O(\log \log n)}$.

Even though in most cases the exact value of constants is not an important problem, I do believe that getting better values for the constants in the above theorems is very interesting. The original construction gave rather poor constants which were not calculated explicitly, but both ϵ and c were at most 10^{-4}. The first major improvement is due to Bellare, Goldwasser, Lund and Russel [6] who by decreasing the number of bits in the answer improved the value of ϵ in Theorem 1 to $1/29$. Later results by Bellare and Sudan [7] use additional insight to improve ϵ (in n^ϵ) for the clique approximation problem to $1/4 - o(1)$ and for chromatic number of $1/10 - o(1)$. They also gave a slight improvement over the value of [6] for the value of c in Theorem 2 to $1/66$. These results depend, however, on the assumption that NP is not contained in probabilistic time which is $2^{O(\log n)^d}$ for some constant d.

5 Future directions

There are several open problems, some that are more or less technical in nature.

- What is the best value for the constant ϵ in Theorem 1? The most important question is whether it can be arbitrarily close to 1. Bellare and Sudan [7] conjecture that this is the case.

- If we unravel the proof of Theorem 2 then it simply takes a 3SAT-formula φ and transforms it do a different 3SAT-formula ψ. The reduction has the property that if φ is satisfiable then so is ψ while if φ is not satisfiable then no assignment satisfies more than a fraction $1 - c$ of the clauses in ψ. The construction of ψ is polynomial time, but extremely complicated. Now that we know that such a construction exists maybe we can find an easy one?

- Of course we would like to base all intractability results for approximation on $NP \neq P$.

References

[1] Arora S., Lund C., Motwani R., Sudan M, and M. Szegedy, "Proof verification and intractability of approximation problems." *Proc. 33rd IEEE Symposium on Foundation of Computer Science*, October 1992, pp. 14-23.

[2] Arora S. and Safra S. "Probabilistic Checkable Proofs: A New Characterization of NP." *Proc. 33rd IEEE Symposium on Foundation of Computer Science*, October 1992, pp. 1-13.

[3] Babai L., Fortnow L, and Lund C. "Non-deterministic Exponential time has Two-Prover Interactive Protocols *Proc. 31st IEEE Symposium on Foundation of Computer Science*, October 1990, pp. 16-25.

[4] Babai L., Fortnow L, Levin L, and Szegedy M. "Checking Computations in Polylogarithmic Time" *Proc 23rd ACM Symposium on theory of computation*, May 1991, pp 21-31.

[5] Babai L. and Moran S. "Arthur-Merlin Games: A Randomized Proof System and a Hierarchy of Complexity Classes", *Journal of Computer and System Sciences*, Vol 36, pp 254-276, 1988.

[6] Bellare M., Goldwasser S., Lund C., and Russell A. "Efficient Probabilistically Checkable Proofs; Applications to Approximation",*Proc 25th ACM Symposium on theory of computation*, May 1993, pp 294-304.

[7] Bellare M. and Sudan M. "Improved Non-Approximability Results", manuscript 1993, to appear at 26th ACM Symposium on theory of computation, May 1994.

[8] Berman P. and Schnitger G. "On the Complexity of approximating the independent set problem", *Proceedings of 6th Annual Symposium on Theoretical Aspects of Computer Science*, pp 256-268, 1989. Springer Verlag, Lecture Notes in Computer Science 349.

[9] Ben-Or M., Goldwasser S., Kilian J. and Wigderson A. "Multi-Prover Interactive Proofs: How to remove Intractability", *Proceeding 20th ACM Symposium on Theory of Computing*, 1988, pp 113-131.

[10] Christofides N. "Worst-case analysis of a new heuristic for the traveling salesman problem", Technical report, Graduate School of Industrial Administration, Carnegie-Mellon University, Pittsburgh, 1976.

[11] Cook S. A. "The complexity of Theorem Proving Procedure." *Proceeding 3rd ACM Symposium on Theory of Computing*, 1971, pp 151-158.

[12] Feige U., Goldwasser S., Lovász L, Safra S. and Szegedy M. "Approximating Clique is Almost NP-complete" *Proc. 32nd IEEE Symposium on Foundation of Computer Science*, October 1991, pp. 2-12.

[13] Fortnow L., Rompel J. and Sipser M. "On the power of Multi-Prover Interactive Protocols" *Proceedings 3rd IEEE Symposium on Structure in Complexity Theory*, pp 156-161, 1988.

[14] Garey M. R. and Johnson D.S. "Computers and intractability; a guide to the theory of NP-completeness", W.H. FREEMAN, 1979.

[15] Goldreich O., Micali S., and Wigderson A. "Proofs that Yield Nothing but their Validity or All Languages in NP have Zero-Knowledge Proof System", *Journal of ACM*, Vol 38, 1991, pp 691-729.

[16] Goldwasser S., Micali S. and Rackoff C. "The Knowledge Complexity of Interactive Proof Systems", *SIAM Journal on Computing*, Vol 18, pp 186-208, 1989.

[17] Goldwasser S., and Sipser M. "Private Coins versus Public Coins in Interactive Proof Systems, *Proceeding 18th ACM Symposium on Theory of Computing*, 1986, pp 59-68.

[18] Gurevich Y, "Average Case Completeness", *Journal of Computer and System Sciences*, Vol 42, 1991, pp 346-398.

[19] Johnson D. "Approximation algorithms for combinatorial problems" *Journal of Computer and System Sciences*, Vol 9, 1974, pp 256-278.

[20] Kann V. "On the approximability of NP-complete optimization problem", Ph. D. thesis, 1992, department of numerical analysis and computing science, Royal Institute of Technology.

[21] Karmarkar N. and Karp R. M. "An efficient approximation scheme for one-dimensional bin packing problem", *Proc. 23rd IEEE Symposium on Foundation of Computer Science*, 1982, pp. 312-320.

[22] Levin, L. "Average Case Complete Problems" *SIAM Journal on Computing*, Vol 15, 1986, pp 285-286.

[23] Lovász L. "On the ration of optimal integral and fractional covers" *Discrete mathematics*, Vol 13, 1975, pp 383-390.

[24] Lund C., Fortnow L., Karloff H. and Nisan N. "Algebraic Methods for Interactive Proof Systems" *Proc. 31st IEEE Symposium on Foundation of Computer Science*, October 1990, pp. 2-10.

[25] Lund C. and Yannakakis M. "On the Hardness of Approximating Minimization Problems" *Proceeding 25th ACM Symposium on Theory of Computing*, 1993, pp 59-68.

[26] Papadimitriou C. and Yannakakis M. "Optimization, approximation and complexity classes" *Journal of Computer and System Science*, vol 43 pp 425-440, 1991.

[27] Shamir A. "IP=PSPACE", *Proc. 31st IEEE Symposium on Foundation of Computer Science*, October 1990, pp. 11-15.

The Parallel Hierarchical Memory Model

Ben H.H. Juurlink and Harry A.G. Wijshoff

High Performance Computing Division, Department of Computer Science
Leiden University, P.O. Box 9512, 2300 RA Leiden, The Netherlands
{benj,harryw}@cs.leidenuniv.nl

Abstract. Modern computer systems usually have a complex memory system consisting of increasingly larger and slower memory. Traditional computer models like the Random Access Machine (RAM) have no concept of memory hierarchy, making it inappropriate for an accurate complexity analysis of algorithms on these types of architectures.

Aggarwal et al. introduced the Hierarchical Memory Model (HMM). In this model, access to memory location x requires $f(x)$ instead of constant time. In a second paper they proposed an extension of the HMM called the Hierarchical Memory Model with Block Transfer (HMBT), in which a block of consecutive locations can be copied in unit time per element after the initial access latency.

This paper introduces two extensions of the HMBT model: the Parallel Hierarchical Memory Model with Block Transfer (P-HMBT), and the pipelined P-HMBT (PP-HMBT). Both models are intended to model memory systems in which data transfers between memory levels may proceed concurrently.

Tight bounds are given for several problems including dot product, matrix transposition and prefix sums. Also, the relationship between the models is examined. It is shown that the HMBT and P-HMBT are both strictly less powerful than the PP-HMBT. It is also shown that the HMBT and P-HMBT are incomparable in strength.

Index Terms — Hierarchical memory, data locality, algorithms.

1 Introduction

In order to keep up with the memory request rate of the central processing unit, modern computer systems usually have a complex memory system consisting of a relatively small amount of registers, followed by one or two levels of cache and main memory. The idea behind using such a hierarchical memory organization is that, by the principle of locality, most of the memory accesses can be made from the fast memory, so that the average access time is closer to the access time of the fast memory. Thus it is important in such an environment to reuse data as much as possible before returning it to slower memory. This property of an algorithm is usually called *temporal locality*. Another form of reuse is *spatial locality*. Spatial locality occurs when consecutive memory locations are accessed. This is important because the time to access one word may be long but several consecutive addresses can be transferred rapidly, since the memory can be organized in banks that allow one clock cycle for each access.

An important aspect for memory hierarchies is parallelism in block transfer, meaning that transfers at different levels may proceed concurrently and may be overlapped with processing. Traditionally, this represents one of the early uses of parallelism in computers. In the next section we introduce the Parallel Hierarchical Memory Model with Block Transfer (P-HMBT), that enables parallelism to be exploited between block transfers.

2 Definition of the Model

The most frequently used model of a computer is the *Random Access Machine* or RAM [3]. It consists of an accumulator, a read-only input tape, a write-only output tape, a program and a memory. The memory consists of a set of memory cells $\underline{A} = \{A_n : n \in \mathbb{N}\}$ and each cell is capable of holding an integer of arbitrary size. We will write n for A_n. There is no upper bound on the number of cells that can be used.

In the RAM there is no concept of memory hierarchy; each memory access is assumed to take one unit of time. This may be appropriate in cases where the size of the problem is small enough to fit in the main memory, but in the presence of registers, caches, disks and so on, this model may be inaccurate. In a more realistic model the memory access time is measured by a nondecreasing function $f : \underline{A} \to \mathbb{N}$.

In [1] the *Hierarchical Memory Model* (HMM) was proposed. The control is identical to that of the RAM, but access to location x requires $f(x)$ instead of constant time. In the HMM, however, there is no concept of block transfer to utilize spatial locality in algorithms. The *Hierarchical Memory Model with Block Transfer* (HMBT) proposed in [2], is defined as the HMM, but in addition a block copy operation $BC(x, y, l)$ is present. It copies the locations $x + i$ to $y + i$ for $0 \le i \le l$ and is valid only if the intervals $[x, x + l]$ and $[y, y + l]$ are disjoint. A block copy operation is assumed to take $\max(f(x), f(y)) + l$ time.

In the HMM and the HMBT, the memory can be viewed as consisting of a hierarchy of levels. A *memory level* is defined as a continuous array of memory cells $A_m \ldots A_n$ such that $f(A_i) = f(A_{i+1})$, for $m \le i < n$, and $f(A_{m-1}) < f(A_m)$ and $f(A_n) < f(A_{n+1})$. We are now able to give a description of the P-HMBT model.

Definition 1. The concept of control, costs and block transfer of the P-HMBT are identical to the HMBT, except for the requirement that block transfers cannot cross level boundaries and can only take place between successive levels. In addition, block transfers are allowed to proceed in parallel between levels. Specifically, if $C_1, C_2, \ldots, C_m$ is a sequence of block transfers such that $C_i = BC(x_i, y_i, l_i)$, then these block transfers can be executed in parallel provided that all x_i's are in different levels. Such a parallel block transfer is written as $C_1 \| C_2 \| \ldots \| C_m$ and is assumed to take $\max_i t_i$ time, where $t_i = \max(f(x_i), f(y_i)) + l_i$.

We take $f(x) = \lfloor \log x \rfloor$ as a model for realistic memory systems. The choice for $\lfloor \log x \rfloor$ is technical, but we believe that it resembles real life well enough to give realistic qualitative predictions. In particular, it resembles memory systems consisting of increasingly larger and slower memory.

The function $f(x) = \lfloor \log x \rfloor$ partitions the memory into levels of continuous locations such that the i-th level $L(i)$ contains the memory locations $[2^i, 2^{i+1}-1]$. We will assume that concurrent block moves like $BC(x, y, l) \parallel BC(y, z, l)$ where x, y and z are in successive levels, cause no memory conflicts, since we can alternate between the odd numbered and even numbered levels. This will increase the running time by at most a constant factor. Alternatively, we can double the memory available to an algorithm by moving all data to the next lower level and use the upper half of each level as a buffer, without increasing the running time by more than a constant factor.

3 Bounds for Simple Problems

Consider the following problem which is called the *touch problem*. Initially, n inputs $a_1, a_2, \ldots, a_n$ are stored in memory locations $1, 2, \ldots, n$. An algorithm is said to touch the input a_i if, at some time during the execution of the algorithm, a_i is stored in location 1 of the hierarchical memory. The problem is to touch all inputs. This yields a lowerbound for various algorithms in which the entire input has to be examined. Examples of such algorithms include merging, searching a random sequence and computing the dot product of two vectors.

Theorem 2. *Any algorithm that touches n inputs on the P-HMBT with access cost function $f(x) = \lfloor \log(x) \rfloor$ requires $\Omega(n \log^*(n))$ time.*

Proof. We prove the theorem by using the same technique as in [2]. Let $b_i(t)$ be the least k such that a_i has been stored in memory location k during one of the first t steps of the computation. Define the *potential* at step t as $\phi(t) = \sum_{i=1}^{n} \log^*(b_i(t))$. The initial potential is $\phi(0) = \sum_{i=1}^{n} \log^*(i)$. When the algorithm terminates after T steps, each input has been stored in memory location 1 during one of the first T steps of the computation. Hence the final potential is $\phi(T) = \sum_{i=1}^{n} \log^*(1) = 0$. Thus $\phi(0) - \phi(T) = \sum_{i=1}^{n} \log^*(i) = \Omega(n \log^*(n))$.

We prove the theorem by showing that any move that takes time L on the P-HMBT decreases the potential by at most $O(L)$. From this the theorem follows. W.l.o.g. assume that $L = k + 2^k$ for some integer k. In L time units we can copy one location from level L of the hierarchical memory to level $L-1$, two locations from level $L-1$ to level $L-2$, $\ldots$, and $L - k = 2^k$ locations from level $k+1$ to level k. On level i, $1 \leq i \leq k$, we can only copy 2^{i-1} locations to the next memory level in this move (the size of the next higher level). The decrease in potential is maximal if we copy the last locations of level i to the first locations of level $i-1$. In other words, the following move decreases the potential as much as possible in L time units.

- On level i, $1 \leq i \leq k$, the memory locations $2^i + 2^{i-1} + j$, $0 \leq j \leq 2^{i-1} - 1$, are copied to locations $2^{i-1} + j$.

- On level i, $k+1 \leq i \leq L$, the locations $2^{i+1} - 1 - j$, $0 \leq j \leq L-i$, are copied to locations $2^{i-1} + j$.

It follows that the decrease in potential due to this move is at most

$$
\Delta\phi = \sum_{i=1}^{k} \sum_{j=0}^{2^{i-1}-1} (\log^*(2^i + 2^{i-1} + j) - \log^*(2^{i-1} + j))
$$

$$
+ \sum_{i=k+1}^{L} \sum_{j=0}^{L-i} (\log^*(2^{i+1} - 1 - j) - \log^*(2^{i-1} + j))
$$

$$
\leq \sum_{i=1}^{k} \sum_{j=0}^{2^{i-1}-1} (\log^*(2^{i+1}) - \log^*(2^{i-1})) + \sum_{i=k+1}^{L} \sum_{j=0}^{L-i} (\log^*(2^{i+1}) - \log^*(2^{i-1})).
$$

It can be shown that both terms are bounded by $O(L)$ [8]. $\qquad\square$

Remark When describing algorithms for the P-HMBT model, it is generally assumed that the input is located in memory locations $1, 2, \ldots, n$ or in memory locations $n, n+1, \ldots, 2n-1$. Both initial situations are equivalent in the following sense. We can go from one situation to the other in time $O(n)$ by repeatedly moving each level concurrently to the next higher (lower) level. So, if an algorithm takes at least $\Omega(n)$ time, this does not affect the running time by more than a constant factor.

Theorem 3. *The following problems can be solved in time $O(n \log^*(n))$ on the P-HMBT with access cost function $f(x) = \lfloor \log(x) \rfloor$.*

(i) The touch problem,
(ii) searching a random sequence of size n, and
(iii) computing the dot product of two vectors of length n each.

Proof. (i) Initially, the $n = 2^m$ inputs are stored in locations $n, n+1, \ldots, 2n-1$, i.e. at level $\log(n)$ of the hierarchical memory. Treat it as a sequence of $n/\log(n)$ blocks B_i, $1 \leq i \leq n/\log(n)$, of length $\log(n)$ each. First move B_1 to level $\log(n) - 1$ of the hierarchical memory and then concurrently move B_1 to level $\log(n) - 2$ and B_2 to level $\log(n) - 1$. Continue in this pipelined fashion until B_1 reaches level $\log\log(n)$ and repeat recursively on that block. Then move the next block from level $\log\log(n)+1$ to level $\log\log(n)$ and concurrently move each block on level i, $\log\log(n) + 2 \leq i \leq \log(n)$, to the next higher level. Continue until all blocks have been processed. The running time $T(n)$ of this algorithm obeys the recurrence relation

$$
T(n) = n/\log(n)T(\log(n)) + O((\log(n) - \log\log(n))\log(n)) + O(n)
$$
$$
= n/\log(n)T(\log(n)) + O(n),
$$

which solves to $T(n) = O(n \log^*(n))$. The same idea applies to *(ii)* and *(iii)*. $\square$

4 Relationship Between the Models

In this section we explore the relationship between the HMBT model and the P-HMBT model. First the following definition is needed. A machine model M is said to be hierarchically weaker than M' ($M \preceq M'$) if each problem that can be solved on model M in time T can also be solved on model M' in time $O(T)$. We want to determine if HMBT $\preceq$ P-HMBT or P-HMBT $\preceq$ HMBT.

Theorem 4. *HMBT $\npreceq$ P-HMBT.*

Proof. It can be shown that search operations require at least $\Omega(\log^2(n))$ time on the P-HMBT model, since at least one location must be accessed and access to location x requires $\sum_{i=0}^{\log x} i = \Theta(\log^2(x))$ time on this model. This holds regardless of the data structure used. On the HMBT, however, search operations can be performed in time $O(\log^2(n)/\log\log(n))$ by using a B-tree [4] where the buckets have size $\log(n)$. Each bucket is organized as a perfectly balanced binary search tree that is stored in continuous memory locations. To search the tree $O(\log(n)/\log\log(n))$ buckets must be searched. This can be done in time $O((\log\log(n))^2)$ per bucket by first moving each bucket to the first $O(\log(n))$ memory locations. Each bucket can be moved to fast memory in time $O(\log(n))$. Thus, search operations on this data structure can be performed in time

$$O(\log(n)/\log\log(n)((\log\log(n))^2 + \log(n))) = O(\log^2(n)/\log\log(n))$$

on the HMBT model. $\qquad\square$

Lemma 5. *Any algorithm that transposes a $\sqrt{n} \times \sqrt{n}$ matrix requires $\Omega(n\log^*(n))$ time on a HMBT machine.*

Proof. The proof of this lemma can be found in [2]. $\qquad\square$

Lemma 6. *A $\sqrt{n} \times \sqrt{n}$ matrix A can be transposed on a P-HMBT machine in optimal $O(n)$ time.*

Proof. Assume the input is in memory locations $n, n+1, \ldots, 2n-1$. The algorithm works essentially as follows. First, $A(1, 1\ldots\sqrt{n})$ is moved to memory locations $2n, 2n+1, \ldots, 2n+\sqrt{n}-1$. Then, $A(1, 2\ldots\sqrt{n})$ is moved to locations $4n, 4n+1, \ldots, 4n+\sqrt{n}-2$ and, concurrently, $A(2, 1\ldots\sqrt{n})$ is moved to $2n+1, 2n+2, \ldots, 2n+\sqrt{n}$. After that, $A(1, 3\ldots\sqrt{n})$ is moved to $8n, 8n+1, \ldots, 8n+\sqrt{n}-3$, $A(2, 2\ldots\sqrt{n})$ to $4n+1, 4n+2, \ldots, 4n+\sqrt{n}-1$ and $A(3, 1\ldots\sqrt{n})$ is moved to $2n+2, 2n+3, \ldots, 2n+\sqrt{n}+1$. The algorithm proceeds in this way until $A(i, j)$ is in memory location $2^j \cdot n + i - 1$ for $1 \leq i, j \leq \sqrt{n}$. Then the transposed matrix is transferred back to level $\log(n)$ of the hierarchical memory. A more formal description of this algorithm is given below.

> for $i \leftarrow 0$ to $\sqrt{n} - 1$ do
>   do steps (a) and (b) concurrently
>   (a) $BC(n + i\sqrt{n}, 2n + i, \sqrt{n} - 1)$

(b) **for** $j \leftarrow 1$ **to** i **do concurrently**
$$BC(2^j \cdot n + i - j + 1, 2^{j+1} \cdot n + i - j, \sqrt{n} - 1 - j)$$
od
for $i \leftarrow 1$ **to** $\sqrt{n} - 1$ **do**
    for $j \leftarrow i$ **to** $\sqrt{n} - 1$ **do concurrently**
$$BC(2^j \cdot n + \sqrt{n} - (j - i), 2^{j+1} \cdot n + \sqrt{n} - (j - i) - 1, \sqrt{n} - 1 - j)$$
od
for $i \leftarrow 0$ **to** $\sqrt{n} - 1$ **do**
    do steps (a) and (b) concurrently
    (a) $BC(2n, n + i\sqrt{n}, \sqrt{n} - 1)$
    (b) **for** $j \leftarrow 1$ **to** $\sqrt{n} - 1 - i$ **do concurrently**
$$BC(2^{j+1} \cdot n, 2^j \cdot n, \sqrt{n} - 1)$$
od

The costs are $O(\sqrt{n})$ per block move. Thus the time for all these moves is bounded by $O(n)$.   □

Theorem 7. *P-HMBT $\not\preceq$ HMBT.*

Proof. Follows directly from lemma 5 and lemma 6.   □

Until now we assumed that only one block move per memory level was allowed in each step, and a next block copy operation could not start until all previous block moves had been completed. A more powerful model is obtained if these restrictions are removed. In the *pipelined* P-HMBT model (PP-HMBT), the only condition that must be satisfied is that during each cycle only one data word can be transferred from level i to level $i - 1$ or vice versa. As an example, consider the following sequence of block moves.

$$\text{Initiate at cycle 0: } BC(y, x, 2) \parallel BC(z, y, 2)$$
$$\text{Initiate at cycle 3: } BC(y, x + 3, 2)$$

In figure 1, the content of each level during each cycle is illustrated.

This example shows two important differences between the pipelined P-HMBT model and the "straight" P-HMBT model. First, a block copy operation involving a memory cell x may be started although the previous block move involving cell x has not been completed. Secondly, it is not required for a block move $BC(x, y, l)$ that all memory cells $[y, y + l]$ are free by the time the transmission of $[x, x + l]$ is started. Instead, if a data word d occupies a cell x at cycle k, then that cell must become free at cycle $k - 1$.

Obviously, P-HMBT $\preceq$ PP-HMBT, since any computation for the P-HMBT model is a legal computation on the PP-HMBT model and can be executed on the PP-HMBT in the same amount of time.

Theorem 8. *Any algorithm that takes time T on a HMBT machine, can be modified into an algorithm that takes time $O(T)$ on a pipelined P-HMBT machine. Hence, HMBT $\preceq$ PP-HMBT.*

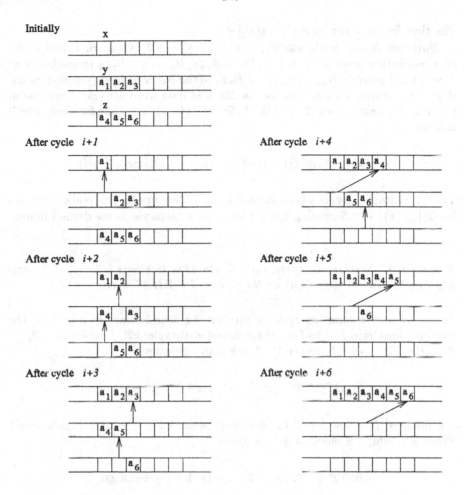

Fig. 1. Pipelining data transfers on the PP-HMBT machine.

Proof. Consider a block move $BC(x, y, l)$ and let $x > y$ (the case $x < y$ is treated similarly). The time for this move on the HMBT is $\lfloor \log(x) \rfloor + l$. This move is simulated as follows. Let $L(i)$ denote the i-th level of the hierarchical memory. The contents of memory locations $2^i, \ldots, 2^{i+1} - 1$ of the HMBT machine are stored in the lower part of $L(i + 2)$ in the pipelined P-HMBT. The upper part of each level is used as a buffer. Let $\alpha = \lfloor \log(y) \rfloor$ and let $\gamma = \lfloor \log(x) \rfloor$. Then, the block $B = [x, x + l]$ is contained in memory locations $2^\gamma \ldots 2^{\gamma+2} - 1$ of the HMBT machine (since $l < x$), and therefore in $L(\gamma + 2) \cup L(\gamma + 3)$. Copy the part of B that is in $L(\gamma + 2)$ to the buffer of $L(\gamma + 1)$ and then back to the lowest numbered locations in the buffer of $L(\gamma + 2)$. Move the part of B that is in $L(\gamma + 3)$ to the buffer of $L(\gamma + 2)$ to obtain a continuous block to be moved.

The time for these moves is $O(\lfloor\log(x)\rfloor + l)$.

Partition B into subblocks B_i, $\alpha + 2 \leq i \leq \gamma + 2$, where B_i contains the data words that have to go to $L(i)$. Move $B_{\alpha+2}, B_{\alpha+3}, \ldots, B_{\gamma+2}$ to the buffer of $L(\gamma+1)$ and *pipeline* $B_{\alpha+2}, B_{\alpha+3}, \ldots, B_{\gamma+1}$ to the buffer of $L(\gamma)$ (pipeline means that the transmission starts as soon as the first data word arrives). Continue in this way until each block B_i is in the buffer of $L(i-1)$. The time for these moves is bounded by

$$\lfloor\log(x)\rfloor + \lfloor\log(x)\rfloor - \lfloor\log(y)\rfloor + l - 1 = O(\lfloor\log(x)\rfloor + l).$$

Then, simultaneously copy each B_i back to $L(i)$. The time for this is also bounded by $O(\lfloor\log(x)\rfloor + l)$. Summing up over these three steps yields the desired bound.
□

Theorem 9. *Touching n inputs, initially stored in locations $1, 2, \ldots, n$ of memory, can be done in time $O(n)$ on the pipelined P-HMBT model.*

Proof. At each computing cycle, a number of transmissions are initiated. The transmissions from $L(i)$ to $L(i-1)$ are initiated at cycles $k2^{i-1}$ for $k = 0, 1, 2, \ldots$. At cycles $k2^{i-1}$ with k is even the block copy operations

$$BC(2^i, 2^{i-1}, 2^{i-1} - 1), \text{ for } 1 \leq i \leq \log n,$$

are initiated (the lower half of each level is moved to the next higher level). When k is odd, the block copy operations

$$BC(2^i + 2^{i-1}, 2^{i-1}, 2^{i-1} - 1), \text{ for } 1 \leq i \leq \log n,$$

are initiated (the upper half of each level is moved to the next higher level). It is easily verified that with this sequence of block moves, exactly one data word is transferred between each pair of successive levels during each cycle. From this observation the time bound follows. It remains to be shown that no data is overwritten during execution. Consider two successive levels $L(i)$ and $L(i + 1)$. The transmissions from $L(i + 1)$ to $L(i)$ are initiated at cycles $k2^i$; location $2^{i+1} + j$, $0 \leq j \leq 2^i - 1$, is transferred to location $2^i + j$ during cycle $k2^i + i + 1 + j$ and location $2^{i+1} + j$, $2^i \leq j \leq 2^{i+1} - 1$, is transferred to location j during cycle $k2^i + i + 1 + j$. Likewise, the transmissions from $L(i)$ to $L(i - 1)$ are initiated at cycles $k2^{i-1}$; location $2^i + j$, $0 \leq j \leq 2^i - 1$, is moved during cycle $k2^{i-1} + i + j$. In other words, if a cell x at level $L(i + 1)$ is copied to a cell y at level $L(i)$ during cycle k, then that cell became free during cycle $k - 1$. □

This theorem shows that the PP-HMBT model is strictly more powerful than the HMBT model and the P-HMBT model.

5 More Results for the P-HMBT and PP-HMBT Model

Prefix Sums Let $\oplus$ be an associative operator over a domain D. The prefix sums problem is defined as follows. Given an array $A = (a_1, a_2, \ldots, a_n)$ of size n. It is required to compute the sums

$$s_i = a_1 \oplus a_2 \oplus \cdots \oplus a_i$$

for $i = 1, \ldots, n$. The prefix sums can be computed in $O(n)$ time on the RAM. From theorem 2, a lowerbound of $\Omega(n \log^*(n))$ follows for computing prefix sums on the P-HMBT. We will show that the prefix sums problem can be solved in time $O(n(\log^*(n))^2)$ on the P-HMBT.

Suppose that initially the array A is stored in memory locations $n, n + 1, \ldots, 2n - 1$, i.e. at level $\log(n)$ of the hierarchical memory. As in theorem 2, treat it as a sequence of $n/\log(n)$ blocks of length $\log(n)$ each.

1. Move these blocks, using the pipelining strategy, to higher levels until the first block reaches level $\log\log(n)$. Repeat recursively on that block and continue until all blocks have been processed.
   At this point we have computed the sums

   $$m_{ij} = a_{(i-1)\log n+1} \oplus a_{(i-1)\log n+2} \oplus \cdots \oplus a_{(i-1)\log n+j}$$

   for $1 \leq i \leq n/\log(n)$ and $1 \leq j \leq \log(n)$.
2. Copy $m_{i,\log n}$, $1 \leq i \leq n/\log(n)$, to memory locations $n/2 + i - 1$, i.e. to level $\log n - 1$. Call this array $Q = (q_1, q_2, \ldots, q_{n/\log n})$.
3. Move the array Q to level $\log(n/\log(n)) = \log(n) - \log\log(n)$ and call prefix sums recursively.
4. Add q_i, $1 \leq i \leq n/\log(n) - 1$ to $m_{i+1,j}$, $1 \leq j \leq \log(n)$. It can be verified that adding a scalar to an array of size m can be done in $O(m \log^*(m))$ time by adapting the algorithm given in the proof of theorem 2.

The running time $T(n)$ of this algorithm fulfills the recurrence

$$T(n) = n/\log(n)T(\log(n)) + T(n/\log(n)) + O(n \log^*(n)).$$

Since $T(2) = constant$, it can be verified that $T(n) = O(n(\log^*(n))^2)$.

On a PP-HMBT machine, prefix sums can be performed in time $O(n)$ as follows. In location 1, the partial sum s_i is stored. The inputs are stored in the lower half of each level and are "fed" to location 1 using a similar scheme as for the touch problem. The outputs (i.e. the partial sums s_i for $1 \leq i \leq n$) are directed along the upper half of each level. It can be shown that this algorithm can be made to run in time $O(n)$ on the PP-HMBT model.

List Ranking Another important problem is *list ranking*. Given a linked list of n elements. The list ranking problem is to determine for each element its distance to the end of the list. Usually, the linked list is represented by two arrays C (the contents or data array) and S (the successor array). The first element in the list (the head) is stored in $C(1)$, and for $1 \leq i \leq n-1$, if the i-th element is stored in $C(j)$, then the $(i+1)$-th element is stored in $C(S(j))$. The list ranking problem can be solved in time $O(n)$ on the RAM.

Many parallel algorithms for the list ranking problem are *shortcut* based, meaning that the essential step in the execution is

$$S(i) \leftarrow S(S(i)).$$

The difficulty is to avoid that two consecutive elements are shortcut in parallel. A solution to this problem has been presented by Kruskal *et al.* [9]. We implement their algorithm where the numbers of processors p equals $\sqrt{n}$. The algorithm partitions the nodes of the list into a $\sqrt{n} \times \sqrt{n}$ array. Each processor is assigned one row of the array. The processors visit the columns of the array synchronously. If the successor of the node visited is not in the same column, the node is compacted with its successor, the node is marked and its successor is removed from the list (marked as deleted). This ensures that no two consecutive elements $C(i)$ and $C(S(i))$ are shortcut simultaneously. This procedure is repeated with the transpose of the array for the nodes that are not yet marked. In this step no further conflicts will occur, because if a node $C(i)$ is not yet marked, then $C(S(i))$ is in the same row as $C(i)$. After finishing this step at most $2/3n + O(1)$ nodes remain, because if a node has not been compacted with one of its successors, then both of its neighbors must have been compacted. Then, the remaining nodes are packed into an array of size $2/3n + O(1)$. This is done by doing a prefix sums with 0 assigned to the nodes that are marked deleted and 1 assigned to the nodes that are not deleted. The whole procedure is then applied recursively to the remaining nodes.

On the P-HMBT model this algorithm is implemented as follows. Assume that each node in the list is represented by a quadruple $V(i) = \langle C(i), S(i), M(i), D(i) \rangle$, where $C(i)$ and $S(i)$ have the usual meaning and where $M(i)$ ($D(i)$) is a single bit indicating whether a node is marked (deleted). Assume also that each quadruple exactly fits into one memory cell (this affects the running time only by a constant factor). Initially, the array V is stored in level $L(\log(n))$ of the hierarchical memory. The algorithm proceeds as follows.

1. Partition V into $\sqrt{n}$ blocks B_i, $1 \leq i \leq \sqrt{n}$, of size $\sqrt{n}$ each. For each of these blocks, repeat step (2).
2. Treat a block B_i as $\sqrt{n}/\log(n)$ blocks E_j of size $\log(n)$ each. For each of these blocks perform the following steps.
   (a) Move the block to level $L(\log\log(n))$ and copy the successor of each element to the array $NEXT$ located at level $L(\log(n) - 1)$.
   (b) Move the array $NEXT$ to level $L(\log\log(n)+1)$ and compact each node with its successor provided that the successor is not in the same block B_i.

    (c) Copy E_j and $NEXT$ back to level $L(\log(n))$.

3. Repeat steps (1) – (2) with the transpose of V.
4. Contract the remaining nodes into an array of size $2/3n + O(1)$. This is accomplished by doing a prefix sums on the nodes with 0 assigned to the nodes that are marked deleted and 1 assigned to the nodes that are not deleted.
5. Solve the resulting smaller list recursively.
6. Extend the solution to all elements of the original list.

This algorithm can be made to run in time

$$T(n) = T(2/3n) + O(n \log(n)),$$

which solves to $T(n) = O(n \log(n))$.

On a PP-HMBT machine, list ranking can be performed in optimal time $O(n)$ as follows. As for the P-HMBT, the input is divided into $\sqrt{n}$ blocks B_i of size $\sqrt{n}$ each, and each block is divided into $\sqrt{n}/\log n$ subblocks E_j of size $\log n$. Each block E_j is pipelined to location 1 of the hierarchical memory and each node is paired with its successor provided that the successor is not in the same block B_i. The cost per block are $O(\log n)$, i.e. in constant time per element. Again the input is permuted and these steps are repeated, which also requires $O(n)$ time. This implies that the running time $T(n)$ obeys the recurrence relation

$$T(n) = T(2/3n) + O(n) ,$$

which solves to $T(n) = O(n)$.

For all problems considered in this paper, the running time on the PP-HMBT equals the RAM time. Whether this holds in general or for a certain class of problems, needs further investigation.

6   Summary and Concluding Remarks

In this paper we introduced two models for hierarchical memory that enables parallelism to be exploited between block transfers, meaning that transfers at different levels may proceed concurrently. Algorithms are given for, e.g., the touch problem, prefix sums and list ranking. Efficient algorithms in these models exhibit both temporal and spatial locality.

We examined the relationship between the proposed models P-HMBT and PP-HMBT and we showed that there is a partial ordering on these models and the HMBT proposed in [2]. This partial ordering adheres to the following diagram.

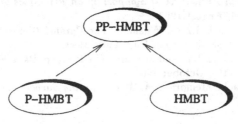

Finally, bounds are given for two important problems, namely prefix sums and list ranking. We noticed that parallel algorithmic techniques can also be used to obtain efficient algorithms for the P-HMBT model. This may not come as a surprise since many parallel algorithms have the following generic scheme (top-down design).

- Divide the problem at hand into p subproblems each taking approximately equal amount of time.
- Solve each subproblem in parallel.
- Combine the results.

Several directions for future research exist. For example, in existing computer systems the block size (cache line size, page size) at each level is usually fixed and the performance is greatly affected by this size [10]. It would be important to understand the behavior of algorithms in this setting. Also, implications for memory organizations in multiprocessor architectures will have to be investigated [5, 7].

Acknowledgement We want to thank Peter Knijnenburg for several useful suggestions.

References

1. A. Aggarwal, B. Alpern, A.K. Chandra, and M. Snir. A Model for Hierarchical Memory. In *19-th Annual ACM Symposium on Theory of Computing*, pages 305–314, May 1987.
2. A. Aggarwal, A.K. Chandra, and M. Snir. Hierarchical Memory with Block Transfer. In *28-th Symposium on Foundations of Computer Science*, pages 204–216, October 1987.
3. A.V. Aho, J.E. Hopcroft, and J.D. Ullman. *The Design and Analysis of Computer Algorithms*. Addison-Wesley, 1974.
4. A.V. Aho, J.E. Hopcroft, and J.D. Ullman. *Data Structures and Algorithms*. Addison-Wesley, 1983.
5. A. Chin. Complexity Models for All-Purpose Parallel Computation. In A. Gibbons and P. Spirakis, editors, *Lectures on parallel computation*, chapter 14. Cambridge University Press, 1993.
6. J.W. Hong and H.T. Kung. I/O Complexity: The Red-Blue Pebble Game. In *Proc. of the 13-th Annual ACM Symp. on Theory of Computing*, pages 326–333, May 1981.
7. B.H.H. Juurlink and H.A.G. Wijshoff. Experiences with a Model for Parallel Computation. In *12th Annual ACM Symposium on Principles of Distributed Computing*, pages 87–96, August 1993.
8. B.H.H. Juurlink and H.A.G. Wijshoff. The Parallel Hierarchical Memory Model. Technical Report 93-33, Leiden University, 1993.
9. C.P. Kruskal, R. Rudolph, and M. Snir. The Power of Parallel Prefix. *IEEE Trans. on Comp.*, C-34(10), October 1985.
10. Alan Smith. Cache Memories. *ACM Computing Surveys*, 14(3), September 1982.

Randomized geometric algorithms (abstract)

Ketan Mulmuley *

The University of Chicago

Since Rabin's seminal paper [1], randomized methods have flourished in several branches of computer science. Computational geometry has been no exception. In the past decade, several simple, yet striking randomized techniques have emerged in this field, and they have found an enormous number of applications. In this talk, we shall give a brief survey of some of the main techniques in this area.

The simplest and the most well known randomized geometric algorithm is Quicksort. Why is Quicksort a geometric algorithm? Because the items to be sorted can be identified with points on the real line. Then the sorting problem becomes equivalent to the problem of constructing the partition of the real line formed by these points. In other words, the problem is to compute the resulting intervals and their adjacencies. Viewed this way, it becomes clear that Quicksort is a one-dimensional geometric algorithm.

A large number of problems in computational geometry can be viewed as higher-dimensional generalizations of the preceding problem solved by Quicksort. In higher dimensions, the objects need not be points anymore. Instead, they can be hyperplanes, hypersurfaces, polyhedra and so forth. The "sorting problem" in this scenario becomes the problem of constructing the resulting partition. More formally, we wish to compute all faces of this partition and the adjacencies among them. For example, when the objects are hyperplanes in R^d, the partition to be constructed could correspond to the arrangement formed by these hyperplanes. As another example, the objects could be sites (points) in R^d and the partition to be constructed could be the Voronoi diagram formed by these sites.

By now, several powerful general techniques have become available for the design of randomized geometric algorithms. We shall discuss a few of them, which include:

1. Random sampling (randomized divide and conquer): This generalizes the randomized divide and conquer view of Quicksort; e.g. see [2, 3].
2. Randomized incremental algorithms: This generalizes the paradigm underlying the iterative (nonrecursive) version of Quicksort; e.g. see [4, 5].
3. Dynamic sampling: This generalizes (e.g. see [6]) the paradigm underlying skip lists [7].
4. Static and dynamic use of history in searching.

We will also discuss some open problems and future promising research directions in this area. Let us mention a few here.

---

* Supported by Packard fellowship.

One limitation of the previous techniques is that they assume that the underlying "configuration space" has bounded dimension. What this means is that generally there is a constant hidden within the big-oh notation which depends exponentially on the dimension d of the problem. Quite often, one can show that such factors are inevitable. In that case, there is not much that one can do. But in quite a few problems, it is not at all clear if such factors are really inevitable. One good example is provided by linear programming. It is well known that linear programming can be done in time that is polynomial in the bit length of the input [8, 9]. An outstanding open problem is whether there exists a strongly polynomial algorithm for linear programming; this means the number of basic operations, ignoring the bit-cost of arithmetic operations, should be polynomial in the number of input parameters. Seidel[10] gave an extremely simple randomized incremental algorithm for linear programming. This algorithm had $O(d!n)$ running time, where n is the number of constraints and d is the dimension. The exponential dependence on d was only to be expected, since the algorithm is based on the techniques in computational geometry, which are generally geared for fixed dimension. The dependence was made subexponential in [11]. Is there a possibility that the dependence could be reduced even further? If the dependence could be made polynomial, then we shall have a strongly polynomial algorithm for linear programming. That is perhaps too much to hope for at this time. But quasi-polynomial dependence may be achievable.

Another important problem is to get high-probability performance bounds, instead of just the expected performance bounds, for the randomized algorithms based on the techniques mentioned above. For search times, this is generally possible. In the case of running times, what one would want to show is that, with high probability, the running time does not deviate from its expected value by more than a constant factor. In a few rare cases, such results are known; e.g. see [12]. But much needs to be done in this area.

Much research needs to be done regarding efficient applicability of the randomized techniques in nonlinear settings. For example, consider the problem when the objects in the input are algebraic hypersurfaces and we are interested in point location among these hypersurfaces. This is a natural generalization of the point location problem for arrangements formed by hyperplanes. Randomized-divide-and conquer paradigm has been applied for this problem [13], but the dependence of the performance bounds on the dimension d seems far from optimal.

Another important area of investigation seems to be the applicability of such techniques in parallel computational geometry. Significant work has been done in this area (e.g. [14]), but much more needs to be done.

Finally, the current derandomization techniques seem to complicate the algorithms considerably. Can one design directly much simpler alternatives to the current algorithms that have been obtained through derandomization?

So far, we have only discussed the "upper bound aspect" of randomized geometric algorithms. Equally important are lower bounds on the performance of such algorithms. We shall briefly touch upon some recent results on lower bounds for randomized geometric algorithms and state some open problems in this area.

References

1. M. O. Rabin. Probabilistic algorithms. In J. F. Traub, editor, *Algorithms and Complexity*, pages 21–30. Academic Press, New York, 1976.
2. K. L. Clarkson. New applications of random sampling in computational geometry. *Discrete & Computational Geometry*, 2:195–222, 1987.
3. D. Haussler and E. Welzl. Epsilon-nets and simplex range queries. *Discrete & Computational Geometry*, 2:127–151, 1987.
4. K. L. Clarkson and P. W. Shor. Applications of random sampling in computational geometry, II. *Discrete & Computational Geometry*, 4:387–421, 1989.
5. K. Mulmuley. A fast planar partition algorithm, I. In *29th Annual IEEE Symposium on the Foundations Of Computer Science*, pages 580–589, 1988.
6. K. Mulmuley and S. Sen. Dynamic point location in arrangements of hyperplanes. *Discrete & Computational Geometry*, 8:335–360, 1992.
7. W. Pugh. Skip lists: A probabilistic alternative to balanced trees. *Commun. ACM*, 33(6):668–676, 1990.
8. N. Karmarkar. A new polynomial-time algorithm for linear programming. *Combinatorica*, 4:373–395, 1984.
9. L. G. Khachiyan. Polynomial algorithms in linear programming. *U.S.S.R. Comput. Math. and Math. Phys.*, 20:53–72, 1980.
10. R. Seidel. Linear programming and convex hulls made easy. In *6th Annual ACM Symposium on Computational Geometry*, pages 211–215, 1990.
11. J. Matoušek, M. Sharir, and E. Welzl. A subexponential bound for linear programming. In *8th Annual ACM Symposium on Computational Geometry*, pages 1–8, 1992.
12. K. Mehlhorn, M. Sharir, and E. Welzl. Tail estimates for the space complexity of randomized incremental algorithms. In *3rd Annual ACM Symposium On Discrete Algorithms*, pages 89–93, 1992.
13. B. Chazelle, H. Edelsbrunner, L. Guibas, and M. Sharir. A singly-exponential stratification scheme for real semi-algebraic varieties and its applications. *Theoret. Comput. Sci.*, 84:77–105, 1991.
14. J. H. Reif and S. Sen. Polling: A new randomized sampling technique for computational geometry. In *21st Annual ACM Symposium on Theory Of Computing*, pages 394–404, 1989.

Connecting the Maximum Number of Grid Nodes to the Boundary with Non-intersecting Line Segments[†]

LEONIDAS PALIOS

The Geometry Center, Univ. of Minnesota

Abstract: We consider the problem of finding the maximum number of nodes in a grid (from a given set of such nodes) that can be connected to the boundary of the grid by means of non-intersecting line segments parallel to the grid axes. The work is motivated from the VLSI/WSI array processor technology, and in particular, the single-track switch model for configurable array processors ([4]). The problem has been investigated by Bruck and Roychowdhury, who described an algorithm to find the maximum number of compatible connections of n given nodes in the grid in $O(n^3)$ time and $O(n^2)$ space ([2]). In this paper, we present methods that take advantage of the dependency of similar configurations and enable us to resolve the problem in $O(n^2 \log n)$ time and $O(n^2)$ space; instrumental in our algorithm is the use of a new type of priority search trees which is of interest in its own right.

1. Introduction.

The work presented in this paper is motivated from the VLSI/WSI array processor technology. An array *processor* is a synchronous parallel computer with a number of *processing elements* that operate in parallel in lockstep fashion (see [3], and [4]). Unfortunately, due to faults during manufacturing, it is often the case that array processors contain faulty processing elements. The design can be made fault-tolerant, however, by incorporating spare processing elements in the array. Then, if each faulty processing element can be substituted by a spare one, the array processor can still be used. The process of finding the appropriate substitutes and establishing the necessary connections is called *reconfiguration*.

One of the models for reconfigurable array processors is the *single-track switch* model, described by Kung, Jean, and Chang ([4]): the array processor consists of a two-dimensional array of processing elements with double-row-column spare ones placed around them; the reconfiguration involves substituting a processing element for the one next to it along a straight-line *compensation path* that connects a faulty

---

[†] Work supported by the National Science Foundation (NSF/DMS-8920161), the Dept. of Energy (DOE/DE-FG02-92ER25137), Minnesota Technology, Inc., and the Univ. of Minnesota.

element to a spare one located in the same row or column, so that no two compensation paths intersect. Then, the problem of determining whether all the faulty processors can be compatibly substituted by spare ones, can be stated as follows:

> Given a set of nodes located at vertices of (a bounded portion of) the grid, determine whether each of these nodes can be connected to the boundary of the grid by means of a single line segment parallel to one of the coordinate axes of the grid, so that no two such line segments intersect.

An instance of this problem where all the nodes can indeed be connected to the grid boundary as described above (along with the corresponding line segments) is shown in Figure 1; however, one can easily produce cases where this is not possible. The problem was first addressed by Kung, Jean, and Chang ([4]), who formulated it as a maximum independent set problem and adapted an algorithm by Bron and Kerbosch to solve it. Later, Roychowdhury and Bruck gave an $O(n^2)$ time algorithm, whe-

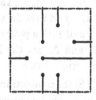

Figure 1

re n is the number of the given nodes in the grid ([8]). A year later, Birk and Lotspiech described an optimal $O(n \log n)$ time algorithm ([1]).

In this paper, we consider the *maximization* version of the above problem, namely:

> Given a set of nodes in the grid, compute a *maximum* size subset S such that each node in S can be connected to the boundary of the grid by means of a single line segment parallel to one of the axes of the grid, and no two such line segments intersect.

An algorithm for this problem was presented by Bruck and Roychowdhury in [2]; it relies on appropriate partitions of the grid into a constant number of rectilinear polygons, whose corresponding optimal connection patterns are independent of each other and are easier to compute as connections are restricted to three, two, or even one of the allowed directions (up, down, left, and right). The combination of the optimal solutions for these polygons gives the optimal solution for the configuration. Taking the best among the optimal solutions of all the configurations considered yields the optimal solution for the entire grid. For n nodes, the algorithm runs in $O(n^3)$ time and requires $O(n^2)$ space.

In this paper, we describe an $O(n^2 \log n)$ time and $O(n^2)$ space algorithm for the above problem. The general approach is the same as that in [2]; we are able, however, to take advantage of dependencies among the optimal solutions of similar configurations. This idea is so powerful that in some cases it leads to methods faster by an order of magnitude. Additionally, instrumental in our algorithm is the introduction of a new type of priority search trees which is of interest in its own right.

Finally, we note that the related problem of computing the maximum size of a *rectilinear wiring* pattern in a bounded portion of the grid is NP-complete, even in the case where the number of bends is restricted to one (*Manhattan wiring*) ([7]).

The paper is structured as follows. In Section 2, we recall the relevant terminology and present our notation. Section 3 outlines the fundamental observations on which the algorithm of Bruck and Roychowdhury as well as our improved version rely. Our algorithm is described and analyzed in Section 4. Section 5 concludes the paper with a summary of results, some final remarks, and open questions.

2. Terminology and Notation.

We begin by mentioning that we assume, without loss of generality, that the given n nodes are all contained in the square with vertices at the grid points $(1,1)$, $(n,1)$, (n,n), and $(1,n)$; this configuration can be obtained by ignoring the rows and columns of the grid that do not contain any nodes, and by renumbering the remaining ones; this normalization process can be performed in $O(\min\{n + \Delta_r + \Delta_c,\ n \log n\})$ time, where Δ_r and Δ_c are the length and width of the smallest rectangle enclosing all n nodes in the grid initially.

Since connections to the grid boundary are restricted to be parallel to the grid axes, a node can be connected to the boundary by means of a line segment along one of four possible directions, namely, *up*, *down*, *left*, or *right*. However, for a given node in the grid, it is likely that not all four directions may lead to feasible connections; think of other nodes located in the same row or column. Let R be a simple rectilinear polygon of the grid, which has been associated with some of the above four directions. Then, by a *connection pattern* in R, we refer to a set of non-intersecting line segments, each corresponding to a different node in R which it connects to the boundary of the grid along one of the associated directions. A connection pattern in R that maximizes the number of such line segments is called *optimal*; the corresponding maximum number (of line segments) is referred to as the *optimal solution* in R. Obviously, there may be more than one optimal connection pattern, but there is only one optimal solution. A connection pattern is characterized as *vertically partitioned* if there is a column c $(1 \le c < n)$ such that no connection (line segment) intersects the *interior* of the vertical strip bounded by the columns c and $c+1$. See also [2]. From the same reference, we borrow the notion of *quadrants* of a node; Figure 2 depicts the four quadrants A, B, C, and D of node p.

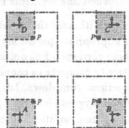

Figure 2

Notation. For clarity in the following discussion, references to optimal solutions are qualified with the associated allowed directions; for example, an "optimal *left-down* solution" involves line segments connecting nodes to the left and bottom boundary sides of the grid only. In figures, the allowed directions are indicated by arrows; for instance, an arrow pointing left indicates that nodes are allowed to be connected to the left side of the grid boundary. Additionally, row indices in figures increase from bottom to top, and column indices from left to right. Finally, we use the expression $R(r_1, r_2; c_1, c_2)$ (where $r_1 \le r_2$ and $c_1 \le c_2$) to denote the rectangle enclosed by and including rows r_1 and r_2, and columns c_1 and c_2. If such a rectangle is defined in terms of a node, say, t, in the grid, we use $t.row$ and $t.column$ to denote t's row and column respectively.

3. Foundations of the Algorithm.

In this section, we summarize the fundamental observations of Bruck and Roychowdhury ([2]), on which their algorithm as well as the improved version we present in this paper rely. To formalize our description, for a node t, and for all $k > t.row$ and $k' > t.column$, we define

$$
\begin{aligned}
left(t)[k] &= \text{optimal left solution in } R(t.row + 1, k; 1, t.column - 1); \\
right(t)[k] &= \text{optimal right solution in } R(t.row + 1, k; t.column + 1, n); \\
down(t)[k'] &= \text{optimal down solution in } R(1, t.row; t.column + 1, k').
\end{aligned}
$$

Bruck and Roychowdhury observed that the optimal connection pattern for the entire grid either has a vertical partitioning or is of one of the two symmetric configurations depicted in Figure 3 (the shown left and right connections indicate a pair of left and right connections in the optimal connection pattern whose horizontal

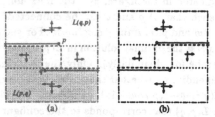

Figure 3

spans overlap, and which have *minimum separation*; this justifies the assigned directions along which connections are allowed in the indicated grid partition). Hence, the maximum number of nodes in the grid (from the given set) that can be connected to the grid boundary by means of non-intersecting line segments is

> max { optimal solution over all vertically partitioned connection patterns,
>
> optimal solution over all connection patterns as shown in Figure 3 }. (1)

A. Vertically Partitioned Connection Patterns. According to its definition, a vertically partitioned connection pattern is characterized by a column c ($1 \leq c < n$) such that the nodes in the rectangle $R(1, n; 1, c)$ are connected up, down, or left, while the nodes in the rectangle $R(1, n; c + 1, n)$ are connected up, down, or right. Since the solutions in the two rectangles are clearly independent of each other, the optimal solution corresponding to a vertically partitioned connection pattern in the grid is

$$
\max_{1 \leq c < n} \{ \quad \text{optimal } up\text{-}down\text{-}left \text{ solution in } R(1, n; 1, c)
$$

$$
+ \quad \text{optimal } up\text{-}down\text{-}right \text{ solution in } R(1, n; c + 1, n) \quad \}. \tag{2}
$$

B. Connection Patterns as shown in Figure 3. Due to the symmetry of the configurations of Figure 3(a) and Figure 3(b), we restrict our attention to the former one. It is immediate from the figure that the solutions in the two L-shaped polygons defined by the line segments associated with p and q are independent; so, if we denote the lower (upper) one as $L(p, q)$ ($L(q, p)$ resp.), the optimal solution among all the connection patterns that correspond to the configuration of Figure 3(a) is

$$
\max_{p, q} \{ \text{ optimal solution in } L(p, q) + \text{ optimal solution in } L(q, p) \}, \tag{3}
$$

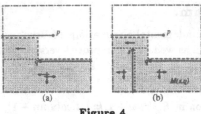

Figure 4   Figure 5

where the optimal solutions in $L(p,q)$ and $L(q,p)$ satisfy the connection restrictions indicated in the figure, and the maximum is computed over all pairs of nodes p and q such that (i) p and q can be connected left and right respectively, and (ii) $p.row >$ $q.row$ and $p.column \geq q.column$. For such a pair (p,q), let us concentrate on how to express the optimal solution in $L(p,q)$ (an expression for the optimal solution in $L(q,p)$ can be found similarly). Only the two configurations shown in Figure 4 are possible, depending on whether the optimal connection pattern contains a node in the rectangle $R(q.row+1, p.row-1; 1, q.column-1)$ which is connected down. The optimal solution in $L(p,q)$ that corresponds to the configuration of Figure 4(a) is precisely

$$\text{optimal down-left-right solution in } R(1, q.row; 1, n) \ + \ left(q)[p.row-1]. \qquad (4)$$

In the configuration of Figure 4(b), s denotes the highest node connected down in the optimal connection pattern in $L(p,q)$, and the corresponding optimal solution is

$$s.a \ + \ \big(left(q)[p.row-1] - left(q)[s.row]\big) \ + \ \text{optimal solution in } M(s,q), \qquad (5)$$

where $s.a$ denotes the optimal down-left solution in the quadrant A of s (see Figure 2), and $left(q)[p.row-1] - left(q)[s.row]$ is equal to the optimal left solution in R(s.row+1, p.row-1; 1, q.column-1). Therefore, the optimal solution in $L(p,q)$ is the maximum among the value of expression (4) and the values of expression (5) over all nodes s in the rectangle $R(q.row+1, p.row-1; 1, q.column-1)$ that can be connected down.

Finally, the optimal solution in $M(s,q)$ under the connection restrictions indicated in Figure 4(b) is equal to

$$\max_t \{ \ down(s)[t.column-1] \ + \ right(t)[q.row] \ + \ t.b \ \}, \qquad (6)$$

where t may be q or any node in the rectangle $R(1, q.row-1; s.column+1, q.column-1)$ that can be connected right (Figure 5); the node t denotes the leftmost node in $M(s,q)$ connected right.

4. The Algorithm.

We now present the techniques and data structures that enable us to compute the optimal solution for n given nodes in the grid in $O(n^2 \log n)$ time. Very briefly, the algorithm begins with a preprocessing phase (Section 4.1), continues with the computation of the optimal solutions for the cases A and B (Sections 4.2 and 4.3 resp.) and concludes by computing and outputting the maximum of the values returned for each of the above two cases (see expression (1)).

4.1. The Preprocessing Phase.

In this phase, we precompute and store in tables (for constant time access) some quantities that will be useful later. First, as expressions (5) and (6) indicate, we need the optimal solutions in the quadrants of all the nodes. Bruck and Roychowdhury also precompute these quantities; their approach involves computing each of these solutions independently in $O(n^{1.5} \log n)$ time per node, for a total of $O(n^{2.5} \log n)$ time. However, the optimal solutions in corresponding quadrants of two nodes are not completely independent, and we can use this fact to our advantage. Indeed, in Section 4.1.1, we show how we can compute the optimal left-down solution in the quadrant A of all the nodes in $O(n^2)$ total time; extending the method to deal with solutions in the three remaining quadrants is trivial. Since there are only four such quadrants, the total time to precompute all these optimal solutions is still $O(n^2)$.

Next, for each node t, we precompute the values of $left(t)[.]$, $right(t)[.]$, $down(t)[.]$, as well as the ones needed for the computation of the optimal solution in $L(q, p)$ and connection patterns as shown in Figure 3(b). For a specific node, each of these arrays can be computed in $O(n)$ time and require $O(n)$ space to store. Therefore, precomputation of all these arrays for all the nodes takes $O(n^2)$ time and $O(n^2)$ space.

Finally, we also compute arrays $leftmost[.]$, $rightmost[.]$, $lowest[.]$ and $highest[.]$: $leftmost[k]$ ($rightmost[k]$ resp.) is equal to the column of the leftmost (rightmost resp.) node located in row k; if no node exists in row k, it is set equal to $n+1$ (0 resp.). The entries for the other two arrays are defined similarly over the columns of the grid. The entries of all four arrays can be computed in $O(n)$ time and space.

4.1.1 Computing the optimal left-down solution in the quadrant A of all the nodes.
Although seemingly counterintuitive, it turns out that computing the optimal left-down solution in the quadrant A of all $O(n^2)$ grid vertices together takes less time than computing the corresponding solution for each node independently. Let $a(i, j)$ ($0 \le i \le n$, $0 \le j \le n$) denote the maximum number of nodes in the rectangle $R(0, i; 0, j)$ that can be connected left or down by means of non-intersecting line segments; clearly, the optimal left-down solution in the quadrant A of a node located at the intersection of row r and column c is precisely the value $a(r, c)$ (see Figure 2).

Let us try to find a recursive definition for $a(i, j)$. First, since no nodes are found in either the 0-th row or the 0-th column, $a(i, 0) = a(0, i) = 0$, for all i such that $0 \le i \le n$. For the $a(i, j)$s, where both i and j are positive, we distinguish the following two cases:

1. *No node in the i-th row of $R(0, i; 0, j)$, if any, is connected down:* Then, the optimal connection pattern in $R(0, i; 0, j)$ consists of an optimal connection pattern in $R(0, i - 1; 0, j)$ plus a connection to the left from the leftmost node among the nodes in the i-th row of $R(0, i; 0, j)$, if any such nodes exist. In other words, the optimal solution is

$$opt_1 \;=\; a(i-1, j) \;+\; \begin{cases} 1, & \text{if } leftmost[i] \le j; \\ 0, & \text{otherwise,} \end{cases}$$

where $leftmost[k]$ is equal to the column index of the leftmost node in row k.

2. *No node in the j-th column of $R(0, i; 0, j)$, if any, is connected left:* Then, the optimal connection pattern in $R(0, i; 0, j)$ consists of an optimal connection pattern in $R(0, i; 0, j - 1)$ plus a connection to the bottom boundary side of the grid from the lowest node among the nodes in the j-th column of $R(0, i; 0, j)$, if any such nodes exist. In other words, the optimal solution is

$$opt_2 = a(i, j - 1) + \begin{cases} 1, & \text{if } lowest[j] \leq i; \\ 0, & \text{otherwise,} \end{cases}$$

where $lowest[k]$ is equal to the row index of the lowest node in column k.

These two cases cover all possible left-down connection patterns in $R(0, i; 0, j)$: a connection pattern that does not fall in either case 1 or 2 must have a connection down from a node in the i-th row and a connection left from a node in the j-th column of $R(0, i; 0, j)$, and thus a pair of intersecting connections; contradiction. Therefore, the optimal left-down solution in $R(0, i; 0, j)$ is

$$a(i, j) = \max\{ opt_1, opt_2 \}.$$

Given that the values of $leftmost[\ ]$ and $lowest[\ ]$ have been precomputed, the computation of $a(i, j)$ takes constant time, if the values of $a(i - 1, j)$ and $a(i, j - 1)$ are already known. This implies that, if the $a(i, j)$s are computed by increasing row and in each row by increasing column, the entire computation will require $O(n^2)$ total time. Moreover, the total space required is only $O(n)$; notice that since $a(i, j)$ depends on $a(i - 1, j)$ and $a(i, j - 1)$, and the computation proceeds by increasing row, we need only maintain the values of $a(i, j)$ in the current row and the row below it (in fact, $n + 1$ entries suffice).

4.2. Vertically Partitioned Connection Patterns.

As before, the idea of computing a quantity in a recursive fashion gives a very efficient method to compute the optimal up-down-left solutions in $R(1, n; 1, c)$ and the optimal up-down-right solutions in $R(1, n; c + 1, n)$ for all columns c. Indeed, in [6], we show that (after the preprocessing phase) the optimal up-down-left solutions in $R(1, n; 1, c)$ for all c can be computed in optimal $O(n)$ time by sweeping the grid from left to right; the optimal up-down-right solutions in $R(1, n; c + 1, n)$ can be computed similarly. Then, expression (2) implies that the optimal solution over all vertically partitioned connection patterns can be found in $O(n)$ time.

4.3. Connection Patterns as shown in Figure 3.

Due to symmetry, we concentrate in the configuration of Figure 3(a). The expression (3) implies that the optimal solution over all connection patterns as shown in Figure 3(a) can be computed in $O(n^2)$ time provided that we have found the optimal solutions in $L(p, q)$ and $L(q, p)$. Again, we restrict our attention to the computation of the optimal solution in $L(p, q)$ for all appropriate pairs (p, q) (the optimal solution in $L(q, p)$ can be found similarly); We show next how to do that in $O(n^2 \log n)$ time, which implies an $O(n^2 \log n)$ overall time complexity for the optimal solution over all connection patterns as in Figure 3.

The basic idea is to compute together all the optimal solutions in the $L(p,q)$s for each specific node q and all ps; to do that, we augment the partial solutions that contribute to the above optimal solutions in such a way that the optimal solution in $L(p,q)$ is equal to the sum of the corresponding optimal (augmented) partial solutions minus a corrective term (which depends only on p and q). Crucial to carrying out this task efficiently is the use of a new type of priority search trees, which we introduce in Section 4.3.1; the space taken by such a tree is $O(n)$, and since we need only one such tree at any given time, the additional space required is $O(n)$.

4.3.1 The Data Structure: A new type of Priority Search Trees. Suppose that we are given an ordered set S of m objects (in our case, nodes in the grid, whose order is defined as a function of their row and column indices). Additionally, the elements of S have been assigned priorities. Our objective is to perform the following operations on the elements of S fast:

MaxUpTo(p):   find the maximum among the priorities of all the elements of S preceding p and including p if it belongs to S.

SubUpTo(k, p):   subtract k from the priority of all the elements of S preceding p.

Remove(p):   remove the element p from the set S.

It is interesting to observe that the description of the elements of S (and to a lesser degree, the desired operations) suggests that the right choice for the data structure to represent the set S may be a priority search tree. Unfortunately, the standard priority search tree described in [5] stores elements at the internal nodes of the tree, which causes the *SubUpTo( )* operation to take $\Omega(\log^2 m)$ time in the worst case.

We therefore use a different type of priority search trees. The basic structure is a balanced binary search tree whose leaves from left to right correspond to the elements of S in order. The novelty lies in the way the heap ordering of the priorities is stored and maintained in the tree: each tree node is associated with two fields *max* and *debt*. So, if we want to subtract k from the priorities of the first i elements of S, we find the (at most $2\log i$) maximal subtrees whose sets of leaves partition the set of the i leftmost leaves of our priority search tree, and add k to the *debt* fields of their roots. In the beginning, all the *debt* fields are initialized to 0; the *max* field of each tree leaf is initialized to the priority of the associated element of S, while the *max* field of an internal tree node t is initialized according to the heap ordering condition

$$t.max \;=\; \max\{\; c_l.max - c_l.debt,\; c_r.max - c_r.debt \;\}, \tag{7}$$

where c_l and c_r are the two children of t. A subtle point is that now the current priority of an element of S associated with a leaf t of the tree equals $t.max$ minus the sum of the *debt* fields of all the tree nodes on the path from the root to t (inclusive). Clearly, building and initializing the tree takes time linear in the size of S.

Let us now see how we can perform each of the above operations fast.

- *MaxUpTo(p)*:   Thanks to the heap ordering, finding the maximum over the priorities of the elements of S preceding and including p if it belongs to S reduces into descending the tree guided by p, and computing the maximum over the quantities

$$c_l.max - c_l.debt - \sum_{\text{ancestor } a \text{ of } c_l} a.debt$$

of the left child c_l at each tree node where the path we are following makes a right turn; note that each such child is the root of a subtree, whose leaves all correspond to elements of S preceding p. Finally, if p is in S, we also take into account p's priority, i.e.,

$$\tau_p.max - \tau_p.debt - \sum_{\text{ancestor } a \text{ of } \tau_p} a.debt,$$

where τ_p is the tree leaf corresponding to p.

- *SubUpTo(k, p):*   Thanks to the *debt* field, we need only descend the tree guided by p, and add k to the *debt* field of the left child of every tree node where we take a right turn. After reaching a leaf, we walk up the same path while making sure that the *max* field of each of the nodes in the path satisfies the heap invariant (7), updating the field appropriately if needed.

- *Remove(p):*   The removal consists of two steps. First, we reduce the priority of p, so that it does not contribute to the heap ordering at all. Unless the entire tree is just a single node, in which case we just discard it and terminate, the tree leaf τ_p corresponding to p has a parent that has another child, say τ_q, corresponding to another element q of S. Then we compute $d = \tau_p.max - \tau_p.debt - \tau_q.max + \tau_q.debt$: if $d \leq 0$, the priority of p is no more than that of q, and we proceed to the second step; otherwise, we add d to $\tau_p.debt$ (which is sufficient to reduce the priority of p so that it equals the priority of q), and then we move up from τ_p to the root, if needed, updating the *max* fields, so that equality (7) holds. In the second step, we remove the tree leaf τ_p, and reduce the 3-node subtree rooted at τ_p's parent, say t, into a single leaf that inherits τ_q's fields, except that its *debt* field is set to $t.debt + \tau_q.debt$. The removal may require rebalancing, and local updating of the *max* fields if a rotation occurs.

The above discussion implies that each of the three operations takes time linear in the height of the priority search tree, that is, time logarithmic in the number of leaves of the tree, since the tree is balanced. Summarizing, we have:

Lemma 4.1. *Given a set S of size m, a priority search tree of the type we introduced in this section enables us to perform each of the MaxUpTo( ), SubUpTo( ), or Remove( ) operation on S in $O(\log m)$ time. The priority search tree requires $O(m)$ time to construct and $O(m)$ space to store.*

We close this section by mentioning that the priority search tree we introduced supports other operations in logarithmic time as well: insertion, computing the maximum over the priorities of all the elements of S *between* two (potential) elements of S (the equivalent of *MinYinXRange( )* described in [5]), and the equivalents of *MinXinRectangle( )* and *MaxXinRectangle( )* described in [5]. The equivalent of *EnumerateRectangle( )*, however, may take as much as $\Theta((k+1)\log m)$ time, if k elements of S are reported.

4.3.2 Computing the optimal solution in $L(p, q)$ for all the appropriate pairs (p, q).

We show next how we can compute the optimal solutions in $L(p, q)$ for all the appropriate pairs (p, q) in $O(n^2 \log n)$ total time. Since the expression (5) relies on the optimal solution in $M(s, q)$ (see Figure 4(b)), we first compute the optimal solution in $M(u, v)$ for all the appropriate pairs (u, v); thanks to the priority search tree described in Section 4.3.1, we can do that in $O(n^2 \log n)$ total time.

The idea is that, for each node u that can be connected down, we compute the optimal solutions in $M(u, v)$ for all appropriate vs together; these are the nodes in the quadrant B of u (except u) that can be connected right (Figure 4(b)). In particular,

0. We build a priority search tree T_u (of the type described in Section 4.3.1) over the set S_u of nodes in the quadrant B of u (except u) that can be connected right; note that there is at most one node per row in S_u. The elements of S_u are ordered by increasing column (and decreasing row on the same column), which can be done in linear time using radix-sort. The priority of an element (node) t in S_u equals the optimal solution in the rectangle $R(1, u.row; u.column, n)$ under the connection restrictions shown in Figure 6 (t denotes the leftmost node connected right), which is

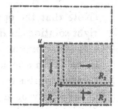

Figure 6

$$init(t) = down(u)[t.column - 1] + right(t)[u.row] + t.b; \qquad (8)$$

the three terms correspond to the optimal solutions in R_1, R_2, and R_3 respectively.

1. We process the nodes w that define a region $M(u, w)$ with u by decreasing row. Since these nodes are precisely the nodes in S_u, we copy the elements of S_u in a list Q, and we radix-sort them by decreasing row in linear time.

2. Then, each node, say v, in Q is processed in order as follows:

a. we perform a $MaxUpTo(v)$ operation in T_u, which, according to the ordering of S_u, returns the maximum over the priorities of v and all the nodes associated with tree leaves preceding v (all the nodes in the rectangle $R(1, v.row - 1; u.column + 1, v.column - 1)$); the value returned is precisely the optimal solution in $M(u, v)$.

b. next, we perform a $SubUpTo(1, v)$ operation in T_u to reduce by 1 the priorities of all the nodes associated with tree leaves preceding v, that is, all the nodes in the rectangle $R(1, v.row - 1; u.column + 1, v.column - 1)$.

c. finally, we remove the node v from the set S_u (actually, the leaf associated with v in T_u) by performing a $Remove(v)$ operation.

(For a proof of correctness, see [6].) From a complexity point of view, the above procedure takes $O(n \log n)$ time to find the optimal solutions in $M(u, v)$ for a fixed u and all the appropriate vs; the tree T_u requires $O(n)$ time to build, and each of the $MaxUpTo(\ )$, $SubUpTo(\ )$, and $Remove(\ )$ operations takes $O(\log n)$ time to execute, since the number of elements of S_u is $O(n)$ initially and always decreases (Lemma 4.1). Repeating the procedure for all nodes u that can be connected down yields the optimal solutions in all the $M(u, v)$ that we need in $O(n^2 \log n)$ total time; all these values are then stored in an $O(n^2)$-size array for constant time access.

We are now ready to compute the optimal solutions in $L(p,q)$ for all the appropriate pairs (p,q). Again, for each node q that can be connected right, we compute the optimal solutions in $L(p,q)$ for all appropriate ps together. We perform the following steps:

0. We build a priority search tree T_q over the set S_q of nodes that lie in the rectangle $R(q.row + 1, n; 1, q.column - 1)$ and can be connected down, plus a dummy node located at $(0,0)$. The elements of S_q are ordered by increasing row (and decreasing column in the same row), which again can be done in linear time using radix-sort. The priority of the dummy node is equal to the optimal solution for the configuration shown in Figure 7(a), that is,

$$\text{optimal down-left-right solution in } R(1, q.row; 1, n) + left(q)[n]. \qquad (9)$$

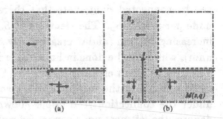

Figure 7

(Note that the optimal down-left-right solutions in $R(1, i; 1, n)$ for all i ($1 \leq i < n$) can be precomputed in $O(n)$ time in a fashion similar to that outlined in Section 4.2.) The priority of any other element (node) s of S_q is equal to the optimal solution in the region shown in Figure 7(b) under the indicated connection restrictions (s denotes the highest node connected down), which is

$$s.a + (left(q)[n] - left(q)[s.row]) + \text{optimal solution in } M(s,q); \qquad (10)$$

the three terms correspond to the optimal solutions in R_1, R_2, and $M(s,q)$.

1. For each node p in the quadrant C of q (except q) that can be connected left, we perform a $MaxUpTo(p)$ operation to compute the maximum among the optimal solutions associated with the nodes in S_q located below the row of p; if the value returned is val, then the optimal solution in $L(p,q)$ is

$$val - (left(q)[n] - left(q)[p.row]).$$

The correctness of the procedure follows from the comparison of Figures 4 and 7, and expressions (4) and (9), and (5) and (10). Since the $MaxUpTo(\ )$ operation takes time logarithmic in the size of S_q, the total time required to find the optimal solution in $L(p,q)$ for each such pair p and q is $O(\log n)$. In other words, after having computed the optimal solutions in all the $M(u,v)$s, additional $O(n^2 \log n)$ time suffices for the computation of the optimal solutions in $L(p,q)$ for all the appropriate pairs (p,q).

5. Concluding Remarks.

Our results are summarized in the following theorem:

Theorem 5.1. *Given n nodes in the grid, the maximum number of them that can be connected to the boundary of the grid by means of non-intersecting line segments (parallel to the grid axes) can be computed in $O(n^2 \log n)$ time and $O(n^2)$ space.*

Although we concentrated on finding the maximum number of non-intersecting connections, the algorithm can be easily modified to yield an optimal connection pattern as well; we simply need to maintain pointers from each optimal solution to the optimal (partial) solutions that contributed to it; the optimal connection pattern can then be computed in $O(n)$ time by tracing back these pointers.

Instrumental in the algorithm is the new type of priority search trees presented in Section 4.3.1; its key feature is that, given an ordered set S, it allows us to increase or decrease the priorities of the elements in an *interval* of S by the same amount in time logarithmic in the size of S. In fact, one can achieve time complexity logarithmic in the size of the interval, provided that we have a "finger" in one of the elements of the interval. This is precisely the case with our algorithm, since the intervals we consider always contain the first element of the current set S; this observation, however, does not improve the asymptotic time complexity of the algorithm in the worst case.

The immediate open question is whether the time or space complexity of the described algorithm can be improved. Additionally, different restrictions in the connection pattern yield interesting variants: one of them involves disallowing *near-misses* (see [1], [4]); in [6], we present an extension of our algorithm that resolves this case in $O(n^2 \log n)$ time and $O(n^2)$ space. A variant that seems more complicated requires that no two line segments (connecting nodes to the grid boundary) intersect in their *interiors*, thus allowing the line segments to abut to each other.

Acknowledgements. I would like to thank Bhaskar Dasgupta of the Computer Science Dept, Univ. of Minnesota for suggesting the problem and for discussions.

6. References.

1. Y. BIRK and J.B. LOTSPIECH, "A Fast Algorithm for Connecting Grid Points to the Boundary with Nonintersecting Straight Lines," *Proc. 2nd Annual Symp. on Discrete Algorithms* (1991), 465–474.

2. J. BRUCK and V.P. ROYCHOWDHURY, "How to Play Bowling in Parallel on the Grid," *Journal of Algorithms* 12 (1991), 516–529.

3. K. HWANG and F.A. BRIGGS, "Computer Architecture and Parallel Processing," McGraw Hill, New York, 1985.

4. S.Y. KUNG, S.N. JEAN, and C.W. CHANG, "Fault-Tolerant Array Processors using Single-Track Switches," *IEEE Trans. on Computers* 38 (1989), 501–514.

5. E.M. MCCREIGHT, "Priority Search Trees," *SIAM Journal on Computing* 14 (1985), 257–276.

6. L. PALIOS, "Connecting Grid Points to the Boundary of the Grid by Means of Non-intersecting Line Segments," *Report GCG56*, The Geometry Center, 1993.

7. R. RAGHAVAN, J. COHOON, and S. SAHNI, "Manhattan and Rectilinear Wiring," *Tech. Report 81-5*, Computer Science Dept., University of Minnesota, 1981.

8. V.P. ROYCHOWDHURY and J. BRUCK, "On Finding Non-intersecting Paths in Grids and its Application in Reconfiguring VLSI/WSI Arrays," *Proc. 1st Annual Symp. on Discrete Algorithms* (1990).

On Self-Stabilizing Wait-Free Clock Synchronization

MARINA PAPATRIANTAFILOU*   PHILIPPAS TSIGAS**

CWI, Amsterdam, The Netherlands
& CTI, Patras, Greece
& Dept. of CE and Informatics, University of Patras, 26500 Patras, Greece

Abstract. Clock synchronization algorithms which can tolerate any number of processors that can fail by ceasing operation for an unbounded number of steps and resuming operation (with or) without knowing that they were faulty are called *Wait-Free*. Furthermore, if these algorithms are also able to work correctly even when the starting state of the system is arbitrary, they are called *Wait-Free, Self-Stabilizing*. This work deals with the problem of *Wait-Free, Self-Stabilizing Clock Synchronization* of n processors in an "in-phase" multiprocessor system and presents a solution with quadratic synchronization time. The best previous solution has cubic synchronization time. The idea of the algorithm is based on a simple analysis of the difficulties of the problem which helped us to see how to "re-parametrize" the cubic previously mentioned algorithm in order to get the quadratic synchronization time solution. Both the protocol and its analysis are intuitive and easy to understand.

AMS Subject Classification (1991): 68M10, 68Q22, 68Q25
CR Subject Classification (1991): D.4.1, D.4.5, D.4.7
Keywords and Phrases: Concurrency, Digital Clocks, Distributed Computing, Fault tolerance, PRAM computation model, Self-Stabilization, Synchronization, Synchronous Systems, Wait-Free Synchronization

1 Introduction

SYNCHRONIZATION among the processors of a multi-processor system is commonly obtained using clocks. In general a clock is implemented in a multi-processor system in one of the following ways: i) using a single clock that is connected to all the processors in the system, ii) using individual clocks for

---

* Partially supported by the ESPRIT II Basic Research Actions Program of the EC under contract no. 7141 (project ALCOM II) and by a Scholarship from the Dutch Organization for International Cooperation in Higher Education (NUFFIC). Email: ptrianta@cti.gr

** Partially supported by the Dutch Science Foundation (NWO) through NFI Project ALADDIN under contract number NF 62-376 and by the ESPRIT II Basic Research Actions Program of the EC under contract no. 7141 (project ALCOM II). Email: tsigas@cti.gr

every processor that are connected to a pulse generator which generates clock pulses stimulating the individual clock, iii) using individual clocks and pulse generators for each processors. It is easy to see that the less centralized the clock implementation is the more resilient to faults it is.

In the past clock synchronization solutions that can tolerate faults have been proposed for the case of arbitrary, or Byzantine, faults [19, 18, 20, 8, 21, 23]. In those system models it has been proven that no algorithm can work unless more than one third of the processors are nonfaulty [8]. In the case of authenticated Byzantine faults the things are not so bad; there exist algorithms that can tolerate any number of faulty processors [12]. The negative results in that model are that: i) the faulty processors can influence the clocks of the non-faulty ones by speeding them up, ii) reaccession of repaired processors is not possible unless more than half of the processors are non-faulty [12]. *Self-stabilizing* algorithms for the clock synchronization problem have also been proposed [11, 1, 6]. An algorithm is called *self-stabilizing* if it can tolerate *transient faults* in the sense that, after a transient fault leaves the system in an arbitrary state, if no further fault occurs for a sufficiently long period of time then the system converges into a consistent global state and can solve the task. A *transient fault* is a fault that causes the state of the system (processes' local states and shared variables) to change arbitrarily. More about self-stabilization can be found in e.g. [7, 2, 4, 5, 22, 9].

So, if we want to sum it all up, the "ideal" clock synchronization algorithm that is highly resilient to failures must have the following features: (i) it must not only tolerate any number of processors' *napping faults* like the authenticated Byzantine model but also guarantee that the nonfaulty processors' clocks remain unaffected by the failures, (ii) faulty processors should be able to rejoin the system and become synchronized in a number of steps k (*synchronization time*) that is independent of the number of the working processors, and (iii) it must work correctly regardless of the system state in which it is started.

Recently Dolev and Welch in [10] presented this highly resilient view of clock synchronization as *Wait-Free, Self-Stabilizing Clock Synchronization*. The assignment of this name to the problem is due to the facts that the first two conditions mentioned in previous paragraph capture the spirit of the *wait-freedom* (cf., e.g., [16, 3, 13]) in the presense of *napping faults* and the third condition captures the spirit of *self-stabilization*. In that paper they present two *Wait-Free, Clock Synchronization* algorithms for n processors which assume a global clock pulse ("in-phase" systems) and nonglobal read/modify/write atomicity. Those solutions guarantee synchronization within $O(n^3)$ and $O(n^2)$ steps; the first solution is also a *Self-Stabilizing* one, while the second depends on the initialization.

In this paper we work on the same problem. By pointing out a simple approach in analyzing its difficulties, we show how to "re-parametrize" the $O(n^3)$ algorithm of [10], thus getting a solution to the Clock Synchronization problem which is both *Wait-Free* and *Self-Stabilizing*, and has synchronization time $O(n^2)$. Moreover, its analysis and proof of correctness are simple and intuitive.

2   The Model

The system consists of n identical processors. A processor p_i is a (possibly infinite) state machine. The processors communicate via a set of single-writer, multi-reader atomic shared variables. Each variable is owned by one processor. The owner of a variable can write it, while all the other processors can read it. In each one of its states, p_i has a pointer to the variables of at most one of the other processors in the system. In each one of its steps, p_i (i) reads its own variables and the variables indicated by its state pointer, (ii) changes state and (iii) updates its own variables. At this point it should be mentioned that, as proven in [10], there can be no wait-free, self-stabilizing clock synchronization algorithm with only *blind* write operations (i.e. updates of its shared variables by each processor without knowing their previous values). This is the reason why p_i has to read its own variables, as well, at each step.

We consider "in-phase" systems, in which all processors share a common clock pulse. Each pulse is a (possibly empty) set of processor names; the set of processors that *make a step* in the pulse. Each processor can make at most one step in one pulse. If a processor does not make a step in some pulse it will be said to *take a pause*.

A *configuration* is a tuple of processors' states and of the values of the shared variables. A system *execution* is a sequence $c_0 \pi_1 c_1 \pi_2 \ldots$ of alternating pulses (denoted by π_x) and configurations (denoted by c_x). Pulses indexed with consecutive numbers will be called *consecutive*. Each configuration c_i in a system execution is derived from its directly preceding configuration c_{i-1} by the state transitions and the shared variable updates of the processors that make a step in the pulse π_i in between these configurations; the shared variable reads by all the processors that make a step in π_i return the respective values of c_{i-1}. An execution is *initialized* if its first configuration is explicitly specified by the protocol. We will refer to a sub-sequence (starting and ending with a configuration) of the sequence which describes a system execution by the term *sub-execution* of that execution. Similarly, a sub-sequence of a sub-execution will be called its sub-execution, as well. The *length* of a sub-execution is the number of pulses in it. In a sub-execution s' (with length greater or equal to l) of a system execution s, a processor p_i will be said to have made l *continuous steps* if it makes steps for l *consecutive* pulses of s'.

This system model, from the theoretical point of view, can be seen as describing the well-known theoretical PRAM model (cf. [15, 17]) with faults. In the real world, it essentially describes existing synchronous multiprocessor systems in which faults may occur, or processors are scheduled independently (cf. [14]). Pause intervals can be interpreted as faults in the connections of the pausing processor or as transient faults, or even as processor crashes.

In a solution to the clock synchronization problem, each processor owns a shared variable which encodes the value of its clock. The *requirement* from a wait-free clock synchronization algorithm is that there should be a positive integer k such that for any execution s of the protocol:

- ADJUSTMENT: For any $l > k$ and for any processor p_i that makes l continuous steps during a sequence of consecutive pulses $\pi_{j+1}, \ldots, \pi_{j+l}$, p_i's clock in c_{j+l} equals its clock in c_{j+l-1} incremented by one.
- AGREEMENT: For any $l \geq k$ and for any two processors p_i and p_j that have both made l continuous steps during any sequence of pulses $\pi_{j+1}, \ldots, \pi_{j+l}$, p_i's and p_j's clocks in c_{j+l} are equal.

If *self-stabilization* should also be guaranteed by the solution, then the above two requirements should be met even in non-initialized executions. This suffices because a subexecution that starts after transient faults have ceased can be viewed as a non-initialized execution.

3 The Protocol

3.1 Informal Description

First we will try to give an insight into the characteristics of the problem by applying an easy strategy: each processor (which has possibly taken a pause) tries to catch up with the maximal clock in the system, by scanning in cyclic order the other processors' clocks and by simply updating its own clock to the maximum clock value it sees in each step incremented by one. In schedules in which for a period of time only one processor (not necessarily the same during that period) holds the maximal clock value in the system, we can think of the maximal value as a "ball" which is "passed" from one processor to the other, under a proper interleaving of their working steps and their pauses. Now, suppose that there exists a processor p_i which tries to find the maximal clock value and which does not take any pauses; it should be able to achieve its goal within a certain number of steps. However, there might be a set S of other processors (more than two) which are scheduled (take pauses or make steps) so that each one p_x of them does not hold the maximal clock value at the pulses when its clock is read by p_i but reads that value from another processor in S immediately after its own value has been read by p_i; then it keeps and increments that value for a number of pulses that are not enough for p_i to complete a cycle and read p_x's clock again; in the meantime another processor p_y can do the same as p_x did. This "game" can be played by all the processors in S scheduled in a way that they cyclically take turns in misleading p_i and preventing it from catching up with the maximal clock in the system. The duration of such a game can be infinite, but the game is also "stop-able" at any time, which implies that at any time p_i will be likely to violate the adjustment requirement.

The protocol presented here —which is a modification of the protocol presented in [10]— protects the correctly working processors by preventing the misbehaving ones from misleading them. This is done in the following way: The idea is still that each processor tries to keep up with the maximal clock value by scanning in cyclic order the other processors' shared variables. Processors have a way to detect their pauses. Each processor p_i owns a shared variable CNT_i

```
var (CLOCK₁, CNT₁), ..., (CLOCKₙ, CNTₙ): (int, int) ;
    /* Shared variables declaration*/

SYNCH(i) /* version for process i */
var j, clock_j, cnt_j, diff, my_clock, my_cnt, susp: int ;
    prev: array [1..n] of int ;
begin
    repeat
        for j = 1 to n (j ≠ i) do
            READ (CLOCKⱼ, CNTⱼ) INTO (clock_j, cnt_j) ;
            my_cnt := my_cnt + 1 ;
            diff := cnt_j − prev[j] ;   prev[j] := cnt_j ;
            if susp ≠ 0 then susp := susp − 1 ;
            if diff > n − 1 then susp := 2n(n − 1) ;
            if susp = 0 then my_clock := max(clock_j, my_clock) + 1 ;
            WRITE (my_clock, my_cnt) TO (CLOCKᵢ, CNTᵢ) ;
        end_for
    forever
end
```

Fig. 1. The Protocol

which it uses in order to count its own total number of steps and a local array $prev$, in whose entry $prev[j]$ it records the value it last read from the CNT_j variable of processor p_j. Following a read of CNT_j, p_i can compute (in its local variable $diff$) the number of steps that p_j made since it was last seen by p_i, by using the recorded information in p_i's local array $prev$. If p_i realises that p_j executed more than $n-1$ steps between these two reads, it concludes that it (p_i) has paused for some step in that interval. Since p_i has paused it might have misled (with the "ball" scenario described previously) a correctly working processor that is trying to fing the maximal clock value. In order to make p_i incapable of further deceit, the protocol suspends it: for a certain number of steps (namely $2n(n-1)$ as can be seen in Figure 1) p_i does not increment its clock. This is implemented with the proper manipulation of p_i's local variable $susp$. In terms of the game described previously, the above mechanism simply bounds its duration: each processor that "catches and passes the ball" will, in at most $n-1$ steps, realize that it has paused in some pulses and, thus, it will become suspended: it will not be allowed to "touch the ball" for a certain number of steps.

In order to complete the informal explanation of the protocol's operations, we consider the processor p_i that has taken some pause and its clock needs adjustment. After the end of its suspension period, if it correctly keeps making continuous steps, it is guaranteed that after it has performed a complete scan of the other processors' variables ($n - 1$ steps) its own clock value will be no less than $n - 1$ units smaller than the maximal clock value of the system at that time. After that point, the protocol via the suspension mechanism ensures that in the following $2n(n-1)$ pulses there will be either (i) $n - 1$ consecutive pulses in which a processor with the maximal clock value will continuously make steps, or (ii) (by the pigeon-hole principle) $n - 1$ pulses, not necessarily consecutive, in which the maximal clock value will not be incremented. Both these cases are convenient for p_i, because, if it makes continuous steps in that interval, it will either (i) actually read the maximal clock value in one of those steps, or (ii) have enough time to catch up with that value, respectively. Once it has the maximal clock value, p_i will continue holding the maximal clock value for as long as it keeps making continuous steps, since it will increment its clock by one at each step.

The formal description of the protocol is given in Figure 1.

3.2 Proof of Correctness

We will first show that the protocol described meets the requirement of a solution to the wait-free clock synchronization problem: for any processor p_i ($1 \leq i \leq n$) which is working correctly (performs continuously steps without taking pauses in between) for at least $k = (4n+1)(n-1)$ pulses, as long as it continues working correctly, its clock will not need adjustment and will agree with the clock of any other processor which has been working correctly for at least k pulses. Towards that we will first prove that p_i after at most k continuous steps will be guaranteed to hold the maximal clock value in the respective system's configuration. Some auxiliary definitions will help the presentation of our arguments.

NOTATION: If c denotes a system *configuration* then
(i) $CLOCK_i(c)$ denotes the value of the respective shared variable in c,
(ii) $MAX\_CLOCK(c)$ denotes $max\{CLOCK_i(c) : 1 \leq i \leq n\}$, and
(iii) d_i^c denotes the difference $MAX\_CLOCK(c) - CLOCK_i(c)$.

DEFINITIONS:
• A processor p_i ($1 \leq i \leq n$) is *suspended* in some configuration in a system execution if its local variable $susp \neq 0$ in that configuration.

• An *adjustment phase* for a processor p_i in a subexecution s' of an execution s is a subexecution $s'' = c_j \pi_{j+1} c_{j+1} \ldots \pi_{j+l} c_{j+l}$ of s', such that these three conditions hold:
(i) p_i makes a step in all the pulses in s'' and its local variable $susp$ equals 0 in all the configurations in s'', and
(ii) either c_j is the first configuration of s' or in pulse π_j p_i makes a step which changes the value of its local variable $susp$ from 1 to 0, and
(iii) either pulse π_{j+l} is the last pulse of s' or in pulse π_{j+l+1} p_i takes a pause.

• A processor p_i performs a *forwarding step* in a particular pulse π_j in some system execution, if p_i makes a step in π_j and $CLOCK_i(c_j) = MAX\_CLOCK(c_j)$

and $MAX\_CLOCK(c_j) = MAX\_CLOCK(c_{j-1}) + 1$. A pulse in an execution is *forwarding* if there exists a processor p_i which makes a forwarding step at that pulse; otherwise we will call the pulse *non-forwarding* (in which case it holds that $MAX\_CLOCK(c_j) = MAX\_CLOCK(c_{j-1})$).

From now on, let s be a system execution (arbitrarily initialized) and let s' be a subexecution of s of length at least $k = (4n + 1)(n - 1)$ such that some processor p_i takes a step at each pulse of s'. We will prove that p_i, at most by the k-th of these steps, will hold the maximal clock value in the system.

Lemma 1. *In the configuration c following the $(2n + 1)(n - 1)$-th pulse of s', p_i's local variable susp will equal 0.*

Proof. In its first $n-1$ steps in s', p_i will load its array $prev$ with the value of the CNT_x shared variable of every other processor p_x. Due to the fact that prior to these steps that array could contain arbitrary values, there is a possibility that p_i becomes suspended in one of these steps, thus assigning to its local variable $susp$ the value $2n(n - 1)$. However, in the next steps, p_i will be finding $diff \leq n - 1$, and, consequently, it will be decrementing the value of $susp$ by one at each pulse. Therefore, by the $(2n + 1)(n - 1)$-th pulse of s', $susp$ will equal 0. $\square$

The above lemma implies that at most after its first $(2n+1)(n-1)$ continuous steps p_i will enter an adjustment phase, which, due to our assumption for p_i, is going to last at least $2n(n - 1)$ pulses. During the adjustment phase and if there are no transient faults in the system, its local variable $susp$ will never become non-zero and the value of $CLOCK_i$ will be incremented by at least 1 at each pulse.

Lemma 2. *In the configuration c following the first $n - 1$ pulses of p_i's adjustment phase in s', it will be $d_i^c \leq n - 1$. Moreover, for any sequence of l ($l \leq 2n(n-1)$) pulses in that phase, if c_j and c_{j+l} are the configurations directly preceding the first and immediately following the last pulse of the sequence, it will hold that $d_i^{c_j} \geq d_i^{c_{j+l}} + l_{nf}$, where l_{nf} is the number of non-forwarding pulses in the specified sequence.*

Proof. From Lemma 1 we have that s' will indeed contain an adjustment phase for p_i of length at least $2n(n - 1)$ pulses.

We first prove the first part of the lemma. Let c^- denote the configuration directly preceding the first step of p_i in the sequence of $n - 1$ pulses specified. It holds that:

$$MAX\_CLOCK(c) - MAX\_CLOCK(c^-) \leq n - 1$$

because at each step the maximal clock of the system can be incremented by at most one. But $MAX\_CLOCK(c^-)$ is the value of $CLOCK_x$ in c^- for some processor p_x in the system, which p_i is going to read in one of these $n-1$ steps. Since the values of the $CLOCK$ variables are never decremented it follows that:

$$CLOCK_i(c) \geq MAX\_CLOCK(c^-) \Rightarrow$$
$$MAX\_CLOCK(c) - CLOCK_i(c) \leq MAX\_CLOCK(c) - MAX\_CLOCK(c^-)$$

which, combined with the first inequality, implies that

$$MAX\_CLOCK(c) - CLOCK_i(c) \leq n - 1$$

The inequality of the second part of the lemma can be derived by addition of the following two relations:

$$CLOCK_i(c_{j+l}) \geq CLOCK_i(c_j) + l$$
$$MAX\_CLOCK(c_{j+l}) = MAX\_CLOCK(c_j) + l - l_{nf}$$

The former holds because p_i is not suspended and, thus, it increments its clock by at least one in each step. The latter holds because the system's maximal clock is incremented by one in each pulse, unless the pulse is non-forwarding. $\square$

The previous lemma states that once p_i enters the adjustment phase, after its first scanning of the other processor variables, it is guaranteed to have a clock value which differs by at most $n-1$ from the maximal clock value of that configuration and that this difference can only decrease in the following steps of p_i. Hence, we have the following:

Lemma 3. *Consider the subexecution which starts with the configuration following the $(n-1)$-th pulse of p_i's adjustment phase in s' and ends with the configuration following the $2n(n-1)$-th pulse of that phase. If in this subexecution there are $n-1$ or more non-forwarding pulses, then it will hold that $d_i^c = 0$, where c is the last configuration of the subexecution.*

Proof. It follows from Lemma 2 and from a fact that is directly derived from the rules of the protocol: if p_i at some step reads the maximal clock value of that configuration then, as long as it continues working correctly it will still hold the maximal clock value in the system and will increment it (by incrementing its own clock) by one at each pulse. $\square$

Lemma 4. *At the configuration c following the $2n(n-1)$-th pulse of p_i's adjustment phase in s' it will be $CLOCK_i(c) = MAX\_CLOCK(c)$.*

Proof. Assume, towards a contradiction, that $CLOCK_i(c) < MAX\_CLOCK(c)$. Let A denote the subexecution specified by the first $2n(n-1)$ pulses of this adjustment phase. Also, consider any processor p_x ($x \neq i$) which makes steps during A. We make two crucial remarks:
(i) Under our assumption, p_x cannot perform $n-1$ continuous forwarding steps during A. Otherwise, we already have a contradiction: Since $CLOCK_x$ is read by p_i every $n-1$ steps and because p_i's steps in the specified interval are continuous

by definition, p_i would have adjusted its own clock to $CLOCK_x$ and, hence to the maximal clock of the system during one of these $n-1$ steps of p_x.

(ii) Once p_x performs its first $n-1$ steps (not necessarily continuous) in A, it will load its local variable $prev[i]$ with a correct value of CNT_i written by p_i during A; thus, p_x will have a consistent reference time-point for detecting its pauses thereafter. After that point, due to our assumption, p_x cannot make more than $n-1$ forwarding steps in A: if it does, we know from (i) that these steps will not be continuous. But then, by at most the $(n-1)$-th such step it will detect its pause, and, as a result it will become suspended. Since the length of a subexecution in which a processor is continuously suspended is at least equal to the length of A ($2n(n-1)$ pulses), p_x will not increment its clock again during A.

What (ii) essentially implies is that the number of forwarding steps of each processor p_x ($x \neq i$) in A is at most $2(n-1)$, which means that the total number of forwarding pulses in A is at most $2(n-1)^2$. The latter in turn implies that the number of non-forwarding pulses during A is at least $2(n-1)$ and, in particular after p_i's first scanning ($n-1$ steps) in A it is at least $n-1$. But then, by Lemma 3, p_i should hold the maximal clock value at c, which contradicts our assumption. □

Theorem 5. *The construction correctly implements a self-stabilizing wait-free clock synchronization solution with $k = (4n+1)(n-1)$.*

Proof. After a processor p_i has worked correctly for at least $k = (4n+1)(n-1)$ steps, it is guaranteed by Lemma 4 that it will hold the maximal clock value in the system. After that, it can be directly derived from the rules of the protocol, that as long as p_i continues working correctly it will still hold the maximal clock value in the system and that it will increment its clock by one at each pulse, thus satisfying the adjustment requirement. The same will hold with any other processor that has been working continuously and correctly for at least k pulses concurrently with p_i. This implies that its clock value will agree with the clock value of p_i, thus, the agreement requirement is satisfied, as well.

The self-stabilizing property of the protocol is due to the fact that no initialization conditions were used in the analysis. □

Conclusions

In this work we show a wait-free and self-stabilizing protocol that achieves clock synchronization among n processors in at most $(4n+1)(n-1)$ steps, and which improves the previously known solution which had synchronization time $O(n^3)$ steps. The best known non-self-stabilizing solution to the same problem also has synchronization time $O(n^2)$. Given these two facts, two points that deserve consideration are (i) whether the requirement for self-stabilization imposes any overhead in the complexity of the problem, and (ii) whether the problem can be solved with a linear time algorithm.

Acknowledgments

We are thankful to Moti Yung for his help in the first steps of this work. We would also like to thank the anonymous referees for their accurate and useful remarks.

References

1. A. ARORA, S. DOLEV, AND M. GOUDA. Maintaining Digital Clocks in Step. *Parallel Processing Letters 1*, 1, 1991, pp. 11-18.
2. Y. AFEK, S. KUTTEN AND M. YUNG. Memory–Efficient Self Stabilization on General Networks. In *Proceedings of the 4th International Workshop on Distributed Algorithms*, volume 486 of *Lecture Notes in Computer Science*, Springer-Verlag 1990, pp. 15–28.
3. H. ATTIYA, N. A. LYNCH AND N. SHAVIT. Are Wait-Free Algorithms Fast? In *Proceedings of the 31st IEEE Symposium on Foundations of Computer Science*, 1990, pp. 55–64.
4. B. AWERBUCH, B. PATT-SHAMIR AND G. VARGHESE. Local Checking and Correction. In *Proceedings of the 32nd IEEE Symposium on Foundations of Computer Science*, 1991, pp. 268–277.
5. B. AWERBUCH AND G. VARGHESE. Distributed Program Checking: A Paradigm for Building Self-Stabilizing Distributed Protocols, In *Proceedings of the 32nd IEEE Symposium on Foundations of Computer Science*, 1991, pp. 258–267.
6. J. M. COURVER, N. FRANCEZ AND M. GOUDA. Asynchronous Unison. In *Proceedings of the 12th IEEE Conference on Distributed Computing Systems*, 1992, pp. 486–493.
7. E.W. DIJKSTRA. Self Stabilizing Systems in Spite of Distributed Control. *Communication of the ACM 17*, 1974, pp. 643–644.
8. D. DOLEV, J.Y. HALPERN AND H.R. STRONG. On the Possibility and Impossibility of Achieving Clock Synchronization. *Journal of Computer Systems Science 32*, 2, 1986, pp. 230–250.
9. S. DOLEV, A. ISRAELI AND S. MORAN. Self Stabilization on Dynamic Systems Assuming Only Read/Write Atomicity. *Distributed Computing 7*, 1, 1993, pp. 3–16.
10. S. DOLEV AND J.L. WELCH. Wait-Free Clock Synchronization. In *Proceedings of the 12th ACM Symposium on Principles of Distributed Computing*, 1993. pp. 97–108.
11. M.G. GOUDA AND T. HERMAN. Stabilizing Unison. *Information Processing Letters 35*, 1990, pp. 171–175.
12. J. HALPERN, B. SIMONS, R. STRONG AND D. DOLEV. Fault-Tolerant Clock Synchronization. In *Proceedings Of the 3rd ACM Symposium on Principles of Distributed Computing*, 1984, pp. 89–102.
13. HERLIHY, M. Wait-free synchronization. *ACM Transactions on Programming Languages and Systems 13*, 1 (Jan. 1991), pp. 124–149.
14. K. HWANG. *Advanced Computer Architectures, Parallelism, Scalability, Programmability*. McGraw-Hill, Inc. 1993.
15. R. KARP AND V. RAMACHANDRAN. Parallel Algorithms for Shared Memory Machines. In J.van Leeuwen, ed., *Handbook of Theoretical Computer Science, Volume A: Algorithms and Complexity* Elsevier, Amsterdam 1990. Also in: Technical

Report UCB/CSD 88/408, Computer Science Division, University of California, March 1988.

16. L. LAMPORT. On Interprocess Communication. *Distributed Computing 1*, 1, 1986, pp. 86–101.

17. F.T. LEIGHTON. *Introduction to Parallel Algorithms and Architectures: Arrays, Trees, Hypercubes*. Morgan Kaufmann Publishers, Inc., 1992.

18. L. LAMPORT AND P.M. MELLIAR-SMITH. Synchronizing Clocks in the Presence of Faults. *Journal of the ACM 32*, 1, 1985, pp. 1–36.

19. K. MARZULLO. *Loosely-Coupled Distributed Services: A Distributed Time Service*, Ph.D. Thesis, Stanford University, 1983.

20. S. MAHANEY AND F. SCHNEIDER. Inexact Agreement: Accuracy, Precision and Graceful Degradation. In *Proceedings of the 4th ACM Symposium on Principles of Distributed Computing*, 1985, pp. 237–249.

21. T.K. SRIKANTH AND S. TOUEG. Optimal Clock Synchronization. *Journal of the ACM 34*, 3, 1987, pp. 626–645.

22. G. VARGHESE. *Self-Stabilization by Local Checking and Correction*. Ph.D. Thesis, MIT Laboratory for Computer Science, 1992.

23. J.L. WELCH AND N. LYNCH. A New Fault-Tolerant Algorithm for Clock Synchronization. *Information and Computation 77*, 1, 1988, pp. 1–36.

Hard Graphs for Randomized Subgraph Exclusion Algorithms

Marcus Peinado

Department of Computer Science
Boston University
Boston, MA 02215, USA

Abstract. A randomized version of the CLIQUE approximation algorithm by Boppana and Halldórsson is analyzed. The Boppana Halldórsson algorithm is currently the only approximation algorithm for the CLIQUE problem with a non-trivial performance guarantee. This paper presents a class of graphs on which the performance ratio of the randomized version of the algorithm is not better than $\Omega(\sqrt{n})$ with probability greater than $1 - 1/n^{\omega(1)}$.

1 Introduction

Unlike many other NP-hard problems, the CLIQUE problem has resisted attempts to find efficient approximation algorithms. Indeed, the well known result of [2] proves that no deterministic polynomial time algorithm can approximate the maximum clique in a graph to within a factor of n^c for some (very small) $c > 0$ unless $P = NP$. Recently, Boppana and Halldórsson [3], [4] have found a subgraph exclusion algorithm with a performance guarantee of $O(n/\log^2 n)$. This algorithm is currently the only one known with a non-trivial performance guarantee. (For CLIQUE the *performance guarantee* of an algorithm is the maximum over all inputs of the ratio of the size of the largest clique in the input graph and the clique size found by the algorithm.) Boppana and Halldórsson also show, non-constructively, that graphs on which the performance of their algorithm is not better than $\Theta(n/\log^2 n)$ have to exist. Indeed, it is not too difficult to construct graphs explicitly on which the performance of the algorithm is bad. These simple constructions rely on the fact that the algorithm selects the vertices in one particular ordering (e.g. lexicographic order). Thus, one might reasonably expect that if the algorithm selects the vertices *at random*, its performance might with high probability improve. A similar idea is the basis of most successful randomized algorithms: bad worst-case performance of the deterministic algorithm on few inputs is traded off against a very small probability of bad performance of the randomized algorithm on many inputs. Kučera [8] investigates this idea in the context of the greedy heuristic for the graph coloring problem.

This paper shows that randomization can only have very limited success when applied to the Boppana-Halldórsson algorithm. It displays a class of graphs which contain cliques of size n^α (given any constant $\alpha < 1/2$) and proves that the size of the clique found by the randomized version of the Boppana-Halldórsson algorithm is smaller than n^δ (for all $\delta > 0$) with probability greater than $1 - n^{-\omega(1)}$. Consequently, even with polynomial amplification, the probability of finding a larger clique is less than $n^{-\omega(1)}$. As an intermediate step in the proof, it is shown that the central subprocedure of the algorithm – which in itself is a generalization of the randomized greedy method – performs worse than n^α for all $\alpha < 1$.

The critical component of the graphs discussed in this paper are random graphs with a forced clique of size n^α ($0 < \alpha < 1$). These graphs have been used by Jerrum [7] to show that the Metropolis process cannot approximate CLIQUE to within a factor n^α for any $\alpha < 1/2$.

The analysis of the algorithm's performance on random graphs is complicated by the fact that during its execution, the algorithm destroys (excludes) certain parts of the graph. As parts of the originally random graph are removed, it can no longer be assumed that the remaining edges are independent. Therefore, the well known techniques for the analysis of random graphs cannot be applied. For this reason, the graphs presented here are somewhat more complicated. They retain their basic random graph structure even after a limited number of subgraph exclusions.

The remaining parts of the introduction contain definitions, a description of the algorithm and an outline of the paper.

1.1 Definitions and Notation

Let $G = (V, E)$ be a graph. For $v \in V$ let $\mathcal{N}_G(v) = \{u \in V | \{v, u\} \in E\}$ and for $S \subseteq V$ let $\mathcal{N}_G(S) = \bigcap_{v \in S} \mathcal{N}_G(v)$. Similarly define $\bar{\mathcal{N}}_G(v) = \{u \in V | \{v, u\} \notin E$ and $u \neq v\}$ and $\bar{\mathcal{N}}_G(S) = \bigcap_{v \in S} \bar{\mathcal{N}}_G(v)$. If the graph G can be infered from the context we will simply write $\mathcal{N}(v)$, $\bar{\mathcal{N}}(v)$. $\mathcal{G}(n, p)$ denotes the class of random graphs with n vertices and edge probability p. $\mathcal{P}(S)$ denotes the power set of the set S. Throughout this paper, n denotes the number of vertices in the given graph. All logarithms have base 2.

1.2 The Algorithm

We give a brief summary of the presentation in [4]. The algorithm consists of a subgraph exclusion procedure and a recursive subprocedure (RAMSEY) which is motivated by Ramsey theory and which, given an input graph, returns a clique and an independent set. The subgraph exclusion procedure calls RAMSEY, stores the clique returned, and removes the independent set from the graph. This is repeated until the graph has become empty.

The RAMSEY subprocedure is a generalization of the greedy method which builds a clique by repeatedly selecting a vertex (*pivot vertex*) and deleting its non-neighbors. RAMSEY improves and generalizes the greedy method by making an additional call to search the non-neighborhood of the pivot vertex. Thus, each recursive call has two cliques to chose from: the clique found in the neighborhood of the pivot together with the pivot and the clique found in the non-neighborhood. RAMSEY returns the larger one.

Clearly, the same idea can be used to find an independent set by interchanging the terms neighborhood and non-neighborhood. RAMSEY returns both an independent set and a clique in the input graph.

RAMSEY$((V, E))$:
    IF (V, E) is empty THEN return $(\emptyset, \emptyset)$
    ELSE choose a vertex $v \in V$
        $(C_1, I_1) := $ RAMSEY$(\mathcal{N}(v))$
        $(C_2, I_2) := $ RAMSEY$(\bar{\mathcal{N}}(v))$
        return (larger of $(C_1 \cup \{v\}, C_2)$, larger of$(I_1, I_2 \cup \{v\})$)

Using Ramsey theory, Boppana and Halldórsson [4] show for the clique C and independent set I returned by RAMSEY(G) that $|C| \cdot |I| \geq \log^2 n/4$. This bound in itself does not guarantee a minimum size of C since $|I|$ can be large.

The purpose of the subgraph exclusion algorithm is to modify the graph such that, eventually, $|I|$ will be small. This is achieved by repeatedly calling RAMSEY and excluding (removing) the returned independent sets:

IS Removal(G):
    $i := 1$
    $(C_i, I_i) := $ RAMSEY(G)
    WHILE $G \neq \emptyset$
        $G := G \setminus I_i$
        $i := i + 1$
        $(C_i, I_i) := $ RAMSEY(G)
    return $max_{j \leq i} C_j$

A clique in G can loose at most one vertex per iteration because a clique and an independent set can share at most one vertex. If the graph has a large enough clique, a constant fraction of the graph will be left even if all independent sets of a certain minimum size k are excluded. If RAMSEY is run on the resulting graph, the size of I can be at most k. This implies a lower bound on $|C| \geq \log^2 n/(4k)$. If the largest clique is small, the performance of the algorithm on the graph is trivially good. The result of this analysis is a performance guarantee of $O(n/\log^2 n)$ (cf. [4] for details). Furthermore, [4] shows non-constructively that this performance guarantee is tight, i.e. that graphs have to exist on which the performance of the algorithm is not better than $O(n/\log^2 n)$.

A useful concept in the analysis of RAMSEY which was used in [4] is the tree of recursive calls made by RAMSEY, with each node labeled by the corresponding pivot vertex. The cliques and independent sets found by RAMSEY are closely related to the paths in this tree. Given any path from the root to a leaf, the leaf and the parents of all left edges form a clique. Similarly, the leaf and the parents of all right edges form an independent set.

The algorithm as described by Boppana and Halldórsson is deterministic. The pivots are selected according to some predefined ordering (e.g. lexicographic). It is relatively easy to construct graphs (together with an ordering of the vertices) on which the algorithm performs badly. This simple construction which depends on the fact that the algorithm chooses the vertices in the order given, breaks down if the pivot vertices are chosen at random. In this paper, we analyze a randomized version of the algorithm in which RAMSEY chooses the pivots at random, i.e. in each recursive call the pivot is chosen uniformly at random from the vertex set of the input graph to the recursive call. We call the randomized RAMSEY subprocedure R-RAMSEY. Furthermore, we allow polynomial amplification, i.e. we analyze a procedure PaR-ramsey, which calls R-RAMSEY $n^{O(1)}$ times and returns the largest clique and the largest independent set found in all runs. Thus, if R-RAMSEY finds a clique of a certain size with probability at least n^{-k} for some $k \in \mathbb{N}$, then PaR-ramsey will return a clique of that size with probability arbitrarily close to one. Finally, let PaR-IS-exclusion denote the subgraph exclusion procedure which calls PaR-ramsey instead of RAMSEY.

1.3 Roadmap

In section 2, we define a Markov chain whose state space was chosen to model the size of the clique found by the algorithm as it progresses. The transition probabilities are closely related to the probability that the algorithm finds a useful vertex in the next step. We go on to show that it is extremely unlikely that the Markov chain will, within the given number of steps, reach a state which would correspond to a non-negligible clique size.

In section 3, we define a graph property and show that if the input graph possesses this property, RAMSEY will, with very high probability, find only a negligibly small clique. The central part of the corresponding proof shows the correspondence between the behavior of the algorithm and the Markov chain and uses the result of section 2.

What remains to be done is to construct graphs which contain large cliques and which have the property defined in section 3. Section 4 defines a class of graphs and shows that it has these properties. This completes the analysis for the randomized ramsey subprocedure. Finally, section 5 shows that the constructed graphs continue to be hard even if the subgraph exclusion procedure is added.

2 The Markov chain

The vertices of the graph we construct are of two kinds: those which are members of a large clique of size $\sqrt{n}$ (*C-vertices*) and those which are not members of any clique of non-negligible size (*non-C-vertices*). The probability that the algorithm finds a C-vertex when it chooses a pivot is the fraction of C-vertices in the graph. We have to guarantee that whenever the algorithm randomly selects a vertex, this probability is sufficiently small.

The intuition behind our strategy to ensure this is as follows: Initially, the fraction of C-vertices in the graph should be small ($n^{-\epsilon}$ for some $\epsilon \in (0.5, 1)$). Whenever a pivot is chosen, this fraction should not increase by too much in the input graphs to the next recursive call. More precisely, if the pivot is a C-vertex, the fraction should not increase by more than a constant factor. If the pivot is a non-C-vertex, the fraction should remain approximately constant, i.e. the C-vertices and the non-C-vertices should be split by approximately the same fraction into neighborhood and non-neighborhood. After a certain number of non-C-vertices have been selected, the number of remaining C-vertices (i.e. portion of any big clique) should be sufficiently small.

Definition 1 *Given $\epsilon > 0$ and a constant $f > 1$, let $(X_i)_{i \in \mathbb{N}}$ be $(0,1)$-random variables whose distribution is bounded by*

$$P(X_i = 1 | \sum_{j=1}^{i-1} X_j = k) \leq q_k = \begin{cases} (1 + n^\epsilon/f^k)^{-1} & \text{if } k \leq \log^{1-\epsilon/4} n \\ 1 & \text{otherwise} \end{cases} \quad (1)$$

Let SP_f denote this class of stochastic processes.

Consider any path in the computation tree of RAMSEY and interpret the event $X_i = 1$ as the i-th vertex in the path being a C-vertex and $X_i = 0$ as it being a non-C-vertex, for i smaller than the path length. For the graphs we will construct, $P(X_i = 1 | \sum_{j=1}^{i-1} X_j = k)$ represents the probability that the i-th vertex in the path will be a C-vertex given that there are k C-vertices in the path before i. The q_k have been chosen so as to be manageable upper bounds on this probability.

Definition 2 *For $i, j \in \mathbb{N}$ let*

$$p_{ij} = \begin{cases} q_i & \text{if } j = i + 1 \\ 1 - q_i & \text{if } j = i \\ 0 & \text{otherwise} \end{cases}$$

The state space $\mathbb{N}$ together with the transition probabilities $(p_{ij})_{i,j \in \mathbb{N}}$ define a Markov chain $\mathcal{MC}_f$. Let the initial distribution be concentrated on state 0 and let the random variables $(Y_i)_{i \in \mathbb{N}}$ denote the state of the Markov chain after i transitions.

We simplify the analysis by approximating the sums of X_i by the Markov chain just defined. The intuition is the same. The state of the Markov chain corresponds to the number of C-vertices in the path so far. It can be shown by induction on i that for all $x \in \mathbb{R}$ and $i \in \mathbb{N}$

$$P(\sum_{j=1}^{i} X_j > x) \leq P(Y_i > x) \quad (2)$$

The following lemma is the key in the proof that the number of C-vertices in any path is likely to remain small.

Lemma 1. *Let $f > 1$ (f constant) and $\epsilon \in (0,1)$. If $T \leq n^{\epsilon/3}$ and $(X_i)_{i \in \mathbb{N}} \in SP_f$ then there exists an $N_0 \in \mathbb{N}$ such that for all $n \geq N_0$:*

$$P(\sum_{i=1}^{T} X_i > \log^{1-\epsilon/4} n) < \frac{1}{n^h} \quad \text{for all } h \in \mathbb{N} \quad (3)$$

Proof. Because of (2), it is sufficient to consider Y_T instead of $\sum_{i=1}^T X_i$. Define the random variables T_i ($i \in \mathbb{N}$) as the number of steps the Markov chain spends in state i provided that it reaches that state. Note that for $m > 0$, $P(T_i < m) = 1 - p_{ii}^m$. Now, for $k = 1 + \log^{1-\epsilon/4} n$

$$P(Y_T \geq k) = P(\sum_{i=1}^{k-1} T_i < T) \leq P(\bigcap_{i=1}^{k-1}\{T_i < T\}) = \prod_{i=1}^{k-1} P(T_i < T) = \prod_{i=0}^{k-1}(1 - p_{ii}^T)$$

$$\leq \prod_{i=1}^{k-1}(1 - \frac{1}{(1+\frac{l^i}{n^\epsilon})^T}) \leq \prod_{i=1}^{k-1}(1 - e^{-\frac{Tl^i}{n^\epsilon}}) \leq (1 - e^{-\frac{Tl^k}{n^\epsilon}})^{k-1}$$

$$\leq (1 - e^{-\frac{1}{n^{\epsilon/2}}})^{\log^{1-\epsilon/4} n} < \left(\frac{2}{n^{\epsilon/2}}\right)^{\log^{1-\epsilon/4} n} \leq n^{-\epsilon/4 \log^{1-\epsilon/4} n}$$

As n grows, the exponent goes toward minus infinity. The last three steps are valid for sufficiently large n. The step from $1 - e^{-\frac{1}{n^{\epsilon/2}}}$ to $\frac{2}{n^{\epsilon/2}}$ follows by considering the limit as $n \to \infty$ of the quotient of the two functions and applying l'Hospital's rule. $\square$

3 The Graph Property

We return now to a level of detail which includes the properties of the RAMSEY algorithm. Consider any graph $G = (V, E)$ and $L \subseteq V$. For $C, D \subseteq V$, $C \cap D = \emptyset$, let $\mathcal{N}_{CD}^G = \mathcal{N}(C) \cap \bar{\mathcal{N}}(D)$. We call the induced subgraph of G whose vertex set is $\mathcal{N}_{CD}^G$ the (C, D)-*induced subgraph of* G. Furthermore, let

$$\mathcal{C}_G = \{(C, D) \subseteq V^2 : C \cap D = \emptyset \text{ and } |L \cap (C \cup D)| < \log^{1-\epsilon/4} n\} \qquad (4)$$

If G and L are clear from the context, we write $\mathcal{C}$ and $\mathcal{N}_{CD}$ instead of $\mathcal{C}_G$ and $\mathcal{N}_{CD}^G$. Assume RAMSEY is run on G. Consider any node x in the computation tree and the vertices on the path that leads to node x. Let C be the set of those vertices in the path at which it turns to the left (neighborhood) and let D be those vertices at which the path turns to the right (non-neighborhood). The input graph to the recursive call corresponding to node x consists of the vertices which are adjacent to all vertices in C and nonadjacent to all vertices in D. This graph is exactly the (C, D)-induced subgraph of G.

Before we explicitly construct a class of graphs on which the algorithm has a bad performance ratio, we define hardness in terms of a set of criteria on a graph. It will be proved using lemma 1 that these conditions are sufficient to guarantee that PaR-ramsey is very likely to find only very small cliques.

Definition 3 *Let* $\mathcal{P} = \{(G_1, L_1), (G_2, L_2), \ldots\}$ *be an infinite collection of pairs* (G_i, L_i), *where* $G_i = (V_i, E_i)$ *is a graph and* $L_i \subseteq V_i$. $\mathcal{P}$ *is called* **hard** *if for some constant* $\epsilon \in (0, 1)$ *and almost all* i,

1. *the size of the largest clique plus the size of the largest independent set in* G_i *restricted to* $V_i \setminus L_i$ *is less than* $n^{\epsilon/3} - 2\log^{1-\epsilon/4} n$, *and*

2. *for all* $(C, D) \in \mathcal{C}_{G_i}$

$$|L_i \cap \mathcal{N}_{CD}| < g(n) \quad or$$
$$|L_i \cap \mathcal{N}_{CD}|n^\epsilon/f^k < |\mathcal{N}_{CD} \setminus L_i|$$

where $k = |L_i \cap (C \cup D)|$ *and* $g(n)$ *is some function such that* $g(n) < n^{o(1)}$.

In the pairs (G, L) we construct, L will contain the only large cliques in the graph. Thus, $L \cap \mathcal{N}_{CD}$ are the C-vertices in the (C, D)-induced subgraph and $\mathcal{N}_{CD} \setminus L$ are the non-C-vertices. The term *hard* refers to the difficulty for RAMSEY of finding a large subset of L. Intuitively, the first condition in the definition implies that the graph becomes empty after relatively few steps of the algorithm. The second condition states that either there are negligibly few (less than g) vertices from L or there are many more vertices not from L than there are from L in the (C, D)-induced subgraph. k denotes the number of C-vertices in the path and f is the constant factor by which the ratio of non-C-vertices to C-vertices decreases each time a C-vertex is found.

Lemma 2. *Let* $\mathcal{P}$ *be a class of pairs* (G, L). *If* $\mathcal{P}$ *is hard, there exists a function* $h(n) < n^{o(1)}$ *such that, as n increases, for* $(G, L) \in \mathcal{P}$

$$P(\text{PaR-ramsey}(G) \text{ finds more than } h(n) \text{ vertices } v \in L) \to 0 \tag{5}$$

Proof. Since $\mathcal{P}$ is hard, there exists a function $g(n)$ as given in definition 3. Consider $(G, L) \in \mathcal{P}$. Any path in any computation tree of RAMSEY(G) has a point T_0 at which the number of C-vertices falls below $g(n)$. This means that T_0 is the first node in the path at which $|L \cap V_i| \le g(n)$, where (V_T, E_T) is the graph associated with T_0. The path can contain at most $g(n)$ vertices $v \in L$ after T_0 and we will show now, that with high probability, it does not contain more than $\log^{1-\epsilon/4} n$ vertices $v \in L$ before T_0.

We do so by interpreting the vertices on the path as a stochastic process and applying lemma 1. Let X_i be the indicator random variable of the event $v_i \in L$, where v_i is the i-th vertex in the path, $i \in \{1, \ldots, T_0\}$. Let V_i be the vertex set of the graph associated with position i and let k_i be the number of $v \in L$ in the path before position i. Clearly, $V_i = \mathcal{N}_{CD}$, where C and D are the sets of neighboring and non-neighboring vertices in the path before position i as defined at the beginning of this section. If $k_i < \log^{1-\epsilon/4} n$ then $(C, D) \in \mathcal{C}$ and since T_0 is the first point in the path at which $|V_i \cap L| < g(n)$, the inequality $|L \cap V_i|n^\epsilon/f^{k_i} < |V_i \setminus L|$ has to hold for all path positions i before T_0 by point 2 of definition 3. Hence, we have for all induced subgraphs (V_i, E_i) at path positions i before T_0 with $k_i \le \log^{1-\epsilon/4} n$:

$$P(X_i = 1 | \sum_{j=1}^{i-1} X_j = k_i) = P(v \in L | k_i \text{ vertices } v \in L \text{ before } i) = \frac{|L_i|}{|V_i|}$$

$$= \frac{|L_i|}{|L_i| + |V_i \setminus L|} \le \frac{|L_i|}{|L_i| + |L_i|n^\epsilon/f^k} = q_{k_i}$$

where $L_i = L \cap V_i$ and q_{k_i} is defined in (1). If $k_i > \log^{1-\epsilon/4} n$, the probability is trivially bounded above by $q_{k_i} = 1$. Thus, the X_i are as in definition 1 and lemma 1 can be applied to bound the number of C-vertices before T_0:[1]

$$P(\sum_{i=1}^{T_0} X_i > \log^{1-\epsilon/4} n) = P(\sum_{i=1}^{T_0} X_i > \log^{1-\epsilon/4} n | T_0 > n^{\epsilon/3}) P(T_0 > n^{\epsilon/3}) +$$

$$P(\sum_{i=1}^{T_0} X_i > \log^{1-\epsilon/4} n | T_0 \leq n^{\epsilon/3}) P(T_0 \leq n^{\epsilon/3})$$

$$\leq P(\sum_{i=1}^{n^{\epsilon/3}} X_i > \log^{1-\epsilon/4} n) + P(T_0 > n^{\epsilon/3}) \leq 2P(\sum_{i=1}^{n^{\epsilon/3}} X_i > \log^{1-\epsilon/4} n) < n^{-k}$$

for all $k \in \mathbb{N}$. The third step follows because $T_0 > n^{\epsilon/3}$ implies $\sum_{i=1}^{n^{\epsilon/3}} X_i > \log^{1-\epsilon/4} n$. To see this, note that the first condition in definition 3 implies that the graph is exhausted after at most $k = n^{\epsilon/3} - 2\log^{1-\epsilon/4} n$ vertices $v \notin L$ have been selected, because all vertices in the path at which it turns left form a clique and all those where it turns right form an independent set. If $T_0 > n^{\epsilon/3}$ there have to be more than $n^{\epsilon/3}$ vertices in the path and, since at most $k = n^{\epsilon/3} - 2\log^{1-\epsilon/4} n$ of them can be in $V_i \setminus L$, at least $2\log^{1-\epsilon/4} n$ of the $n^{\epsilon/3}$ vertices must be in L.

The last step follows directly from lemma 1. Therefore, with high probability, the total number of $v \in L$ in the path is bounded by $h(n) = g(n) + \log^{1-\epsilon/4} n < n^{o(1)}$. Considering all $O(n)$ paths and using polynomial amplification can increase the probability only by a polynomial factor. □

Lemma 2 decouples the construction of a hard graph from the algorithm itself. We can now construct a hard graph for RAMSEY only in terms of the conditions stated in definition 3.

4 A Class of Hard Graphs

We will now describe a class of graphs which contain large (n^α, $0 < \alpha < 1$) cliques which PaR-ramsey will not find with high probability. More formally, we describe a class $\mathcal{P}$ of hard pairs (G, L) in which L contains cliques of size n^α. With high probability, the largest clique PaR-ramsey finds is smaller than n^δ for all $\delta > 0$. It can be shown that the class $\mathcal{G}(n, p, n^\alpha)$ of random graphs with a built-in clique of size n^α (as described in [7]) has this property. Here, we describe a somewhat more complicated class of graphs which in addition can be shown to preserve these properties even if a limited number of independent sets is excluded from the graph (as is done by PaR-IS-exclusion).

---

[1] Formally, there is a slight problem, because (X_i) has only been defined for $i \in \{1, \ldots, T_0\}$ while $\mathcal{SP}_f$ is a sequence of X_i for $i \in \mathbb{N}$. However, this is irrelevant since lemma 1 depends only on X_i for $i \in \{1, \ldots, T_0\}$. To make the statement formally correct one can append any sequence (X_j) ($j > T_0$) which satisfies (1) to the (X_i) ($i \leq T_0$) defined above.

Definition 4 *Given the number of vertices n, a constant α $(0 < \alpha < 1)$ and a graph $G_s = (V_s, E_s)$ with $2n^\alpha$ vertices (called the **skeletal graph** of G) construct a graph G with vertex set $V = \{1, \ldots, n\}$ as follows.*

1. *Partition V into two disjoint sets L and $T = V \backslash L$ such that $|L| = 3n^\alpha \log^2 n$.[2] Partition L into n^α disjoint sets L_i $(i \in \{1, \ldots, n^\alpha\})$ of equal size $|L_i| = 3 \log n$. Similarly, partition T into n^α disjoint sets T_i of equal size. We will call the T_i and L_i **segments**.*

2. *The edges within each segment are determined by making each segment L_i an independent set and each segment T_i a random graph.*

3. *The edges between vertices in different segments are determined by G_s. We will only be considering skeletal graphs G_s with $2n^\alpha$ vertices half of which are labeled t_j and l_j, respectively $(j = 1, \ldots, n^\alpha)$. Let vertex t_i of G_s correspond to segment T_i and let l_i correspond to L_i.*
   If $\{u, w\} \in E_s$ then let there be edges in E between all vertices of the segment corresponding to u and all vertices of the segment corresponding to w. If $\{u, w\} \notin E_s$ then let there be no edge between any vertex in the segment corresponding to u and any vertex in the segment corresponding to w.

G consists of $2n^\alpha$ segments. The segments L_i are independent sets and the segments T_i are random graphs. If each segment were collapsed into one vertex, the result would be the skeletal graph G_s.

The following algorithm will be used to generate the skeletal graphs.

Definition 5 *Given an even n, generate $G_s = (V_s, E_s)$ as follows. Let $T_s = \{t_1, \ldots, t_{n/2}\}$, $L_s = \{l_1, \ldots, l_{n/2}\}$ and $V_s = T_s \cup L_s$. Furthermore, let $\{u, w\} \in E_s$ if $u, w \in L_s$ and $u \neq w$. Determine all other edges of G_s by independent random coin flips with probability 0.5.*

G_s is essentially a random graph of size n with a built-in clique of size $n/2$.

Definition 6 *Given $0 < \alpha < 1$, let $\mathcal{MG}_{n,\alpha}$ be the set of all pairs (G, L) constructed by the procedure in definition 4 for $n \in \mathbb{N}$ (such that the segment sizes are integers) and for the skeletal graphs of def 5. Let $\mathcal{MG}_\alpha = \bigcup_n \mathcal{MG}_{n,\alpha}$.*

Consider a pair $(G, L) \in \mathcal{MG}_\alpha$ as generated in def 4. Each segment L_i consists of $3 \log^2 n$ C-vertices and each segment T_i consists of significantly more ($n^\epsilon - 3 \log^2 n$) non-C-vertices. L forms a complete n^α-partite graph. Hence, every C-vertex is a member of cliques of size n^α. All vertices of the same segment are connected to the same T_j and L_j. Thus, whenever a vertex is selected, the graph is split along segment lines. If the pivot is a C-vertex ($v \in L_i$) then the set of T segments is split between the neighborhood and non-neighborhood whereas all L segments will be in the neighborhood. It is crucial to ensure that there are

---

[2] The fact that these numbers might not be integers is only a minor technical issue which can be resolved by rounding up or down appropriately or by defining the graph only for those n for which the segment sizes are integers.

enough T segments in the neighborhood. This is achieved by the random skeletal graph.

The proof that $\mathcal{MG}_\alpha$ is hard for all $\alpha < 1$ is contained in the next three lemmas which make up the rest of this section. Lemma 3 states a general random graph property. Lemma 4 states the corresponding property in the graph we have constructed and lemma 5 uses this property to show that the constructed graphs are hard with high probability. The random graph property in question is a tight bound on the size of the neighborhood $\mathcal{N}(S)$ of any sufficiently small set S of vertices of the skeletal graph. Due to the construction, this implies a bound on the number of T_i and L_i still present in the graph after a certain number of pivots have been selected. Due to space limitations, the proofs of the following lemmas are only sketched or omitted.

4.1 Neighborhood Sizes in the Graphs

For the rest of this section, let β, γ be constants such that $0 < \beta < 1 < \gamma$. Let $r = r(n) = (1 - 3 \log\log n / \log n) \log n$. Given a graph $G_s = (V_s, E_s)$ with n vertices, let

$$\mathcal{C}_{r(n)} = \{(C, D) \subseteq V_s^2 : C \cap D = \emptyset \text{ and } |C \cup D| < r(n)\} \qquad (6)$$

and, for $C, D \subseteq V_s$ let $Y_{CD}^s = |\mathcal{N}_{CD}^{G_s} \cap T_s|$ and $Z_{CD}^s = |\mathcal{N}_{CD}^{G_s} \cap L_s|$

Lemma 3. *Let* (V_s, E_s), $|V_s| = n$, *be a graph generated as in definition 5. Then, as* $n \to \infty$,

$$P(\forall (C, D) \in \mathcal{C}_r(n) : Y_{CD}^s > \beta \frac{n}{2} 2^{-|C \cup D|}) \to 1 \text{ and} \qquad (7)$$

$$P(\forall (C, D) \in \mathcal{C}_r(n) : Z_{CD}^s < \gamma \frac{n}{2} 2^{-|(C \cup D) \setminus L_s|}) \to 1 \qquad (8)$$

Proof. (sketch): The probability of the complementary event is bounded above by $|\mathcal{C}_{r(n)}| \cdot \max_{\mathcal{C}_{r(n)}} p_{CD}$, where p_{CD} stands for the probability that Y_{CD}^s, (Z_{CD}^s respectively) does not meet the required bound. It is elementary to bound $|\mathcal{C}_{r(n)}|$. A bound on p_{CD} can be obtained using Chernoff bounds [6]. $\qquad\square$

In the graphs in $\mathcal{MG}_\alpha$, the connections between different segments are determined by a the edges of the skeletal graph G_s. Thus, $|\mathcal{N}_{G_s}(v)|$ corresponds to the number of segments which are adjacent to segment v. The next lemma maps the random graph property just proved to the graphs in $\mathcal{MG}_\alpha$.

Given $(G = (V, E), L) \in \mathcal{MG}_\alpha$ and Gs skeletal graph $G_s = (V_s, E_s)$, define the function $f : \mathcal{P}(V) \to \mathcal{P}(V_s)$ as follows:

$$f(S) = \{t_i \in V_s : S \cap T_i \neq \emptyset\} \cup \{l_i \in V_s : S \cap L_i \neq \emptyset\} \qquad (9)$$

The function f maps a set S of vertices of G to the segments intersected by S. For $C, D \subseteq V$ let $Y_{CD} = |\{i : T_i \subseteq \mathcal{N}_{CD}\}|$ and $Z_{CD} = |\{i : L_i \subseteq \mathcal{N}_{CD}\}|$ be the number of segments T_i (L_i respectively) contained in $\mathcal{N}_{CD}$. Let

$$\mathcal{C}_f = \{(C, D) \subseteq V^2 : f(C) \cap f(D) = \emptyset \text{ and } |f(C \cup D)| < r(n^\alpha)\} \qquad (10)$$

$\mathcal{C}_f$ is the class of pairs (C, D) which intersect less than $r(n^\alpha)$ segments.

Lemma 4. *For $(G, L) \in \mathcal{MG}_\alpha$:*

$$P(\forall (C, D) \in \mathcal{C}_f : Y_{CD} > \beta \frac{n^\alpha}{2} 2^{-|f(C \cup D)|}) \to 1 \ and \qquad (11)$$

$$P(\forall (C, D) \in \mathcal{C}_f : Z_{CD} < \gamma \frac{n^\alpha}{2} 2^{-|f((C \cup D) \backslash L)|}) \to 1 \qquad (12)$$

Proof. (omitted): The proof proceeds by showing that the events in (7) and (11) ((8) and (12), respectively) are equivalent and applying lemma 3. $\qquad\square$

Lemma 5. *For all $0 < \alpha < 1$, the class $\mathcal{MG}_\alpha$ is hard with high probability and the L of each pair contains a clique of size n^α.*

Proof. omitted. $\qquad\square$

This concludes the proof that $\mathcal{MG}_\alpha$ is *hard* for PaR-ramsey. It remains to show that $\mathcal{MG}_\alpha$ is hard even for PaR-IS-exclusion.

5 The Subgraph Exclusion Algorithm

Theorem 6. *For $\alpha < 1/2$ and for almost every graph G from $\mathcal{MG}_\alpha$, there exists a function $g(n) < n^\delta$ (for all $\delta > 0$) such that*

$$P(PaR\text{-}IS\text{-}exclusion(G) \text{ finds more than } g(n) \text{ vertices } v \in L) \to 0 \qquad (13)$$

Proof. The proof is based on the following fact: In each of the first n^α exclusions, exactly one entire L_i (and no vertex of any other L_j) will be removed from the graph. Therefore, after the first n^α independent sets have been excluded, L will have disappeared completely from the graph.

To see this, note first that each L_i is an independent set which is larger than any independent set in L^c. In almost every G the largest independent set in the subgraph induced by T will be smaller than $2 \log^2 n$. Since each L_i is an independent set larger than $3 \log^2 n$, any independent set containing an L_i is larger than any independent set not containing an L_i.

Second, note that if there is an L_i in the graph, RAMSEY will find all $v \in L_i$ as an independent set. This is the case simply because all the vertices of each L_i are in the neighborhood of exactly the same set of vertices. Therefore, in the computation tree of RAMSEY(G), there will be one path for each L_i which contains all $v \in L_i$ as parents of right (non-neighbor) edges.

Third, each independent set contains vertices from at most one L_i because all vertices from different L_i are connected by an edge. This completes the proof of the fact.

Now, we note that the only property of the T_i used in the proof of lemma 5 is $|T_i| > n^\epsilon$ for some constant $\epsilon > 0$. Since $\alpha < 1/2$, there exists a constant $\epsilon > 0$ such that initially, $|T_i| = n/n^{1/2-\epsilon} - 3\log^2 n \geq \frac{1}{2} n^{1/2+\epsilon}$. The largest independent set in each T_i has less than $3 \log n$ vertices (for almost every G). Therefore, after n^α exclusions, i.e. after L has been completely removed from the graph, there will be more than $n^{1/2+\epsilon} - 3\log^2 n - 3n^\alpha \log n = \Theta(n^{1/2+\epsilon})$ vertices left in each

T_i. This means that as long as not more than n^α exclusions have occurred, we can prove hardness like we did in lemma 5 and then apply lemma 2. After n^α exclusions, L has been removed and the graph does not contain any clique larger than $O(\log^2 n)$. □

This implies that the *performance ratio* of PaR-IS-exclusion is not better than $O(\sqrt{n})$. The class of graphs which was described in definition 6 is a particular example of graphs on which the algorithm shows this behavior.

6 Conclusion

We have constructed a class of graphs on which the Boppana and Halldórsson algorithm has a performance ratio of $\Omega(\sqrt{n})$. We have shown that not even randomizing the algorithm and allowing polynomial amplification can improve the performance ratio on this class of graphs.

Several open problems remain: Can the graphs be constructed deterministically, i.e. without having to rely on random graph properties? One of the two random graphs used in the construction (T_i) can easily be replaced by a deterministic graph due to Frankl and Wilson [5]. However, finding a deterministic version of the skeletal graph G_s used in step 3 of the construction would imply deterministically constructing a graph with no clique and no independent set larger than $O(\log n)$. Finding a polynomial time procedure for this problem has been a long standing open problem [1]. It appears that while this problem is not solved, major changes to our construction would be needed to make it fully deterministic (and polynomial time).

What is the true performance guarantee of the randomized and polynomially amplified version of the algorithm? Our result gives only a lower bound. Is it better than the performance guarantee of $O(n/\log^2 n)$ of the original algorithm?

References

1. Noga Alon, Joel Spencer, and Paul Erdös. *The Probabilistic Method*. Wiley, 1992.
2. S. Arora, C. Lund, R. Motwani, M. Sudan, and M. Szegedy. Proof verification and hardness of approximation problems. In *33rd FOCS* , pages 14–23, 1992.
3. R. Boppana and M. Halldórsson. Approximating maximum independent sets by excluding subgraphs. In *SWAT*, pages 11–25. Springer Verlag, 1990.
4. R. Boppana and M. Halldórsson. Approximating maximum independent sets by excluding subgraphs. *BIT*, 32:180–196, 1992.
5. P. Frankl and R. M. Wilson. Intersection theorems with geometric consequences. *Combinatorica*, 1(4):357–368, 1981.
6. Torben Hagerup and Christine Rüb. A guided tour of Chernoff bounds. *Information Processing Letters*, 33:305 – 308, 1989.
7. Mark Jerrum. Large cliques elude the metropolis process. *Random Structures and Algorithms*, 3(4):347–360, 1992.
8. Ludek Kučera. The greedy coloring is a bad probabilistic algorithm. *Journal of Algorithms*, 12:674–684, 1991.

Task Scheduling in Networks[*]

(Extended Abstract)

Cynthia Phillips [**,1], Clifford Stein[***,2] and Joel Wein [†,3]

[1] Sandia National Labs, Albuquerque, NM, USA
[2] Department of Computer Science, Dartmouth College, Hanover, NH, USA
[3] Department of Computer Science, Polytechnic University, Brooklyn, NY, USA

Abstract. Scheduling a set of tasks on a set of machines so as to yield an efficient schedule is a basic problem in computer science and operations research. Most of the research on this problem incorporates the potentially unrealistic assumption that communication between the different machines is instantaneous. In this paper, we remove this assumption and study the problem of *network scheduling*, where each job originates at some node of a network, and in order to be processed at another node must take the time to travel through the network to that node.

Our main contribution is to give approximation algorithms and hardness proofs for many of the fundamental problems in network scheduling. We consider two basic scheduling objectives: minimizing the makespan, and minimizing the average completion time. For the makespan we prove small constant factor hardness-to-approximate and approximation results for the most general forms of the problem. For the average completion time, we give a log-squared approximation algorithm; the techniques used in this approximation are somewhat general and have other applications. For example, we give the first non-trivial approximation algorithm to minimize the average completion time of a set of jobs with release dates on identical parallel machines.

Another contribution of this paper is to introduce an interesting class of questions about the design of networks to support specific computational tasks, and to give a polylogarithmic approximation algorithm for one of those problems; specifically, we give approximation algorithms to determine the minimum cost set of machines with which to augment a network so as to make possible a schedule of a certain target length.

---

[*] Much of this work was done while the second and third authors were visiting Sandia National Labs, the third author was visiting DIMACS, and the second author was visiting Polytechnic University.

[**] caphill@cs.sandia.gov. This work was performed under U.S. Department of Energy contract number DE-AC04-76DP00789.

[***] cliff@cs.dartmouth.edu. Research partly supported by NSF Award CCR-9308701, a Walter Burke Research Initiation Award and a Dartmouth College Research Initiation Award.

[†] wein@mem.poly.edu. Research partially supported by NSF Research Initiation Award CCR-9211494 and a grant from the New York State Science and Technology Foundation, through its Center for Advanced Technology in Telecommunications.

1 Introduction

Scheduling a set of tasks on a set of machines so as to yield an efficient schedule is a basic problem in computer science and operations research. It is also a difficult problem and hence, much of the research in this area has incorporated a number of potentially unrealistic assumptions. One such assumption is that communication between the different machines is instantaneous. In many application domains, however, such as a network of computers or a set of geographically-scattered repair shops, decisions about when and where to move the tasks are a critical part of achieving efficient resource allocation. In this paper we remove the assumption of instantaneous communication from the traditional parallel machine models and study the problem of *network scheduling*, where each job originates at some node of a network, and in order to be processed at another node must take the time to travel through the network to that node.

Until this work, network scheduling problems had either loose [2, 5] or no approximation algorithms. Our main contribution is to give approximation algorithms and hardness proofs for many of the fundamental problems in network scheduling. Our upper bounds are robust, as they depend on general characteristics of the jobs and the underlying network. In particular, our algorithmic techniques to optimize average completion time yield other results, such as the first nontrivial approximation algorithms for a combinatorial scheduling question: the minimization of the average completion time of a set of jobs with release dates on identical parallel machines. (To differentiate our network scheduling models from the traditional parallel machine models, we will refer to the latter as *combinatorial* scheduling models.)

Our results not only yield insight into the network scheduling problem, but also demonstrate contrasts between the complexity of certain combinatorial scheduling problems and their network variants.

An instance $\mathcal{N} = (G, \ell, \mathcal{J})$ of the network scheduling problem consists of a network $G = (V, E)$, $|V| = m$, with non-negative edge lengths ℓ; we define $\ell_{\max}$ to be the maximum edge length. At each vertex v_i in the network is a machine M_i. We are also given a set of n jobs, $J_1, \ldots, J_n$. Each job J_j originates, at time 0, on a particular *origin machine* M_{o_j} and has a processing requirement p_j; we define $p_{\max}$ to be $\max_{1 \le j \le n} p_j$. Each job must be processed on one machine without interruption. Job J_j is not available to be processed on a machine M' until time $d(M', M_{o_j})$, where $d(M_i, M_k)$ is the length of the shortest path in G between M_i and M_k. We assume that the M_i are either identical (J_j takes time p_j on every machine) or that they are *unrelated* (J_j takes time p_{ij} on M_i, and the p_{ij} may all be different). The identical and unrelated machine models are fundamental in traditional parallel machine scheduling and are relatively well understood [3, 12, 13, 14, 15, 18, 24]. Unless otherwise specified, in this abstract the machines in the network are assumed to be identical.

We study algorithms to minimize the two most basic objective functions. One is the *makespan* or *maximum completion time* of the schedule; that is, we would like all jobs to finish by the earliest time possible. The second is the *average completion time*. When we consider an instance of network scheduling, the minimum makespan or average completion time that could be achieved if we ignore the communication constraints will be called the *combinatorial* schedule length or *combinatorial* average completion time.

Another contribution of this paper is to introduce a class of questions about the design of networks to support specific computational tasks, and to give a polylogarithmic approximation algorithm for one of those problems. There is much work on designing networks to support specific *communication* requirements [9, 16], but the design of scheduling environments to support high quality schedules at lowest resource cost is a relatively unexplored area, especially in the context of a network.

We define a ρ-approximation algorithm to be an algorithm that gives a solution of cost no more than ρ times optimal. Unless specified all approximation algorithms are also polynomial time algorithms.

Summary of Results

• *We give a 2-approximation algorithm for minimizing the makespan and show that, unless $\mathcal{P} = \mathcal{NP}$, no ρ-approximation algorithm exists with $\rho < 4/3$.* The techniques we use for the upper bound can be extended to much more general models. For example, we can still find a 2-approximation in the case when the machines are *unrelated*. We can also allow the edge lengths, or transit times, to be dependent on which job is using the edge. Further, we can incorporate a cost for the traversal of edges or for being processed by machines into the model. Note that this 2-approximation algorithm matches the best known approximation algorithm for scheduling unrelated machines with no underlying network [18]; in fact we use the algorithm of [18] as a subroutine. Our hardness result provides an interesting contrast to the combinatorial problem, for which Hochbaum and Shmoys [12] gave a polynomial approximation scheme.

• *We show that busy schedules can be as much as a $\Omega\left(\sqrt{\frac{\log m}{\log\log m}}\right)$ factor longer than the optimal schedule but no more than an $O\left(\frac{\log(m\ell_{\max})}{\log\log(m\ell_{\max})}\right)$ factor longer than the optimal schedule.* Simple strategies, such as constructing schedules with no unforced idle time, provide schedules of length a small constant factor times optimal, at minimal computational cost, for a variety of scheduling problems [7, 8, 15, 23]. We call such schedules *busy schedules*, and show that for the network scheduling problem their quality degrades significantly. This is in striking contrast to the combinatorial model (for which Graham showed that a busy strategy yields a 2-approximation algorithm [7]), and provides further evidence that the introduction of a network makes these problems qualitatively harder. However, such schedules are of some quality; the busy scheduling strategy does better than previously known approximation algorithms for identical machines in a network, which achieve an $O(\log m\ell_{\max})$ bound [2, 5, 19].

• *We give an $O(\log^2 n)$-approximation algorithm for the $\mathcal{NP}$-hard problem of minimizing the average completion time.* It formulates the problem as a hypergraph matching integer program and then approximately solves a relaxed version of the integer program. We can then find an integral solution to this relaxation, employing as a subroutine either the techniques of Raghavan [21] and Raghavan and Thompson [22]; or Plotkin, Shmoys and Tardos [20]. In combinatorial scheduling, a schedule with minimum average completion time can be found in polynomial time, even if the machines are unrelated.

• *We give other applications of our algorithmic technique for average completion time.* The techniques for the average completion time algorithm are somewhat general and can be extended to yield other results. One consequence is the first $O(\log^2 n)$-approximation algorithm for minimizing the average completion time of jobs with release dates on identical parallel machines. No previous approximation algorithms were known, even for the special case of *just one machine* [15]. For unrelated machines with release dates we give $O(\log^2 n)$-approximation algorithms as well; however these are pseudopolynomial-time algorithms and are only polynomial-time bounded if the job sizes are bounded by a polynomial in the number of jobs and machines. Note that this case is still of some interest, as even the one machine version is strongly $\mathcal{NP}$-hard [15]. In the special case in which the job sizes are polynomially bounded by the number of jobs and machines we give $O(\log^2 n)$-approximation algorithms for minimizing the average *weighted* completion time of jobs with release dates on unrelated machines. No prior approximation algorithms were known.

We also give a $\frac{5}{2}$-approximation for a variant of network scheduling in which each job has not only an origin, but also a *destination*. Finally, we introduce a network design problem and give a polylogarithmic approximation algorithm.

We note that although the focus of this paper is on the off-line and centralized versions of these problems, the algorithms to approximate minimum makespan can be converted into *on-line* centralized algorithms using the techniques in [25].

Previous Work. The problem of network scheduling has received some attention, mostly in the distributed setting. Deng et. al. [5] considered a number of variants of the problem. In the special case in which each edge in the network is of unit length, all job processing times are the same, and the machines are identical, they showed that the off-line problem is in $\mathcal{P}$. It is not hard to see that the problem is $\mathcal{NP}$-Complete when jobs are allowed to be of different sizes; they give a $O(\log(m\ell_{\max}))$ off-line approximation algorithm for this problem. They also give a number of results on the distributed problem when the network topology is completely connected, a ring or a tree.

Awerbuch, Kutten and Peleg [2] considered the distributed version of the problem under a novel notion of on-line performance, which subsumes the minimization of both average and maximum completion time. They give distributed algorithms with polylogarithmic performance guarantees in general networks. They also characterize the performance of feedback-based approaches. In addition they derived off-line approximation results similar to those of Deng et. al [2, 19]. Alon et. al. [1] proved an $\Omega(\log m)$ lower bound on the performance of any distributed scheduler that is trying to minimize schedule length. Fizzano et. al. [6] give a distributed 4.3-approximation algorithm for schedule length in the special case in which the network is a ring.

Our work differs from these papers by focusing on the centralized off-line problem and by giving approximations of higher quality. In addition, our approximation algorithms work in a more general setting, that of unrelated machines.

2 Makespan

2.1 A 2-Approximation Algorithm For Makespan

Let $\mathcal{U}' = (G, \ell, \mathcal{J}')$ be an instance of the unrelated network scheduling problem with optimal schedule length D. Assuming that we know D, we will show how to construct a schedule of length at most $2D$. It is simple to convert this into a 2-approximation algorithm for the problem in which we are not given D [12].

Job J_j is able to run on machine M_i in the optimal schedule when:

$$d(M_{o_j}, M_i) + p_{ij} \leq D. \tag{1}$$

For a given job J_j, we will denote by $Q(J_j)$ the set of machines that satisfy (1). If we restrict each J_j to only run on the machines in $Q(J_j)$, the length of the optimal schedule remains unchanged.

Form unrelated machines scheduling problem ($\mathcal{Z}$) as follows:

$$p_{ij} = \begin{cases} p'_{ij} & \text{if } M_i \in Q(J_j) \\ \infty & \text{otherwise} \end{cases} \tag{2}$$

If the optimal schedule for the unrelated network scheduling problem has length D, then the optimal solution to the unrelated parallel machine scheduling problem (2) is at most D. We will use the 2-approximation algorithm of Lenstra, Shmoys and Tardos [18] to assign jobs to machines. The following theorem is easily inferred from [18], and characterizes assignment of jobs to machines.

Theorem 1 Lenstra, Shmoys, Tardos [18]. *Let $\mathcal{Z}$ be an unrelated parallel machine scheduling problem with optimal schedule of length D. Then there exists a polynomial-time algorithm that finds a schedule S of length $2D$. Further, S has the property that no job starts after time D.*

Theorem 2. *There exists a polynomial-time 2-approximation algorithm for makespan minimization in the unrelated network scheduling problem.*

Proof. Given an instance of the unrelated network scheduling problem, with shortest schedule of length D, form the unrelated parallel machine scheduling problem $\mathcal{Z}$ defined by (2) and use the algorithm of [18] to produce a schedule S of length $2D$ in which each machine M_i is assigned a set of jobs S_i. Let $|S_i|$ be the sum of the processing times of the jobs in S_i and let $S_i^{\max}$ be the job in S_i with largest processing time on machine i. By Theorem 1 and the fact that the last job run on machine i is no longer than the longest job run, $|S_i| - S_i^{\max} \leq D$. Let S'_i denote the set of jobs $S_i - S_i^{\max}$. We form the schedule for each machine i by running job $S_i^{\max}$ at time $D - |S_i^{\max}|$, followed by the jobs in S'_i.

In this schedule the jobs assigned to any machine clearly finish by time $2D$; it remains to be shown that all jobs can be routed to the proper machines by the time they need to run there. Since job $S_i^{\max}$ starts at time $D - |S_i^{\max}|$, by (1), it arrives in time. The remaining jobs only need to arrive by time D, so by (1) and (2), they have arrived in time.   □

Observe that the techniques presented here are quite general and can be applied to any problem that can be characterized in a condition such as (2). In particular, even if the time for a job to travel over an edge is dependent on that job, we can still have a 2-approximation algorithm. Using [24] we can also handle costs on a job traversing an edge or being processed by a machine.

2.2 Nonapproximability

Theorem 3. *It is $\mathcal{NP}$-complete to determine if an instance of the network scheduling problem has a schedule of length 3, even in a network with $\ell_{\max} = 1$. Therefore, no ϵ-approximation algorithm for the network scheduling problem can have $\epsilon < 4/3$ unless $\mathcal{P} = \mathcal{NP}$, even in a network with $\ell_{\max} = 1$.*

It is not hard to see, via matching techniques, that it is polynomial-time decidable whether there is a schedule of length 2. We can show that this is not the case when the machines in the network can be unrelated. This is already known [18] if we allow multiple machines at one node. If no zero length edges are allowed, i.e. each machine is forced to be at a different network node, their proof does not work, but we can give a different proof.

Theorem 4. *There does not exist an ϵ-approximation algorithm for the unrelated network scheduling problem with $\epsilon < 3/2$ unless $\mathcal{P} = \mathcal{NP}$, even in a network with $\ell_{\max} = 1$.*

2.3 Naive Strategies

The algorithms in Section 2.1 give reasonably tight bounds on the approximation of the schedule length. Although these algorithms run in polynomial time, they may be rather slow [20, 21, 22]. We thus explore whether a simpler strategy might also yield good approximations.

A natural candidate is a *busy* strategy: construct a *busy schedule*, in which, at any time t there is no idle machine M_i and idle job J_j so that job J_j can be started on M_i at time t. Busy strategies and their variants have been analyzed in a large number of scheduling problems (see [15]) and have been quite effective in many of them. For combinatorial identical machine scheduling, Graham showed such strategies yield a $(2 - \frac{1}{m})$ approximation guarantee [7]. In this section we analyze the effectiveness of busy schedules for network scheduling. Part of the interest of this analysis lies in what it reveals about the relative hardness of scheduling with and without an underlying network; namely, the introduction of an underlying network can make simple strategies much less effective.

A Lower Bound. We construct a family of instances of the network scheduling problem, and demonstrate, for each instance, a busy schedule which is significantly longer than the shortest schedule for that instance. The network $G = (V, E)$ consists of $\mathcal{L}$ levels of nodes, with level $i, 1 \leq i \leq \mathcal{L}$, containing ρ^{i-1} nodes. Each node in level $i, 1 \leq i < \mathcal{L} - 1$, is connected to every node in level $i + 1$. Each machine in levels $1, \ldots, \mathcal{L} - 1$ receives ρ jobs of size 1 at time 0. The machines in level $\mathcal{L}$ initially receive no jobs. The optimal schedule length for this instance is 2 and is achieved by each machine in level $i, 2 \leq i \leq \mathcal{L}$ taking exactly one job from level $i - 1$.

Theorem 5. *For the family of instances of the network scheduling problem defined above, there exist busy schedules a factor $\Omega(\sqrt{\frac{\log m}{\log \log m}})$ longer than optimal.*

An alternative view of the network scheduling model is that each job J_j has a release date, a time before which it is unavailable for processing, and that J_j's release date can be different on different machines. It is known that when release dates are introduced into the identical machine scheduling problem and each job's release date is the same on all machines, busy strategies still give a $(2 - \frac{1}{m})$-approximation guarantee [10, 11]. Our result shows that when the release dates of the jobs are allowed to be different on different machines, busy scheduling degrades significantly as a scheduling strategy.

An Upper Bound. In contrast to the lower bounds of the previous subsection, we can prove that busy schedules are of some quality. Given an instance $\mathcal{I}$, we define $C_{\max}^* = C_{\max}^*(\mathcal{I})$ to be the length of an optimal schedule for $\mathcal{I}$ and $C_{\max}^A = C_{\max}^A(\mathcal{I})$ to be the length of the schedule produced by algorithm A.

Consider a busy schedule S for an instance $\mathcal{I}$ of the network scheduling problem. We define $p_j(t)$ to be the number of units of J_j remaining to be processed at time t, and $W_t = \sum_{k=1}^{n} p_k(t)$ be the total work remaining to be processed at time t.

Lemma 6. *Consider a busy schedule S for an instance $\mathcal{I}$ of the network scheduling problem. $W_{iC_{\max}^*} \leq \frac{1}{2}(W_0/i!)$ for $i \geq 1$.*

Proof. (Sketch) We partition S into consecutive blocks B_i of length $C_{\max}^*$ and compare each block to an optimal schedule S^* for $\mathcal{I}$ of length $C_{\max}^*$.

Consider a job J_j that does not get started by time $C_{\max}^*$ in S, and let M_j be the machine on which J_j is processed in S^*. This means that in block B_1, machine M_j is busy for p_j units of time during J_j's "slot" in S^*. Hence for every job j that is not started by $C_{\max}^*$ there is an equal amount of unique work which we can identify that *is* processed in B_1, implying that $W_{C_{\max}^*} \leq W_0/2$. Successive applications of this argument yields $W_{iC_{\max}^*} \leq W_0/2^i$ for $i \geq 1$.

To obtain the stronger bound, we increase the amount of processed work which we identify with each unstarted job. If work w_{jk} takes job J_j's slot in block B_k, then job J_j can run in w_{ij}'s slot by block B_{k+2}, thus adding another slot dedicated to job J_j. $\square$

We assume $W_0 \leq m^2 \ell_{\max}$; if not one can show that any busy scheduling algorithm is an $O(1)$-approximation algorithm. Combining this observation with Lemma 6 we have:

Theorem 7. *Let A be any busy scheduling algorithm and $\mathcal{I}$ an instance of the network scheduling problem. Then $C_{\max}^A(\mathcal{I}) = O(\frac{\log(\ell_{\max}m)}{\log\log(\ell_{\max}m)} C_{\max}^*(\mathcal{I}))$.*

2.4 Scheduling with Origins and Destinations

In this subsection we consider a variant of the network scheduling problem in which each job, after being processed, has a *destination machine* to which it must travel. Specifically, in addition to having an originating machine M_{o_j}, job J_j also has a terminating machine M_{t_j}. Job J_j begins at machine M_{o_j}, travels distance $d(M_{o_j}, M_{d_j})$ to machine M_{d_j}, the machine it gets processed on, and then proceeds to travel for $d(M_{d_j}, M_{t_j})$ units of time to machine M_{t_j}. We call this problem the *point-to-point scheduling problem*.

Theorem 8. *There exists a polynomial-time $\frac{5}{2}$-approximation algorithm to minimize makespan in the point-to-point scheduling problem.*

Proof. (Sketch) We get an assignment of jobs to machines as we did in the proof of Theorem 2, but we replace (1) with

$$d(M_{o_j}, M_i) + p_{ij} + d(M_i, M_{t_j}) \leq D. \tag{3}$$

This gives us an assignment of jobs to machines. Pick any machine M_i and let $\mathcal{J}_i$ be the set of jobs that run on machine M_i. Note that this set of jobs has the properties described in Theorem 1. We partition the set of jobs $\mathcal{J}_i$ into three groups, and place each job into the lowest numbered group which is appropriate:

1. $\mathcal{J}_i^0$ contains the job in $\mathcal{J}_i$ with the longest processing time,
2. $\mathcal{J}_i^1$ contains jobs for which $d(M_{o_j}, M_i) \leq D/2$,
3. $\mathcal{J}_i^2$ contains jobs for which $d(M_{o_j}, M_i) \geq D/2$.

Let $p(\mathcal{J}_i^k)$ be the sum of the processing times of the jobs in group $\mathcal{J}_i^k$, $k = 1, 2$. By arguments similar to those in Theorem 2, $p(\mathcal{J}_i^1) + p(\mathcal{J}_i^2) \leq D$. We will always schedule $\mathcal{J}_i^1 \cup \mathcal{J}_i^2$ in a block B of D consecutive time steps. The first $p(\mathcal{J}_i^1)$ time steps will contain jobs in $\mathcal{J}_i^1$ while the last $p(\mathcal{J}_i^2)$ time steps will contain jobs in $\mathcal{J}_i^2$. There may be idle time in the interior of the block.

We consider two possible scheduling strategies.

Case 1: $(p(\mathcal{J}_i^1) \leq p(\mathcal{J}_i^2))$ In this case we first run the long job in $\mathcal{J}_i^0$. We then run block B from time D to $2D$. Since $p(\mathcal{J}_i^1) \leq D/2$, the jobs in $\mathcal{J}_i^1$ all finish by time $3D/2$ and reach their destinations by time $5D/2$. It is straightforward that this holds for the rest of the jobs.

Case 2: $(p(\mathcal{J}_i^1) \geq p(\mathcal{J}_i^2))$ We first run block B from time $D/2$ to $3D/2$. We then start the long job in $\mathcal{J}_i^0$ at time $3D/2$; it arrives at its destination by time $5D/2$. Since $p(\mathcal{J}_i^2) \leq D/2$, the jobs in $\mathcal{J}_i^2$ don't start until time D and hence we are guaranteed that they have arrived at machine M_i by that time. It is straightforward that the rest of the jobs all work out correctly. □

3 Average Completion Time

Given a schedule S, let C_j be the time that job J_j finishes running. Then the average completion time is just $\sum_j C_j/n$, whose minimization is equivalent to the minimization of $\sum_j C_j$.

3.1 Algorithms

Polynomial-time algorithms for minimizing the average completion time of a combinatorial scheduling problem are typically based on a bipartite matching formulation [15]. Consider first the case of unit-sized jobs. One side of the bipartition has a node for every job and the other side has a node for every machine at every time: The cost of running job J_j on machine M_i at time k is $k+1$, because the completion time of that job is $k+1$. We can adapt this approach for the network variant of the problem by excluding edges that correspond to times when job J_j could not run on machine M_i because $d(M_i, M_{o_j})$ is too large.

Theorem 9. *There exists a polynomial-time algorithm to solve the network scheduling problem with unit-size jobs on identical machines with the objective of minimizing the average completion time.*

We now describe an $O(\log^2 n)$-approximation algorithm for the general problem of arbitrary processing times on identical machines, by formulating the scheduling problem as a *hypergraph matching problem*. This formulation, and the method of achieving an approximate solution for it, constitute an interesting general technique; in Section 3.2 we describe some other applications of it. In the hypergraph matching formulation, the vertices consist of two sets, J and M, and the hyperedges will be denoted by F. J will have n nodes, one for each job. M will have mT nodes, where T is an upper bound on the makespan of any schedule, such as $mp_{\max}$. M will have a node for each (machine, time) pair; we will denote the node that corresponds to machine M_i at time t as (i, t). A hyperedge $e \in F$ represents scheduling a job J_j on machine M_i from time t_1 to t_2 by including nodes $j, (i, t_1), (i, t_1 + 1), \ldots, (i, t_2)$. The cost of an edge e, denoted c_e, will be the completion time of job J_j if it is scheduled using that edge. There will be one edge in the hypergraph for each feasible placement of a job on a machine; we exclude edges that would violate the network constraints. The problem of whether a hypergraph matching problem has value at most C is an integer program $(\mathcal{I})$, a feasible solution of which represents a schedule of cost at most C. Decision variable $x_e \in \{0, 1\}$ denotes whether hyperedge e is in the matching.

$$\sum_{j \in e} x_e = 1 \quad j = 1 \ldots, n$$

$$\sum_{(i,t) \in e} x_e \leq 1 \quad \forall (i, t) \in M \tag{4}$$

$$\sum_e x_e c_e \leq C$$

We can now employ the techniques of Plotkin, Shmoys and Tardos [20] to find a solution to a relaxed version of $(\mathcal{I})$. With additional work, we are able to convert the solution to a schedule and show that:

Theorem 10. *There exists an $O(\log^2 n)$-approximation algorithm for the minimization of average completion time in the network scheduling problem.*

3.2 Applications of the Algorithmic Technique

As this is a fairly general formulation of the scheduling problem, it yields other results. If each job J_j has a weight w_j, we can use these techniques to approximate another basic optimization criterion, the average *weighted* completion time, $\sum_j w_j C_j / n$. In addition, we have already noted that the network scheduling model can be characterized by a set of machines and a set of jobs, with each job having a *release date*, which can be different on different machines. The traditional notion of release dates is that the release date of a job is the same on all machines; our formulation handles this as well. Most of the generalization of Theorem 10 to these cases are not polynomial time algorithms, however, but rather only polynomial in the maximum job size. We generalize Theorem 10 as follows.

Theorem 11. *There exists a $O(\log^2 n)$-approximation algorithm for minimizing the average weighted completion time of n jobs with release dates on m unrelated machines (whether or not the release dates are traditional or are different on different machines). The running time of these algorithms is polynomial in n, m and $p_{\max}$, the maximum job size. For the special case of minimizing the average completion time of n jobs with release dates on identical machines, the algorithm is a polynomial-time approximation algorithm.*

The problem of minimizing average completion time on unrelated parallel machines when the jobs have release dates is strongly $\mathcal{NP}$-hard even when there is *just one machine* [17]; even for this special case, however, no approximation algorithms were known. Therefore a pseudopolynomial-time algorithm is of interest. Further, our technique yields the first polynomial-time approximation algorithm for the setting of one machine or identical machines.

The technique can also be applied to a problem motivated by satellite communication systems; the details are left for the full version of the paper.

4 Network Augmentation To Improve Schedule Quality

We have shown that the introduction of a network into parallel machine scheduling increases the algorithmic complexity of the problem. It can also significantly increase the length of the schedule over the combinatorial schedule length. Given a set of jobs $\mathcal{J}$, let $\mathcal{L}_m(\mathcal{J})$ denote the length of the shortest combinatorial schedule of those jobs on m identical machines. Now consider the introduction of a network G of m nodes, and an initial assignment of the jobs in $\mathcal{J}$ to the nodes of the network at time $t = 0$. Define $\mathcal{L}_m^G(\mathcal{J})$ to be the length of the shortest schedule of the jobs in G. We can consider the following problem.

The Network Augmentation Problem: Given a network $G = (V, E)$ and an assignment of a set of jobs $\mathcal{J}$ to the nodes of the network at time 0. Given $\mathcal{L}'$, $p_{\max} \leq \mathcal{L}' \leq \mathcal{L}_m^G(\mathcal{J})$. Find the minimum cardinality set of machines to add at the nodes of G so that $\mathcal{L}_m^G(\mathcal{J}) \leq \mathcal{L}'$.

Of course we can not tell in polynomial time if $\mathcal{L}' \leq \mathcal{L}_m^G(\mathcal{J})$; however we can tell within a small constant factor, which is all that is necessary for our bounds. In particular, we can show that:

Theorem 12. *Consider an instance of the network augmentation problem, with schedule length target $\mathcal{L}'$, and let C be the size of the minimum augmentation. There exists a polynomial-time algorithm that yields an augmentation of size $O((C+1)\log m)$ and enables a schedule for $\mathcal{J}$ in G of length $O(\log^2 m)\mathcal{L}'$, and a polynomial-time algorithm that yields an augmentation of size $O((C+1)\log^2 m)$ and enables a schedule for $\mathcal{J}$ in G of length $O(\log m)\mathcal{L}'$.*

The algorithm proceeds by constructing a well-separated-cover of the network, as defined in [2]. This decomposes the network into neighborhoods, and the existence of a schedule of a certain length can be related to a condition on the ratio of (amount of work) to (number of machines) in each neighborhood. Determining the set of machines to add so as to satisfy this load condition reduces to a generalization of set-cover, which can be approximated within a logarithmic factor [4].

Acknowledgments. We are grateful to Phil Klein for several helpful discussions early in this research, to David Shmoys for several helpful discussions, especially about the upper bound for average completion time, to David Peleg and Baruch Awerbuch for explaining their off-line approximation algorithm to us, and to Perry Fizzano for reading an earlier draft of this paper.

References

1. N. Alon, G. Kalai, M. Ricklin, and L. Stockmeyer. Lower bounds on the competitive ratio for mobile user tracking and distributed job scheduling. In *Proceedings of the 33rd Annual Symposium on Foundations of Computer Science*, pages 334–343, 1992.
2. B. Awerbuch, S. Kutten, and D.Peleg. Competetive distributed job scheduling. In *Proceedings of the 24th Annual ACM Symposium on Theory of Computing*, pages 571–580, 1992.
3. J.L. Bruno, E.G. Coffman, and R. Sethi. Scheduling independent tasks to reduce mean finishing time. *Communications of the ACM*, 17:382–387, 1974.
4. V. Chvátal. A greedy heuristic for the set-covering problem. *Mathematics of Operations Research*, 4(3):233–235, August 1979.
5. X. Deng, H. Liu, J. Long, and B. Xiao. Deterministic load balancing in computer networks. In *Proceedings of 2nd IEEE Symposium on Parallel and Distributed Processing*, 1992.
6. P. Fizzano, D. Karger, C. Stein, and J. Wein. Job scheduling in rings. In *Proceedings of the 1994 ACM Symposium on Parallel Algorithms and Architectures*, 1994.

7. R.L. Graham. Bounds for certain multiprocessor anomalies. *Bell System Technical Journal*, 45:1563–1581, 1966.

8. R.L. Graham. Bounds on multiprocessing anomalies. *SIAM Journal of Applied Mathematics*, 17:263–269, 1969.

9. M. Grotschel, C. L. Monma, and M. Stoer. Design of survivable networks. In *Handbook in Operations Research and Management Science*. 1992. To appear.

10. D. Gusfield. Bounds for naive multiple machine scheduling with release times and deadlines. *Journal of Algorithms*, 5:1–6, 1984.

11. L. Hall and D. B. Shmoys. Approximation schemes for constrained scheduling problems. In *Proceedings of the 30th Annual Symposium on Foundations of Computer Science*, pages 134–141. IEEE, October 1989.

12. D.S. Hochbaum and D.B. Shmoys. Using dual approximation algorithms for scheduling problems: theoretical and practical results. *Journal of the ACM*, 34:144–162, 1987.

13. D.S. Hochbaum and D.B. Shmoys. A polynomial approximation scheme for machine scheduling on uniform processors: using the dual approximation approach. *SIAM Journal on Computing*, 17:539–551, 1988.

14. W. Horn. Minimizing average flow time with parallel machines. *Operations Research*, 21:846–847, 1973.

15. E.L. Lawler, J.K. Lenstra, A.H.G. Rinooy Kan, and D.B. Shmoys. Sequencing and scheduling: Algorithms and complexity. In S.C. Graves, A.H.G. Rinnooy Kan, and P.H. Zipkin, editors, *Handbooks in Operations Research and Management Science, Vol 4., Logistics of Production and Inventory*, pages 445–522. North-Holland, 1993.

16. D. N. Lee, K. T. Medhi, J. L. Strand, R. G. Cox, and S. Chen. Solving large telecommunications network loading problems. *AT&T Technical Journal*, 68(3):48–56, 1989.

17. J.K. Lenstra, A.H.G. Rinnooy Kan, and P. Brucker. Complexity of machine scheduling problems. *Annals of Discrete Mathematics*, 1:343–362, 1977.

18. J.K. Lenstra, D.B. Shmoys, and É. Tardos. Approximation algorithms for scheduling unrelated parallel machines. *Mathematical Programming*, 46:259–271, 1990.

19. D. Peleg, 1992. Private communication.

20. S. Plotkin, D. B. Shmoys, and E. Tardos. Fast approximation algorithms for fractional packing and covering problems. In *Proceedings of the 32nd Annual Symposium on Foundations of Computer Science*, 1991. To appear.

21. P. Raghavan. Probabilistic construction of deterministic algorithms: approximating packing integer programs. *Journal of Computer and System Sciences*, 37:130–143, 1988.

22. P. Raghavan and C. D. Thompson. Randomized rounding: a technique for provably good algorithms and algorithmic proofs. *Combinatorica*, 7:365–374, 1987.

23. D. B. Shmoys, C. Stein, and J. Wein. Improved approximation algorithms for shop scheduling problems. In *Proceedings of the 2nd ACM-SIAM Symposium on Discrete Algorithms*, pages 148–157, January 1991. To appear in *Siam J. Computing*.

24. D. B. Shmoys and E. Tardos. Scheduling parallel machines with costs. In *Proceedings of the 4th ACM-SIAM Symposium on Discrete Algorithms*, pages 448–455, January 1993.

25. D. B. Shmoys, J. Wein, and D.P. Williamson. Scheduling parallel machines on-line. *SIAM Journal on Computing*, 1994. To appear.

Parallel Dynamic Lowest Common Ancestors

Eric Schenk*

Department of Computer Science, University of Toronto, Toronto, Ontario, M5S–1A4, Canada

Abstract. This paper gives a CREW PRAM algorithm for the problem of finding lowest common ancestors in a tree under the insertion and deletion of leaves. For a tree with a maximum of n vertices, the algorithm takes $O(m/p + r \log p + \min(m, r \log n))$ time and $O(n)$ space using p processors to process a sequence of m operations that are presented over r rounds. Furthermore, lowest common ancestor queries can be done in worst case constant time using a single processor. For one processor, the algorithm matches the bounds achieved by the best sequential algorithm known. The new algorithm is somewhat simpler and has smaller constants in the time and space complexity.

1 Introduction

Finding lowest common ancestors in trees is a frequently occurring problem in the literature and has found application in such diverse problems as computing dominators in reducible flow graphs [1], detecting negative cycles in sparse graphs [8], planarity testing [7], and computing weighted matchings [4]. This paper gives a Concurrent Read Exclusive Write Parallel Random Access Machine (CREW PRAM) algorithm for the problem of finding lowest common ancestors in a tree under the insertion and deletion of leaves. As part of the solution to dynamic lowest common ancestors this paper also defines and solves the dynamic restricted range minimum problem.

Both problems considered in this paper are treated as Abstract Data Types with a fixed set of operation types. An external agent is assumed to make rounds of one or more requests, where each round consists of requests of the same type and must be dealt with on-line before the next round. The number of rounds of requests is denoted by r. The number of requests in the ith round is denoted by q_i, and the total number of requests is denoted by m. The maximum size attained by the data structure is denoted by n.

The *dynamic lowest common ancestor problem* is to maintain one or more trees under the following operations.

initialize(T, x): Initialize the tree T to contain the single vertex x.
lca(T, x, y): Find the vertex of greatest depth in T that is an ancestor of both x and y.
add-leaf(T, x, y): Add the new vertex y as a child of the vertex x in T.
delete-leaf(T, x): Remove the leaf x from the tree T.

We place the further restriction that within a round at most one *add-leaf* operation can be specified at each existing vertex. We achieve $O(m/p + r \log p + \min(m, r \log n))$ time and $O(n)$ space on a p processor CREW PRAM. The work is optimal for $p \in O(n/r \log n)$. Also, any *lca* query can be performed in worst case constant time on one processor. This is the first parallel algorithm for the dynamic problem.

Gabow [4] gives a sequential algorithm to process sequences of m *lca* and *add-leaf* operations creating a tree of maximum size n in $O(m)$ time and $O(n)$ space. This matches the sequential complexity of our algorithm and the method used here to deal

* This research was supported in part by an NSERC postgraduate scholarship.

with deletions applies directly to Gabow's construction as well. However, a sequential implementation of our algorithm is somewhat simpler and has smaller constants in both the time and space complexity.

Both Gabow's algorithm and our algorithm are based upon algorithms for the *static lowest common ancestor problem*, which is to preprocess a fixed tree T of size n such that on-line *lca* queries can be answered efficiently. Harel and Tarjan [5] give an algorithm that achieves $O(n)$ time and space for preprocessing and constant time to answer queries. Gabow [4] bases his algorithm for the dynamic lowest common ancestor problem on this algorithm. Berkman and Vishkin [3] use a completely different approach to the static problem to derive a constant time, n processor, preprocessing algorithm on a special variant of the CRCW PRAM. This results in a data structure that allows *lca* queries to be answered in constant time using a single processor. Our solution to the dynamic lowest common ancestor problem is based upon this algorithm.

Our construction for the dynamic lowest common ancestor problem is a reduction to the *dynamic restricted range minimum problem*. This problem is to maintain one or more lists of elements, where each element x has an associated value $value(x)$, under the following operations.

initialize-min$(X, x_1, x_2, \ldots, x_q)$: Create a new list $X = [x_1, x_2, \ldots, x_q]$.

list-insert$(X, x, y, side)$: Insert a new element x into the list X either to the left or right of y as indicated by the value of *side*. The new element x must must have $value(x)$ greater than that of its left neighbor, and no less than that of its right neighbor, if such a neighbor exists.

rmin(X, x, y): Find the rightmost minimum valued element in X between x and y inclusive.

prec(X, x, y): Determine if x precedes y in the list X.

Within a round at most one insertion can be specified at each element currently in a list X. For the purposes of analysis, an initialization operation for a list of q elements is counted as q operations. We achieve $O(m/p + r \log p + \min(m, r \log n))$ time and $O(n)$ space on a p processor CREW PRAM. The work is optimal for $p \in O(n/r \log n)$. Also, any *rmin* or *prec* query can be performed in worst case constant time on one processor.

The remainder of this paper is organized as follows. Section 2 gives the reduction from the dynamic lowest common ancestor problem to the dynamic restricted range minimum problem. Section 3 gives an slightly inefficient algorithm for dynamic restricted range minimum. Section 4 shows how to make this algorithm optimal for very small lists. Section 5 combines the algorithms of section 3 and 4 to obtain a general optimal algorithm. Section 6 gives some concluding remarks.

2 Dynamic Least Common Ancestors

To solve the dynamic least common ancestors problem we reduce it to the dynamic restricted range minimum problem. Our reduction is similar to that used by Berkman and Vishkin [3] in their solution to the static lowest common ancestor problem. Let the sequence $X = [x_1, \ldots, x_n]$, where $n = |V|$, be the preorder tour of a tree T. Let $[x_i, \ldots, x_j]$ denote the sublist of elements between x_i and x_j inclusive. The following lemma is easily established.

Lemma 1. *For any* $i < j$, *let* z *be the rightmost vertex of minimal depth in the sublist* $[x_i, \ldots, x_j]$. *If* $x_i = z$, *then* $lca(T, x_i, x_j) = z$; *otherwise* $lca(T, x_i, x_j)$ *is the parent of* z *in* T.

Provided that the parent and depth in T of each vertex is recorded, this reduces the problem of computing lowest common ancestors in T to the problems of determining the relative order of elements within the list X and of finding the rightmost element with minimal depth among the elements in a sublist of the list X.

The operation *initialize*(T, x) is implemented by initializing the list X to a single element x, where *value*(x) is assigned 0. The operation *add-leaf*(T, x, y) is performed by setting *value*(y) to *value*$(x) + 1$ and inserting y to the immediate right of x in X. The *delete-leaf* operation is dealt with by marking the affected vertex as deleted and incrementing a count of the number of deleted vertices. Whenever the number of deleted vertices exceeds the number of vertices remaining in T, we rebuild the entire data structure by executing *initialize-min* on the undeleted vertices in T. This at most doubles the overall amortized complexity.

The solution to the dynamic restricted range minimum problem presented in the remainder of this paper, together with the above discussion implies the following.

Theorem 2. *For trees of maximum size* n, *any sequence of* m *dynamic lowest common ancestor operations, presented over* r *rounds, can be processed in* $O(m/p + r \log p + \min(m, r \log n))$ *time and* $O(n)$ *space on a* p *processor CREW PRAM. In addition, lca queries can be processed in constant time on a single processor.*

3 A Simple Dynamic Restricted Range Minimum Algorithm

This section presents a simple, but not very efficient, algorithm for the dynamic restricted range minimum problem. This algorithm relies on a solution to a restricted version of the dynamic lowest common ancestor problem, which we solve by taking advantage of some number theoretic properties of a binary tree.

3.1 Preliminaries

Let B be a binary tree and let $|B|$ denote the number of leaves in B. For any vertex v in B let B_v denote the subtree rooted at v, let $\ell(v)$ denote the leftmost leaf and let $r(v)$ denote the rightmost leaf of B_v. The depth of a vertex v is denoted $depth(v)$, where the root has depth 0. Let $lca_B(x, y)$ denote the lowest common ancestor of any two vertices x and y in B. Finally, let $v.parent$, $v.left$ and $v.right$ respectively denote the parent, the left child and the right child of vertex v.

Consider a list in which each element x has a associated value *value*(x). We define a total ordering such that $x < y$ if (1) *value*$(x) < $ *value*(y) or (2) *value*$(x) = $ *value*(y) and y precedes x in the list. Under this ordering, the rightmost minimal valued element in a sublist is the minimum element in that sublist.

By analogy with open intervals on a number line we define open intervals on a list. The notation $[x, \ldots, y)$ denotes the list of elements between x and y excluding y. Similarly $(x, \ldots, y]$ denotes the list of elements between x and y excluding x. In both cases, if $x = y$ the list is empty.

The operators $\vee$, $\wedge$, $\oplus$ and $\neg$ denote the bitwise boolean or, bitwise boolean and, bitwise boolean exclusive or, and bitwise complement operations, respectively.

3.2 The Basic Data Structure

The basic structure for representing a list $X = [x_1, \ldots, x_n]$ is a binary tree B of depth $O(\log |B|)$ whose leaves in left to right order are $x_1, \ldots, x_n$. There are three additional components:

1. Each vertex v in B has a field $v.min$ that points to to the minimum leaf in B_v.
2. Each leaf v of B holds an array $A_v[1, \ldots, depth(v)]$. Let v_i denote the ancestor of v at depth i. For all $1 \leq i \leq depth(v)$, if v_i is a right child, then $A_v[i]$ points to the minimum element in the list $[\ell(v_i), \ldots, v)$, otherwise $A_v[i]$ points to the minimum element in the list $(v, \ldots, r(v_i)]$. In either case, if the list in question is empty, then $A_v[i] = 0$.
3. Each vertex v in B is labelled by a pair of integers $\langle \pi_v, \mu_v \rangle$. The root of the tree is labelled $\langle 0, 1 \rangle$. If vertex v is labelled $\langle \pi_v, \mu_v \rangle$, then its left child is labelled $\langle \pi_v, 2\mu_v \rangle$, and its right child is labelled $\langle \pi_v \vee \mu_v, 2\mu_v \rangle$. Note that if v is at depth d, then $\mu_v = 2^d$. Also note that the d least significant bits of π_v, read from least to most significant bit, describe the path from the root to v, with a 0 bit indicating a left branch and a 1 bit indicating a right branch.

Since B has depth $O(\log |B|)$, this structure takes $O(n \log n)$ words of space, and for each vertex v, the integers used in the label $\langle \pi_v, \mu_v \rangle$ fit into an $O(\log n)$ bit word.

3.3 Computing rmin and prec

For any two leaves x_i, x_j, where $i < j$, element $rmin(X, x_i, x_j)$ can be found as follows. Let $\gamma = \pi_{x_i} \oplus \pi_{x_j}$. Let d be the position of the least significant 1 bit in the binary representation of γ. Observe that this is the position where the paths to x_i and x_j diverge. Thus, d is the depth of $lca_B(x_i, x_j)$ in B. It is easily shown that if $\lambda = \gamma \wedge \neg(\gamma - 1)$, then $\lambda = 2^d$. Let x'_i be the ancestor of x_i at depth $d + 1$ and x'_j be the ancestor of x_j at depth $d + 1$. By definition $A_{x_i}[d + 1]$ contains the minimum element in $(x_i, \ldots, r(x'_i)]$ and $A_{x_j}[d + 1]$ contains the minimum element in $[\ell(x'_j), \ldots, x_j)$. It follows directly that the element $rmin(X, x_i, x_j)$ is $\min(x_i, A_{x_i}[d + 1], A_{x_j}[d + 1], x_j)$.

To compute $rmin(X, x, y)$ for any two leaves x, y, we need to be able to determine if x precedes y in the list (i.e. check if $prec(X, x, y)$ is true). Assume without loss of generality that $x \neq y$. Compute λ as described above. The leaf x precedes y in the list if and only if $\pi_x \wedge \lambda = 0$ (i.e. bit d of π_x is 0). To see this, note that the path from the root to x takes a left branch at $lca_B(x, y)$ exactly when bit d of π_x is 0.

Using the above algorithms both $rmin$ and $prec$ can be computed in constant time by one processor, and therefore any collection of q_i such queries can be processed in parallel in constant time on a q_i processor CREW PRAM.

3.4 Rebalancing and Initialization

As elements are inserted into the list, the tree B must continue to have depth $O(\log |B|)$. We use a brute force rebalancing scheme that replaces any potentially unbalanced subtree with a perfectly balanced subtree. For each vertex v, store the size of B_v in $v.size$ and the size of B_v the last time it was rebalanced in $v.oldsize$. For a given constant $\alpha \in (0, 1)$, a subtree B_v is balanced if and only if $(1 - \alpha)v.oldsize \leq v.size \leq (1 + \alpha)v.oldsize$. (This criteria can be replaced by any of the criteria defined for weight balanced trees. See, for example, Andersson [2].)

To rebalance a subtree B_v, we replace it with a balanced subtree and recompute the data structure on those vertices. Let $q = |B_v|$. Even before rebalancing the subtree B_v will have depth $O(\log q)$. Therefore, the vertices of B_v can be collected into an array in $O(\log q)$ time on q processors by recursively descending the tree. The vertices can be formed into a perfectly balanced tree in constant time on q processors [9]. Let w be the root of the resulting tree. For any vertex z, the label $\langle \pi_z, \mu_z \rangle$ can be computed from z's parent, if any, in constant time. Thus, these labels can be computed for all vertices z in B_w in a total of $O(\log q)$ time on q processors by descending the tree. Similarly, for any vertex z the labels $z.min$, $z.size$ and $z.oldsize$ can be computed from the children of z, if any, in constant time. Thus, these labels can be computed for all vertices z in B_w in a total of $O(\log q)$ time on q processors by ascending the tree. It remains to describe how to compute A_x for each leaf x in B_w. It can be easily shown that the following procedure computes A_x on one processor in $O(\log |B|)$ time.

$R' \leftarrow 0, L' \leftarrow 0, v \leftarrow x$
for i from $depth(x)$ down to 1 do
    loop invariant: $L' = \min[l(v), \ldots, x]$ and $R' = \min[x, \ldots, r(v)]$.
    if v is a right child then $A_x[i] \leftarrow L'$; $L' \leftarrow \min(L', v.parent.left.min)$.
    if v is a left child then $A_x[i] \leftarrow R'$; $R' \leftarrow \min(R', v.parent.right.min)$.
    $v \leftarrow v.parent$

The total time to rebalance the subtree B_v is $O(\log |B|)$ time on $q = |B_v|$ processors. Finally, this procedure can also be used to initialize the data structures for a collection of lists with total of q elements in $O(\log q)$ time on q processors, where $O(\log q)$ time is required to allocate the memory for the data structures [6].

3.5 Insertions

Let $I_1, \ldots, I_q$ be the requested insertions, where the request $I_i = list\text{-}insert(x_i, y_i, s_i)$ specifies that the new leaf x_i is to be inserted to the left or right of y_i depending on whether $s_i = $ "left" or $s_i = $ "right". Insertions are dealt with in two stages. In the first stage the new leaves are inserted into B and the data structure is extended to these new leaves. In the second stage, subtrees of B are rebalanced if necessary.

The insertion algorithm begins by allocating a record for each new leaf x_i, and allocating a corresponding record z_i for the new internal node that will be the parent of x_i and y_i. This can be accomplished in $O(\log q)$ time on q processors [6]. Next, for all requests I_i, we replace the leaf y_i with a three node subtree rooted at z_i and with x_i and y_i as leaves, where the order of the leaves is determined by the value of s_i. This can be performed in constant time on q processors. Having constructed these replacement trees, we compute the labels for all of them using the same method as in the rebalancing procedure. This requires a further $O(\log |B|)$ time on q processors, giving a total of $O(\log |B|)$ time on q processors.

The data structure on the vertices in the replacement subtrees is correct by construction. We must show that the rest of the data structure is still correct. Consider any vertex v not in any replacement subtree. Since the path from v to the root is unchanged, $\langle \pi_v, \mu_v \rangle$ remains correct trivially. Before the insertions $v.min$ was correct. In order for it to now be incorrect it would have to be the case that a leaf was inserted into B_v with a lesser value than $v.min$. But any new leaf x_i inserted into B_v must have a value greater than its corresponding insertion point y_i, which was in the subtree B_v before the insertions. It follows that $v.min$ is correct for all vertices. By a similar argument, if v is a leaf we can show that $A_v[i]$ remains correct for all $1 \le i \le depth(v)$.

It remains to update subtree sizes and carry out rebalancing. For each z_i, it is necessary to recompute $v.size$ at every ancestor v of z_i. This is done by having one processor walk up the tree from each z_i and compute $v.size$ at each ancestor of z_i from the sizes of its children. Some care must be taken in scheduling to avoid write conflicts (for example, by alternately moving up from the left and the right). A similar process is used to find, for each z_i, the highest unbalanced subtree on the path from z_i to the root of B. Let W be the set of unbalanced subtrees found in this search. Since the subtrees in W are disjoint they can be rebalanced in parallel. Updating sizes and collecting W can be done in $O(\log |B|)$ time on q processors. The rebalancing step can be performed in a further $O(\max_{B_v \in W} \log |B|)$ time on $\sum_{B_v \in W} |B_v|$ processors.

3.6 Complexity

Let c_i be the number of vertices in subtrees rebalanced in round i. By the balance criteria, a subtree is rebalanced only if it has changed in size by a constant fraction since its last rebalancing. Thus the size of a rebalanced subtree is proportional to the number of insertions since the last rebalancing. Since each insertion occurs in at most $O(\log n)$ subtrees, it follows that $\sum_{i=1}^{r} c_i \in O(m \log n)$. Using this, together with the complexities derived in the previous subsections, we obtain that using $p \le n$ processors, the time complexity for all r rounds of operations is $O(m \log^2 n/p + r \log n)$.

4 An Optimal Algorithm for Small Lists

The algorithm presented in section 3 performs large amounts of work both in rebalancing, and in computing the A_v arrays for new vertices v. In this section we consider an algorithm that avoids this work for lists of length at most $b \in O((\log \log n)^2)$.

Our construction is based upon the algorithm given in section 3 with modifications motivated by the following observations. First, since $b \in O(\log n)$, rebalancing is not necessary to maintain a depth bound of $O(\log n)$. Second, since $b \log b \in O(\log n)$ we can store the entire contents of an array A_v in a single $O(\log n)$ bit word by keeping the at most b elements in a contiguous block and storing indices into this block in the array. Finally, an examination of the changes in the contents of an array A_y due to the insertion of a new element x beside y reveals that the resulting arrays A_x and A_y are very similar to the original contents of the array A_y. In fact, we can show that they are identical except that one or more positions in A_x must be assigned the value y, and one or more positions in A_y must be assigned the value x. Furthermore we can maintain a list of the positions that must be changed with only a small additional effort.

4.1 A Modified Data Structure

We begin by defining some low level notation and operations. Let $\delta = \lceil \log(b+1) \rceil$. Consider an array C of length b in which each element is in $\{0, \ldots, b\}$. Then C can be represented by a $b\delta$ bit word, in which bits $(i-1)\delta + 1, \ldots, i\delta$ represent the ith entry in C. Copying the entire contents of an array into a new array can be done trivially in constant time. Also, accessing or modifying any single element of an array can be performed in constant time on one processor using bit masking techniques. Finally, given two arrays C and F stored in this manner and a value $a \in \{0, \ldots, b\}$, we can use the same techniques to compute, in constant time on one processor, the array $mask(C, F, a)$ that results from setting $C[i]$ to a for all i such that $F[i] = 1$.

We define the modified data structure. A list $[x_1, \ldots, x_q]$ is represented by a potentially unbalanced binary tree B whose leaves are $x_1, \ldots, x_q$. The number of vertices in the tree is stored in $B.size$. A fixed size array of records $M[1, \ldots, b]$ stores the vertices of B. There are additional components for each vertex v in B:

1. Each vertex v in B has a field $v.min$ that points to the minimum leaf in B_v.
2. Each vertex v in B has a field $v.index$ that indicates the position of v in the array M.
3. Each vertex v in B holds five arrays: A_v, L_v, R_v, $L_v^{\perp}$ and $R_v^{\perp}$, each represented by a single word. Let v_i denote the ancestor of v at depth i. For all $1 \leq i \leq b$, entry i of each array is set to 0 unless otherwise specified. Non zero entries of the array are defined for $1 \leq i \leq depth(v)$ as follows.
    (a) If v_i is a right child, then $A_v[i]$ stores the index of the minimum element in the list $[\ell(v_i), \ldots, \ell(v))$, otherwise $A_v[i]$ stores the index of the minimum element in the list $(r(v), \ldots, r(v_i)]$. In either case, if the list in question is empty, then $A_v[i] = 0$. (Note that A_v is defined for all vertices v, in contrast with section 3.2, where it is defined only for leaves.)
    (b) If v_i is a left child and $(r(v), \ldots, r(v_i)] = \emptyset$ then $L_v^{\perp}[i] = 1$.
    (c) If v_i is a right child and $[\ell(v_i), \ldots, \ell(v)) = \emptyset$ then $R_v^{\perp}[i] = 1$.
    (d) If v_i is a left child and $v.min$ is smaller than every element in the list $(r(v), \ldots, r(v_i)]$, or the list is empty, then $L_v[i] = 1$.
    (e) If v_i is a right child and $v.min$ is smaller than every element in the list $[\ell(v_i), \ldots, \ell(v))$, or the list is empty, then $R_v[i] = 1$.
4. Each vertex v in B is labelled by a pair of integers $\langle \pi_v, \mu_v \rangle$, defined as for the basic data structure (see section 3.2).

The space used is $O((\log \log n)^2)$ words, where each word is of size $O(\log n)$ bits. Note that both $rmin$ and $prec$ can be computed from this data structure using the algorithms described in section 3.3.

4.2 Insertions

Let $I_1, \ldots, I_q$ be the requested insertions, where the request $I_i = list\text{-}insert(x_i, y_i, s_i)$ specifies that the new leaf x_i is to be inserted to the left or right of y_i depending on whether $s_i = $ "left" or $s_i = $ "right".

The insertion algorithm begins by assigning index values to to each new leaf x_i, and to each new internal vertex z_i, where z_i will be the parent of x_i and y_i. This is accomplished as follows. For all requests I_i we set $x_i.index \leftarrow B.size + 2i$, $M[B.size + 2i] \leftarrow x_i$, $z_i.index \leftarrow B.size + 2i + 1$ and $M[B.size + 2i + 1] \leftarrow z_i$. After this, the size of the tree is updated by setting $B.size \leftarrow B.size + 2q$. Finally, for all requests I_i we replace the leaf y_i with a three node subtree rooted at z_i and having leaves x_i and y_i, where the order of the leaves is determined by the value of s_i. These steps can be performed in constant time on q processors.

Having constructed these replacement trees, we must compute the labels for each of them. It follows from the definitions that for any i the correct values for $z_i.min$, $depth(z_i)$, $\langle \pi_{z_i}, \mu_{z_i} \rangle$, A_{z_i}, R_{z_i}, L_{z_i}, $L_{z_i}^{\perp}$ and $R_{z_i}^{\perp}$ are exactly those values found in the record for y_i before the insert, therefore we can simply copy these. It remains to describe the computation of the labels on the leaves of the replacement trees. The following three easily established lemmas will be useful.

Lemma 3. *Let v be any non root vertex in B and let $z = v.parent$. If v is a left child then (1) $L_v^\perp[i] = 1$ if and only if $i = depth(v)$, and (2) $R_v^\perp = R_z^\perp$. If v is a right child then (1) $R_v^\perp[i] = 1$ if and only if $i = depth(v)$, and (2) $L_v^\perp = L_z^\perp$.*

Lemma 4. *If v is any non root vertex in B and $z = v.parent$ then,*

1. *If $v.min = z.min$ and v is a right child then $L_v = L_z$ and $R_v[i] = 1$ if and only if $R_z[i] = 1$ or $i = depth(v)$.*
2. *If $v.min = z.min$ and v is a left child then $R_v = R_z$ and $L_v[i] = 1$ if and only if $L_z[i] = 1$ or $i = depth(v)$.*
3. *If $v.min$ is larger than the right neighbor of $r(v)$, or if $r(v)$ has no right neighbor, then $L_v = L_v^\perp$.*
4. *If $v.min$ is larger than the left neighbor of $\ell(v)$, or if $\ell(v)$ has no left neighbor, then $R_v = R_v^\perp$.*

Lemma 5. *Let v be any non root vertex in B, let $z = v.parent$ and let u be v's sibling. If v is a left child then $A_v = mask(A_z, L_u, u.min)$. If v is a right child then $A_v = mask(A_z, R_u, u.min)$.*

Consider any replacement subtree rooted by a vertex z. For any child v of z, the values of $v.min$, $depth(v)$ and $\langle \pi_v, \mu_v \rangle$ can be computed in constant time on one processor directly from their definitions and the labels for z. Furthermore, using the above three lemmas, the values for $L_v^\perp$, $R_v^\perp$, L_v, R_v, and A_v can also be computed in constant time on one processor from the labels for z. Thus, the labels for all of the replacement trees can be computed in $O(q)$ time on q processors, leading to a total of $O(q)$ time on q processors to perform q insertions.

4.3 Initialization

The input to an initialization operation is a list $[x_1, \ldots, x_q]$ of elements and their associated values. Within a round several initialization operations can be requested, call them $I_1, \ldots, I_k$. The initialization algorithm proceeds in three stages. In the first stage the necessary memory for all lists is allocated. In the second a balanced tree is built, and in the third the labelling of the tree is computed. Only the first stage is performed completely in parallel. In the remaining stages at most one processor performs the initialization of a particular list.

The first stage is straight forward. An array M of records of size b must be allocated for each initialization request. This can be accomplished in $O(k/p + \log p)$ time on p processors [6].

In the second stage we use one processor to build a tree B with the following structure. Let $x_1, \ldots, x_q$ be the records for the q leaves, and let $y_1, \ldots, y_{q-1}$ be the records for the $q - 1$ internal vertices. The root of the tree is y_1. For $1 \le i < q - 1$ we make x_i the left child of y_i and y_{i+1} the right child of y_i. Finally, x_{q-1} is the left child of y_{q-1} and x_q is the right child of y_{q-1}. We use this skewed tree structure because it simplifies the remainder of the initialization computations.

In the third stage we must compute the labels for each vertex v. The labels $depth(v)$,

$\langle \pi_v, \mu_v \rangle$ and $v.min$ can all be computed in $O(q)$ time. By lemma 3 the $L_v^\perp$ and $R_v^\perp$ arrays can also be computed in $O(q)$ time (from the bottom of B up). By lemma 5 the A_v arrays can be computed in $O(q)$ time by a depth first tour of the tree B, provided that the L_v and R_v arrays are available. By definition, for $v \in \{y_1, \ldots, y_{q-1}, x_q\}$, $L_v[i] = 0$ for all i. Also by definition, for $v \in \{x_1, \ldots, x_{q-1}\}$, $L_v[i] = 1$ if and only if $i = depth(v)$. Thus the L_v arrays can be computed in $O(q)$ time. It remains therefore to show how to compute the R_v array.

By definition $R_{y_1}[i] = 0$ for all i. Also, using the results of lemma 4, for $1 < i < q$, R_{y_i} can be computed from $R_{y_{i-1}}$ in constant time, and furthermore, R_{x_q} can be computed from $R_{y_{q-1}}$. Thus the R_{y_i} and R_{x_q} arrays can be computed in $O(q)$ time. Finally, it can be shown that the following procedure correctly computes R_{x_i} for $1 \le i \le q - 1$ in $O(q)$ time.

```
Set the array C to all 0's.
Set the stack S to empty.
for i = q - 1 down to 1 do
    if S is not empty then
        loop
            set y to the top item on stack S
            if x_i < y then
                pop the top item on stack S; D ← R_y; R_y = C; R_y ← mask(R_y, D, 0)
            end if
        until S is empty or y < x_i.
    push x_i onto stack S; R_{x_i} ← C; C[i] ← 1
```

This gives a total of $O(q)$ time to initialize a single list $[x_1, \ldots, x_q]$, omitting the cost of allocating memory. Give a collection of k lists of combined size q_i, initialization can be carried out in $O(\min(q_i, k/p + (\log \log n)^2)) \in O(q_i/p + \min(q_i, (\log \log n)^2))$ time on p processors by assigning roughly equal size tasks to each of the p processors.

4.4 Complexity

Either q_i parallel $rmin$ or $prec$ queries, or q_i insertions can be done in $O(q_i/p)$ time on p processors. Performing q_i parallel initialization requests requires $O(q_i/p + \min(q_i, (\log \log n)^2))$ time. Summing these times we obtain a total of $O(m/p + \min(m, r(\log \log n)^2))$ time on p processors to perform m operations over r rounds on one or more trees that reach size at most n. As well, the space required for each tree is $O((\log \log n)^2)$ words, regardless of the number of elements actually stored in the tree.

5 An Optimal Algorithm

Examining the algorithm of section 3 shows that much of the work performed arises from always rebalancing an entire subtree, despite the fact that the subtrees near the leaves are already balanced. This cost can be circumvented by successively reducing the restricted range minimum problem to smaller versions of itself.

5.1 A Self Reduction

Assume we have two algorithms, A_1 and A_2, to solve the dynamic restricted range minimum problem. Let $T_1(n, m, r, p)$ be the time to solve a sequence of m operations on one or more lists, constructing lists of size at most n, presented over r rounds, on a p processor CREW PRAM using algorithm A_1. Similarly, let $T_2(n, m, r, p)$ be the time for A_2 to solve the same problem. We construct an algorithm that, for $f(n) \in (\log n)^{O(1)}$, uses A_1 and A_2 to solve this problem in time $T_1(n/f(n), m/f(n), \min(r, m/f(n)), p) + T_2(f(n), m, r, p) + O(m/p + r \log p + \min(m, r \log n) + \min(m, rd(n)))$, where $d(n)$ is the maximum depth of a tree representing a list in algorithm A_2.

We begin by describing the data structure for a fixed list $X = [x_1, \ldots, x_q]$. Partition X into sublists $G_1, \ldots, G_{\lceil q/f(n) \rceil}$ each of size $O(f(n))$. For each sublist G we maintain three pointers: $G.head$, $G.tail$ and $G.min$, that point to the head, the tail, and the minimum element of the sublist G respectively. Each element x_i also has a pointer $g(x_i)$ that points to the sublist it is contained in. The sublists are maintained using algorithm A_2. A second list $B = [b_1, \ldots, b_{\lceil q/f(n) \rceil}]$ is maintained by algorithm A_1, where $b_i = G_i.min$. Note that each b_i occurs in two lists, G_i and B. Also, the list B is doubly linked so that for each b_i, $b_i.next$ and $b_i.prior$ point to the next and prior elements in the list.

A request $prec(x, y)$ can be satisfied as follows. If $g(x) = g(y)$ then it can be computed in the sublist $g(x)$ using algorithm A_2. Otherwise $prec(x, y)$ is the same as $prec(g(x).min, g(y).min)$ computed in list B using algorithm A_1.

A request $rmin(x_i, x_j)$ can be satisfied as follows. Without loss of generality, suppose $i < j$. If x_i and x_j are in the same sublist, i.e. $g(x_i) = g(x_j)$, then the request can be satisfied directly using algorithm A_2 on sublist $g(x_i)$. Otherwise, there are three possibilities. The minimum could be in the list $g(x_i)$ between x_i and $g(x_i).tail$, it could be in the list $g(x_j)$ between x_j and $g(x_j).head$, and it could be the minimum element in the list B between $g(x_i).min.next$ and $g(x_j).min.prior$. The first two of these can be found using algorithm A_2, the last by using algorithm A_1.

Processing q_i queries on q_i processors can be done in the sum of the times to process q_i queries on q_i processors with algorithms A_1 and A_2.

To process an initialization request for a list $[x_1, \ldots, x_q]$, divide the list into sublists of size $f(n)$ in constant time on q processors, initialize each sublist using algorithm A_2, construct the list B in a further constant time on q processors, and initialize B using algorithm A_1.

Processing a set of q insertions is more involved. First we must group together insertions that are in the same sublist. In the next section we describe how to solve this problem in $O(\min(q, q/p + \log n))$ time on p processors. Once the grouping has been done, the insertions in each sublist are carried out using algorithm A_2. Next any sublists that have become larger than $f(n)$ elements must be identified; this can be done in $O(q/p + \log p)$ time on p processors. Let q' be the total size of these sublists. Each of these sublists must be split in half. Toward this end, each sublist is mapped into an array, then divided into two sublists of size at most $f(n)$ each, and then the initialization procedure of algorithm A_2 must be run on each new sublist. Mapping the sublists into arrays can be done in $O(\min(q', q'/p + d(n)))$ time on p processors. The sublists can be split in half in $O(q'/p)$ time on p processors. For each sublist that is split, one of the new sublists will have a global minimum equal to the global minimum of the sublist being replaced. That sublist is already represented in list B. The other new sublist must have its global minimum inserted into the list B. This is done using algorithm A_1. The total time required for the q insertions is $O(q/p + \log p + \min(q, q/p + \log n) + \min(q', q'/p + d(n))) =$

$O(q/p+q'/p+\log p+\min(q,\log n)+\min(q',d(n)))$ plus q insertions and q' initializations of elements using A_2 and at most q insertions using A_1.

5.2 The Grouping Problem

In the algorithm we describe above, it is necessary to group together insertions that occur in the same sublist. This problem is made somewhat easier by the fact that the number of insertions into a given sublist cannot exceed the size of the sublist. Furthermore, within a sublist, each insertion point can be assigned a unique index. Using these facts, the problem that must be solved is equivalent to the following problem.

The Grouping Problem: Given a list $X = [x_1, \ldots, x_q]$ of distinct elements from the set $S_1 \times S_2$, where $|S_2| \in O(f(n))$, construct a list $Y = [x_{\sigma(1)}, \ldots, x_{\sigma(q)}]$ where σ is a permutation on $\{1, \ldots, q\}$ such that all elements in X that have the same value in the first coordinate are contiguous.

We sketch a solution for $f(n) \in (\log n)^{O(1)}$. First, sort X by the second coordinate. Reif [10] gives an algorithm that sorts q elements drawn from a set of size $(\log n)^{O(1)}$ in $O(\log n)$ time on a $q/\log n$ processor EREW PRAM. Let L_i denote the set of elements with i in the second coordinate. The sets $L_1, \ldots, L_{|S_2|}$ can be determined from the sorted list in constant time using q processors. We first describe how to complete the algorithm for $f(n) \in O(\log n)$. Observe that for each i, L_i contains no elements that have the same value in the first coordinate. Thus, by considering the elements of S_2 one at a time we can construct linked lists of elements that match on the first coordinate. Furthermore, the length of each list and the rank of each element in its list are determined. Using p processors this can be done in $O(q/p + \log n)$ time. Prefix sums space can be used to determine the allocation of space for each list in $O(q/p + \log n)$ time. The elements can be packed into an array in a further $O(q/p)$ time.

The algorithm can be used recursively for larger values of $f(n)$. Sort the input on the second coordinate and divide the input into subsets by dividing S_2 into $\log n$ equal sized parts. The problem can be solved recursively in parallel on these subsets, assigning to each subset a number of processors proportional to its size. The solutions thus obtained can then be merged using the above algorithm by treating contiguous elements that match on the first coordinate as a single element. For $f(n) = (\log n)^{O(1)}$ only a constant number of levels recursions are required.

The total time to solve the grouping problem is $O(q/p + \log n)$ time on p processors. If $q < \log n$ we can instead perform a bucket sort on one processor in $O(q)$ time. Combining these techniques we get a final complexity of $O(\min(q, q/p + \log n))$ time to perform grouping on p processors.

5.3 Complexity

We examine the complexity of the construction. Recall that q_i is the number of operations requested in round i. Let q'_i denote the number of elements that are in sublists that are split in round i. We can count the total number of elements involved in splits by charging 2 to each element inserted since the last split. Since $\sum_{i=1}^{r} q_i = m$ we have that $\sum_{i=1}^{r} q'_i = 2m$. Now, excluding the cost of running algorithms A_1 and A_2, processing a round requires $O(q_i/p+q'_i/p+\log p+\min(q_i,\log n)+\min(q'_i,d(n)))$ time on p processors. Each of the sublists maintained by algorithm A_2 has size at most $f(n)$, and the list B

maintained by algorithm A_1 has size at most $n/f(n)$. Furthermore, at most $m/f(n)$ operations are performed on the list B. Summing the times for the various components of the algorithm we obtain $T(n, m, r, p) = T_1(n/f(n), m/f(n), \min(r, m/f(n)), p) + T_2(f(n), m, r, p) + O(m/p + r \log p + \min(m, r \log n) + \min(m, rd(n)))$. Let $S_1(n)$ and $S_2(n)$ be the space used by algorithms A_1 and A_2 to represent a single list of size n respectively. The construction given in this section requires $O(n)$ space to represent the division into sublists, a further $S_1(n/f(n))$ space is used to represent list B, and $nS_2(f(n))/f(n)$ space is used to represent the $O(n/f(n))$ sublists. Thus the total space used by the construction is $S(n) = S_1(n/f(n)) + nS_2(f(n))/f(n) + O(n)$.

Setting both A_1 and A_2 to the algorithm of section 3, and setting $f(n)$ to $\log^2 n$ and $d(n)$ to $O(\log \log n)$, we obtain an algorithm with time complexity $O(m(\log \log n)^2/p + r(\log \log n)^2 + r \log p + \min(m, r \log n))$ on p processors, and space complexity $O(n \log \log n)$. Setting A_1 to the algorithm thus obtained and setting A_2 to the algorithm of section 4, and setting $f(n)$ to $(\log \log n)^2$ and $d(n)$ to $O((\log \log n)^2)$, we obtain an algorithm with time complexity $O(m/p + r \log p + \min(m, r \log n))$ on p processors and space complexity $O(n)$.

6 Concluding Remarks

With some additions the range minimum algorithms presented here can support the deletion of local maxima. Also, the lowest common ancestors algorithm can be extended to allow the insertion and deletion of roots in the tree. Finally, all of the algorithms presented in this paper only perform well in an amortized sense. It remains open whether or not the amortization can be eliminated.

References

1. Alfred V. Aho, John E. Hopcroft, and Jeffrey D. Ullman, *On computing least common ancestors in trees*, SIAM J. Comput. **5** (1976), 115–132.
2. Arne Andersson, *Improving partial rebuilding by using simple balance criteria*, Proceedings of the Workshop on Algorithms and Data Structures (F. Dehne, J.-R. Sack, and N. Santoro, eds.), Springer Verlag, 1989, pp. 393–402.
3. Omer Berkman and Uzi Vishkin, *Recursive *-tree parallel data-structure*, Proceedings of the 30th Annual IEEE Symposium on the Foundations of Computer Science, 1989, pp. 196–202.
4. Harold N. Gabow, *Data structures for weighted matching and nearest common ancestors with linking*, Proceedings of the 1st Annual ACM-SIAM Symposium on Discrete Algorithms, 1990, pp. 434–443.
5. Dov Harel and Robert Endre Tarjan, *Fast algorithms for finding nearest common ancestors*, SIAM J. Comput. **13** (1984), no. 2, 338–355.
6. Lisa Higham and Eric Schenk, *PRAM memory allocation and initialization*, Parallel Processing Letters **3** (1993), no. 3.
7. J. Já Já and J. Simon, *Parallel algorithms in graph theory: Planarity testing*, SIAM J. Comput. **11** (1982), 314–328.
8. D. Maier, *An efficient method for storing ancestor information in trees*, SIAM J. Comput. **8** (1979), 559–618.
9. A. Moitra and S. S. Iyengar, *Derivation of a parallel algorithm for balanced binary trees*, IEEE Trans. Software Engrg. **SE-12** (1986), no. 3, 442–449.
10. J. H. Reif, *An optimal parallel algorithm for integer sorting*, Proceedings of the 26th Annual IEEE Symposium on the Foundations of Computer Science, 1985, pp. 335–344.

An $O(\log \log n)$ Algorithm to Compute the Kernel of a Polygon*

S. Schuierer[†]

Abstract

The kernel of a polygon **P** is the set of all points that see the interior of **P**. It can be computed as the intersection of the halfplanes that are to the left of the edges of **P**. We present an $O(\log \log n)$ time CRCW-PRAM algorithm using $n/\log \log n$ processors to compute a representation of the kernel of **P** that allows to answer point containment and line intersection queries efficiently. Our approach is based on computing a subsequence of the edges that are sorted by slope and contain the "relevant" edges for the kernel computation.

1 Introduction

Visibility problems play an important role in computational geometry. Given a simple polygon **P** in the plane we say two points p and q in **P** are *visible* from each other or *see* each other if the line segment $\overline{pq}$ is contained in **P**. The *kernel of* **P** *kernel*(**P**) is then defined as the set of all points that see all the other points in **P**. It can be easily shown that *kernel*(**P**) is the intersection of all the halfplanes that lie to the left of the polygon's edges given a counterclockwise orientation of **P** [6].

Although $\Omega(n \log n)$ time is required to compute the intersection of n arbitrary halfplanes, the fact that the halfplanes correspond to the edges of a simple polygon can be exploited to obtain a linear time sequential algorithm [6] or an $O(\log n)$ time algorithm for a CREW-PRAM with $n/\log n$ processors [4]. Other visibility related algorithms for parallel models of computation that have been previously studied are an optimal $O(\log n)$ time CREW-PRAM algorithm to find the visibility polygon of a point [1] and an optimal $O(\log n)$ time CREW-PRAM algorithm for detecting weak visibility of a simple polygon [3].

Our approach is very similar to the one used in [4]. In a first step the halfplanes which are irrelevant for the kernel computation are filtered out and, then, a representation of the intersection of the remaining set of halfplanes is computed. This representation of the kernel allows to answer point containment and line intersection queries in time $O(\log n/\log p)$ if p processors are available.

As it turns out this idea can be implemented on a parallel random access machine where concurrent write and read accesses to one cell are allowed (*CRCW-PRAM*) in time $O(\log \log n)$ using $n/\log \log n$ processors. The particular model of CRCW-PRAM we make use of here is the *COMMON* CRCW-PRAM where concurrent write is only allowed if all processors write the same value to one cell.

The paper is organized as follows. After introducing some notation we describe in the second section which edges can be filtered out. The result are two sequences of halfplanes which are sorted by slope and whose intersection yields the kernel. In Section 4 we consider the problem of computing these sequences of halfplanes in parallel. In Section 5 we describe how to use

---

*This work was supported under a Deutsche Forschungsgemeinschaft Grant, Project "Datenstrukturen", Ot 64/5-4.

[†]Institut für Informatik, Universität Freiburg, Rheinstr. 10, D-79110 Freiburg, FRG, email: schuierer@informatik.uni-freiburg.de

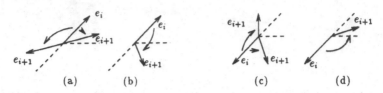

Figure 1: *The definition of* $incr(e_i, e_{i+1})$.

dualization to find a representation of the intersection of these halfplanes that allows efficient point-containment and line-intersection queries.

2 Definitions and Notation

As was pointed out above the kernel of a simple polygon **P** is the intersection of all the halfplanes that are to the left of its edges given a counterclockwise orientation of **P**. Since the approach we present works for closed and open curves we assume in the following that we are given a simple oriented polygonal chain $C = (e_1, \ldots, e_n)$ where edge e_i starts at vertex v_i and ends at vertex v_{i+1}.

We denote the halfplane with e_i on its boundary and interior to the left of e_i by $h^+(e_i)$ and the halfplane with e_i on its boundary and interior to the right of e_i by $h^-(e_i)$. The *kernel* of C is defined as $\bigcap_{1 \leq i \leq n} h^+(e_i)$. Hence, if C is closed curve with a counterclockwise orientation, then the kernel of C is the set of all points that see the interior of C.

One main step in our algorithm is to produce a sequence of edges that are sorted according to turning angle. By turning angle we mean the angle of the edge plus the number of times the chain has spiraled around itself. Here, the *angle* of edge e_i is defined as the angle between the directed line through e_i and the x-axis and is denoted by $\Theta(e_i)$.

The turning angle can be defined incrementally by considering the difference between the angles of two consecutive edges. If e_i and e_{i+1} are two consecutive edges of C, then we define the *incremental angle* between e_i and e_{i+1} to be the signed angle between e_i and e_{i+1} if they are considered as vectors and denote it by $incr(e_i, e_{i+1})$. $incr(e_i, e_{i+1})$ is positive if e_{i+1} turns to the left from e_i and negative if e_{i+1} turns to the right.

More precisely, $incr(e_i, e_{i+1})$ is defined as follows (see Figure 1):

$$incr(e_i, e_{i+1}) = \begin{cases} \text{(a) } \Theta(e_{i+1}) - \Theta(e_i) & \text{if } \Theta(e_i) \leq \pi \text{ and } \Theta(e_{i+1}) < \Theta(e_i) + \pi; \\ \text{(b) } \Theta(e_{i+1}) - \Theta(e_i) - 2\pi & \text{if } \Theta(e_i) \leq \pi \text{ and } \Theta(e_{i+1}) > \Theta(e_i) + \pi; \\ \text{(c) } \Theta(e_{i+1}) - \Theta(e_i) & \text{if } \Theta(e_i) > \pi \text{ and } \Theta(e_{i+1}) > \Theta(e_i) - \pi; \\ \text{(d) } \Theta(e_{i+1}) - \Theta(e_i) + 2\pi & \text{if } \Theta(e_i) > \pi \text{ and } \Theta(e_{i+1}) < \Theta(e_i) - \pi. \end{cases}$$

We can now define the *turning angle* $\Theta_C(e)$ of an edge e w.r.t. C inductively by

$$\begin{aligned} \Theta_C(e_1) &= \Theta(e_1) \\ \Theta_C(e_{i+1}) &= \Theta_C(e_i) + incr(e_i, e_{i+1}) \\ &= \Theta(e_1) + \sum_{j=1}^{i} incr(e_j, e_{j+1}). \end{aligned}$$

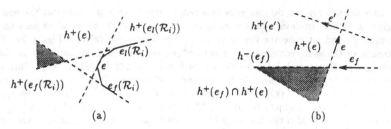

Figure 2: *An edge of a reflex or convex chain is irrelevant.*

3   Computing the Kernel of a Polygonal Chain

In this section we show that there exist sorted sequences of edges S_1 and S_2 that contain all of the edges that are "relevant" for the computation of $kernel(C)$. We say edge e is *relevant* if the line through e intersects $kernel(C)$ in more than one point. Our approach is analogous to the one presented in [4] though the proof we present is different.

Consider two consecutive edges e_i and e_{i+1} of C that meet in vertex v_{i+1}. Vertex v_{i+1} is said to be *convex* if e_{i+1} is in $h^+(e_i)$, otherwise it is called *reflex*. A subchain C' of C is called *convex* (*reflex*) if all the vertices of C' except the first and last are convex (reflex).

We assume that C consists of the maximal reflex chains $\mathcal{R}_1, \mathcal{R}_2, \ldots, \mathcal{R}_m$ and the maximal convex chains $C_1, C_2, \ldots, C_m$ starting with $\mathcal{R}_1$ which is followed by C_1 which in turn is followed by $\mathcal{R}_2$ and so on till we reach the last chain C_m. Clearly, it is possible to decompose any polygonal chain into such a sequence of convex and reflex chains where the first and/or last chain $\mathcal{R}_1$ resp. C_m may be empty. For a chain C, we denote the first edge by $e_f(C)$ and the last edge by $e_l(C)$. Note that we always have $e_l(\mathcal{R}_i) = e_f(C_i)$ and $e_l(C_i) = e_f(\mathcal{R}_{i+1})$.

We start with the following observation about edges of reflex chains.

Observation 3.1 *If e is an edge of $\mathcal{R}_i$ between $e_f(\mathcal{R}_i)$ and $e_l(\mathcal{R}_i)$ and $\Theta_C(e_f(\mathcal{R}_i)) - \Theta_C(e_l(\mathcal{R}_i)) < \pi$, then e is not relevant (see Figure 2a).*

Next we consider the relevant edges among the convex chains.

Lemma 3.2 *If C is a polygonal chain consisting of reflex chains $\mathcal{R}_1, \mathcal{R}_2, \ldots, \mathcal{R}_m$ and convex chains $C_1, C_2, \ldots, C_m$ with $e_f = e_f(C)$ such that*

(i) *there is no edge $e \in C_1 \cup C_2 \cup \cdots \cup C_m$ with $\Theta_C(e) > \Theta_C(e_f)$ and*

(ii) *there is no edge $e \in C$ with $\Theta_C(e_f) - \Theta_C(e) \geq \pi$,*

then $kernel(C)$ equals the intersection of $h^+(e_f)$ with the halfplanes corresponding to the last edges of the reflex chains $\mathcal{R}_i$, $1 \leq i \leq m$.

Proof: W.l.o.g. we can assume that $\Theta(e_f) = \pi$. Our first claim is that C is y-monotone. In order to show this we prove that $\Theta_C(e)$ is in $(0, \pi)$, for all $e \in C$. Since the angles of the edges along a convex chain are increasing, the maximum turning angle is achieved at some edge e_{max} with $e_{max} = e_l(C_{i_1})$, $1 \leq i_1 \leq m$; the angles along a reflex chain are decreasing, so the minimum is achieved at some edge e_{min} with $e_{min} = e_l(\mathcal{R}_{i_2})$, $1 \leq i_2 \leq m$. By Assumption (i) $\Theta_C(e_{max})$ is less than $\Theta_C(e_f) = \pi$ and, furthermore, by Assumption (ii) $\Theta_C(e_{min}) > \Theta_C(e_f) - \pi = 0$. This immediately implies that C is y-monotone.

We have already argued that an edge of a reflex chain $\mathcal{R}$ that is neither at the end nor at the beginning of $\mathcal{R}$ is irrelevant for the computation of $kernel(C)$. So let e' be an edge of the convex chain C_j and let e be the first edge of C_j. For illustration refer to Figure 2b. Since e' and e belong to the same convex chain, e is in $h^+(e')$ and since C is monotone, e is above the line through e_f. Hence, e belongs to the wedge formed by $h^+(e')$ and $h^-(e_f)$. Furthermore, $0 < \Theta_C(e) \leq \Theta_C(e')$ and, therefore, the line ℓ' through e' intersects the line through e_f in $h^-(e)$ and ℓ' does not intersect $h^+(e) \cap h^+(e_f)$. Since $h^+(e) \cap h^+(e') \cap h^+(e_f)$ is non-empty, $h^+(e')$ contains $h^+(e) \cap h^+(e_f)$ completely. Hence, $h^+(e) \cap h^+(e') \cap h^+(e_f) = h^+(e) \cap h^+(e_f)$ and e' is irrelevant. Since e is also the last edge of $\mathcal{R}_j$, this implies that all edges of C are irrelevant except for e_f and the end edges of the reflex chains. $\qquad\Box$

With the help of the above lemma we can now define a sequence S_1 of edges that contains "most" of the relevant edges of C. To this end let $\theta_i = \max_{1 \leq j \leq i} \Theta_C(e_j)$ and let S_1 be the set of edges e_i with $\Theta_C(e_i) = \theta_i > \theta_{i-1}$, where $\theta_0 = -\infty$, i.e., S_1 is the set of edges whose turning angle is larger than the turning angle of any previous edge.

Lemma 3.3 *If C is a polygonal chain consisting of reflex chains $\mathcal{R}_1, \mathcal{R}_2, \ldots, \mathcal{R}_m$ and convex chains $C_1, C_2, \ldots, C_m$ such that there is no edge e of C between two consecutive edges e' and e'' in S_1 with $\Theta_C(e') - \Theta_C(e) \geq \pi$, then $kernel(C)$ is the intersection of $h^+(e')$, with $e' \in S_1$, and $h^+(e_l(\mathcal{R}_i))$, $1 \leq i \leq m$.*

Proof: We prove by induction on the edges of C that $\bigcap_{1 \leq j \leq i} h^+(e_j)$ contains the intersection of the halfplanes $h^+(e_r)$, with $e_r \in S_1$ and $r \leq i$, and $h^+(e_l(\mathcal{R}_k))$, with $1 \leq k \leq m$. The claim is obvious if we consider only e_1. Hence, consider edge e_i and assume that the claim has been shown for the subchain of C consisting of the edges $e_1, \ldots, e_{i-1}$. As e_i is irrelevant if it doesn't belong to a convex chain, we can assume that e_i is part of convex chain C_j. If $\Theta_C(e_i) > \theta_{i-1}$, then e_i belongs to S_1 and we are done. Otherwise, let e_k be the edge in S_1 with largest $k < i$. The induction hypothesis implies that

$$\bigcap_{1 \leq j \leq i} h^+(e_j) \supseteq \bigcap_{e_j \in S_1, j < k} h^+(e_j) \cap \bigcap_{1 \leq j \leq m} h^+(e_l(\mathcal{R}_j)) \cap \bigcap_{k \leq j \leq i} h^+(e_j).$$

By Lemma 3.2 $\bigcap_{k \leq j \leq i} h^+(e_j)$ contains

$$h^+(e_k) \cap \bigcap_{1 \leq j \leq m} h^+(e_l(\mathcal{R}_j))$$

since $\Theta_C(e_j) \leq \Theta_C(e_k)$, for $k < j \leq i$, and there is no edge e_j, $k < j \leq i$, with $\Theta_C(e_k) - \Theta_C(e_j) \geq \pi$ which completes the induction step and, hence, the proof. $\qquad\Box$

Lemma 3.4 *Let e_i and $e_{i'}$ be two consecutive edges in S_1. If there is an edge e_j with $i < j < i'$ such that $\Theta_C(e_i) - \Theta_C(e_j) \geq \pi$, then $kernel(C)$ is empty.*

Proof: Let j' be the smallest index $j' > i$ with $\Theta_C(e_i) - \Theta_C(e_{j'}) \geq \pi$. Hence, the part of C from e_i to $e_{j'-1}$ is monotone w.r.t. $\Theta(e_i) + \pi$ and $h^+(e_i) \cap h^+(e_{j'-1}) \cap h^+(e_{j'}) = \emptyset$ (see **Figure 3a**). $\qquad\Box$

Corollary 3.5 *If C is a polygonal chain consisting of reflex chains $\mathcal{R}_1, \mathcal{R}_2, \ldots, \mathcal{R}_m$ and convex chains $C_1, C_2, \ldots, C_m$ and $\Theta_C(e_f(C)) - \Theta_C(e_l(\mathcal{R}_i)) \geq \pi$, for some $1 \leq i \leq m$, then $kernel(C)$ is empty.*

Now suppose we mirror C at the y-axis and revert the orientation of the edges and call this new chain C'. Note that the kernel of C' is the image of the kernel of C and a relevant edge of C' is the image of a relevant edge of C. Similarly, an edge belongs to a reflex (convex) chain of C' if

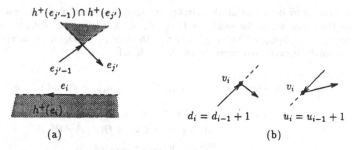

Figure 3: *The kernel of C is empty if $\Theta_C(e_i) - \Theta_C(e_{j'}) \geq \pi$.*

and only if its preimage belongs to a reflex (convex) chain of C. Hence, we can define a sequence S_2 of edges of C' analogously to the sequence S_1 of C only now starting with the image of ϵ_n. But the image of the *last* edge of a reflex chain of C is now the *first* edge of a reflex chain of C' and vice versa. If we take into account that we have $\bigcap_{e \in S_1} h^+(e) \subseteq \bigcap_{1 \leq i \leq m} e_f(\mathcal{R}_i)$, then, if we mirror back and again change the orientation, we have analogously $\bigcap_{e \in S_2} h^{\mp}(e) \subseteq \bigcap_{1 \leq i \leq m} e_l(\mathcal{R}_i)$. Therefore, we obtain

$$ kernel(C) \;=\; \bigcap_{e \in S_1} h^+(e) \cap \bigcap_{1 \leq i \leq m} h^+(e_l(\mathcal{R}_i)) = \bigcap_{e \in S_1} h^+(e) \cap \bigcap_{e \in S_2} h^+(e). \qquad (1) $$

4 Highly Parallel Computation of the Kernel

In this section we present an $O(\log \log n)$ algorithm for a CRCW-PRAM with $O(n/\log \log n)$ processors to compute the kernel of a simple polygonal chain. With the observations from the previous section it seems that the following approach to computing the kernel is a reasonable try.

1. Compute S_1 and S_2;
2. Merge S_1 and S_2 into a single sequence S;
3. Intersect the halfplanes in S;

In the following we will address each of the four steps.

4.1 Computing S_1

The computation of S_1 consists of two steps. First we have to compute $\Theta_C(e_i)$, and then θ_i, for $1 \leq i \leq n$. Recall that $\theta_i = \max_{1 \leq j \leq i} \Theta_C(e_j)$. Computing $\Theta_C(e_i)$ essentially amounts to computing the sums $\sum_{j=1}^{i} incr(e_j, e_{j+1})$, for $1 \leq i \leq n$. Unfortunately, there is a lower bound of $\Omega(\log n / \log \log n)$ for the computation of the parity function of n bits on a CRCW-PRAM which, in particular, implies a lower bound for the computation of the sum of n numbers.

In order to beat this lower bound, we make use of a different representation of $\Theta_C(e_i)$. We say vertex v_{i+1} with adjacent edges e_i and e_{i+1} is a *downturn* if $\Theta(e_i) \leq \pi$ and $\Theta(e_{i+1}) > \Theta(e_i) + \pi$. Similarly, we say vertex v_{i+1} is an *upturn* if $\Theta(e_i) > \pi$ and $\Theta(e_{i+1}) < \Theta(e_i) - \pi$ (see Figure 3b). Let the number of upturns up to and including vertex v_i be u_i and the number of downturns d_i.

Lemma 4.1 *If C is a polygonal chain, then*

$$ \Theta_C(e_i) = \Theta(e_i) + (u_i - d_i) 2\pi. $$

Proof: The proof is by induction on the number of edges. The claim obviously holds for $i = 1$. Now suppose the claim is true for some $i > 1$, i.e., $\Theta_C(e_i) = \Theta(e_i) + (u_i - d_i)2\pi$. We want to show that $\Theta_C(e_{i+1}) = \Theta(e_{i+1}) + (u_{i+1} - d_{i+1})2\pi$ with $\Theta_C(e_{i+1}) = \Theta_C(e_i) + incr(e_i, e_{i+1})$. In order to do so we distinguish four cases depending on $\Theta(e_i)$.

1. $\Theta(e_i) \leq \pi$ and $\Theta(e_{i+1}) < \Theta(e_i) + \pi$.
   By definition we have $incr(e_i, e_{i+1}) = \Theta(e_{i+1}) - \Theta(e_i)$, $d_{i+1} = d_i$ and $u_{i+1} = u_i$. Hence,

$$\begin{aligned} \Theta_C(e_{i+1}) &= \Theta_C(e_i) + incr(e_i, e_{i+1}) \\ &= \Theta(e_i) + (u_i - d_i)2\pi + \Theta(e_{i+1}) - \Theta(e_i) \\ &= \Theta(e_{i+1}) + (u_{i+1} - d_{i+1})2\pi. \end{aligned}$$

2. $\Theta(e_i) \leq \pi$ and $\Theta(e_{i+1}) > \Theta(e_i) + \pi$.
   We have $incr(e_i, e_{i+1}) = \Theta(e_{i+1}) - \Theta(e_i) - 2\pi$, $d_{i+1} = d_i + 1$, and $u_{i+1} = u_i$. Hence,

$$\begin{aligned} \Theta_C(e_{i+1}) &= \Theta_C(e_i) + incr(e_i, e_{i+1}) \\ &= \Theta(e_i) + (u_i - d_i)2\pi + \Theta(e_{i+1}) - \Theta(e_i) - 2\pi \\ &= \Theta(e_{i+1}) + (u_{i+1} - (d_i + 1))2\pi \\ &= \Theta(e_{i+1}) + (u_{i+1} - d_{i+1})2\pi. \end{aligned}$$

3. The cases 3. $\Theta(e_i) > \pi$ and $\Theta(e_{i+1}) > \Theta(e_i) - \pi$ and 4. $\Theta(e_i) > \pi$ and $\Theta(e_{i+1}) < \Theta(e_i) - \pi$ can be handled analogously;

$\square$

Hence, the computation of $\Theta_C(e_i)$ can be reduced to computing $t_i = u_i - d_i$. Assume that t_i is in the interval $[-k, k]$, $1 \leq k \leq n$. In the following we show that if $kernel(C)$ is non-empty, then $k = 1$.

In order to compute t_i, for all $1 \leq i \leq n$, we make use of an optimal $O(\log \log n)$ CRCW-PRAM algorithm that solves the *all nearest smaller values (ANSV)* problem which is defined as follows [2]. Given a sequence of n elements $(a_1, \ldots, a_n)$ from a totally ordered domain. Find, for each element a_i, $1 \leq i \leq n$, the left and right closest smaller element, i.e., find the element with maximum $j < i$ such that $a_j < a_i$ and find the minimum $k > i$ with $a_k < a_i$. The element a_j is called the *left match* of a_i and a_k is called the *right match*.

We now apply the ANSV-algorithm to the up- and downturns. The first step is that for each downturn we find the right closest up- or downturn. In order to do so, processor i writes $n - i$ into array cell a_i if v_i is an up- or downturn and $n + 1$ otherwise. Clearly, an application of the ANSV-algorithm yields the closest right up- or downturn, for each vertex. If v_i is a downturn and its right match v_j an upturn, then v_i and v_j are called *matched*. If there is no up- or downturn to the right of downturn v_i, then we consider v_i as matched nevertheless. Now we compute, for each matched downturn v_i, its left closest unmatched up- or downturn. Each of these steps can be carried out in $O(\log \log n)$ steps with $n/\log \log n$ processors [2].

We claim that for the remaining (unmatched) vertices t_i is contained in the interval $[-k+1, k]$. The proof is by contradiction. If there is no unmatched downturn, then $t_i \geq 0$ and the claim holds. Otherwise, we observe that the minima of the sequence (t_i) are achieved at downturns. So assume that there is an unmatched downturn v_i such that $t_i = -k$. Its closest up- or downturn v_j to the right exists and is a downturn since otherwise v_i and v_j were matched. But this implies that $t_j = -k - 1$ in contradiction to our assumption that $t_j \in [-k, k]$.

The above procedure is now iterated one more time, matching upturns with downturns and finding the closest left unmatched up- or downturn, for each matched downturn. With this iteration the possible range for the t_i values of the remaining up- or downturns is again reduced by one. If there are any unmatched up- or downturns left after the iteration, then this implies that either there is an i with $t_i > 1$ or we started out with a $t_i < -1$. In the first case we have $\Theta_C(e_i) = \Theta(e_i) + t_i 2\pi > 4\pi$ which implies that $kernel(C)$ is empty [4, 6]. In the second case we have $\Theta_C(e_i) = \Theta(e_i) - t_i 2\pi < -2\pi$ and, hence, $\Theta_C(e_f(C)) - \Theta_C(e_i) > 2\pi$ which implies by Corollary 3.5 that $kernel(C)$ is also empty in this case.

Hence, we can assume that all t_i's are contained in $[-1, 1]$ and the above procedure is executed at most two times. Any downturn that is matched in the second step is known to have a t_i value of 0 and any upturn is known to have a t_i value of 1. Since each matched pair P' in the first step has a pointer to an unmatched pair P the t-value of P' can be computed in constant time per vertex. In order to obtain the t_i values for all vertices we just have to apply the ANSV-algorithm one more time so that each vertex can find its left closest up- or downturn. All of these operations can be carried out in time $O(\log \log n)$.

Given the t_i values for each vertex, $\Theta_C(e_i)$ can be computed in time $O(\log \log n)$, for all $1 \leq i \leq n$, by Lemma 4.1. In order to obtain S_1 it is necessary to compute $\theta_i = \max_{1 \leq j \leq i} \Theta_C(e_j)$ which is a prefix maxima operation on $\Theta_C(e_i)$ and can be carried out in time $O(\log \log n)$ with $n/\log \log n$ processors [2].

Finally, we have to check if there is an edge e between two consecutive edges e' and e'' of S_1 with $\Theta_C(e') - \Theta_C(e) \geq \pi$ by Lemma 3.3. This can be done by yet another application of the ANSV-algorithm in order to find, for each edge, the closest edge in S_1 to the left and then compare the two turning angles. If there is one edge e with left closest neighbour e' in S_1 such that $\Theta_C(e') - \Theta_C(e) \geq \pi$, then $kernel(C)$ is empty.

The sequence S_2 can be computed and checked analogously.

4.2 Merging S_1 and S_2

The important property of S_1 and S_2 is that the edges in both sequences are sorted by turning angles. Hence, an application of an optimal $O(\log \log n)$ merging algorithm to obtain the merged sequence S seems possible. Unfortunately, the edges of S_1 and S_2 are scattered among the edges of C which we assume to stored in an array A. In order to compact the edges of S_1 and S_2 into a contiguous part of A we need at least $\Omega(\log |S_1|/\log \log n)$ time [7] which we cannot afford since S_1 (or S_2) can be as large as $\Omega(n)$. The solution is to apply the ANSV-algorithm in order to find, for each edge e not in S_1, its left closest match $lcm_{S_1}(e)$ in S_1 and the same for edges not in S_2. We create two new arrays A_1 and A_2 with $A_i[j] = e_j$ if e_j is in S_i and $A_i[j] = lcm_{S_i}(e_j)$ otherwise. Note that $\bigcap_{e \in A_i} h^+(e) = \bigcap_{e \in S_i} h^+(e)$, for $i = 1, 2$. Hence, we can merge the two arrays A_1 and A_2 into one array A' of size $2n$ in time $O(\log \log n)$.

4.3 Intersecting S_1 and S_2

Since the *turning angles* of the edges in A vary at most between -2π and 4π, we can split A' into $O(1)$ contiguous parts such that the *angles* of the edges in these parts are in the intervals $[-\pi/2, \pi/2)$ or $[\pi/2, 3\pi/2)$ and merge these parts again into two sequences E_+ and E_- such that the angles of all edges in E_+ are sorted and in the range $[-\pi/2, \pi/2)$ and the angles of the edges in E_- are also sorted and in the range $[\pi/2, 3\pi/2)$. In order to compute the intersection of the halfplanes associated with the edges in $E_+ \cup E_-$ we apply a dualization method as described below.

5 Halfplane Intersection and the Convex Hull

In this section we show how to reduce the intersection of halfplanes that are sorted according to angle to the computation of the convex hull of certain point sets that are sorted on one coordinate. To this end we apply the concept of *geometric duality* which maps points onto lines and vice versa while keeping incidence relations. More precisely we apply the following dual map D see [5]

$$D(p) = \{(x,y) \in I\!\!R^2 \mid y = p_1 x - p_2\} \text{ with } p = (p_1, p_2) \text{ and}$$
$$D(\ell) = (a_1, -a_2) \text{ with } \ell = \{(x,y) \in I\!\!R^2 \mid y = a_1 x + a_2\}$$

D has the special property that it preserves order which means in this case that if p is above the line ℓ, then $D(\ell)$ is above $D(p)$. In order to compute the intersection of the halfplanes $\mathcal{H} = \{h^+(e) \mid e \in A\}$ we now proceed as follows. $\mathcal{H}$ is split into the two sets $\mathcal{H}_+ = \{h^+(e) \mid e \in E_+\}$ and $\mathcal{H}_- = \{h^+(e) \mid e \in E_-\}$. Hence, $\bigcap \mathcal{H} = \bigcap \mathcal{H}_+ \cap \bigcap \mathcal{H}_-$. Note that all the halfplanes in $\mathcal{H}_+$ have their interior above their boundary and all the halfplanes in $\mathcal{H}_-$ have their interior below their boundary. Further, let $\mathcal{L}_+$ be the set of directed boundary lines of the halfplanes in $\mathcal{H}_+$ and $\mathcal{L}_-$ be defined analogously for $\mathcal{H}_-$. Now $p \in \bigcap \mathcal{H}_+$ if and only if p is above all lines ℓ in $\mathcal{L}_+$; in the dual plane this corresponds to the fact that for all $\ell \in \mathcal{L}_+$, the point $D(\ell)$ is above the line $D(p)$, i.e., that the lower convex hull of $D(\mathcal{L}_+)$ is above $D(p)$. The *lower convex hull* of $D(\mathcal{L}_+)$ is defined as the set of points that are above the convex hull of $D(\mathcal{L}_+)$. Similarly, the *upper convex hull* of $D(\mathcal{L}_-)$ is defined as the set of points that are below the convex hull of $D(\mathcal{L}_-)$. We denote the boundary of the lower convex hull of $D(\mathcal{L}_+)$ by $\mathcal{H}_+(\mathcal{L}_+)$ and the boundary of the upper convex hull of $D(\mathcal{L}_-)$ by $\mathcal{H}_-(\mathcal{L}_-)$.

In the same way we get that $p \in \bigcap \mathcal{H}_-$ if and only if, for all $\ell \in \mathcal{L}_-$, the point $D(\ell)$ is below the line $D(p)$, i.e., if the convex hull of $D(\mathcal{L}_-)$ is below $D(p)$. The angles of the lines in $\mathcal{L}_+$ are sorted and contained in the interval $[-\pi/2, \pi/2)$ and the angles of the lines in $\mathcal{L}_-$ are sorted and contained in $[\pi/2, 3\pi/2)$. If the equation of the line bounding a halfplane with angle θ is $y = a_1 x + a_2$, we have the relation $a_1 = \tan \theta$. Hence, $D(\mathcal{L}_+)$ and $D(\mathcal{L}_-)$ are point sets that are sorted by x-coordinate, and the optimal $O(\log \log n)$ time algorithm of Wagener [8] can be used to compute the convex hull of $D(\mathcal{L}_+)$ and $D(\mathcal{L}_-)$.

Unfortunately, there is a lower bound of $\Omega(\log n/\log \log n)$ for the computation of an array representation of the convex hull of a sorted point set; therefore, a special data structure—called the *bridge tree*—is used in [8]. The bridge tree is defined on point sets and is not designed to be redualized. Nevertheless, we can use the bridge tree representation of the convex hull of $D(\mathcal{L}_+)$ and $D(\mathcal{L}_-)$ to answer several queries about $\bigcap \mathcal{H}_+$ and $\bigcap \mathcal{H}_-$.

A bridge tree supports the following queries in time $O(\log n/(\log p + 1) + 1)$ if p processors are assigned to the query:

(i) Given a point q, test whether q is contained in the lower hull of $D(\mathcal{L}_+)$ (or the upper hull of $D(\mathcal{L}_-)$).

(ii) Given a point q, report the tangents to the lower hull of $D(\mathcal{L}_+)$ (or the upper hull of $D(\mathcal{L}_-)$) passing through q.

(iii) Given a line ℓ, report the intersection of ℓ with the lower hull of $D(\mathcal{L}_+)$ (or the upper hull of $D(\mathcal{L}_-)$).

(iv) Given a direction d, report the extremal points of the lower hull of $D(\mathcal{L}_+)$ (or the upper hull of $D(\mathcal{L}_-)$) in direction d.

In the following it is our aim to show that the same types of queries can be handled for $\mathcal{K} = \bigcap \mathcal{H}_+ \cap \bigcap \mathcal{H}_-$ within the same time bounds. A query of type (i) can be answered if we make use of the observation that the points contained in $\bigcap \mathcal{H}_+$ are mapped to lines that do not intersect the lower hull of $D(\mathcal{L}_+)$. Hence, to test if a point p is contained in $\mathcal{K}$ can be mapped to the query whether the dual line $D(p)$ does neither intersect the lower hull of $D(\mathcal{L}_+)$ nor the upper hull of $D(\mathcal{L}_-)$.

In order to answer a query of type (iii) we observe the following.

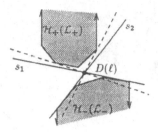

Figure 4: *The tangents to the lower hull of $D(\mathcal{L}_+)$ and the upper hull of $D(\mathcal{L}_-)$ correspond to the intersection points of ℓ.*

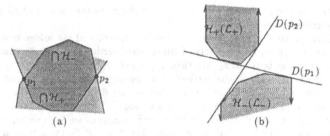

Figure 5: *The duality between $\bigcap \mathcal{H}$ and the lower hull of $D(\mathcal{L}_+)$ and the upper hull of $D(\mathcal{L}_-)$.*

Lemma 5.1 *The line ℓ intersects $\mathcal{K}$ if and only if two of the four tangents of $D(\ell)$ to $\mathcal{H}_+(\mathcal{L}_+)$ and $\mathcal{H}_-(\mathcal{L}_-)$ do not intersect the interior of $\mathcal{H}_+(\mathcal{L}_+) \cup \mathcal{H}_-(\mathcal{L}_-)$.*

Proof: Omitted. □

Since there are only four tangents through $D(\ell)$ to the lower hull of $D(\mathcal{L}_+)$ and the upper hull of $D(\mathcal{L}_-)$ and these tangents can be found and tested for intersection with the interiors of the hulls in time $O(\log n/(\log p + 1) + 1)$ with the help of a bridge tree, we can report the intersection points of a line ℓ with $\mathcal{K}$, if they exist, within the same time bound (see Figure 4).

5.1 Computing the Intersection Points of $\mathcal{H}_+(\mathcal{L}_+)$ and $\mathcal{H}_-(\mathcal{L}_-)$

In order to answer queries of type (ii) and (iv) for $\mathcal{K}$, we first have to find the intersection points p_1 and p_2 of the boundary of $\bigcap \mathcal{H}_+$ with the boundary of $\bigcap \mathcal{H}_-$ since, for example, p_1 is the extremal point of $\mathcal{K}$ in direction π. The lines $t_1 = D(p_1)$ and $t_2 = D(p_2)$ are simultaneously tangents to the lower hull of $D(\mathcal{L}_+)$ and the upper hull of $D(\mathcal{L}_-)$ (see Figure 5).

Note that the lower hull of $D(\mathcal{L}_+)$ is on a different side of t_1 (t_2) as the upper hull of $D(\mathcal{L}_-)$. In order to compute t_1 and t_2 we make use of the internal structure of the bridge tree for the the upper hull of $D(\mathcal{L}_-)$ which can be described as follows. On the first level of the bridge tree the points of $D(\mathcal{L}_-)$ are split into $n^{1/6}$ contiguous sequences $\mathcal{P}_i$ of points, each consisting of at most $O(n^{5/6})$ points. For each $\mathcal{P}_i$, at most two *bridges* are stored where a bridge is an edge of the upper hull of $D(\mathcal{L}_-)$ such that the x-range of the bridge intersects the x-range of $\mathcal{P}_i$. A bridge b is called a *proper* bridge of $\mathcal{P}_i$ if one of the end points of b belongs to $\mathcal{P}_i$; otherwise it is called

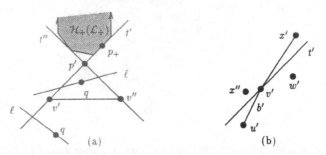

Figure 6: *q is not contained in W.*

a *passing* bridge. In the latter case no point of $\mathcal{P}_i$ belongs to the upper hull of $D(\mathcal{L}_-)$. This structure is recursively applied to all $\mathcal{P}_i$.

Our algorithm proceeds according to the recursive structure of the bridge tree. In the first step it is our aim to sort out the two sets of consecutive vertices $\mathcal{P}_{i_0}$ and $\mathcal{P}_{i_1}$ that contain the vertices that are incident to t_1 and t_2. We then apply the procedure recursively to $\mathcal{P}_{i_0}$ and $\mathcal{P}_{i_1}$.

Let $v_1, \ldots, v_k$, $k \leq 2n^{1/6}$, be the vertices that belong to the bridges on the first level of the bridge tree. With $n^{1/6} \cdot n^{2/3} = n^{5/6}$ processors we can find, for each of the k vertices, the tangents to the lower hull of $D(\mathcal{L}_+)$ in time $O(\log n/(\log n^{2/3} + 1) + 1) = O(1)$ with the help of the bridge tree for the lower hull of $D(\mathcal{L}_+)$. With $n^{1/3}$ processors we can check in constant time which tangents have all the k vertices to one side. Let t' and t'' be these two tangents and v' and v'' be the vertices incident to t' and t'', respectively, with v' to the left of v''. The lines t' and t'' form four wedges. Let p' be the intersection point of t' and t'' and let W be the wedge that contains v' and v'' and W' the wedge opposite to W.

Lemma 5.2 *The support points of the tangents t_1 and t_2 at the upper hull of $D(\mathcal{L}_-)$ are not contained in the interior of W.*

Proof: To see this let q be first a point in the interior of the triangle T spanned by p', v' and v''. Since the line segment from v' to v'' is contained in the upper hull of $D(\mathcal{L}_-)$, any line ℓ that is incident to q and that is the dual of a point in $\mathcal{K}$ does not intersect $\overline{v'v''}$ and, hence, intersects the line segments from v' to p' and from v'' to p'. This implies that W' is entirely above ℓ and, hence, ℓ is not a tangent to $D(\mathcal{L}_+)$ (see Figure 6a).

Now consider a point q in W below the line segment from v' to v''. We claim that neither t_1 nor t_2 intersects q. For consider a line ℓ through q. Either ℓ intersects t' below v' or t'' below v''. In the first case v' and the intersection point p_+ of t' with $D(\mathcal{L}_+)$ are on the same side of ℓ and, hence, ℓ cannot be t_1 or t_2. An analogous statement holds in the second case (see again Figure 6a). Therefore, the intersection points of t_1 and t_2 and $D(\mathcal{L}_-)$ are not in the interior of W. □

A consequence of Lemma 5.2 is that we only have to consider points that are outside W. Let b' be the bridge incident to v'. Suppose that b' starts at vertex u' and ends at v'. Let w' be the next vertex after v' in the top level of the bridge tree that belongs to a bridge.

Lemma 5.3 *If p is a point above t', then p_x is between v'_x and w'_x.*

Proof: Let x' be a vertex of $\mathcal{H}_-(\mathcal{L}_-)$ after w'. If x' is above t', then w' is entirely below the line segment from v' to x' and, hence, does not belong to the upper hull of $D(\mathcal{L}_-)$ in contradiction of

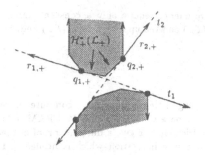

Figure 7: *Defining $q_{i,+}$, $r_{i,+}$, and $\mathcal{H}_+^*(\mathcal{L}_+)$.*

our choice of w' (see Figure 6b). A similar argument holds if x' is before u'. And if there exists a vertex x'' between u' and v' that is above t', then b' is below x'' and, hence, does not belong to the upper hull of $D(\mathcal{L}_-)$ (see again Figure 6b). □

Of course, we can argue in a similar way if b' starts in v'. Furthermore, b'', v'', and t'' can be dealt with analogously.

If v' belongs to $\mathcal{P}_{i'}$, then our observations imply that we can recur on the $O(n^{5/6})$ vertices of $\mathcal{P}_{i'}$ in order to find t_1 and similarly for t_2. Since we reduce the exponent of the number of vertices we have to look at with each recursion step by a factor of 5/6, the total time needed to find t_1 and t_2 is $O(\log \log n)$. If we stop the recursion after, say, ten levels, then the number of vertices to be considered has been reduced to $O(n^{1/6})$ and the problem can be solved directly in constant time as in the first step. Hence, in fact, we need only constant time to compute t_1 and t_2. If we do not find the two tangents t_1 and t_2, then the lower hull of $D(\mathcal{L}_+)$ intersects the upper hull of $D(\mathcal{L}_-)$ and the kernel of C is empty. In particular, this implies that we can check in time $O(\log \log n)$ if C is starshaped or not.

With the help of t_1 and t_2 we also obtain a more accurate representation of $\mathcal{K}$ in the dual plane. To this end let $q_{i,+}$ be the intersection point of $\mathcal{H}_+(\mathcal{L}_+)$ with t_i and $q_{i,-}$ be the intersection point of $\mathcal{H}_-(\mathcal{L}_-)$, for $i = 1, 2$. We denote the part of the boundary of $\mathcal{H}_+(\mathcal{L}_+)$ between $q_{1,+}$ and $q_{2,+}$ by $\mathcal{H}_+^*(\mathcal{L}_+)$ and the part of the boundary of $\mathcal{H}_-(\mathcal{L}_-)$ between between $q_{1,-}$ and $q_{2,-}$ by $\mathcal{H}_-^*(\mathcal{L}_-)$. Let $r_{i,+}$ be the ray that is contained in t_i, starts at $q_{i,+}$, and that does not contain $q_{i,-}$, for $i = 1, 2$; similarly, let $r_{i,-}$ be the ray that is contained in t_i, starts at $q_{i,-}$, and that does not contain $q_{i,+}$, for $i = 1, 2$ (see Figure 7).

Lemma 5.4 *The points in $\mathcal{K}$ are duals of lines that do not intersect $\mathcal{H}_+^*(\mathcal{L}_+)$, $\mathcal{H}_-^*(\mathcal{L}_-)$, $r_{1,+}$, $r_{2,+}$, $r_{1,-}$, or $r_{2,-}$.*

Proof: Omitted. □

We denote the region above $r_{1,+} \cup \mathcal{H}_+^*(\mathcal{L}_+) \cup r_{2,+}$ by $D_+(\mathcal{K})$·and the region below $r_{1,-} \cup \mathcal{H}_-^*(\mathcal{L}_-) \cup r_{2,-}$ by $D_-(\mathcal{K})$. $\mathcal{K}$ is the set of points whose dual line do not intersect the interior of $D_+(\mathcal{K}) \cup D_-(\mathcal{K})$.

In order to answer a query of type (ii), that is in order to compute the tangent to $\mathcal{K}$ through a given point p, we just note that the intersection points q_1 and q_2 of $D(p)$ with the boundary of $D_+(\mathcal{K}) \cup D_-(\mathcal{K})$ are the duals of the tangents to $\mathcal{K}$. To see this just note that $D(q_1)$ is a tangent to $\mathcal{K}$ through p since q_1 is on the boundary of $D(\mathcal{K})$ and $D(p)$ is incident to q_1.

Finally, in order to answer a query of type (iv) we note that the set of parallel lines in the primal plane with slope d is represented by the vertical line ℓ_d through the point $(d, 0)$ in the dual plane. Hence, we have to intersect ℓ_d with the boundaries of $D_+(\mathcal{K})$ and $D_-(\mathcal{K})$.

Clearly, all of the above queries require only a constant number of queries to the bridge tree of the lower hull of $D(\mathcal{L}_+)$ and the upper hull of $D(\mathcal{L}_-)$ and, hence, the query time is still $O(\log n/(\log p + 1) + 1)$.

6 Conclusions

We have presented an $O(\log \log n)$ time algorithm to compute the kernel of a polygonal chain with $O(n/\log \log n)$ processors on a COMMON CRCW-PRAM. Our approach is based on the algorithm by Cole and Goodrich [4]. We give a different proof of their main theorem and show how to avoid parallel prefix sum computation which is needed in their algorithm. Since our intersection of halfplanes algorithm makes use of dualization and the new optimal $O(\log \log n)$ convex hull algorithm of Wagener [8], we also address the question of how to answer the following four queries for the kernel of a polygon $\mathbf{P}$ in the primal plane given a query structure in the dual plane.

(i) Given a point p, test whether p is in the $kernel(\mathbf{P})$.

(ii) Given a point p, report the tangents to the $kernel(\mathbf{P})$ passing through p.

(iii) Given a line ℓ, report the intersection of ℓ with $kernel(\mathbf{P})$.

(iv) Given a direction d, report the extremal points of $kernel(\mathbf{P})$ in direction d.

All of these queries can be answered in time $O(\log n/(\log p + 1) + 1)$ if p processors are available.

References

[1] M. J. Atallah and D. Z. Chen. Optimal parallel algorithm for visibility of a simple polygon from a point. In *Proc. 5th ACM Symp. on Computational Geometry*, pages 114–123, 1989.

[2] O. Berkman, B. Schieber, and U. Vishkin. *Some Doubly Logarithmic Optimal Parallel Algorithms Based on Finding All Nearest Smaller Values*. Technical Report UMIACS-TR-88-79, Institute for Advance Computer Studies, University of Maryland, 1988.

[3] D. Z. Chen. An optimal parallel algorithm for detecting weak visibility of a simple polygon. In *Proc. 8th ACM Symp. on Computational Geometry*, pages 63–72, 1992.

[4] R. Cole and M. Goodrich. Optimal parallel algorithms for polygon and point-set problems. In *Proc. 4th ACM Symp. on Computational Geometry*, pages 201–210, 1988.

[5] Herbert Edelsbrunner. *Algorithms in Combinatorial Geometry*. Springer Verlag, EATCS monographs on theoretical computer science, 1987.

[6] D. T. Lee and F. P. Preparata. An optimal algorithm for finding the kernel of a polygon. *Journal of the ACM*, 26(3):415–421, July 1979.

[7] P. Ragde. The parallel simplicity of compaction and chaining. In *Proc. 17th Intern. Colloq. on Automata, Languages, and Programming*, pages 744–751, 1990.

[8] H. Wagener. Optimal parallel hull construction for simple polygons in $O(\log \log n)$ time. In *Proc. 33rd IEEE Symp. on Foundations of Computer Science*, pages 593–599, 1992.

Computing the L_1-Diameter and Center of a Simple Rectilinear Polygon in Parallel*

Sven Schuierer[†]

Abstract

The diameter of a set S of points is the maximal distance between a pair of points in S. The center of S is the set of points that minimise the distance to their furthest neighbours. The problem of finding the diameter and center of a simple polygon with n vertices for different distance measures has been studied extensively in recent years. There are algorithms that run in linear time if the geodesic Euclidean metric is used and $O(n \log n)$ time if the link metric is used.

In this paper we consider the L_1-metric inside a simple rectilinear polygon P, i.e. the distance between two points in P is defined as the length of a shortest rectilinear path connecting them. We give an $O(\log n)$ time algorithm to compute the L_1-diameter and center on a EREW-PRAM with $n/\log n$ processors if a triangulation of the polygon is provided.

1 Introduction

Rectilinear paths and rectilinear obstacles play an important role in many applications. VLSI design is one of the most prominent examples. A natural metric to measure the distance between two points in such a setting is the L_1-metric since it corresponds to the length of a shortest rectilinear path. In this paper we address the problem of finding the maximal distance, i.e., the maximum of the length of a shortest rectilinear path between two points inside a simple rectilinear polygon, also known as the *diameter problem*. A closely related problem is that of locating the set of points inside a simple rectilinear polygon whose maximal L_1-distance to any other point in the polygon is minimized. This set is known as the *center* of the polygon.

Center and diameter problems have been considered in various other contexts and a number of results have been obtained. Pollack et al. [18] give an $O(n \log n)$ algorithm to compute the geodesic center of a polygon if the Euclidean metric is used. It is not known up to date if the time complexity of this algorithm is optimal or whether it can be improved upon. The Euclidean diameter of a simple polygon can be computed in linear time by a very recent result of Hershberger and Suri [10] based on matrix searching. If the L_1-metric is used, a prune-and-search technique leads to a linear time diameter and center algorithm [19].

A metric with a somewhat different flavour is the *link metric* which counts the number of turns of a path. The problem of computing the *link center* has received a lot of attention [4, 11, 12] and again an $O(n \log n)$ upper bound could be achieved. This bound can be improved if we restrict ourselves to rectilinear paths. Nilsson and Schuierer show that in this case both the rectilinear link diameter and the rectilinear link center can be computed in linear time [16, 17]. These algorithms require a triangulated polygon as input. In combination with the recent triangulation algorithm by Chazelle [2] they are now optimal.

In the parallel setting the search for an optimal triangulation algorithm still continues but in the light of the triangulation algorithms by Clarkson et al. [3] and Goodrich [6] there is a growing need for parallel algorithms which are optimal in a triangulated polygon. While the computation of shortest

---

*This work was supported by the Deutsche Forschungsgemeinschaft under Grant No. Ot 64/8-1.

[†]Institut für Informatik, Universität Freiburg, Rheinstr. 10–12, D-79104 Freiburg, Fed. Rep. of Germany; email: schuierer@informatik.uni-freiburg.de

paths in various metrics has been studied intensively [1, 7, 8, 15] the diameter and center problem have not received as much attention with the exception of the link diameter and the link center problem. Ghosh and Maheshwari developed an $O(\log^2 n \log \log n)$ time algorithm to compute the link center of a simple polygon on a CREW-PRAM using $O(n^2)$ processors [5]. Lingas *et al.* considered the rectilinear link distance and gave among many other algorithms an $O(\log n \log^* n)$ time algorithm to compute the rectilinear link diameter and center of a simple rectilinear polygon on a CREW-PRAM using $O(n/\log n \log^* n)$ processors [14]. For details on models of the PRAM see [9, Section 1.3].

In this paper we present an $O(\log n)$ time algorithm to compute the L_1-diameter and center of a simple polygon using $n/\log n$ processors on the weakest PRAM model, the EREW-PRAM given a triangulation. Our algorithm is based on two simple observations. The first observation is that if we partition the polygon into a number of simpler parts, called histograms, then a path that spans the diameter will pass through at least one of the histograms. Hence, if we compute a candidate path for each histogram, then we only have to compute the maximum of the lengths of the candidate paths to find the diameter. The second observation is that enough distance information in order to compute a candidate path can be localized at each histogram without having to do too many operations. Our algorithm is time and work optimal, that is, the time as well as the total number of operations used is optimal since it can easily be seen that computing the sum of n numbers can be reduced to the computation of the diamter of a simple polygon.

The paper is organized as follows. Sections 2 and 3 introduce some notation and a number of basic observations. The main part of the paper is contained in Section 4 which presents the algorithm to compute the L_1-diameter. Section 5 then deals with the computation of the L_1-center.

2 Preliminaries

In this paper we are only concerned with rectilinear paths and rectilinear polygons. A *rectilinear path* $\mathcal{P}$ is the concatenation of axes-parallel line segments, called *links*, such that no two consecutive links are collinear; a *simple rectilinear polygon* **P** is a simple and closed rectilinear path. For convenience we identify **P** with the region it encloses. If we want to refer to the enclosing path, we speak of the *boundary of* **P**.

If we are given a rectilinear path $\mathcal{P}$, we define its *length* to be the sum of the lengths of its links. Given two points p and q in **P**, the L_1-*distance* between them is the length of a shortest rectilinear path in **P** from p to q. We denote it by $L_1(p, q)$. The *eccentricity of* p *in* **P**, denoted by $\varepsilon_{\mathbf{P}}(p)$, is defined as the maximum distance $L_1(p, q)$ of p to a point q in **P**. If $L_1(p, q) = \varepsilon_{\mathbf{P}}(p)$, we call q a *furthest neighbour* of p.

The L_1-*diameter* of **P** is the maximum eccentricity of a point in **P**. We denote it by $D(\mathbf{P})$. The L_1-*radius* of **P**, denoted by $R(\mathbf{P})$, is the minimum eccentricity of a point in **P**. The set of points p with $\varepsilon_{\mathbf{P}}(p) = R(\mathbf{P})$ forms the *center* of **P** and is denoted by *cen*(**P**).

2.1 The Histogram Partition of a Simple Rectilinear Polygon

One of the main structures used in this paper is the histogram partition of a simple rectilinear polygon **P**. It is the decomposition of **P** into simpler parts, called *histograms*. A polygon **Q** is called *monotone* w.r.t. to an orientation θ if all lines that are orthogonal to θ intersect in a line segment or the empty set. A *histogram* is a monotone rectilinear polygon where one of the monotone chains is a single straight line segment, called the *base* of the histogram. A *maximal histogram* **H** is a histogram in **P** such that there is no histogram in **P** with the same base that properly contains **H**. A *window* of the maximal histogrom **H** is a maximal line segment on the boundary of **H** that is not part of the boundary of **P**. The base is not considered as a window.

The *histogram partition of* **P** w.r.t. an edge e, denoted by $\mathcal{H}(\mathbf{P}, e)$, can now be defined as follows. Let $\mathbf{H}_e$ be the maximal histogram of e in **P**. If $\mathbf{P} = \mathbf{H}_e$, then $\mathcal{H}(\mathbf{P}, e)$ is defined to be $\{\mathbf{H}_e\}$. Otherwise, let w_i, $1 \leq i \leq k$, be the windows of $\mathbf{H}_e$. Each window w_i splits **P** into two subpolygons one of which does not contain $\mathbf{H}_e$. We denote this subpolygon by $\mathbf{P}_{w_i}$. The histogram partition is

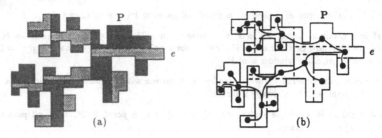

Figure 1: The tree associated to the histogram partition of **P**.

now given by $\mathcal{H}(\mathbf{P}, e) = \bigcup_{1 \leq i \leq k} \mathcal{H}(\mathbf{P}_{w_i}, w_i) \cup \{\mathbf{H}_e\}$. For illustration see Figure 1a. If a triangulation of the polygon is given [3, 6], a histogram partition can be computed in time $O(\log n)$ on a EREW-PRAM with $n/\log n$ processors [14]. From now on we will assume that any histogram we refer to is an element of $\mathcal{H}(\mathbf{P}, e)$ if not explicitly stated otherwise.

There is a natural tree structure associated to the histogram partition. We call it the *histogram tree of* **P** *and* e and denote it by $\mathcal{T}(\mathbf{P}, e)$ (see Figure 1b). It is defined as follows. The nodes of $\mathcal{T}(\mathbf{P}, e)$ are the histograms of $\mathcal{H}(\mathbf{P}, e)$. There is an edge between histogram $\mathbf{H}_1$ and $\mathbf{H}_2$ in $\mathcal{H}(\mathbf{P}, e)$ if the base of $\mathbf{H}_1$ is a window of $\mathbf{H}_2$. The root of $\mathcal{T}(\mathbf{P}, e)$ is the histogram with base e.

It is easy to see that the subtree of $\mathcal{T}(\mathbf{P}, e)$ that is rooted at histogram H with base b contains all and only those histograms that form the subpolygon $\mathbf{P}_b$ of b.

3 Gates in a Simple Rectilinear Polygon

A second important concept that we need in order to deal with L_1-shortest paths inside **P** are *gates*.

Definition 3.1 *Let* **Q** *be a subpolygon of* **P** *and* p *a point in* **P** *but outside* **Q**. *If* g *is a point in* **Q** *such that, for all points* $q \in$ **Q**, $L_1(p, q) = L_1(p, g) + L_1(g, q)$, *then* g *is called the* gate *of* p *to* **Q**. *The map* $\Gamma_{\mathbf{Q}} : \mathbf{P} \longrightarrow \mathbf{Q}$ *which maps a point to its gate is called the* gate map *of* **Q**. *If a gate exists for all points outside* **Q**, *then* **Q** *is called* gated.

It can shown that a gate is always unique which justifies the above definition [20].

Lemma 3.1 ([20, Example 5.10.1]) *If* $\Gamma_{\mathbf{Q}}(p) = g$, *for* p *in* **P**, *then* g *is the point of* **Q** *that is closest to* p.

A gate captures the influence of a point outside a gated set **Q**. In a certain sense which we will make precise later gated sets allow to localize the distance information of the whole polygon.

Often the class of gated sets consists only of trivial sets. One main characteristic of the L_1-metric in simple polygons is that geodesically convex sets are gated. A set is *geodesically convex* if it is a closed subset C of **P** such that all shortest paths between any two points of C are contained in it.

Lemma 3.2 ([20, Theorem 5.13]) *All geodesically convex sets of* **P** *are gated.*

We are in particular interested in gates for *subpolygons of chords*. A *chord* of **P** is an axis-parallel line segment in **P** that touches the boundary of **P** in exactly two points. Note that, for instance, windows are chords. A chord c splits **P** into two subpolygons $\mathbf{P}_1$ and $\mathbf{P}_2$. We call $\mathbf{P}_1$ and $\mathbf{P}_2$ the *subpolygons of* c.

Corollary 3.3 *Let* **P** *be a simple rectilinear polygon and* c *a chord in* **P**.

(i) If P_1 is a subpolygon of c, then P_1 is gated, all gates to P_1 are on c, and $\Gamma_{P_1} = \Gamma_c$.

(ii) If H is a maximal histogram in P with base c, then H is gated and all gates to H belong either to a window or the base of H, that is, for a point $p \in P \setminus H$ we have $\Gamma_H(p) = \Gamma_c(p)$ or $\Gamma_H(p) = \Gamma_w(p)$, for a window w of H.

Another important observation which can be easily proven with the help of gates is that a furthest neighbour of a point p in P is always a vertex.

Lemma 3.4 *Let P be a simple rectilinear polygon. If p is a point in P, then all points q with $L_1(p,q) = \varepsilon_P(p)$ are vertices.*

An immediate consequence of the above lemma is that the diameter of P is always spanned by a pair of vertices and we need only to consider the vertices of P in the following.

3.1 Gates in the Histogram Partition

Since the histograms of $\mathcal{H}(P,e)$ are maximal histograms of a window, the results of the previous section apply to them. Recall that if w is a window of histogram H in $\mathcal{H}(P,e)$, then P_w denotes the subpolygon of w that does not contain H.

Lemma 3.5 *Let P be a simple rectilinear polygon and H a histogram of $\mathcal{H}(P,e)$. Let w_1 and w_2 be two different windows of H. If p_1 is a point in P_{w_1} and p_2 a point in P_{w_2}, then*

$$L_1(p_1, p_2) = L_1(p_1, \Gamma_{w_1}(p_1)) + L_1(\Gamma_{w_1}(p_1), \Gamma_{w_2}(p_2)) + L_1(\Gamma_{w_2}(p_2), p_2).$$

By Lemma 3.5 gates can be used to calculate the distance between two vertices if their shortest path passes through H. How many different gates can a histogram have? We only consider gates of windows since only these will play a role in our algorithm.

Lemma 3.6 *Let H be a histogram in $\mathcal{H}(P,e)$, w a window of H, and H_w the maximal histogram with base w. If v is a vertex in P_w and g_w is the gate of v on w, then there is a vertex v_w of H_w with $\Gamma_w(v_w) = g_w$.*

An important consequence of Lemma 3.6 is that there is only a very limited number of different gates. Each gate is the orthogonal projection of a vertex of a histogram onto its base; hence, there are exactly half as many different gates as there are vertices of the histograms in $\mathcal{H}(P,e)$. Half as many since there are always two vertices that have the same x-coordinate or the same y-coordinate. If P consists of n vertices, then the sum of the number of vertices of the histograms in $\mathcal{H}(P,e)$ is at most $3n$ [13] and we obtain the following result.

Corollary 3.7 *The sum of the number of different gates for the vertices of P in all histograms is linear in the number of vertices of P.*

Clearly, the orthogonal projection of the vertices of a histogram H onto its base can be computed in constant time with m processors on a EREW-PRAM if H consists of m edges. Hence, the computation of all gates in $\mathcal{H}(P,e)$ can be carried out in time $O(\log n)$ on a EREW-PRAM with $n/\log n$ processors.

3.2 The Weight of Gates

Let H be a histogram and w a window of H. We denote the set of points $\{\Gamma_w(v) \mid v \in P_w\}$ by $\Gamma_w(P_w)$ and define the *weight* of a gate $g \in \Gamma_w(P_w)$ to be the maximum distance of g to a vertex v in P_w with $\Gamma_w(v) = g$. The weight of g is denoted by $w(g)$. A *maximum weight neighbour* $wn(g)$ of g is a vertex v in $\Gamma_w(P_w)$ with $L_1(v,g) = w(g)$. The computation of the weight and a maximum weight neighbour of for each gate is an important building block of our algorithm to compute the diameter of P.

We make use of the above observation that gates are orthogonal projections of the vertices and the window end points onto the base of H. If w is a window of H, we call the end point of w that is closer to the base of H the *designated end point* of w.

Lemma 3.8 *Let* H *be a histogram in* $\mathcal{H}(\mathbf{P}, e)$ *with base* b. *If* g *is a gate on* b, *then a maximum weight neighbour of* g *is either a vertex of* H *or a furthest neighbour of a designated end point of a window* w.

Lemma 3.8 implies that in order to compute maximum weight neighbours of gates we need to compute furthest neighbours of designated end points of windows. Once these are known, a furthest neighbour of a gate g is a furthest neighbour of the designated end point p with $\Gamma_b(p) = g$ which maximizes the sum $\varepsilon_{\mathbf{P}_w}(p) + L_1(p, g)$ or a vertex of H.

So we are left with the computation of a furthest neighbour of the designated end point of a window. In order to do so we define the following tree structure T_{dp}. Let H be a histogram of $\mathcal{H}(\mathbf{P}, e)$ with base b and p a vertex of H or a designated end point of a window of H. The parent of p is defined as the designated end point of b. We identify the point p with the node in the tree T_{dp}. The representation of T_{dp} we choose is a data structure where the children of a designated end point p are stored in a doubly linked list and p points to the first element in the list. Since the vertices and windows of a histogram are stored in a counterclockwise order around the boundary such a representation can be easily obtained. We define the *length* of an edge in T_{dp} between two nodes p and q as the L_1-distance from p to q. Computing the weight of p now amounts to computing the longest path which starts at p and which is contained in the subtree of T_{dp} that is rooted at p. This problem can be solved in time $O(\log n)$ on a EREW-PRAM with $n/\log n$ processors using standard techniques [9, Section 3.2.2].

4 An Algorithm to Compute the L_1-Diameter of a Polygon

Our algorithm to compute the L_1-Diameter is based on the following simple observation about shortest paths between two vertices that span the diameter. Let v_1 and v_2 be two vertices that span the diameter of P and $\mathcal{P}$ a shortest path between them. Let H_1 and H_2 be the histograms v_1 and v_2 belong to. The path $\mathcal{P}$ induces a shortest path $\mathcal{P}_T$ in $T(\mathbf{P}, e)$ from H_1 to H_2. The path $\mathcal{P}_T$ passes through the lowest common ancestor H_{lca} of H_1 and H_2 in $T(\mathbf{P}, e)$. Note that H_{lca} may also coincide with one of H_1 or H_2. In any case we observe that there is one histogram in $\mathcal{H}(\mathbf{P}, e)$ such that a shortest path between a pair of vertices that span the diameter passes through this histogram.

In other words, one approach to computing the diameter is to compute, for each histogram H, a pair of vertices such that a shortest path between the vertices intersects H and the distance between the vertices is as large as possible. We call such a pair of vertices a H-*restricted diametral pair* and the distance between them the H-*restricted diameter* of $\mathbf{P}_b$ if b is the base of H.

Lemma 4.1 *Let* H *be a histogram in* $\mathcal{H}(\mathbf{P}, e)$ *with base* b *and windows* $w_1, \ldots, w_k$. *If* (v_1, v_2) *is a* H-*restricted diametral pair, then*

(i) v_1 *and* v_2 *belong to* H *or to different subpolygons* $\mathbf{P}_{w_{j_1}}$ *and* $\mathbf{P}_{w_{j_2}}$ *of* P *and*

(ii) *if* g_i *is the gate of* v_i *in* H, *then* v_i *is a maximum weight neighbour of* g_i, *for* $i = 1, 2$.

Lemma 4.1 suggests that in order to compute the H-restricted diameter of $\mathbf{P}_b$ we only need to consider the weighted gates of H. If we define the weight of a vertex v of H to be 0 and a maximum weight neighbour of v to be v itself, then Lemmata 4.1 and 3.5 imply that the H-restricted diameter of P is given by

$$\max\{w(p) + L_1(p, q) + w(q) \mid p \text{ and } q \text{ are gates or vertices of H.}\} \qquad (*)$$

Furthermore, if (p, q) is a pair of points that achieves the above maximum, then $(wn(p), wn(q))$ is a H-restricted diametral pair of P by Lemma 4.1. Recall that $wn(p)$ is the vertex v with the largest distance to p among all vertices whose gate to H is p.

In the light of ($*$) the following conventions seem to be useful. We define the *weighted distance* between two points p and q as $w(p) + L_1(p, q) + w(q)$ and denoted it again by $L_1(p, q)$. The *weighted diameter* of a subset Q or H is the maximal weighted distance between two points and denoted by $D_w(Q)$. A different way to express the relation between ($*$) and the H-restricted diameter of P_b is to say that the weighted diameter of H equals the H-restricted diameter of P_b.

4.1 Computing the Weighted Diameter of H

Let H be a horizontal histogram with windows $w_1, \ldots, w_k$. In order to compute the weighted diameter of H we need the *horizontal vertex visibility map* of H. The horizontal vertex visibility map is the planar partition of H we obtain if we extend, at each vertex or gate, a horizontal line segment in the interior of H until the boundary is reached. These extensions can be computed by the algorithms of Goodrich [6] or Clarkson *et al.* [3]. We consider the intersection points of the extensions and the boundary as vertices of H. The planar partition we obtain consists of rectangles.

Given the horizontal visibility map there is a natural tree structure $T(H)$ associated to it. A rectangle R is defined to be a parent of rectangle R' if R is below R' and R and R' share some part of their boundary. $T(H)$ can be represented by a data structure where every parent points to a linear list that contains the children.

Given the tree $T(H)$ it is our aim to collapse it in a one-node tree while keeping track of the diameter of larger and larger subtrees of it until finally the diameter of H is the diamter of the one-node tree. We proceed in two phases. In the first phase we contract long chains of nodes with only one child. In the second phase we continually contract a constant fraction of the leaves of the resulting tree until we obtain the desired one-node tree. We now describe the phases in more detail.

4.2 Combining Pyramids

Consider a chain C of rectangles where each rectangle has only one child. We first apply list ranking to C and then we merge repeatedly nodes that have an odd rank with nodes that have an even rank until every node of the new tree has at least two children.

If we take a closer look at the above described algorithm, we note that do not only deal with rectangles in the combination process. Instead we will obtain *pyramids* by combining two or more adjacent rectangles. A pyramid is a histogram which is monotone w.r.t. both axes. The boundary of a pyramid Δ can be split into a left staircase L and a right staircase R and the top and the bottom horizontal edge of Δ which we call the *lids of* Δ. Note that weighted gates and vertices are only part of L and R.

To describe the combination process more precisely we assume that we are given a list L of pyramids $\Delta_1, \ldots, \Delta_m$ where pyramid i has rank i in the list L and the x-range of pyramid Δ_{i+1} is contained in the x-range of pyramid Δ_i.

We need a number of definitions and notations to describe the algorithm. We denote the bottom lid of pyramid Δ_i by b_i and the top lid by t_i, the left staircase by L_i, and the right staircase by R_i. We denote the left end point of b_i by b_i^l and the right end point by b_i^r. Similarly, we denote the left end point of t_i by t_i^l and the right end point by t_i^r.

In order to talk about distances and the diameter the following definitions are useful. For a point p_i in Δ_i we define the *left eccentricity of p* to be the largest weighted distance from p to a point on L_i and the *right eccentricity of p* to be the largest weighted distance from p to a point on R_i. The left eccentricity of p is denoted by $\varepsilon_L(p_i)$ and the right eccentricity by $\varepsilon_R(p_i)$. Finally, we need the *weighted eccentricity of p_i* which is the maximum of $\varepsilon_L(p_i)$ and $\varepsilon_R(p_i)$ and denoted by $\varepsilon_w(p_i)$.

We execute the following algorithm which keeps track of the weighted diameter $D_w(\Delta_i)$ of Δ_i and the weighted eccentricities of the end points of the bottom lid b_i of Δ_i and the left and right eccentricities of the end points of the top lid t_i of Δ_i. For illustration refer to Figure 2. It is based on a number of observations which we will present in its proof of correctness.

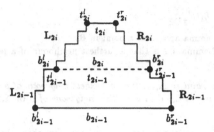

Figure 2: The combination of two pyramids.

Algorithm *Pyramid Combination*
Input: A list $\mathcal{L}$ of adjacent pyramids $\Delta_1, \ldots, \Delta_m$ together with the diameter and the eccentricities of the base and lid end points of Δ_i
Output: One pyramid $\Delta = \Delta_1 \cup \cdots \cup \Delta_m$ and the diameter and the eccentricities of the base and lid end points of Δ

    while $m > 1$ **do**
        for $i := 1$ **to** $\lfloor m/2 \rfloor$ **do in parallel**
            $\Delta_{new,i} := \Delta_{2i-1} \cup \Delta_{2i}$;
            $D_w(\Delta_{new,i}) := \max\{\varepsilon_L(t_{2i-1}^l) + \varepsilon_w(b_{2i}^l), \varepsilon_R(t_{2i-1}^r) + \varepsilon_w(b_{2i}^r), D_w(\Delta_{2i}), D_w(\Delta_{2i-1})\}$;
            $\varepsilon_L(t_{new,i}^l) := \max\{\varepsilon_L(t_{2i}^l), \varepsilon_L(t_{2i-1}^l) + L_1(t_{2i-1}^l, t_{2i}^l)\}$;
            Analogously for $\varepsilon_R(t_{new,i}^r)$;
            $\varepsilon_w(b_{new,i}^l) := \max\{\varepsilon_w(b_{2i-1}^l), \varepsilon_w(b_{2i}^l) + L_1(b_{2i}^l, b_{2i-1}^l)\}$;
            Analogously for $\varepsilon_w(b_{new,i}^r)$;
        endfor
        $m := m/2$;
    endwhile
End *Pyramid Combination*

Algorithm *Pyramid Combination* as presented above only computes the weighted diameter of the pyramids and does not output a pair of points that span the diameter; but it can be easily modified in order to keep track of the points associated with the eccentricities of the end points of the lids and a pair of points that spans the diameter.

4.2.1 Correctness of Algorithm *Pyramid Combination*

In order to see that the above algorithm is correct we need a few observations which are given in the lemmata that follow. These lemmata are stated in terms of histograms rather than pyramids since we need to consider the case of histograms later. Hence, we assume that Δ_i and Δ_{i+1} are two histograms rather than pyramids and that Δ_i and Δ_{i+1} are adjacent, that is, the top lid t_i of Δ_i contains the base b_{i+1} of Δ_{i+1}.

In the above algorithm we have $t_i^l = b_{i+1}^l$ and $t_i^r = b_{i+1}^r$ which is not true anymore if we consider histograms. L_i and R_i are now defined to be the part of Δ_i to the left and right, resp., of t_i. We .art off with an observation about how L_1-shortest paths between two points in Δ_i and Δ_{i+1} can be selected.

Lemma 4.2 *If Δ_i and Δ_{i+1} are two adjacent histograms, p a point in L_i, and q in Δ_{i+1}, then there is a L_1-shortest path from p to q that goes through t_i^l and b_{i+1}^l. In particular, $L_1(p,q) =*

$L_1(p, t_i^l) + L_1(t_i^l, b_{i+1}^l) + L_1(b_{i+1}^l, q).$

Of course, a similar lemma applies to a point in R_i.

The importance of Lemma 4.2 is that a furthest neighbour of a point p in Δ_{i+1}, L_i, or R_i is independent of p.

Corollary 4.3 *If p is a point in Δ_{i+1}, then a furthest neighbour of p in L_i is a furthest neighbour of t_i^l in L_i. If p is a point in L_i, then a furthest neighbour of p in Δ_{i+1} is a furthest neighbour of b_{i+1}^l in Δ_{i+1}.*

With the help of the above considerations we can now show that the weighted L_1-diameter computation of Algorithm *Pyramid Combination* is correct.

Lemma 4.4 *If Δ_{2i-1} and Δ_{2i} are two adjacent histograms and $\Delta_{new,i} = \Delta_{2i-1} \cup \Delta_{2i}$, then $D_w(\Delta_{new,i}) = \max\{\varepsilon_L(t_{2i-1}^l) + \varepsilon_w(b_{2i}^l), \varepsilon_R(t_{2i-1}^r) + \varepsilon_w(b_{2i}^r), D_w(\Delta_{2i}), D_w(\Delta_{2i-1})\}.$*

In addition to showing that the weighted L_1-diameter computation is correct we also have to show that the other values are correctly updated.

Lemma 4.5 *If Δ_{2i-1} and Δ_{2i} are two adjacent histograms and $\Delta_{new,i} = \Delta_{2i-1} \cup \Delta_{2i}$, then $\varepsilon_L(t_{new,i}^l) = \max\{\varepsilon_L(t_{2i}^l), \varepsilon_L(t_{2i-1}^l) + L_1(t_{2i-1}^l, t_{2i}^l)\}.$*

A similar lemma holds for $\varepsilon_R(t_{new,i}^r)$. Finally, we consider the last two steps of the algorithm.

Lemma 4.6 *If Δ_{2i-1} and Δ_{2i} are two adjacent histograms and $\Delta_{new,i} = \Delta_{2i-1} \cup \Delta_{2i}$, then $\varepsilon_w(b_{new,i}^l) = \max\{\varepsilon_w(b_{2i-1}^l), \varepsilon_w(b_{2i}^l) + L_1(b_{2i}^l, b_{2i-1}^l)\}.$*

Again a similar lemma holds for $\varepsilon_w(b_{new,i}^r)$.

It is easy to see that Algorithm *Pyramid Combination* can be carried out in $O(\log m)$ time on EREW-PRAM with m processors if the histogram H consists of m vertices and gates. If we define the *work* as the total number of operations performed, then the algorithm also needs only $O(m)$ work since the number of processors can be halved at each step. Brent's lemma now implies that Algorithm *Pyramid Combination* can be carried out in $O(\log m)$ time on a EREW-PRAM with $m/\log m$ processors [9, Section 1.4].

4.3 Combining Histograms

Hence, we obtain a tree $T(\Delta)$ where the nodes correspond to pyramids such that each pyramid is either a leaf or an internal node with at least two children. For the moment we assume that each internal node of $T(\Delta)$ has exactly two children. We call the bases of the children of a pyramid Δ the *top lids* of Δ. Obviously, a leaf of $T(\Delta)$ has no top lids and its boundary consists only of the boundary of H except for its base.

We now apply a combination operation to $T(\Delta)$ which consists of combining a leaf with its parent and its sibling into one node that then represents the union of all pyramids that correspond to these three nodes. Thus, we obtain a sequence of trees $T(\Delta) = T_1(\Delta), \ldots, T_k(\Delta)$ where $T_i(\Delta)$ is derived from $T_{i+1}(\Delta)$ by the simultaneous application of a number of independent combination operations and $T_k(\Delta)$ consists of a single node that represents the whole histogram H. The result of a combination operation are nodes that correspond to histograms rather than pyramids; so we assume from now on that we are given histograms as input to the combination step.

Since the combination step is very similar to the standard *rake*-operation on binary trees [9, Section 3.3], we can avoid that one node is involved in two concurrent combination operations by using a special numbering of the nodes of $T(\Delta)$ [9, Section 3.3]. This standard numbering technique also allows to apply a combination operation to at least half of the leaves. As a consequence the number of nodes decreases geometrically and $k = O(\log m)$ if $T(\Delta)$ consists of m nodes in the beginning.

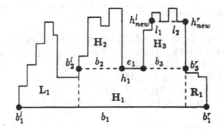

Figure 3: The combination operation on three histograms.

In the following we describe one combination step in detail. Let H_2 be a leaf with parent H_1 and sibling H_3. W.l.o.g. we assume that H_2 is to the left of H_3. Since H_2 and H_3 are adjacent to H_1, their bases b_2 and b_3 are the only lids of H_1 and there is a horizontal edge e_1 of H with two reflex vertices that connects b_2 and b_3. $b_2 \cup e_1 \cup b_3$ forms an edge h_1 of the boundary of H_1. Since H_2 is a leaf, it has no top lids. If H_3 is an internal node, it also has two top lids l_1 and l_2 and there is a horizontal edge e_3 of H with two reflex vertices that connects l_1 and l_2 such that $l_1 \cup e_3 \cup l_2$ forms a horizontal edge h_3 of H_3. Let b_i^l and b_i^r be the left and right end point of base b_i, for $i = 1, 2, 3$; furthermore, by h_i^l and h_i^r we denote the left and right end point of edge h_i, for $i = 1, 3$. Note that $h_1^l = b_2^l$ and $h_1^r = b_3^r$. L_i (R_i) is now defined as the part of H_i to the left (right) of h_i^l (h_i^r). For illustration refer to Figure 3.

As before let $D_w(H_i)$ be the weighted diameter of histogram H_i, $\varepsilon_w(b_i^l)$ and $\varepsilon_w(b_i^r)$ the maximal weighted distance of a point in H_i to the left and right end point of b_i, respectively; $\varepsilon_L(p)$ ($\varepsilon_R(p)$) is the maximal weighted distance of a point p in H_i to a point in L_i (R_i).

As is shown later the diameter of the union of H_1, H_2, and H_3 is given as the maximum of the following values.

(i) The maximum distance D_{L_1} of a point in L_1 to a point in H_2 or H_3;

(ii) The maximum distance D_{R_1} of a point in R_1 to a point in H_2 or H_3;

(iii) The maximum distance $D_{H_2 H_3}$ between a point in H_2 and a point in H_3;

(iv) The maximum D_{123} of the weighted diameters of H_1, H_2, and H_3.

A more precise formulation of the algorithm is as follows.
The following invariants are maintained by Algorithm *Histogram Combination*.

(i) If H is a histogram that corresponds to an internal node, then H has two top lids which are separated by a reflex edge e of P.

(ii) If H is a histogram that corresponds to a leaf, then it has no lids.

(iii) If H is a histogram with base b, then $D(H)$, $\varepsilon_w(b^l)$, and $\varepsilon_w(b^r)$ are correctly computed for H.

(iv) If H corresponds to an internal node with lids t_1 and t_2, then $\varepsilon_L(t_1^l)$, $\varepsilon_R(t_2^r)$, are correctly computed for H.

In order to see that the first and second invariant are maintained we observe that if we merge histogram H_2 with its parent H_1 and its sibling H_3, then we obtain a new histogram H_{new} whose lids are the lids of H_2 and H_3 and whose base is the base of H_1. If H_2 and H_3 are leaves in $T_i(\Delta)$, then the newly created node in $T_{i+1}(\Delta)$ which corresponds to H_{new} has no children and is a leaf.

```
Algorithm Histogram Combination
    H_new := H_1 ∪ H_2 ∪ H_3;
    D_L_1 := max{ε_L(h_1^l) + L_1(h_1^l, b_2^l) + ε_w(b_2^l), ε_L(h_1^l) + L_1(h_1^l, b_3^l) + ε_w(b_3^l)};
    D_R_1 := max{ε_R(h_1^r) + L_1(h_1^r, b_2^r) + ε_w(b_2^r), ε_R(h_1^r) + L_1(h_1^r, b_3^r) + ε_w(b_3^r)};
    D_H_2H_3 := ε_w(b_2^r) + L_1(b_2^r, b_3^l) + ε_w(b_3^l);
    D_123 := max{D_w(H_1), D_w(H_2), D_w(H_3)};
    D(H_new) := max{D_L_1, D_R_1, D_H_1H_2, D_123};
    ε_w(b_new^l) := max{ε_w(b_1^l), ε_w(b_2^l) + L_1(b_1^l, b_2^l), ε_w(b_3^l) + L_1(b_1^l, b_3^l)};
    Similarly for ε_w(b_new^r);
    if H_3 is an internal node
        then h_new^l := h_3^l; h_new^r := h_3^r;
             L_new := L_1 ∪ H_2 ∪ L_3; R_new := R_1 ∪ R_3;
             D_1 := ε_L(h_1^l) + L_1(h_1^l, h_3^l); D_2 := ε_w(b_2^l) + L_1(b_2^l, h_3^l);
             ε_L(h_new^l) := max{ε_L(h_3^l), D_1, D_2}; Similarly for ε_R(h_new^r);
    endif
End Histogram Combination
```

Since H_2 and H_3 are leaves, they contain no top lids and, hence, so does H_{new}. If, on the other hand, only H_2 is a leaf, then H_{new} is an internal node and contains exactly the two top lids of H_3.

Next we show that the L_1-diameter of H_{new} is correctly computed.

Lemma 4.7 Let H_1, H_2, and H_3 be three horizontal histograms such that the bases of H_2 and H_3 are lids of H_1 and H_2 is to the left of H_3. If $H_{new} = H_1 ∪ H_2 ∪ H_3$, then $D(H_{new})$ is given as in Algorithm Histogram Combination.

The weighted eccentricities of the base end points b_{new}^l and b_{new}^r are also correctly computed.

Lemma 4.8 Let H_1, H_2, and H_3 be three horizontal histograms such that the bases of H_2 and H_3 are lids of H_1 and H_2 is to the left of H_3. If $H_{new} = H_1 ∪ H_2 ∪ H_3$, then $ε_w(b_{new}^l) = \max\{ε_w(b_1^l), ε_w(b_2^l) + L_1(b_1^l, b_2^l), ε_w(b_3^l) + L_1(b_1^l, b_3^l)\}$.

Since the computation of $ε_w(b_{new}^r)$ is completely analogous to the computation of $ε_w(b_{new}^l)$ we dispense with a proof of correctness for its computation.

Finally, we have to consider the case that H_3 is an internal node and the left eccentricity of h_{new}^l and the right eccentricity of h_{new}^r in H_{new} have to be computed. Clearly, the part of H_{new} to the left of h_{new}^l consists of L_1, H_2, and L_3 since H_2 is to the left of H_3 and there are no vertices of H_1 between h_1^l and h_1^r. The part of H_{new} to the right of h_{new}^r consists of R_1 and R_3.

Lemma 4.9 Let H_1, H_2, and H_3 be three horizontal histograms such that the bases of H_2 and H_3 are lids of H_1 and H_2 is to the left of H_3. If $H_{new} = H_1 ∪ H_2 ∪ H_3$, then $ε_L(h_{new}^l) = \max\{ε_L(h_3^l), ε_L(h_1^l) + L_1(h_1^l, h_3^l), ε_w(b_2^l) + L_1(b_2^l, h_3^l)\}$.

A similar argument applies to the computation of $ε_R(h_{new}^r)$. This completes the proof of correctness of Algorithm Histogram Combination.

The analysis of Algorithm *Histogram Combination* is very simple. It needs constant time. As we mentioned before, a standard numbering of the nodes allows to apply Algorithm *Histogram Combination* to at least half of the leaves of $\mathcal{T}(\Delta)$ concurrently [9, Section 3.3]. If a EREW-PRAM with m processors is used, $O(\log m)$ combination steps are needed to reduce $\mathcal{T}(\Delta)$ to a one node tree since we can discard a constant fraction of the nodes of each $\mathcal{T}_i(\Delta)$ in one step. Since each step takes constant time and reduces the number of used processors by a constant fraction, the algorithm takes $O(\log m)$ time and uses $O(m)$ work. Brent's Lemma implies that $O(\log m)$ time is also sufficient to reduce $\mathcal{T}(\Delta)$ to a one node tree on a EREW-PRAM with $m/\log m$ processors.

Above we assume that each internal node H of $T(\Delta)$ has exactly two children. If no edges of H are collinear, then this condition always holds. But in rectilinear geometry degeneracies frequently arise and have to be dealt with. So consider an internal node H of $T_i(\Delta)$ with more than two children; we transform H into a number of nodes with exactly two children by conceptually prolonging each vertical edge of H by some very small amount $\varepsilon > 0$ except for the rightmost vertical edge which is prolonged by $(k/2-1)\varepsilon$ if H consists of k edges. We denote the histogram with the prolonged edges by H_ε. If we compute the horizontal vertex visibility map of H_ε, then it will be identical to the horizontal vertex visibility map of H except for a number of rectangles which have a height that is a multiple of ε. Each rectangle of the partition induced by the horizontal vertex visibility map has now at most two children. By treating ε as a number that is smaller than any number greater than 0 we can apply the above algorithm and the weighted diameter of H is correctly computed. The transformation can be easily carried out in $O(\log k)$ time with $k/\log k$ processors on the EREW-PRAM.

5 Computing the L_1-center

In this section we are concerned with the computation of the L_1-center of P. Recall that the L_1-radius $R(P)$ of P is the smallest value r such that there exists a point p in P with $L_1(p,v) \leq r$, for all vertices v of P, and the center is the set of all points that have this property, i.e., $cen(P) = \{p \in P \mid L_1(p,v) \leq R(P), \text{ for all vertices } v \in P\}$. There is a close relationship between the radius and the diameter of a simple polygon.

Lemma 5.1 *If* P *is a simple polygon, then* $R(P) = D(P)/2$.

Let v_1 and v_2 be a diametrical pair of vertices as given by the diameter algorithm. We compute a shortest path $\mathcal{P}$ from v_1 to v_2 and find the point p in the middle of $\mathcal{P}$. Let l be the link of $\mathcal{P}$ to which p belongs and c the chord incident to p which is orthogonal to l. W.l.o.g. we assume that c is vertical. The chord c splits P into two subpolygons P_1 and P_2 such that v_1 is in P_1 and v_2 is in P_2. Let P_1 be locally to the left of c and P_2 to the right. Furthermore, we assume that $g_1 = \Gamma_c(v_1)$ is below $g_2 = \Gamma_c(v_2)$. The location of all points that have distance $R(P)$ to v_1 and v_2 is a diagonal d that goes through p and has a slope of $-45°$. Clearly, d contains the center of P and, hence, the center is a diagonal line segment. The following observation is crucial.

Lemma 5.2 *The square spanned by* $cen(P)$ *is completely contained in* P.

A square S such that d contains one diagonal of S and S contains p is called a *center square*. Hence, $cen(P)$ is contained in the largest center square S_c that is contained in P. S_c can be computed in the following way. For each edge e of P we compute the largest square S_e that does not intersect e and that contains p. The intersection of all these squares is the largest center square S_c contained in P.

Let the four sides of S_c be $s_1, \ldots, s_4$ and the chords in P through s_i be denoted by c_i. Furthermore, let T_i be the set of all the vertices that are on the opposite side of c_i as is the square S_c. We now compute $\Gamma_{s_i}(T_i)$, for $1 \leq i \leq 4$. For each gate g_i on the boundary of S_c, we then compute the line segment on the diagonal of S that has distance $R(P)$ to g_i. The intersection of all these line segments then yields the L_1-center.

The above algorithm can be carried out in $O(\log n)$ time on a EREW-PRAM with $n/\log n$ processors. To see this just note that a shortest path between v_1 and v_2 can be computed within this time bound [15]. Finding the middle point and computing the gates also clearly takes logarithmic time and linear work. Finding the center squares and computing their intersections as well as computing the gates on $s_1, \ldots, s_4$ is also possible in $O(\log n)$ time on a EREW-PRAM with $n/\log n$ processors. Finally, we have to compute the lines segments on the diagonal of S_c from the gates and intersect these which again needs $O(\log n)$ time and linear work.

6 Conclusions and Open Problems

We have presented an optimal $O(\log n)$ time algorithm on a EREW-PRAM to compute the L_1-diameter and the L_1-center of a simple rectilinear polygon. Our approach is based on the histogram decomposition of the polygon. Unfortunately, it seems that the techniques employed here do not generalize in an obvious way to polygons with holes. It is also unclear how to attack the problem in higher dimensions.

A different open problem is the complexity of computing the diameter and radius of a polygon in parallel if a metric that is a linear combination of the L_1 and link metric is used.

References

[1] V. Chandru, S.K. Ghosh, A. Maheshwari, V. T. Rajan, and S. Saluja. *NC-Algorithms for Minimum Link Paths and Related Problems*. TR, Tata Institute of Fundamental Research, Bombay, 1992.

[2] Bernard Chaselle. Triangulating a simple polygon in linear time. In *Proc. 31st IEEE Symp. on Foundations of Computer Science*, pages 220–230, 1990.

[3] K.L. Clarkson, R. Cole, and R.E. Tarjan. Randomized parallel algorithms for trapezoidal decomposition. In *Proc. 7th ACM Symp. on Computational Geometry*, pages 152–161, 1991.

[4] H.N. Djidjev, A. Lingas, and J.-R. Sack. An $O(n \log n)$ algorithm for computing the link center of a simple polygon. *Discrete and Computational Geometry*, 8:131–152, 1992.

[5] S.K. Ghosh and A. Maheshwari. Parallel Algorithms for All Minimum Link Paths and Link Center Problems. In O. Nurmi, E. Ukkonnen, editors, *Proc. 3rd Scand. Workshop on Algorithm Theory*, pages 106–117, LNCS 621, 1992.

[6] M.T. Goodrich. Triangulating a polygon in parallel. *Journal of Algorithms*, 10:327–351, 1989.

[7] M.T. Goodrich, S.B. Shauck, and S. Guha. Parallel methods for visibility and shortest path problems in simple polygons. In *Proc. 7th ACM Symp. on Computational Geometry*, pages 73–82, 1991.

[8] J. Hershberger. Optimal parallel algorithms for triangulated simple polygons. In *Proc. 8th ACM Symp. on Computational Geometry*, pages 33–42, 1992.

[9] J. JáJá. *An Introduction to Parallel Algorithms*. Addison-Wesley, 1992.

[10] J.Hershberger and S. Suri. Matrix searching with the shortest path metric. In *25th ACM Symp. on Theory of Computing*, pages 485–494, 1993.

[11] Yan Ke. An efficient algorithm for link-distance problems. In *Proc. 5th Symp. on Computational Geometry*, pages 69–78, 1989.

[12] W. Lenhart, R. Pollack, J. Sack, R. Seidel, M. Sharir, S. Suri, G. Toussaint, S. Whitesides, and C. Yap. Computing the link center of a simple polygon. *Discrete and Computational Geometry* 3:281–293.

[13] Christos Levcopoulos. *Heuristics for Minimum Decompositions of Polygons*. PhD thesis, University of Linköping, Linköping, Sweden, 1987.

[14] A. Lingas, A. Maheshwari, and J.-R. Sack. Optimal parallel algorithms for rectilinear link distance problems. *Algorithmica*, to appear.

[15] K. M. McDonalds and J. G. Peters. Smallest paths in simple rectilinear polygons. *IEEE Transactions on Computer-Aided Design*, 11(7):864–875, 1992.

[16] B. J. Nilsson and S. Schuierer. Computing the rectilinear link diameter of a polygon. In H. Bieri, editor, *Proc. Workshop on Computational Geometry*, pages 203–216, LNCS 553, 1991.

[17] B.J. Nilsson and S. Schuierer. An optimal algorithm for the rectilinear link center of a rectilinear polygon. In N. Santoro F. Dehne, J.-R. Sack, editor, *Proc. 2nd Workshop on Algorithms and Datastructures*, pages 249–260, LNCS 519, 1991.

[18] R. Pollack, M. Sharir, and G. Rote. Computing the geodesic center of a simple polygon. *Discrete and Computational Geometry*, 4(6):611–626, 1989.

[19] S. Schuierer. *An Optimal Algorithm to Compute the L_1-Diameter and Center of a Simple Rectilinear Polygon*. Technical Report 49, Institut für Informatik, Universität Freiburg, 1994.

[20] M. van de Vel. *Theory of Convex Structures*. Mathematical Library, North-Holland, 1993.

Exploiting Locality in LT-RAM Computations[*]

Jop F. Sibeyn[†]   Tim Harris[‡]

Abstract

As processor speeds continue to increase the primary assumption of the RAM model, that all memory locations may be accessed in unit time, becomes unrealistic. In the following we consider an alternative model, called the Limiting Technology or LT-RAM, where the cost of accessing a given memory location is dependent on the size of the memory module. In general, computations which are performed on an LT-RAM with a memory of size $n \times n$ will result in execution times which are n times slower than a comparable RAM, if no special precautions are taken. Here we provide a general technique by which, for a class of algorithms, this slow-down can be reduced to $\mathcal{O}(2^{6 \cdot \log^{1/2} n})$ for sequential memory access, or to just $\mathcal{O}(1)$ if the memory access can be pipelined.

1 Introduction

The past twenty years the speed of computer systems has constantly increased. Through progress in chip fabrication, memory chips have become both smaller and faster, even as the storage capacity of the chips has increased dramatically. However, since data cannot be packed infinitely dense, eventually a larger memory will result in a higher access time. The point will be reached where a very fast processor can no longer perform a memory access within a constant number of clock-cycles. At that point the basic assumption of the RAM-model: *the processor can access all data in $\mathcal{O}(1)$ time independent of the size of the memory*, becomes unrealistic. Work related to the practicality of constant time assumptions in VLSI can be found in [2].

If we now assume a planar lay-out of memory and charge memory access with time proportional to the distance a signal must travel, then accessing data in

---

[*]This research was partially supported by the Leonardo Fibonacci Institute for the Foundations of Computer Science, and by EC Cooperative Action IC-1000 (project ALTEC: *Algorithms for Future Technologies*).

[†]Max-Planck-Institut für Informatik, Im Stadtwald, 66123 Saarbrücken, Germany. E-mail: jopsi@mpi-sb.mpg.de

[‡]Department of Computer Science, JCMB, The King's Buildings, University of Edinburgh, Scotland, EH9 3JZ. Email: tjh@dcs.ed.ac.uk.

a memory of size $n \times n$ costs $\Theta(n)$ steps. For example, without special care, multiplying two $n \times n$ matrices takes $\mathcal{O}(n^4)$. We refer to this model as the **Limiting-Technology RAM** or **LT-RAM**, as the model finds its motivation in the limiting technologies of the future. Such a model was originally proposed in [3], though it was considered there primarily in the parallel case. Efficient exploitation of such a model requires the copying of blocks from the large, slow memory into a smaller, faster one near the processor, and hence reflects the frequent use of memory hierarchies in current architectures. Hence the LT-RAM allows the consideration of topics such as caching and register reuse, but from a high level. For a more practical consideration of such topics see [4]. We focus primarily on memory which is layed out in two-dimensions, though our results are fundamentally unchanged if we consider higher dimensional layouts.

As an alternative to designing algorithms for the LT-RAM itself, we suggest to apply the following paradigm, which allows a separation of concerns:

Paradigm LTRAM

1. Design an optimal algorithm for the $n \times n$ mesh.

2. Transfer the mesh algorithm to the LT-RAM.

This paper deals with the second step: we develop a technique by which mesh computations can be simulated on an LT-RAM with minimal slow-down. This technique then precludes the necessity of developing LT-RAM algorithms explicitly, as such algorithms would have limited utility in a more general context. The reason for targeting mesh algorithms is that, by their nature, such algorithms, when well designed, fully exploit the inherent locality of a problem (a precise definition of locality is given below), which we are then able to exploit within the very different context of the LT-RAM. Furthermore, many good mesh algorithms already exist. Excellent examples of problems that can be solved efficiently on a two-dimensional mesh, and hence with very little loss on an LT-RAM, are matrix multiplication and graph problems that are related to the computation of a transitive closure (see [5]). If we were to simulate algorithms designed for the RAM on the LT-RAM, then in general it would be impossible to exploit any locality, and the simulation would run no better than the worst case.

In our technique for the above step 2, a block of the memory is fetched to a reserved area close to the processor, comparable to a cache. Then the processor performs the operations that would have been performed on this block by the processors of the mesh. The idea of fetching a block to a location close to the processor is applied recursively for simulating the operations on the block. A problem with this simulation technique is that on a real mesh this block of memory does not live in isolation, but may be modified by the processors surrounding it over time. After simulating t steps of the mesh on a block we are therefore forced to assume that any data in the block which is closer than t steps away from the boundary may now be invalid. We can then only use the central core of the block which does not risk being invalidated in this way. This fact has far reaching consequences for the effectiveness of the recursive application of our technique.

Our simulation works best on problems that can be implemented on an $n \times n$ mesh with almost optimal speed-up. In such cases it results in LT-RAM algorithms that are almost a factor of n faster than would be achieved by running an algorithm on the LT-RAM in a simplistic manner. For example, by applying our simulation, two $n \times n$ matrices can be multiplied in only slightly more than $\mathcal{O}(n^3)$ time. We define 'locality' within this context in the following way:

Definition 1 *We consider a problem* P. *Let* $T_N(P)$ *be the time required for the fastest solution for* P *running on a RAM with* $\mathcal{O}(N)$ *memory. Let* $T'_{n,d}(P)$ *be the number of steps required by the fastest algorithm for* P *running on a d-dimensional* $n \times \cdots \times n$ *mesh, in which the processors are equipped with* $\mathcal{O}(1)$ *memory each. The* (N, d)-*locality of* P *is the number* $L_d(N, P) = T_N(P)/(N \cdot T'_{N^{1/d},d}(P))$.

As an example, the highly local problem of matrix multiply has $L_2(n^2) = 1$. On the other hand, a less inherently local problem such as the sorting of N numbers has $L_2(N) = \log N/\sqrt{N}$. In general the slow-down due to the use of an LT-RAM in comparison with a hypothetical RAM, equals $f(n)/L_2(n^2)$, where $f(n)$ is the additional loss by the transition from a mesh to the LT-RAM. In this paper we show that $f(n)$ is a slowly growing function of n, depending on the memory access model: $f(n) = \mathcal{O}(2^{6 \cdot \log^{1/2} n})$ in the case of sequential memory access, and $f(n) = \mathcal{O}(1)$ in the case of pipelined memory access.

The remainder of this paper is organized as follows: We begin with a more detailed description of the problem. Then, in Section 3 and Section 4, we consider the cases of sequential and pipelined memory access, respectively. Subsequently we give a short evaluation of our method. In Section 6, we extend our approach to the simulation of higher dimensional meshes on the LT-RAM. Finally we consider meshes in which each processor has a memory, which is so large that it takes more than $\mathcal{O}(1)$ time to access it.

2 Simulating a Mesh on an LT-RAM

In this section we analyze how efficiently an algorithm running on a $n \times n$ mesh, in which each processor is equipped with $\mathcal{O}(1)$ of memory, can be simulated on an LT-RAM with a planar memory of capacity $\mathcal{O}(n^2)$. For simplicity we assume that the processor of the LT-RAM is placed somewhere on the outside of the memory. Putting the processor in the middle of the memory would only give a gain in the access-time by a factor two. We provide the processor with a memory hierarchy in the form of additional free memory for storing the blocks and subblocks that will be fetched from the main memory. Let the level i blocks, $i \geq 0$, have size $s_i \times s_i$, with $s_0 = n$. Hence, our LT-RAM can be represented as depicted in Figure 1, though we also need some extra room for storing intermediate results of the computation. We now describe how the memory which is distributed across the mesh is decomposed into blocks and subblocks which are then copied into the small memories located near the LT-RAM. This decomposition provides the basic framework of the simulation.

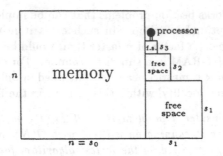

Figure 1: The processor, the memory, and the reserved blocks.

For the moment, we assume that the simulation of the steps of an $s_i \times s_i$ submesh is performed for $\alpha \cdot s_i$ steps on the LT-RAM, where α is a constant between 0 and 1/2 (though later α will be a function of s_i). After these $\alpha \cdot s_i$ steps, only the data in the memory cells that lie $\alpha \cdot s_i$ away from a boundary can be verified as valid. As we described earlier, this is due to possible modifications of boundary elements by neighboring processors of the submesh. We call the set

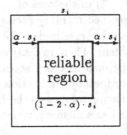

Figure 2: The reliable region for simulating $\alpha \cdot s_i$ steps of the mesh on an $s_i \times s_i$ submesh.

of these cells (and the corresponding submesh) the **reliable region**. The area of the reliable regions is $(1 - 2 \cdot \alpha_i)^2 \cdot s_i^2$ for an internal submesh and larger for a submesh on the boundary. See Figure 2.

Thus, for simulating $\alpha \cdot s_i$ steps on an $s_{i-1} \times s_{i-1}$ mesh, we must make a tessellation of partially overlapping $s_i \times s_i$ submeshes such that after $\alpha \cdot s_i$ steps each mesh processor is represented by a memory element which is somewhere still in a reliable region. Only with these overlaps can we ensure that all mesh computations will take place on the LT-RAM and use valid data. In Figure 3 we give an example, showing that for $\alpha = 1/4$, nine tiles are sufficient. Each $s_i/2 \times s_i/2$ block provides enough data, such that $s_i/4$ calculations may be performed, while maintaining the central subblock of size at least $s_i/4 \times s_i/4$ in the reliable region. In the general case, by selecting appropriately located submeshes, the whole $s_{i-1} \times s_{i-1}$ mesh can be covered by the reliable regions of $g_i = s_{i-1}^2/((1 -$

Figure 3: The tessellation of an $s_i \times s_i$ mesh with nine partially overlapping $s_i/2 \times s_i/2$ submeshes, such that all processors of the mesh are contained in the reliability region of one of the submeshes for $\alpha = 1/4$.

$2 \cdot \alpha_i)^2 \cdot s_i^2)$ $s_i \times s_i$ submeshes (assuming that $(1 - 2 \cdot \alpha_i) \cdot s_i$ divides s_{i-1}). Hence, the time $T_{i-1}(\alpha \cdot s_i)$, for simulating $\alpha_i \cdot s_i$ steps on a $s_{i-1} \times s_{i-1}$ mesh, equals g_i times the time for simulating $\alpha \cdot s_i$ steps on $s_i \times s_i$ submeshes, plus the time required for moving these submeshes close to the LT-RAM processor. This transportation cost depends on the memory-access model (pipelined or not) and s_i. Let M_i be the transportation cost for $s_i \times s_i$ submeshes. We get the following general expression:

$$T_{i-1}(\alpha \cdot s_{i-1}) = \frac{s_{i-1}^3}{(1 - 2 \cdot \alpha)^2 \cdot s_i^3} \cdot (T_i(\alpha \cdot s_i) + M_i). \tag{1}$$

3 Sequential Memory Access

We now consider the case where memory access is not pipelined, i.e. when performing multiple accesses each access costs the same. We would like to determine the value of $T_0(\alpha \cdot n)$, the cost of performing the entire mesh computation on the LT-RAM. Implicitly it is assumed that the original mesh computation requires at least $\alpha \cdot n$ steps, which is clearly true of all but the most trivial computations. The value of $T_0(\alpha \cdot s_1)$ depends on the choices of the s_i, where $i \geq 1$, and the correct choices depend on M_i. If the memory is accessed sequentially, then $M_i = \theta(s_{i-1} \cdot s_i^2)$. Assuming that the constant is one, and with $\alpha = 1/4$, substitution in (1) gives

$$T_{i-1}(s_{i-1}) = 4 \cdot s_{i-1}^3 / s_i^3 \cdot (T_i(s_i) + 4 \cdot s_{i-1} \cdot s_i^2). \tag{2}$$

We now consider the recurrence relation of (2) in detail. Given that simulating s_i steps of a computation on an $s_i \times s_i$ mesh on a RAM requires at least s_i^3 time, we introduce the function $f(s_i) = T_i(s_i)/s_i^3$. $f(n)$ can be interpreted as the extra delay when going from a mesh to an LT-RAM which is attributable to the assumption that the memory access is not instantaneous, and the goal of our simulation is to minimize $f(n)$. To simplify the notation, we consider the case $i = 1$, and use $n = s_0$ and $s = s_1(n)$, which allows (2) to be rewritten as

$$f(n) = 4 \cdot (f(s) + 4 \cdot n/s). \tag{3}$$

Obviously, $s = n^{1-\epsilon}$ and $f(x) = x^{2 \cdot \epsilon}$, satisfies (3) for all $\epsilon > 0$, for sufficiently large n. We derive a sharper bound on f:

Theorem 1 *Choosing $s = n^{1-\log^{-1/2} n}$, $f(n) \leq n^{6 \cdot \log^{-1/2} n}$.*

Proof: Let $s = n^{1-\log^{-1/2} n}$, and $g(x) = x^{6 \cdot \log^{-1/2} x}$. By induction, we may assume that $f(s) \leq g(s)$. Also $4 \cdot n/s = 4 \cdot n^{\log^{-1/2} n} \leq s^{6 \cdot \log^{-1/2} s} = g(s)$, for sufficiently large n. So, considering (3), in order to obtain $f(n) \leq g(n)$, it is sufficient to show that

$$8 \cdot g(s) \leq n^{6 \cdot \log^{-1/2} n} = g(n).$$

We first analyze part of the exponent of $g(s)$:

$$\begin{aligned}
\log^{-1/2} s &= \log^{-1/2} n^{1-\log^{-1/2} n} \\
&= (1 - \log^{-1/2} n)^{-1/2} \cdot \log^{-1/2} n \\
&\simeq \log^{-1/2} n + 1/2 \cdot log^{-1} n.
\end{aligned}$$

Now substitution gives

$$\begin{aligned}
g(s) &= s^{6 \cdot \log^{-1/2} s} \\
&\simeq n^{6 \cdot (1-\log^{-1/2} n) \cdot (\log^{-1/2} n + 1/2 \cdot \log^{-1} n)} \\
&< n^{6 \cdot (\log^{-1/2} n - 1/2 \cdot \log^{-1} n)} \\
&= n^{6 \cdot \log^{-1/2} n}/8 \\
&= g(n)/8.
\end{aligned}$$

$\square$

Theorem 1 shows that the loss in performance when using the LT-RAM for a highly local algorithm is smaller than any power of n and larger than polylogarithmic: for $n = 2^k$, $n^{\log^{-1/2} n} = 2^{\sqrt{k}}$, and $f(n) \leq 2^{6 \cdot \sqrt{k}}$. From the proof of the theorem, it follows that a substantially smaller estimate for $f(x)$ ($f(x) = n^{\alpha \cdot \log^{-\epsilon} x}$, for some $\epsilon > 1/2$) would not satisfy (3).

4 Pipelined Memory Access

In the previous section we analyzed the simulation of a mesh on an LT-RAM under the assumption that every memory access in an $n \times n$ memory takes $\Omega(n)$ steps, even when a whole block is accessed. However, many existing computer systems support the ability to pipeline the fetch of a block of memory. Using such a technique, moving an $s \times s$ square of data may be assumed to cost only $\mathcal{O}(n + s^2)$ time instead of $\mathcal{O}(n \cdot s^2)$. A related parallel model can be seen in [1].

Now we assume that $M_i = s_{i-1} + s_i^2$, and this M_i is substituted in (1). We also choose to replace $\alpha \cdot s_i$ by $t_i = t(s_i)$, to allow a more flexible choice of the time for which a simulation is performed. Thus we find a modified recurrence relation:

$$T_{i-1}(t_{i-1}) = \frac{t_{i-1} \cdot s_{i-1}^2}{t_i \cdot (s_i - 2 \cdot t_i)^2} \cdot (T_i(t_i) + s_{i-1} + s_i^2). \tag{4}$$

This relation has a very different character from (2): as surely $T_i(t_i) \geq t_i \cdot s_i^2$, the factor $T_i(t_i) + s_{i-1} + s_i^2$ is dominated by this term, except in the case where $t_i \cdot s_i^2 \leq s_{i-1}$.

Theorem 2 *Choosing $s_i = s_{i-1}^{2/3}$ and $t_i = s_{i-1}^{1/3}$, for all $i > 0$, $T_0(t) = \mathcal{O}(t \cdot n^2)$, for all $t \geq t_1$.*

Proof: Let $f_i = T_i(t_i)/(t_i \cdot s_i^2)$, and let s_i and t_i be as indicated, for $i \geq 1$ (until $s_i = \mathcal{O}(1)$). Then substitution gives

$$
\begin{aligned}
f_{i-1} &= (f_i + s_{i-1}/(t_i \cdot s_i^2) + 1/t_i)/(1 - 2 \cdot t_i/s_i)^2 \\
&\simeq (f_i + s_{i-1}^{-1/3}) \cdot (1 + 4 \cdot s_{i-1}^{-1/3}).
\end{aligned}
$$

It follows that f_i develops as the sum of a kind of geometric series: supposing that $s_\infty = 2$ and $f_\infty = 1$, we find

$$
f_0 \lesssim 2 \cdot \prod_{i=0} (1 + 4 \cdot s_i^{-1/3}) \simeq 2 + 4 \cdot \sum_{i=0} 2^{-(3/2)^k/3} \simeq 12.
$$

$\square$

The precise value of the constant involved in the estimate for T_0 depends on the omited constant in M_i and the value of s_0. In any case it is small. Hence, the sketched approach is practical for reasonable values of n, and could be advantageous even in present-day-technology memory management.

Notice that for the reduction of $T_0(t)$ to $\mathcal{O}(t \cdot n^2)$, we do not need the full strength of a pipelined memory access. It is sufficient if we have **weakly-pipelined memory access**, in the sense that M_i may have the following form: $M_i = s_{i-1} \cdot s_i^{2-\epsilon} + s_i^2$, for some constant $0 < \epsilon \leq 2$.

Theorem 3 *Choosing $s_i = s_{i-1}^{1-\epsilon/6}$ and $t_i = s_{i-1}^{1-\epsilon/3}$, $T_0(t) = \mathcal{O}(t \cdot n^2)$, for all $t \geq t_1$.*

Proof: Let $f_i = T_i(t_i)/(t_i \cdot s_i^2)$, and let s_i and t_i be as indicated. Then, for $\epsilon \leq 3 - \sqrt{3}$, $f_{i-1} = T_{i-1}(t_{i-1})/(t_{i-1} \cdot s_{i-1}^2) \lesssim (1 + 4 \cdot s_{i-1}^{-\epsilon/6}) \cdot (f_i + 2 \cdot s_{i-1}^{-\epsilon/3})$; and for $\epsilon \geq 3 - \sqrt{3}$, $f_{i-1} \lesssim (1 + 4 \cdot s_{i-1}^{-\epsilon/6}) \cdot (f_i + 2 \cdot s_{i-1}^{\epsilon/3-1})$. Hence, for all $0 < \epsilon \leq 2$,

$$
f_{i-1} \lesssim (1 + 4 \cdot s_{i-1}^{-\epsilon/6}) \cdot (f_i + 2 \cdot s_{i-1}^{-\epsilon/6}).
$$

Supposing that $s_\infty = 2$ and $f_\infty = 1$, we find

$$
f_0 \lesssim 3 + 6 \cdot \sum_{i=0} 2^{-((1/(1-\epsilon/6))^k) \cdot \epsilon/6} = \mathcal{O}(1).
$$

$\square$

5 Evaluation

Folling the paradigm we have called LTRAM, an LT-RAM algorithm may be constructed by simulating an existing mesh algorithm. We now consider how efficient this approach is. Obvious positive features of LTRAM are its generality and its simplicity. But how about the resulting run time in comparison with a trivial simulation of RAM algorithms?

We use the notation of Definition 1. Applying LTRAM on a two-dimensional mesh for a problem P results in a delay $\Theta(f(n)/L_2(n^2, P))$ when compared with the best RAM algorithm. Trivially performing the RAM algorithm on the LT-RAM gives a delay of $\Theta(n)$. As $L_2(n^2, P) \leq 1$, we gain at most a factor $n/f(n)$. We would like to prove that, for all problems P, $L_2(n^2, P) \geq 1/n$, or, formulated differently, that

$$T'_{n,2}(P) = \mathcal{O}(T_{n^2}(P)/n). \tag{5}$$

This would imply that we loose at most a factor $f(n)$.

Of course (5) does not hold without further conditions on P: a problem like multiplying two numbers that are stored at remote positions, takes one step on a RAM and $\Theta(n)$ steps on a mesh. But this an abstruse problem: why would one carry around a memory with $N = n^2$ entries if we are going to multiply only two numbers out of it? So we restrict ourselves to the following, more logical, class of problems:

Definition 2 *A problem P requiring N storage is **non-trivial**, if $T_N(P) = \Omega(N)$.*

A central question is whether (5) holds for all non-trivial problems. If we allow an additional polylog factor, then there is a strong relation with the "NC = P question" from PRAM computation. But, whereas it is generally believed that NC $\neq$ P, the status of (5) remains obscure. For some problems the following proposition may be helpful:

Proposition 1 *If a problem P can be represented by a directed acyclic graph (dag) of width $N = n^2$, and depth d, then P can be solved on an $n \times n$ mesh in $\mathcal{O}(d \cdot n)$ time.*

Proof: Every 'layer' of the dag involves at most n^2 operations. Without loss of generality, we assume that all operations are binary. Performing a binary operation requires that its operands are brought together. We solve this through two rounds of routing: in the first (second) round the first (second) argument of operation i is transferred to the processor with index i, PU_i. Because many operations may use the same arguments, this is not a permutation routing problem, but more like a partial broadcasting by many processors: a processor may send its data element to many processors, and every processor is the destination of precisely one data element. Such a routing can be performed in $\mathcal{O}(n)$ steps. This can be seen easiest by considering the reversed routing problem, which has the following specifications:

- Every processor holds one packet with a key i, $0 \leq i < n^2$, not necessarily different.

- If two packets with the same key reside in the same processor, then one of the packets may be deleted.

- All packets with key i that are not deleted, must be routed to PU_i, for all $0 \leq i < n^2$.

This problem can be solved by: (1) sorting the packets on their keys; (2) wiping the packets with the same keys together while deleting any surplus packets; (3) routing the remaining packets to their destinations (a partial permutation routing problem). The involved subproblems can be solved in $\mathcal{O}(n)$ time (see [7] for the currently best sorting algorithm and references). Hence the reversed problem, and consequently also the original problem, can be solved in $\mathcal{O}(n)$ time. □

Corollary 1 *If a problem P can be represented by a dag of width $N = n^2$ and depth $T_N(P)/n^2$, then P can be solved on an $n \times n$ mesh in $\mathcal{O}(T_N(P)/n)$ time.*

Notice that the condition in Corollary 1 is not a necessary condition: for sorting n^2 numbers the shortest possible dag has length $\mathcal{O}(\log n)$. And yet there are algorithms that sort n^2 numbers in $\mathcal{O}(n)$ steps on an $n \times n$ mesh. A bigger difference might be found for the maximum-flow problem. For a general graph with n nodes and $\Omega(n^2)$ edges, this problem requires n^2 storage, and $\mathcal{O}(n^3)$ time [6, 8]. There is no known algorithm that satisfies the condition of Corollary 1 even when allowing an additional polylog factor. Yet it is imaginable that there is an algorithm for an $n \times n$ mesh that solves the maximum-flow problem in $\mathcal{O}(n^2)$ time.

The problem of determining the parity of the sum of all stored numbers is non-trivial. On a RAM it takes $\Theta(n^2)$ time, on a mesh $\Theta(n)$ time. This shows that in any case (5) cannot be sharpened.

6 Higher Dimensional Memories

We now briefly consider the more general case of a memory layed out in d dimensions, such that the access time of a memory of size N grows as $N^{1/d}$ for some $d > 2$. At a transitory stage of technology, between the actual RAM model and the LT-RAM model, this may give a realistical estimate of the memory access cost. In analogy to our earlier techniques, we are going to simulate the action of a d-dimensional $n \times \cdots \times n$ mesh ($n = N^{1/d}$) on the LT-RAM. In comparison with an algorithm running on a RAM, for a problem P, we now loose a factor $f_d(N)/L_d(N,P)$ in performance, where f_d is defined analogously to the f that we considered before, the additional loss resulting from the transition from the d-dimensional mesh to the LT-RAM. $L_d(N,P)$ is the (N,d)-locality of P, see Definition 1.

It is easy to see that for sequential memory access, we get the following analogue of (2):

$$T_{i-1}(s_{i-1}/4) = 2^d \cdot s_{i-1}^{d+1}/s_i^{d+1} \cdot (T_i(s_i/4) + d \cdot s_{i-1} \cdot s_i^d). \qquad (6)$$

Substituting $f_d(s_i) = T_i(s_i)/s_i^{d+1}$ in (6) gives

$$f_d(s_{i-1}) = 2^d \cdot (f_d(s_i) + 4 \cdot d \cdot s_{i-1}/s_i). \qquad (7)$$

Añalogously to Theorem 1 we can derive that

Theorem 4 *Choosing* $s_i = s_{i-1}^{1-\log^{-1/2} s_{i-1}}$, *for all* $i > 0$, $f_d(n) \leq n^{2 \cdot (d+1) \cdot \log^{-1/2} n}$.

For pipelined memory access, we find

$$T_{i-1}(t_{i-1}) = \frac{t_{i-1} \cdot s_{i-1}^d}{t_i \cdot (s_i - 2 \cdot t_i)^d} \cdot (T_i(t_i) + d \cdot s_{i-1} + s_i^d). \qquad (8)$$

From this it follows that

Theorem 5 *Choosing* $s_i = s_{i-1}^{2/3}$ *and* $t_i = s_{i-1}^{1/3}$, *for all* $i > 0$, $T_0(t) = \mathcal{O}(t \cdot d \cdot n^d)$, *for all* $t \geq t_1$.

7 Meshes with a Large Memory

Assume that associated with an $n \times n$ memory we have m^2 processors. After the previous discussion, clearly it is optimal to arrange these processors in an $m \times m$ grid, each on the lower-right corner of an $n/m \times n/m$ block of the memory. See Figure 4. How can we optimize the use of this **LT-mesh**? Again, a trivial

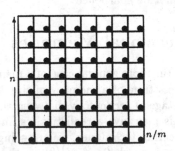

Figure 4: Schematical representation of the memory, with the processors indicated by •.

approach does not bring satisfactory results: we would get a slow-down of $(n/m)^3$ in comparison with an $n \times n$ mesh in which every processor is equipped with $\mathcal{O}(1)$

memory. We therefore once again use as input to our simulation well designed algorithms for the $n \times n$ mesh.

Every processor is responsible for the computation of the correct values in its 'own' submesh. In principle we now can apply immediately the earlier technique of recursively fetching subblocks of memory. Only on the boundaries between the blocks we must take some care: the fetched $s_1 \times s_1$ blocks must stretch over the boundary into the adjacent block in order to ensure that if the simulation is performed for $\alpha \cdot n/m$ steps, that then also the memory cells close to a boundary lie within the reliability region of some block. It is important to notice that, though different processors may access data from the same memory cell, this never happens in the same step, provided that the processors fetch the memory blocks in the same order. One possibility is to fetch the blocks of size $s_1 \times s_1$ in row-major order, starting with the blocks in the highest row, and working its way through the rows of blocks from left to right. We illustrate this in Figure 5.

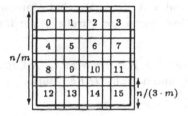

Figure 5: The sixteen $s_1 \times s_1$ blocks that must be accessed by a processor for $\alpha = 1/8$, $s_1 = n/(3 \cdot m)$.

There are no conflicts for memory accesses despite the overlapping of blocks, and we do not need a sophisticated memory organization scheme. Furthermore, whatever mechanism we use to fetch the blocks towards the processor, their paths are distinct. We immediately get the following analogous of Theorem 1 and Theorem 2:

Theorem 6 *For sequential memory access, the simulation of t steps of an $n \times n$ mesh on an $m \times m$ LT-mesh, can be performed in $\mathcal{O}(t \cdot (n/m)^2 \cdot (n/m)^{6 \cdot \log^{-1/2}(n/m)})$ steps. This is achieved for all $t \geq t_1$, with $s_0 = n/m$, $s_i = s_{i-1}^{1 - \log^{-1/2} s_{i-1}}$ and $t_i = s_i/4$, for all $i \geq 1$.*

Theorem 7 *For pipelined memory access, the simulation of t steps of an $n \times n$ mesh on an $m \times m$ LT-mesh, can be performed in $\mathcal{O}(t \cdot (n/m)^2)$ steps. This is achieved for all $t \geq t_1$, with $s_0 = n/m$, $s_i = s_{i-1}^{2/3}$ and $t_i = s_{i-1}^{1/3}$, for all $i \geq 1$.*

8 Conclusion

Our results demonstrate that a good approach to designing an algorithm for the LT-RAM, is to first solve the problem on the mesh, and then to simulate the mesh on the LT-RAM. In this way one retains some portability and pays only a small price in performance on the LT-RAM. However, we have not found optimal results in the case of sequential memory access, which suggests it may be worthwhile to investigate the existence of non-trivial lower-bounds for this problem.

9 Acknowledgments

The authors acknowledge that the simulation problem discussed in this paper was originally proposed by G. Bilardi and F.P. Preparata during a Fibonacci Institute Summer School, held in Trento, Italy, in July 1993.

References

[1] Aggarwal, A., Chandra, A., Snir, M, 'On Communication Latency in PRAM Computations,' *Proc. Symp. on Parallel Algorithms and Architectures*, pp. 11-21, 1989.

[2] Bilardi, G., Pracchi, M., F.P. Preparata, 'A Critique of Network Speed in VLSI Models of Computation,' *IEEE Journal of Solid-State Circuits*, Vol. 17, pp. 696-702, 1982.

[3] Bilardi, G., F.P. Preparata, 'Horizons of Parallel Computation,' *Proc. Int. Conf. for 25th Anniversary of INRIA*, Bensoussan, Verjus (Eds.), Paris, France, 1992.

[4] Goodman, J., 'Using Cache Memory to Reduce Processor-Memory Traffic,' *Proc. of 10th Int. Symp. on Computer Architecture*, pp. 124-131, 1983.

[5] Leighton, T., *Introduction to Parallel Algorithms and Architectures: Arrays-Trees-Hypercubes*, Morgan-Kaufmann Publishers, San Mateo, California, 1992.

[6] Karzanov, A.V., 'Determining the Maximal Flow in a Network with the Method of Preflows,' Soviet Math. Dokl. 15, pp. 434-437, 1974.

[7] Kaufmann, M., J.F. Sibeyn, T. Suel, 'Derandomizing Routing and Sorting Algorithms for Meshes,' *Proc. xth Symposium on Discrete Algorithms*, pp. 669-679, ACM-SIAM, 1994.

[8] Malhotra, V.M., M.P. Kumar, S.N. Maheshwari, 'An $\mathcal{O}(|V|^3)$ Algorithm for Finding Maximum Flows in Networks,' Inf. Proc. Letters, 7, no. 6, pp. 277-278, 1978.

Efficient Preprocessing of Simple Binary Pattern Forests

Mikkel Thorup

University of Copenhagen, Department of Computer Science, Universitetsparken 1, 2100 Kbh. Ø, Denmark; e-mail: mthorup@diku.dk.

Abstract. Multi-pattern matching in trees is fundamental to a variety of programming language systems. A bottleneck in this connection has been the combinatorial problem of constructing the immediate subsumption tree for a simple binary pattern forest. We reduce the complexity of this problem from $O(n^2)$ time and $O(n^2)$ space to $O(n \log n)$ time and $O(n)$ space. Such a result was conjectured possible in 1982 by Hoffmann and O'Donnell, and in 1992 finding it was called a main open problem by Cai, Paige, and Tarjan.

1 Introduction

This paper is concerned with simple multi-pattern matching in trees. We will solve a combinatorial sub-problem which has been considered a bottleneck in this context. More precisely, our result implies an efficient preprocessing of simple multi-patterns for bottom-up parsing in trees. This preprocessing resembles the one done by the Knuth-Morris-Pratt algorithm [6] for string matching. Using the terminology from [5], which will be summarized in Section 2, consider a simple binary pattern forest F and let *patsize* $= p$ denote the size of F, that is, the sum of the sizes of the individual patterns in F. The main result of this paper is

Theorem 1. *The immediate subsumption tree (the transitive irreflexive reduction of the subsumption relation) G_S for F can be computed in $O(p \log p)$ time and $O(p)$ space. Within the same bounds it can be checked that, indeed, the pattern forest is simple.*

The previous best bounds were $O(p^2)$ time and space [5].

In 1982, as a comment to their Algorithm A for construction of the immediate subsumption tree G_S, Hoffmann and O'Donnell [5, p. 82]: *Algorithm A is quadratic in patsize since it constructs $\overline{G}_S$, the transitive closure of G_S, rather than G_S. It seems there should be an algorithm for computing G_S for simple pattern forests which requires $O(patsize)$ steps only. So far, we have not found an algorithm this efficient, ...*

In those days Algorithm A was not a bottle-neck, but ten years later, with the exception of the construction of the immediate subsumption tree, Cai, Paige, and Tarjan [1], succeeded in getting all steps in the bottom-up multi-pattern matching in trees with simple binary pattern forests down to pseudo-linear time and linear space. Algorithm A was left unchanged, hence left as a severe bottleneck. In their conclusion they write [1, p. 58]: *The main open problem in the method of match set elimination is to compute the subsumption tree* T [$= G_S$ above] *in better time and space than Hoffmann and O'Donnell's Algorithm A.*

Let s denote the size of some subject tree S. Moreover, denote by m the number of matches with F in S, i.e. m denotes the size of desired output. Merging our result with Theorem 8.5 in [1, p. 57], we get

Corollary 2. *With an* $O(p \log p)$ *time and* $O(p)$ *space preprocessing of* F, *we can do each step in a bottom-up multi-pattern matching with* F *in* S *in* $O(\log p)$ *time and* $O(p)$ *space. Afterwards, the matches can be reported in* $O(m)$ *time and space. Thus the total matching is done in* $O(p \log p + s \log p + m)$ *time and* $O(p + s + m)$ *space.*

In Theorem 8.5 from [1, p. 57], they have $O(p^2)$ time and $O(p^2)$ space for the preprocessing, hence $O(p^2 + s \log p + m)$ time and $O(p^2 + s + m)$ space for the total matching. In [1, p. 57], as a comment to Theorem 8.5, they mention that the time for the bottom-up step can be improved to $O((\log \log p)^2)$ using more advanced data structures. Exactly the same improvement can be applied to our result. It should be mentioned that the measure for p in [1] differs from the one used in [5]. The measure from [1] is the smaller, but with regards to worst case complexity, the two measures are of the same order. In this paper we use the measure from the original article [5], which, as mentioned in the beginning, is the size of the representation of the input F. We shall discuss the exact measure from [1] at the end of this paper.

For applications of Corollary 2 within various programming language systems, the reader is referred to [1,4,5]. For example, in [1] there is a discussion of the class of very simple pattern forests. A very simple pattern forest is defined as a simple unbounded pattern forests for which the simplicity is preserved under some binary encoding. Hence Theorem 1 and Corollary 2 holds for all very simple pattern forests F.

The paper is divided as follows. Section 2 gives a formal statement of the problem of Theorem 1. Section 3 presents an algorithm for the problem and Section 4 proves that the algorithm is efficient. In the concluding remarks we will briefly discuss the special complexity measure from [1].

It is worthwhile noticing that our algorithm is simple and easy to implement. The hardness of the paper lays in a potential function argument showing that the algorithm is efficient.

2   The problem

This section summarizes the relevant definitions from [5], so as to formally define the problem treated in this paper; namely, the problem of constructing an immediate subsumption tree for a simple binary pattern forest.

We are given an alphabet Σ of function symbols together with a distinguished symbol $*$ ($=v$ in [5]). A Σ-*pattern*, or just *pattern*, is either $*$, or of the form

$a(t_1, t_2)$ where $a \in \Sigma$ and t_i is a Σ-pattern. Thus a Σ-pattern corresponds to an ordered rooted binary tree where all leaves are labeled with $*$ and where all internal nodes have degree 2 and are labeled with symbols from Σ (In [5] the symbols in Σ have designated arities. This is not needed here in the at most binary case, for if we want a to be unary or a constant, we just always use a as $a(t, *)$ or $a(*, *)$, respectively).

The patterns are partially ordered by $\sqsupseteq$ ($=\geq$ in [5]) defined as follows: let t and u be patterns. Then $t \sqsupseteq u$ if either $t = u$, $t = *$, or $t = a(t_1, t_2)$ and $u = a(u_1, u_2)$ where $t_i \sqsupseteq u_i$. Thus, seen as trees, $t \sqsupseteq u$ if t can be obtained from u by first repeatedly removing pairs of leaves with the same father (all leaves in patterns are in such pairs), second re-labeling with $*$ all internal nodes of u that have been turned into leaves by the pruning. If $t \sqsupseteq u$ we say that t *subsumes* u. If $t \sqsupseteq u$ and $t \neq u$, we write $t \sqsupset u$, and say that t *strictly* subsumes u. Finally, we say that two patterns are *incomparable* if neither $t \sqsupseteq u$ nor $u \sqsupseteq t$.

A *sub-pattern* relation between patterns is defined as the least reflexive transitive ordering such that if $t = a(t_1, t_2)$ then both t_1 and t_2 are sub-patterns of t. Thus, in terms of trees, s is a sub-pattern of t if s is a sub-tree of t sharing all its leaves with t.

Although it is not needed for this text, we remark that a pattern t *matches* a pattern u at node v in u if $t \sqsupseteq u^v$ where u^v is the sub-pattern of u descending from v.

Let $\mathcal{F}$ ($=F$ in [5]) denote a fixed non-empty set of Σ-patterns – a *pattern forest* and set p ($=patsize$ in [5]) to be the sum of the sizes (number of nodes) of the individual patterns in $\mathcal{F}$. Let $\mathcal{P}$ ($=PF$ in [5]) denote the set of sub-patterns of $\mathcal{F}$. Notice that (in contrast to in [5] we always have that) $* \in \mathcal{P}$ since $\mathcal{F}$ is non-empty and all leaves are labeled with $*$.

We say that $\mathcal{F}$ is *simple* if for every two incomparable patterns $t, u \in \mathcal{P}$, there is no Σ-pattern s such that $t \sqsupseteq s$ and $u \sqsupseteq s$. In this paper, unless stated otherwise, we will assume that $\mathcal{F}$ is simple. Theoretically, this is a very strong restriction, but it seems not to be a serious problem for many practical applications [1,4,5].

Let t, u be patterns in $\mathcal{P}$. We say that t *immediately subsumes* u if $t \sqsupseteq u$ and there is no $s \in \mathcal{P}$ such that $t \sqsupset s \sqsupset u$. As observed in [5], the simplicity of $\mathcal{F}$ implies that for any $u \in \mathcal{P} \setminus \{*\}$ there is exactly one $t \in \mathcal{P}$ immediately subsuming u; if $u = *$, there are none.

Consider the directed graph $\mathcal{T}$ ($=G_S$ in [5]) with vertex set $\mathcal{P}$ and where (t, u) is an arc if and only if t immediately subsumes u. From the above considerations, it follows that $\mathcal{T}$ is a rooted tree with root $*$. We call this rooted tree $\mathcal{T}$ *the immediate subsumption tree* for $\mathcal{F}$.

This paper presents an efficient way of computing $\mathcal{T}$ and establish thereby Theorem 1.

3   The algorithm

In this section we present an algorithm constructing the immediate subsumption
tree for a simple pattern forest. It should be noticed that the algorithm itself
is obvious and easy to implement. The hardness of this paper is the proof that
the algorithm is as efficient as announced in Theorem 1. This proof is differed to
the next section. Also in the next section, it will be discussed how we can verify
that a pattern forest is simple.

As in the last section, let $\mathcal{F}$ denote a fixed simple pattern forest. Let p denote
the sum of the sizes (number of nodes) of the individual patterns in $\mathcal{F}$, and let
$\mathcal{P}$ denote the set of sub-patterns of $\mathcal{F}$. Moreover, let $\mathcal{T}$ denote the immediate
subsumption tree for $\mathcal{F}$ that we wish to construct.

First, like in [1,5], we have a preliminary step turning $\mathcal{F}$ into a shared forest
by identification of equal sub-patterns. More precisely, in $O(p)$ time, we construct an acyclic directed graph $\mathcal{F}^s$ whose vertices correspond to the elements
from $\mathcal{P}$. The vertex corresponding to $*$ has no outgoing arcs. The vertex corresponding to an element of the form $a(t_1, t_2)$ has two outgoing arcs; namely to
the vertices corresponding to t_1 and t_2. Now $\mathcal{F}^s$ allows us to carry out various
simple operations on the elements from $\mathcal{P}$ in constant time. Assuming $\mathcal{F}^s$, we
have the following algorithm for the construction of $\mathcal{T}$.

Algorithm A Builds the immediate subsumption tree $\mathcal{T}$.
A.1.Let $\mathcal{T}$ consist of the root $*$.
A.2.For all $a(t_1, t_2) \in \mathcal{P}$ in size-wise increasing order:
A.2.1.   Set $s := \mathtt{imm\text{-}sub}(a, t_1, t_2)$.
A.2.2. $\mathtt{attach}(a(t_1, t_2), s)$.

Procedure 1 $\mathtt{imm\text{-}sub}(a, t_1, t_2)$, *where* $a(t_1, t_2) \in \mathcal{P}$, *returns the pattern* $s \in \mathcal{P}$
immediately subsuming $a(t_1, t_2)$.

Procedure 2 $\mathtt{attach}(t, s)$ *where* s *but not* t *is a node in* $\mathcal{T}$, *adds the arc* (s, t)
to $\mathcal{T}$, *i.e. makes* t *a son of* s *in* $\mathcal{T}$.

For any $t \neq *$ in the current version of $\mathcal{T}$, let $\mathtt{father}(t)$ denote the father of t in
$\mathcal{T}$.

Algorithm B Implements $\mathtt{imm\text{-}sub}$ for use in Algorithm A. Thus, the input is
(a, t_1, t_2) where $a(t_1, t_2) \in \mathcal{P}$ and where both t_1 and t_2 have been placed in $\mathcal{T}$.
We want to find a $\sqsubseteq$-wise minimal pattern $s \sqsupseteq a(t_1, t_2)$. Notice for $i = 1, 2$ and
$s_i \in \mathcal{P}$ that since t_i has been placed in $\mathcal{T}$, we have $s_i \sqsupseteq t_i$ if and only if s_i is an
ancestor to t_i in $\mathcal{T}$ (t_i is considered an ancestor to itself). Also notice that s is
unique by the simplicity of $\mathcal{F}$. Thus we return $s = *$ if there are no patterns s_1
and s_2 such that $a(s_1, s_2) \in \mathcal{P} \setminus \{a(t_1, t_2)\}$ and s_i is an ancestor to t_i. Otherwise
we return $s = a(s_1, s_2)$ where s_1 is the nearest ancestor to t_1 such that for some
ancestor s_2' to t_2 we have $a(s_1, s_2') \in \mathcal{P} \setminus \{a(t_1, t_2)\}$, and where s_2 is the least
such s_2'.

B.1.Set $s_1 := t_1$.   —first we aim at solution of the form $s = a(s_1, s_2)$.

B.2.If $t_2 \neq *$,   —aiming at solution with $s_1 = t_1$ and $s_2 \sqsupseteq t_2$.

B.2.1. Set $s_2 := $ left-match$(a, s_1, $father$(t_2))$.

B.3.Otherwise, set $s_2 := \bot$.

B.4.While $s_2 = \bot$ and $s_1 \neq *$: —aiming at solution with $s_1 \sqsupset t_1$ and $s_2 \sqsupseteq t_2$.

B.4.1. Set $s_1 := $ father(s_1).

B.4.2. Set $s_2 = $ left-match(a, s_1, t_2).

B.5.If $s_2 \neq \bot$, return $a(s_1, s_2)$;

B.6.Otherwise return $*$.   —no solution of the form $s = a(s_1, s_2)$.

Procedure 3 left-match(a, s_1, t_2'), *where* $s_1, t_2' \in \mathcal{P}$, *returns* $\bot$ *if there is no* $s_2 \sqsupseteq t_2'$ *such that* $a(s_1, s_2) \in \mathcal{P}$; *otherwise it returns the* $\sqsubseteq$-*wise least* $s_2 \sqsupseteq t_2'$ *such that* $a(s_1, s_2) \in \mathcal{P}$.

This completes our construction of the immediate subsumption tree $\mathcal{T}$. In the next section we will discuss its efficiency, including a description of how left-match and attach can be implemented by standard dynamic data structures.

4 Efficiency

In this section we present a potential function argument for the efficiency of the algorithm from the last section. Moreover, at the end of this section, we mention how it can be verified that a pattern forest is simple.

Theorem 3. *Algorithm A makes a total of* $O(p)$ *calls to* left-match.

Proof: The first two calls to left-match from each call to imm-sub are accounted for in the at most $|\mathcal{P} \setminus \{*\}| < p$ calls to imm-sub. The rest of the proof is devoted to bounding the remaining calls to left-match. The key is to relate these remaining calls to the depth of the patterns in the immediate subsumption tree. For any $t \in \mathcal{P}$, let depth(t) denote the depth of t in the final version of $\mathcal{T}$, i.e. in the immediate subsumption tree (depth$(*) = 0$ and depth(father$(t)) + 1 = $ depth(t)).

CLAIM 1. depth$(t) \leq (|t| - 1)/2$ *where* $|t|$ *is the number of vertices in* t.

PROOF: We settle the claim by induction on depth(t). By definition the root $*$ of $\mathcal{T}$ has size 1 settling the base case. Now if s is the father, hence a strict ancestor to t, then we can get from t to s by repeatedly replacing a sub-pattern of the form $b(*, *)$ with $*$. Every time we do this, we decrease the size with at 2, and we have to do it at least once. Thus, depth$(t) = $ depth$(s) + 1 \leq (|s| - 1)/2 + 1 \leq ((|t| - 2) - 1)/2 + 1 = (|t| - 1)/2$. $\quad\square$

Notice that $|t|$ is always odd, so our bound on depth(t) is always an integer.

CLAIM 2. *If* $t = a(t_1, t_2)$ *then* depth$(t) \leq $ depth$(t_1) + $ depth$(t_2) + 1$.

PROOF: We prove the result by induction on $\mathrm{depth}(t)$. Since $t \neq *$, t has some father s in $\mathcal{T}$. By definition $\mathrm{depth}(t) = \mathrm{depth}(s) + 1$. For the base case, if $s = *$, we have $\mathrm{depth}(s) = 0$. Hence $\mathrm{depth}(t) = 1$ which is an obvious lower-bound for the right hand side. We may therefore assume that s is of the form $a(s_1, s_2)$ where $s_i \sqsubseteq t_i$ and $(s_1, s_2) \neq (t_1, t_2)$. Hence either $\mathrm{depth}(s_1) < \mathrm{depth}(t_1)$ and $\mathrm{depth}(s_2) \leq \mathrm{depth}(t_2)$ or $\mathrm{depth}(s_1) \leq \mathrm{depth}(t_1)$ and $\mathrm{depth}(s_2) < \mathrm{depth}(t_2)$. In either case $\mathrm{depth}(s_1) + \mathrm{depth}(s_2) < \mathrm{depth}(t_1) + \mathrm{depth}(t_2)$. Thus, by the induction hypothesis, we have $\mathrm{depth}(t) = 1 + \mathrm{depth}(s) \leq 1 + \mathrm{depth}(s_1) + \mathrm{depth}(s_2) + 1 \leq \mathrm{depth}(t_1) + \mathrm{depth}(t_2) + 1$, as desired. $\square$

CLAIM 3. *If during a call* $\mathrm{imm\text{-}sub}(a, t_1, t_2)$, *we make* $k + 2$ *calls to* $\mathrm{left\text{-}match}$ *where* $k > 0$, *then* $\mathrm{depth}(a(t_1, t_2)) \leq \mathrm{depth}(t_1) + \mathrm{depth}(t_2) + 1 - k$.

PROOF: Assume during a call $\mathrm{imm\text{-}sub}(a, t_1, t_2)$ that we make $k + 2$ calls to $\mathrm{left\text{-}match}$ where $k > 0$. Then $\mathrm{depth}(t_1) \geq k + 1$. Hence the inequality is trivially satisfied if $*$ is the father of $a(t_1, t_2)$, for then the left hand side is 1 while the right hand side is at least 2. Otherwise the father is of the form $a(s_1, s_2)$ where $\mathrm{depth}(s_1) = \mathrm{depth}(t_1) - k - 1$. Thus by Claim 2, we get

$$
\begin{aligned}
\mathrm{depth}(a(t_1, t_2)) &= 1 + \mathrm{depth}(a(s_1, s_2)) \\
&\leq 1 + \mathrm{depth}(s_1) + \mathrm{depth}(s_2) + 1 \\
&\leq 1 + \mathrm{depth}(t_1) - k - 1 + \mathrm{depth}(t_2) + 1 \\
&\leq \mathrm{depth}(t_1) + \mathrm{depth}(t_2) + 1 - k,
\end{aligned}
$$

as desired. $\square$

Now, for a potential function counting the remaining calls to $\mathrm{left\text{-}match}$, we introduce a counter d_t for each $t \in \mathcal{P}$. Our potential function will be the sum $\sum_{t \in \mathcal{F}} d_t$—not $\sum_{t \in \mathcal{P}} d_t$.

- Initially, that is, before the first call to $\mathrm{imm\text{-}sub}$, we set $d_t := (|t| - 1)/2$ for all $t \in \mathcal{P}$.
- Suppose in connection with a call $\mathrm{imm\text{-}sub}(a, t_1, t_2)$ that we have $k + 2$ calls to $\mathrm{left\text{-}match}$ where $k > 0$. Then for all $u \in \mathcal{P}$, if $a(t_1, t_2)$ occurs l times as a sub-pattern u, we set $d_u := d_u - kl$.

Initially, we have $\sum_{t \in \mathcal{F}} d_t = (p - |\mathcal{F}|)/2 < p/2$. Moreover, any $t \in \mathcal{P}$ is a sub-pattern of at least one $u \in \mathcal{F}$. Thus, in connection with a call to $\mathrm{imm\text{-}sub}$ using $k + 2$ calls to $\mathrm{left\text{-}match}$ we subtract at least k from $\sum_{t \in \mathcal{F}} d_t$. Hence the theorem follows if we can prove that $\sum_{t \in \mathcal{F}} d_t$ never turns negative.

CLAIM 4. *The following conditions are invariant:*

(i) $d_t \geq \mathrm{depth}(t)$ *for all* $t \in \mathcal{P}$.
(ii) *Let* $t = a(t_1, t_2) \in \mathcal{P}$. *Then* $d_t \leq d_{t_1} + d_{t_2} + 1$ *with equality in all cases where* $\mathrm{imm\text{-}sub}(a, t_1, t_2)$ *has not been called.*

PROOF: Initially (i) is satisfied by Claim 1. Moreover, (ii) is initially satisfied since $d_t = (|t| - 1)/2 = ((1 + |t_1| + |t_2|) - 1)/2 = (|t_1| - 1)/2 + (|t_2| - 1)/2 + 1 = d_{t_1} + d_{t_2} + 1$.

Now, focus on some specific call to imm-sub(b, s_1, s_2), and assume that conditions (i) and (ii) are satisfied before this call. We need to prove that the conditions also are satisfied after the call. Suppose we during our call to imm-sub(b, s_1, s_2) make ≤ 2 calls to left-match. Then no counter d_t is changed, so neither (i) nor (ii) can become violated. Hence we may assume that for some $k > 0$, we make $k + 2$ calls to left-match.

For all $t \in \mathcal{P}$, we let $\overline{d}_t$ denote the old value of d_t.

Clearly, our conditions (i) and (ii) are only affected for patterns containing $s = b(s_1, s_2)$ as a sub-pattern. We will apply induction over these patterns. For the base case, assume that $t = s$. Trivially (ii) is satisfied since imm-sub(b, s_1, s_2) has now been called and $d_{t_i} = \overline{d}_{t_i}$ $(i = 1, 2)$ while $d_t = \overline{d}_t - k$. Moreover, by Claim 3 we have

$$\text{depth}(t) \leq \text{depth}(t_1) + \text{depth}(t_2) - k + 1 \leq d_{t_1} + d_{t_2} - k + 1 = \overline{d}_t - k = d_t,$$

which is the statement of (i). Hence both invariants are satisfied for $t = s$.

For the induction step, let $t = a(t_1, t_2)$ be any pattern from $\mathcal{P}$ having s as a strict sub-pattern. Since imm-sub(a, t_1, t_2) has not been called, we need to prove equality in (ii).

For $i = 1, 2$, denote by l_i the number of times that s occurs as a sub-pattern of t_i. Then $l = l_1 + l_2$ is the number of times that s occurs as a sub-pattern of t. Before the call, by (ii), we have $\overline{d}_t = \overline{d}_{t_1} + \overline{d}_{t_2} + 1$. Moreover, we have $d_{t_1} = \overline{d}_{t_1} - l_1 k$, $d_{t_2} = \overline{d}_{t_2} - l_2 k$ and $d_t = \overline{d}_t - lk$. Thus,

$$d_t = \overline{d}_t - lk = \overline{d}_{t_1} + \overline{d}_{t_2} + 1 - lk = \overline{d}_{t_1} + \overline{d}_{t_2} + 1 - (l_1 + l_2)k$$
$$= \overline{d}_{t_1} - l_1 k + \overline{d}_{t_2} - l_2 k + 1 = d_{t_1} + d_{t_2} + 1,$$

so, indeed, we have equality in (ii) after the call.

By the induction hypothesis on (i), we have $d_{t_1} \geq \text{depth}(t_1)$ and $d_{t_2} \geq \text{depth}(t_2)$, so our equality in (ii) together with Claim 2 settles (i), completing the induction. $\quad\square$

By definition $\text{depth}(t) \geq 0$ for all $t \in \mathcal{P}$, so in particular this is the case for all $t \in \mathcal{F}$. Hence, by Claim 4 (i) we may conclude that $\sum_{t \in \mathcal{F}} d_t$ never turns negative. Summing up, we have shown that

- Initially $\sum_{t \in \mathcal{F}} d_t < p/2$.
- In connection with a call to imm-sub using $k + 2$ calls to left-match we subtract at least k from $\sum_{t \in \mathcal{F}} d_t$.
- $\sum_{t \in \mathcal{F}} d_t$ does not turn negative.

Hence $p/2$ bounds the total number of calls to left-match which are not among the first two calls in a call to imm-sub. There are at most p calls to imm-sub, so we may conclude that in total, we have no more than $5p/2$ calls to left-match. $\quad\blacksquare$

For each $t_1 \in \mathcal{P}$, let $\omega(t_2)$ denote the number of patterns $a(t_1, t_2) \in \mathcal{P}$

Theorem 4. *There is a dynamic $O(p)$ space data structure for the rooted tree $\mathcal{T}$ which supports* attach(t, s) *in* $O(\omega(s) \log p)$ *amortized time and supports* left-match(a, t_1, t_2) *in* $O(\log p)$ *amortized time.*

Proof: With each vertex t_2 in $\mathcal{T}$ we associate the set $\{(a, t_1) \mid a(t_1, t_2) \in \mathcal{P}\}$. Thus $\omega(t_2)$ is the cardinality of the set associated with t_2. Now, we compute left-match(a, t_1, t_2) by checking if there is an ancestor s_2 to t_2 in the current version of $\mathcal{T}$ such that (a, t_1) is a member of the set associated with s_2. If so, we return the least such ancestor s_2; otherwise we return $\bot$. This is an instance of the retrieval problem for context trees [7], so Theorem 4 follows from the solution to the retrieval problem in [2] combined with the improvements to the sub-problem of maintaining order in lists in [3]. ∎

Proof of Theorem 1: The correctness of Algorithms A and B is immediate. Hence, since $\sum_{t_2 \in \mathcal{P}} \omega(t_2) = |\mathcal{P}| \leq p$, the result follows directly from Theorems 3 and 4. ∎

Even if $\mathcal{F}$ is not simple, the above algorithms will still produce a rooted tree within the same time bounds. However, this rooted tree will only represent part of the immediate subsumption relation. In order to check that, indeed, $\mathcal{F}$ is simple, we simply run the symmetric algorithm with a procedure "right-match" instead of left-match and check that the resulting trees are identical. Clearly, this does not change our complexities.

5   Concluding remarks

Above we have seen a simple algorithm which constructs the immediate subsumption tree for a simple pattern forest in time $O(p \log p)$ time and $O(p)$ space, improving the previous best bounds of $O(p^2)$ time and $O(p^2)$ space [5]. This settles the conjecture of Hoffmann and O'Donnel [5] quoted in the introduction within a log factor, and hence also the "main" open problem of Cai, Paige, and Tarjan [1].

Some remarks should be made in connection with the complexity measures in [1]. In [1] they make the nice observation that if we ignore the initial $O(p)$ processing at page 4 of $\mathcal{F}$ to the shared forest $\mathcal{F}^s$, then Hoffmann and O'Donnell's algorithm constructs the immediate subsumption tree in $O(q^2)$ time and $O(q^2)$ space where $q = |\mathcal{P}| \leq p$. Notice that the size of $\mathcal{F}^s$ is $O(q)$, so q corresponds to some kind of local complexity measure. Now q can be of the same order as p, so the observation does not improve the general worst case. However, it improves the complexity in the special case where the sub-patterns have high multiplicity, i.e. where $q \ll p$. The same observation does not quite carry through with our algorithm. It is true that our counting benifits from high mulitiplicity, but only if the sub-patterns of high multiplicity to some extent coincide with those for which we make extra calls left-match. However, something general can be said about our construction of T from $\mathcal{F}^s$ in terms of q. First of all our space requirement is $O(q)$ since it is the sum of the sizes of $\mathcal{F}^s$ and T, both of which are $O(q)$. Also, there are at most $q - 1$ calls to imm-sub. Since T is of height at most $q - 1$, no call to imm-sub can result in more than q calls to left-match. Thus, in total we make $O(\min\{q^2, p\})$ calls to left-match. The amortized time for calls to attach and left-match turns out to be $O(\log q)$, so we may conclude that our algorithm constructs T from $\mathcal{F}^s$ in time $O(\min\{q^2, p\} \log q)$. Thus, for the local construction of T from $\mathcal{F}^s$, our algorithm might be up to a log factor slower than Hoffmann and O'Donnell's if $q \ll p$. On the other hand our algorithm is nearly a linear factor faster if $q \approx p$ and concerning the space requirement our algorithm is always better by a linear factor. As stated in Theorem 1, the real result of this paper is the strong improvement of the worst case complexity for the full construction of the immediate sub-sumption tree T from the input pattern forest $\mathcal{F}$. As in the original paper [5] on this problem, our worst case complexity is measured in the standard way in terms of the size p of the representation of the input $\mathcal{F}$.

References

1. CAI, J., PAIGE, R, AND TARJAN, R. More efficient bottom-up multi-pattern matching in trees. *Theor. Comp. Sci.* **106** (1992) 21–60.
2. DIETZ, P.F. Maintaining order in a linked list. In: *Proc. 14th STOC* (ACM, 1982) 122–127.
3. DIETZ, P.F. AND SLEATOR, D.D. Two algorithms for maintaining order in a list. In: *Proc. 19th STOC* (ACM, 1987) 365–372.
4. HOFFMANN, C. AND O'DONNELL, J. An interpreter generator using tree pattern matching. In: *Proc. 6th. POPL* (ACM, 1979) 169–179.
5. HOFFMAN, C. AND O'DONNELL, J. Pattern matching in trees. *J. ACM* **29**:1 (1982) 68–95.
6. KNUTH, D., MORRIS, J. AND PRATT, V. Fast pattern matching in strings. *SIAM J. Comput.* **6**:2 (1977) 323–350.
7. WEGBREIT, B. Retrieval from context trees. *Inf. Proc. Lett.* **3**:4 (1975) 119–120.

A Parallel Algorithm for Edge-Coloring Partial k-Trees

Xiao Zhou, Shin-ichi Nakano and Takao Nishizeki *

Department of System Information Sciences

Graduate School of Information Sciences

Tohoku University, Sendai 980-77, Japan

Abstract

Many combinatorial problems can be efficiently solved for partial k-trees (graphs of treewidth bounded by k). The edge-coloring problem is one of the well-known combinatorial problems for which no NC algorithms have been obtained for partial k-trees. This paper gives an optimal and first NC parallel algorithm which finds an edge-coloring of a given partial k-tree using a minimum number of colors.

1 Introduction

This paper deals with the edge-coloring problem which asks to color all edges of a given simple graph G, using a minimum number of colors, so that no two adjacent edges are colored with the same color. The minimum number is called the *chromatic index* $\chi'(G)$ of G. Vizing showed that $\chi'(G) = \Delta$ or $\Delta + 1$ for any simple graph G where Δ is the maximum degree of G [9]. The edge-coloring problem arises in many applications, including various scheduling and partitioning problems [9]. The problem is NP-complete [12], and hence it is very unlikely that there exists a sequential algorithm which edge-colors a given graph G with $\chi'(G)$ colors in polynomial time. On the other hand, there exist sequential algorithms which edge-color G with $\Delta + 1$ colors in polynomial time [11, 15, 18]. However, no NC parallel algorithms for edge-coloring G with $\Delta + 1$ colors have been obtained except for the case when Δ is bounded [14].

It is known that many combinatorial problems can be solved very efficiently for partial k-trees or series-parallel graphs [1, 2, 5, 6, 19]. Such a class of problems has been characterized in terms of "forbidden graphs" or "extended monadic logic of second order" [1, 2, 5, 6, 19]. The edge-coloring problem does not belong to such a class of the "maximum (or minimum) subgraph problems," and is indeed one of the "edge-covering problems" which does not appear to be efficiently solved for partial k-trees [5]. However, Bodlaender gave a sequential

---

*E-mail:(zhou|nakano|nishi)@ecei.tohoku.ac.jp

algorithm which solves the edge-coloring problem on a partial k-tree G in time $O(n\Delta^{2^{2(k+1)}})$ where n is the number of vertices in G [3]. Furthermore the current authors have recently obtained a linear-time sequential algorithm [20]. On the other hand, NC parallel algorithms for the edge-coloring problem have not been obtained for partial k-trees, although NC parallel algorithms have been obtained for the following three restricted classes of graphs: planar graphs with maximum degree $\Delta \geq 9$ [8]; outerplanar graphs [7, 10]; and series-parallel multigraphs [21]. Note that outerplanar graphs and series-parallel simple graphs are partial 2-trees.

In this paper we give an optimal and first NC parallel algorithm which solves the edge-coloring problem for partial k-trees with fixed k. Given a partial k-tree G with its decomposition tree, the algorithm finds an edge-coloring of G using $\chi'(G)$ colors in $O(\log n)$ time with $O(n/\log n)$ processors. It is known that a decomposition tree of G can be found in $O(\log^2 n)$ time with $O(n/\log n)$ processors [4, 17]. Our idea is to decompose G of large Δ into several subgraphs $G_1, G_2, \cdots, G_s$ of small maximum degree such that $\chi'(G) = \sum_{j=1}^{s} \chi'(G_j)$, and to extend optimal edge-colorings of these subgraphs to an optimal edge-coloring of G. We construct such a surprising decomposition by partitioning the vertex set into $O(\log n)$ subsets so that the "forward degree" of each vertex is bounded. The parallel computation model we use is a concurrent-read exclusive-write parallel random access machine (CREW PRAM).

2 Terminology and definitions

In this chapter we give some definitions. Let $G = (V, E)$ denote a graph with vertex set V and edge set E. We often denote by $V(G)$ and $E(G)$ the vertex set and the edge set of G, respectively. The paper deals with *simple* graphs without multiple edges or self-loops. An edge joining vertices u and v is denoted by (u, v). The *degree* of vertex $v \in V(G)$ is denoted by $d(v, G)$ or simply by $d(v)$. The *maximum degree* of G is denoted by $\Delta(G)$ or simply by Δ. The graph obtained from G by deleting all vertices in $V' \subseteq V(G)$ is denoted by $G - V'$. The subgraph of G induced by the edges in a subset $E' \subseteq E(G)$ is denoted by $G[E']$.

The class of *k-trees* is defined recursively as follows:

(a) A complete graph with k vertices is a k-tree.

(b) If $G = (V, E)$ is a k-tree and k vertices $v_1, v_2, \cdots, v_k$ induce a complete subgraph of G, then $G' = (V \cup \{w\}, E \cup \{(v_i, w) | 1 \leq i \leq k\})$ is a k-tree where w is a new vertex not contained in G.

(c) All k-trees can be formed with rules (a) and (b).

A graph is a *partial k-tree* if and only if it is a subgraph of a k-tree. Thus partial k-trees are simple graphs. Figure 1(a) illustrates a process of generating a 3-tree, and Figure 1(b) depicts a partial 3-tree. In this paper we assume that k is a constant.

A *decomposition tree* of a graph $G = (V, E)$ is a tree $T = (V_T, E_H)$ with V_T a family of subsets of V satisfying the following properties:

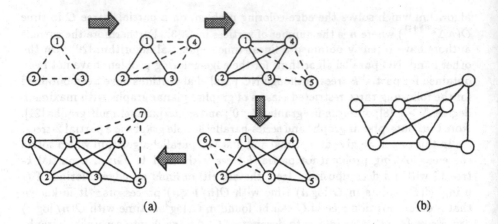

Figure 1: (a) 3-trees and (b) a partial 3-tree.

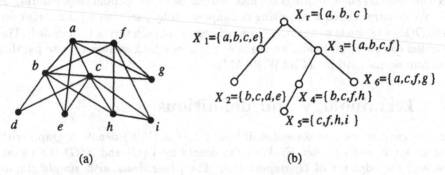

Figure 2: (a) a partial 3-tree and (b) its decomposition tree.

- $\bigcup_{X_i \in V_T} X_i = V$;

- for every edge $e = (v, w) \in E$, there is a node $X_i \in V_T$ with $v, w \in X_i$; and

- if node X_j lies on the path in T from node X_i to node X_l, then $X_i \cap X_l \subseteq X_j$.

Figure 2 illustrates a partial 3-tree and its decomposition tree. The *treewidth* of a decomposition tree $T = (V_T, E_T)$ is $\max_{X_i \in V_T} |X_i| - 1$. The treewidth of G is the minimum treewidth of a decomposition tree of G, taken over all possible decomposition trees of G. It is known that every graph with treewidth $\leq k$ is a partial k-tree, and conversely, that every partial k-tree has a decomposition tree with treewidth $\leq k$.

3 Optimal parallel algorithm

In this section we give an optimal parallel algorithm which edge-colors a partial k-tree G with $\chi'(G)$ colors if a decomposition tree of G is given. Our algorithm runs in $O(\log n)$ time with $O(n/\log n)$ processors.

Bodlaender has given a sequential algorithm which solves the edge-coloring problem for partial k-trees in time $O(n\Delta^{2^{2(k+1)}})$ [3]. A standard dynamic programming algorithm solves the edge-coloring problem for partial k-trees in time $O(n\{\Delta 2^{2(k+1)(\Delta+1)} + (\Delta+1)^{k(k+1)/2}\})$. [1] Such a sequential algorithm can be modified to a parallel algorithm as in the following theorem.

Theorem 3.1 *Let G be a partial k-tree with maximum degree Δ given by its decomposition tree. Then there is a parallel algorithm which solves the edge-coloring problem for G in $O(\log n)$ time using $O(n\{\Delta 2^{6(k+1)(\Delta+1)} + (\Delta+1)^{k(k+1)/2}\}/\log n)$ processors.*

We omit the proof of Theorem 3.1 in this paper for lack of space, but remark that our algorithm applies an extended version of the "tree contraction" technique to the decomposition tree of G.

If Δ is bounded, say $\Delta < 6k$, then $\{\Delta 2^{6(k+1)(\Delta+1)} + (\Delta+1)^{k(k+1)/2}\}$ is also bounded althought it is very large even for small k, and hence the algorithm in Theorem 3.1 is an optimal parallel algorithm. Therefore it suffices to give an optimal parallel algorithm only for the case when $\Delta(G) \geq 6k$.

Our idea is to decompose $G = (V, E)$ to several subgraphs $G_1, G_2, \cdots, G_s$ such that $\chi'(G) = \sum_{j=1}^{s} \chi'(G_j)$ when $\Delta(G) \geq 6k$. Let c be a fixed positive integer, and let $E_1, E_2, \cdots, E_s$ be a partition of E. We say that $E_1, E_2, \cdots, E_s$ is a (Δ, c)-partition of E if $G_j = G[E_j]$, $1 \leq j \leq s$, satisfy

(a) $\Delta(G_j) = c$ for each j, $1 \leq j \leq s - 1$, and $c \leq \Delta(G_s) < 2c$; and

(b) $\Delta(G) = \sum_{j=1}^{s} \Delta(G_j)$.

Since G is a partial k-tree, $G_1, G_2, \cdots, G_s$ are partial k-trees, too. We proved in [20] that $\chi'(G) = \Delta(G)$ if a partial k-tree G has maximum degree $\geq 2k$. Since $\Delta(G) \geq 6k$, $\chi'(G) = \Delta(G)$. Choose $c = 3k$, then $\Delta(G_j) \geq c = 3k$ and hence $\chi'(G_j) = \Delta(G_j)$ for each j, $1 \leq j \leq s$. Since $\Delta(G_j) < 2c = 6k$ for $1 \leq j \leq s$, the algorithm in Theorem 3.1 can find an edge-coloring of G_j with $\Delta(G_j)$ colors in $O(\log n)$ time with $O(n/\log n)$ processors. On the other hand, the condition (b) above implies

$$\chi'(G) = \sum_{j=1}^{s} \chi'(G_j).$$

Therefore edge-colorings of G_j with $\chi'(G_j)$ colors, $1 \leq j \leq s$, can be immediately extended to an edge-coloring of G with $\chi'(G)$ colors. Thus we have the following algorithm.

---

[1] We would like to thank Dr. Petra Scheffler for pointing out this fact.

EDGE-COLOR(G)
{ assume that a decomposition tree T of G is given }
begin

    if $\Delta(G) < 6k$ **then**
1   find an edge-coloring of G with $\chi'(G)$ colors   { Theorem 3.1 }
    else
    begin
2   find a $(\Delta, 3k)$-partition $E_1, E_2, \cdots E_s$ of E;
3   find decomposition trees of $G_1, G_2, \cdots, G_s$ from T;
4   find an edge-coloring of G_j with $\chi'(G_j) = \Delta(G_j)$ colors for each j, $1 \leq j \leq s$;
5   extend these edge-colorings of $G_1, G_2, \cdots, G_s$ to an edge-coloring of G
        with $\chi'(G)$ colors
    end

end;

We have the following two lemmas.

Lemma 3.2 *If a partial k-tree $G = (V, E)$ has the maximum degree $\Delta(G) \geq 6k$, then a $(\Delta, 3k)$-partition of E can be found in $O(\log n)$ time using $O(n/\log n)$ processors.*

Lemma 3.3 *Let T be is a decomposition tree of a partial k-tree G, and let $E_1, E_2, \cdots, E_s$ be a (Δ, c)-partition of E. Then one can find decomposition trees T_j, $1 \leq j \leq s$, of G_j having $O(|E_j|)$ nodes in $O(\log n)$ time using $O(n/\log n)$ processors.*

The proof of Lemma 3.2 is given later, but the proof of Lemma 3.3 is omitted in this paper for lack of space. From Theorem 3.1 and Lemmas 3.2 and 3.3 we have the following theorem.

Theorem 3.4 *The edge-coloring problem can be solved in $O(\log n)$ time using $O(n/\log n)$ processors for a partial k-tree G given by its decomposition tree.*

Proof. Clearly the algorithm above correctly finds an edge-coloring of a partial k-tree G with $\chi'(G)$ colors. Therefore it suffices to prove the time and processor complexities. By Theorem 3.1 and Lemmas 3.2 and 3.3, lines 1, 2 and 3 can be done in $O(\log n)$ time with $O(n/\log n)$ processors. Since $\Delta(G_j) < 6k$, by Theorem 3.1 one can obtain an edge-coloring of G_j with $\Delta(G_j)$ colors in $O(\log|E_j|)$ time and $O(|E_j|)$ computational operations. Hence line 4 requires $O(\log n)$ time and $O(n)$ computational operations in total. Therefore line 4 can be done in $O(\log n)$ time using $O(n/\log n)$ processors. Since $\Delta(G) = \sum_{j=1}^{s} \Delta(G_j)$, one can immediately extend these edge-colorings of $G_1, G_2, \cdots, G_s$ to an edge-coloring of G with $\Delta(G)$ colors in $O(\log n)$ time using $O(n/\log n)$ processors. $Q.\mathcal{E}.\mathcal{D}.$

In the remaining of this section we prove Lemma 3.2. We construct a (Δ, c)-partition of E from a partition $S_1, S_2, \cdots, S_l$ of vertex set V. For each $v \in S_i$, $1 \leq i \leq l$, let

$$
\begin{aligned}
E_b(v, G) &= \{(v, w) \in E \mid w \in S_j \text{ and } j < i\}, \\
E_f(v, G) &= \{(v, w) \in E \mid w \in S_j \text{ and } j \geq i\}, \\
d_b(v, G) &= |E_b(v, G)|, \text{ and} \\
d_f(v, G) &= |E_f(v, G)|.
\end{aligned}
$$

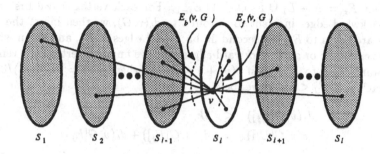

Figure 3: Illustration for notations.

Thus $d(v, G) = d_b(v, G) + d_f(v, G)$. We call the edges in $E_b(v, G)$ *backward edges*, those in $E_f(v, G)$ *forward edges*, $d_b(v, G)$ the *backward degree* of v, and $d_f(v, G)$ the *forward degree* of v. (See Figure 3.) We say that $S_1, S_2, \cdots, S_l$ is a (d_f, c)-*partition* of V if every vertex $v \in V$ satisfies $d_f(v, G) < c$ for a fixed positive integer c. Then we have the following two lemmas.

Lemma 3.5 *More than one third of the vertices in any partial k-tree have degree* $< 3k$.

Proof. Let n' be the number of vertices of a partial k-tree $G = (V, E)$ having degree $< 3k$. Then clearly $3k(|V| - n') \leq 2|E| < 2k|V|$, and consequently $n' > |V|/3$. $\mathcal{Q.E.D.}$

Lemma 3.6 *A $(d_f, 3k)$-partition $S_1, S_2, \cdots, S_l$ of V can be found for every partial k-tree $G = (V, E)$ in $O(\log n)$ time using $O(n/\log n)$ processors, and $l = O(\log n)$.*

Proof. One can easily find a $(d_f, 3k)$-partition of V as follows. Let S_1 be the set of vertices having degree $< 3k$, and delete all vertices in S_1 from G. The resulting graph is a partial k-tree since G is a partial k-tree. Let S_2 be the set of vertices having degree $< 3k$ in the resulting graph, and delete all vertices in S_2. Repeating the same operation, one can find a $(d_f, 3k)$-partition

$S_1, S_2, \cdots, S_l$ of V. By Lemma 3.5 the operation is repeated $O(\log n)$ times, and hence $l = O(\log n)$. Clearly one needs $O(n)$ computational operations to find the partition. Thus the partition can be found in $O(\log n)$ parallel time using $O(n/\log n)$ processors. $\qquad\qquad\qquad\qquad\qquad\qquad\qquad\qquad\qquad\qquad\qquad\mathcal{Q.E.D.}$

We then present a parallel algorithm EDGE-PARTITION to find a $(\Delta, 3k)$-partition $E_1, E_2, \cdots, E_s$ of E, where $s = \lfloor \Delta/3k \rfloor$. We now outline the algorithm before describing it formally. We first construct $E_1, E_2, \cdots, E_{s-1}$ as follows, and then let $E_s = E - E_1 \cup E_2 \cup \cdots \cup E_{s-1}$. For each vertex v, we first number the backward edges incident to v from 1 to $d_b(v, G)$; we then insert the first $3k$ backward edges to E_1, the second $3k$ backward edges to E_2, and so on whenever there remain $3k$ or more backward edges; and we finally insert all the remaining backward edges to $E_{p(v)+1}$ if $p(v) + 1 \le s - 1$, where $p(v) = \lfloor d_b(v, G)/3k \rfloor \le s$. For each j, $1 \le j \le s - 1$, we have

$$d_f(v, G[E_j]) \le 3k,$$
$$d(v, G[E_j]) = d_f(v, G[E_j]) + d_b(v, G[E_j]),$$

and hence

$$d(v, G[E_j]) - 3k \le d_b(v, G[E_j]).$$

If $d(v, G[E_j]) > 3k$, then we adjust the degree $d(v, G[E_j])$ to exactly $3k$ by deleting $d(v, G[E_j]) - 3k$ backward edges incident to v from E_j. More precisely, we first adjust the degrees of vertices in S_l, then vertices in S_{l-1}, and so on. Since a deleted backward edge incident to v in S_i is not incident to any vertex w in $S'_{i'}$, $i' \ge i+1$, the deletion does not change $d(w, G[E_j])$ which has been adjusted not to exceed $3k$. Thus, when the algorithm terminates, we have $\Delta(G[E_j]) \le 3k$ for every j, $1 \le j \le s - 1$. Since $s = \lfloor \Delta/3k \rfloor$ and $E_s = E - E_1 \cup E_2 \cup \cdots \cup E_{s-1}$, we have $3k \le \Delta(G[E_s])$.

We are now ready to describe EDGE-PARTITION formally.

EDGE-PARTITION(G)
{ $G = (V, E)$ is a partial k-tree with maximum degree $\Delta \ge 6k$ }
begin
1   find a $(d_f, 3k)$-partition $S_1, S_2, \cdots, S_l$ of V;   { Lemma 3.6 }
2   $s := \lfloor \Delta/3k \rfloor$;
3   **for each** $v \in V$ **in parallel do**
4   **begin**
5   number the backward edges in $E_b(v, G)$ from 1 to $d_b(v, G)$;
6   let $E_b(v, G) = \{e_{v_q} \mid 1 \le q \le d_b(v, G)\}$
7   **end**;
8   **for each** j, $1 \le j \le s$, **in parallel do** $E_j := \phi$;   { initialize edge-sets E_j }
9   **for each** $v \in V$ **in parallel do** $p(v) := \lfloor d_b(v, G)/3k \rfloor$;
    { construct $E_1, E_2, \cdots, E_{s-1}$ }
10 **for each** $v \in V$ and each j, $1 \le j \le \min\{p(v), s - 1\}$, **in parallel do**

11     $E_j := E_j \cup \{e_{v_q} \in E_b(v,G)|\ 3k(j-1) < q \le 3kj\};$
      { insert to E_j exactly $3k$ backward edges incident to v, and hence $d_b(v, G[E_j]) = 3k$ although probably $d_f(v, G[E_j]) > 0$ and $d(v, G[E_j]) = d_f(v, G[E_j]) + d_b(v, G[E_j]) > 3k$ }

12 **for** each $v \in V$ with $p(v) + 1 \le s - 1$ **in parallel do**

13     $E_{p(v)+1} := E_{p(v)+1} \cup \{e_{v_q} \in E_b(v,G)|\ 3kp(v) < q \le d_b(v,G)\};$
      { insert to $E_{p(v)+1}$ the remaining backward edges incident to v, and hence $d_b(v, G[E_{p(v)+1}]) < 3k$ }

    { adjust $d(v, G[E_j])$ not to exceed $3k$ }

14 **for** $i := l$ **downto** 1 **do**

15   **for** each $v \in S_i$ and each j, $1 \le j \le s - 1$, **in parallel do**
      { $d(v, G[E_j]) - 3k \le d_b(v, G[E_j])$ }

16   **if** $d(v, G[E_j]) > 3k$ **then**

17   delete $d(v, G[E_j]) - 3k$ backward edges in $E_b(v, G[E_j])$ from E_j;
      { $d(v, G[E_j]) = 3k$ }
      { $\Delta(G[E_j]) \le 3k$ }

    { construct E_s }

18   $E_s := E - E_1 \cup E_2 \cup \cdots \cup E_{s-1}$

end

We now prove that $E_1, E_2, \cdots, E_s$ found above is a $(\Delta, 3k)$-partition of E. We first claim that each graph $G_j = G[E_j]$, $1 \le j \le s-1$, satisfies

$$\Delta(G_j) = 3k. \tag{1}$$

Since $\Delta(G_j) \le 3k$, $1 \le j \le s-1$, it suffices to prove that for each j, $1 \le j \le s-1$, there exists a vertex $w \in V(G_j)$ such that $d(w, G_j) = 3k$. By the construction above we have

$$d(v, G_j) = 3k \tag{2}$$

for any vertex v and j, $1 \le j \le \min\{p(v), s-1\}$. In particular, let w be any vertex with $d(w) = \Delta$, then we have

$$p(w) = \left\lfloor \frac{d_b(w, G)}{3k} \right\rfloor = \left\lfloor \frac{\Delta - d_f(w, G)}{3k} \right\rfloor \ge \left\lfloor \frac{\Delta - 3k}{3k} \right\rfloor = \left\lfloor \frac{\Delta}{3k} \right\rfloor - 1 = s - 1 \tag{3}$$

since $d_f(w, G) \le 3k$. By Eqs. (2) and (3) we have $d(w, G_j) = 3k$ for each j, $1 \le j \le s - 1$, completing a proof of Eq. (1).

We then claim that $3k \le \Delta(G_s) < 6k$ and $\Delta(G) = \sum_{j=1}^{s} \Delta(G_j)$. Let w be any vertex with $d(w) = \Delta$. Since $d(w, G_s) = \Delta - 3k(s - 1)$ and $s = \lfloor \Delta/3k \rfloor$, we have $3k \le d(w, G_s) < 6k$. If $d(w, G_s) = \Delta(G_s)$, then

$$3k \le \Delta(G_s) < 6k$$

and

$$\Delta(G) = d(w, G) = 3k(s-1) + \Delta(G_s) = \sum_{j=1}^{s} \Delta(G_j),$$

completing a proof of the claim. Thus it suffices to prove that $d(w, G_s) = \Delta(G_s)$, that is, $d(w, G_s) \geq d(v, G_s)$ for any vertex v. Consider first the case $p(v) + 1 \leq s - 1$. In this case all the edges in $E_b(v, G)$ are once distributed to sets $E_1, E_2, \cdots, E_{p(v)+1}$ at lines 10-13, but some of them are deleted from $E_1, E_2, \cdots, E_{p(v)+1}$ at line 17 and hence by the construction at line 18 $E_b(v, G_s)$ consists of only these deleted backward edges. Furthermore the number of backward edges deleted from E_j satisfies $d(v, G_j) - 3k \leq d_f(v, G_j)$ since $d_b(v, G_j) \leq 3k$. Therefore we have

$$
\begin{aligned}
d(v, G_s) &= d_f(v, G_s) + d_b(v, G_s) \\
&\leq d_f(v, G_s) + \sum_{j=1}^{s-1} d_f(v, G_j) \\
&= d_f(v, G) \\
&\leq 3k \\
&\leq d(w, G_s).
\end{aligned}
$$

Consider next the case $p(v) \geq s - 1$. In this case by Eq. (2) $d(v, G_j) = 3k$ for each j, $1 \leq j \leq s - 1$. By line 18 we have

$$
d(v, G_s) = d(v, G) - \sum_{j=1}^{s-1} d(v, G_j) \leq \Delta - \sum_{j=1}^{s-1} 3k = d(w, G_s).
$$

We now analyze the time and processor complexities. By Lemma 3.6 line 1 can be done in $O(\log n)$ parallel time. Trivially lines 2, 8, 9 and 18 can be done in $O(1)$ parallel time. Lines 3 – 7 and 10 – 13 can be done total in $O(\log n)$ parallel time. Lines 15, 16 and 17 can be easily done in $O(1)$ parallel time. Therefore the **for**-loop at line 14 can be done total in $O(\log n)$ parallel time since $l = O(\log n)$. Thus the algorithm above can be executed total in $O(\log n)$ parallel time. The total number of required computational operations is $O(n)$. Therefore the algorithm above runs in $O(\log n)$ parallel time using $O(n/\log n)$ processors.

Thus we have completed a proof of Lemma 3.2.

The sequential algorithm in [20] uses similar partitions of E and V. However they correspond to a $(\Delta, 2k)$-partition of E and a $(d_f, 2k)$-partition of V in this paper, and l cannot be bounded by $O(\log n)$. Furthermore the construction of $(\Delta, 2k)$-partition of E is sequential and cannot be parallelized.

4 Conclusion

In this paper we gave an optimal and first NC parallel algorithm which solves the edge-coloring problem for partial k-trees with fixed k. Given a partial k-tree G with its decomposition tree, the algorithm runs in $O(\log n)$ time with $O(n/\log n)$ processors where n is the number of vertices in G.

Our algorithm solves a single particular problem, that is, the edge-coloring problem. However the methods which we developed in this paper appear to be useful for many other problems, especially for the "edge-partition problem with respect to property π" which asks to partition the edge set of a given graph into a minimum number of subsets so that the subgraph induced by each subset satisfies the property π. For the edge-coloring problem, π is indeed a matching.

Consider for example a property π: the degree of each vertex v is not greater than $f(v)$, where $f(v)$ is a positive integer assigned to v. Clearly the edge-partition problem with respect to such a property π is the same as the so-called f-coloring problem [13, 16]. Our algorithm can be generalized to solve the f-coloring problem on partial k-trees in parallel.

Acknowledgment

We would like to thank Dr. Hitoshi Suzuki for various discussions and Dr. Petra Scheffler for suggesting us to improve the coefficient. This research is partly supported by Grant in Aid for Scientific Research of the Ministry of Education, Science, and Culture of Japan under a grant number: General Research (C) 05650339.

References

[1] S. Arnborg, B. Courcelle, A. Proskurowski, and D. Seese, *An algebraic theory of graph reduction,* Journal of the Association for Computing Machinery, 40, 5, pp.1134-1164, 1993.

[2] S. Arnborg and J. Lagergren, *Easy problems for tree-decomposable graphs,* Journal of Algorithms, 12, 2, pp.308-340, 1991.

[3] H. L. Bodlaender, *Polynomial algorithms for graph isomorphism and chromatic index on partial k-trees,* Journal of Algorithms, 11, 4, pp.631-643, 1990.

[4] H. L. Bodlaender, *A linear time algorithm for finding tree-decompositions of small treewidth,* Proc. of the 25th Ann. ACM Symp. on Theory of Computing, pp.226-234, San Diego, CA, 1993.

[5] R.B. Borie, R.G. Parker and C.A. Tovey, *Automatic generation of linear-time algorithms from predicate calculus descriptions of problems on recursively constructed graph families,* Algorithmica, 7, pp.555-581, 1992.

[6] B. Courcelle, *The monadic second-order logic of graphs I: Recognizable sets of finite graphs,* Information and Computation, 85, pp.12-75, 1990.

[7] Y. Caspi and E. Dekel, *A near-optimal parallel algorithm for edge-coloring outerplanar graphs,* manuscript, Computer Science Program, University of Texas at Dallas, Richardson, Tx., 1992.

[8] M. Chrobak and T. Nishizeki, *Improved edge-coloring algorithms for planar graphs*, J. Algorithms, 11, pp.102-116, 1990.

[9] S. Fiorini and R.J. Wilson, *Edge-Colouring of Graphs*, Pitman, London, 1977.

[10] A. Gibbons and W. Rytter, *Efficient Parallel Algorithms*, Cambridge Univ. Press, Cambridge, 1988.

[11] H. N. Gabow, T. Nishizeki, O. Kariv, D. Leven and O. Terada, *Algorithm for edge-coloring graphs*, Tech. Rep. TRECIS-8501, Tohoku Univ., 1985.

[12] I.J. Holyer, *The NP-completeness of edge-coloring*, SIAM J. on Computing, 10, pp.718-720, 1981.

[13] S. L. Hakimi and O. Kariv, *On a generalization of edge-coloring in graphs*, Journal of Graph Theory, 10, pp.139-154, 1986.

[14] H. J. Karloff and D. B. Shmoys, *Efficient parallel algorithms for edge-coloring problems*, Journal of Algorithms, 8, pp.39-52, 1985.

[15] T. Nishizeki and N. Chiba, *Planar Graphs: Theory and Algorithms*, North-Holland, Amsterdam, 1988.

[16] S. Nakano, T. Nishizeki and N. Saito, *On the f-coloring multigraphs*, IEEE Transactions on Circuits and Systems, Vol. 35, No. 3, pp. 345-353, 1988.

[17] B.A. Reed, *Finding approximate separators and computing tree-width quickly*, Proc. of the 24th Ann. ACM Symp. on Theory of Computing, pp. 221-228, 1992.

[18] O. Terada and T. Nishizeki, *Approximate algorithms for the edge-coloring of graphs*, Trans. Inst. of Electronics and Communication Eng. of Japan, J65-D, 11, pp. 1382-1389, 1982.

[19] K. Takamizawa, T. Nishizeki, and N. Saito, *Linear-time computability of combinatorial problems on series-parallel graphs*, Journal of the Association for Computing Machinery, 29, 3, pp. 623-641, 1982.

[20] X. Zhou, S. Nakano and T. Nishizeki, *A linear algorithm for edge-coloring partial k-trees*, Proc. of First Europian Symposium on Algorithms, Lect. Notes in Computer Science, Springer-Verlag, 726, pp. 409-418, 1993.

[21] X. Zhou, H. Suzuki and T. Nishizeki, *Sequential and parallel algorithms for edge-coloring series-parallel multigraphs*, Proc. of Third Conf. on Integer Programming and Combinatorial Optimization, pp. 129-145, 1993.

Dominating Cliques in
Distance-Hereditary Graphs *

Feodor F. Dragan

Dept. of Math. and Cybern. Moldova State University
A. Mateevici str. 60 Chişinău 277009 Moldova

Abstract. A graph is distance-hereditary if and only if each cycle on five or more vertices has at least two crossing chords. We present linear time algorithms for the minimum r-dominating clique and maximum strict r-packing set problems on distance-hereditary graphs. Some related problems such as diameter, radius, central vertex, r-dominating by cliques and r-dominant clique are investigated too.

1 Introduction

A subset $D \subset V$ is a *dominating set* in graph $G = (V, E)$ iff for all vertices $v \in V \setminus D$ there is a vertex $u \in D$ with $uv \in E$. It is a *dominating clique* in G iff D is a dominating set in G and a clique (i.e. for all $u, v \in D$ $uv \in E$).

There are many papers investigating the problem of finding minimum dominating sets in graphs with (and without) additional requirements to the dominating sets. The problems are in general NP-complete. For more special graphs the situation is sometimes better (for a bibliography on domination cf. [12], for a recent survey on special graph classes cf. [4]).

Opposite to dominating set for a given graph G a dominating clique does not necessarily exist. As was shown in [6] the problem whether a given graph has a dominating clique is NP-complete even for weakly chordal graphs (a graph G is *weakly chordal* iff G does not contain induced cycles C_k of length $k \geq 5$ and no complements $\overline{C_k}$ of such cycles).

[14] and [15] investigate the dominating clique problem on strongly chordal and chordal graphs. A graph G is *chordal* iff it does not contain any induced cycle C_k of length $k \geq 4$. A graph G is *strongly chordal* iff it is chordal and each cycle in G of even length at least 6 has an odd chord (a chord x_i, x_j in a cycle $C = (x_1, \ldots, x_{2k})$ of even length $2k$ is an *odd chord* if the distance in C between x_i and x_j is odd). Although for chordal graphs there is a simple criterion for the existence of dominating cliques the problem of finding a minimum (cardinality) dominating clique is NP-complete. In the case of strongly chordal graphs the last problem is already polynomial-time solvable.

*This work was partially supported by the VW–Stiftung Project No. I/69041
e-mail address: dragan@university.moldova.su

Here we study the following generalized domination (*r-domination*) problem: Let $(r(v_1), \ldots, r(v_n))$ be a sequence of non-negative integers which is given together with the input graph. For any two vertices u, v denote by $dist(u, v)$ the length (i.e. number of edges) of a shortest path between u and v in G. A subset $D \subset V$ is an *r-dominating set* in G iff for all $v \in V \setminus D$ there is a $u \in D$ with $dist(u, v) \leq r(v)$. It is an *r-dominating clique* in G iff D is additionally a clique.2

[9] investigates the *r*-dominating clique problem on Helly, chordal and dually chordal graphs. For the definition of dually chordal graphs see [5]. The condition for the existence of dominating cliques known from [15] on chordal graphs is shown to be valid also in more general case of *r*-domination in Helly graphs and in chordal graphs. Again the problem of finding a minimum *r*-dominating clique is *NP*-complete in Helly graphs and is linear-time solvable in dually chordal graphs (a superclass of strongly chordal graphs and subclass of Helly graphs).

In this paper we investigate the *r*-dominating clique problem on distance-hereditary graphs. A *distance-hereditary graph* is a connected graph in which every induced path is isometric. That is, the distance of any two vertices in an induced path equals their distance in the graph. These graphs were introduced by E.HOWORKA [13], who gave first characterizations of distance-hereditary graphs. For instance, a connected graph G is distance-hereditary if and only if every circuit in G of length at least 5 has a pair of chords that cross each other. Evidently every distance-hereditary graph is weakly chordal. We show that the condition for the existence of *r*-dominating cliques known from [9] on Helly graphs and chordal graphs is still valid in the case of distance-hereditary graphs. Also we give efficient algorithms for *the minimum r-dominating clique problem* and some related problems (such as *diameter, radius, central vertex, r-dominating by cliques* and *r-dominant clique*) on these graphs.

As we already mentioned the minimum *r*-dominating clique problem is *NP*-complete on chordal graphs and linear-time solvable in dually chordal graphs. Opposite to this problem the problems of *r*-dominating by cliques and *r*-dominant clique are *NP*-complete on dually chordal graphs and polynomial time solvable in chordal graphs (see [9]). So, the obtained results show that the distance-hereditary graphs possess advantages of both chordal and dually chordal classes of graphs.

2 Terminology and Basic Properties

We shall consider finite, simple loopless, undirected and connected graph $G = (V, E)$, where $V = \{v_1, \ldots, v_n\}$ is the vertex set and E is the edge set of G, and we shall use more-or-less standard terminology from graph theory [10].

Let v be a vertex of G. We denote the *neighborhood* of v, consisting of all vertices adjacent to v, by $N(v)$, and the *closed neighborhood* of v, the set $N(v) \cup \{v\}$, by $N[v]$. The *k-th neighborhood* of v, denoted by $N^k(v)$, is defined as the set of all vertices of distance k to v, that is, $N^k(v) = \{u \in V : dist(u, v) = k\}$. The *disk* centered at v with radius k is the set of all vertices having distance at most k to v: $D^k(v) = \{u \in V : dist(u,v) \leq k\}$. Let also $N^k(v,u) = N^k(v) \cap N^{dist(u,v)-k}(u)$.

A vertex v of G is a *leaf* if $|N(v)| = 1$. Two vertices v and u are *twins* if they have the same neighborhood $(N(v) = N(u))$ or the same closed neighborhood $(N[v] = N[u])$. *True twins* are adjacent, *false twins* are not. We denote with $< S >$ the subgraph of G induced by the vertices of $S \subset V$. A *cograph* is a graph that does not contain any induced path of length three.

Several interesting characterizations of distance-hereditary graphs in terms of existence of particular kinds of vertices (leaves, twins) and in terms of metric and neighborhood properties, and forbidden configurations were provided by BANDELT and MULDER [2], and by D'ATRI and MOSCARINI [8]. Some algorithmic aspects are considered in [11] and [8]. The following propositions list the basic information on distance-hereditary graphs that is needed in the sequel.

Proposition 1 ([2],[8]) *For a graph G the following conditions are equivalent:*

(1) *G is distance-hereditary.*

(2) *The house, domino, fan (see Fig. 1) and the cycles C_k of length $k \geq 5$ are not induced subgraphs of G.*

(3) *Every induced subgraph of G contains a leaf or a pair of twins.*

(4) *For arbitrary vertex x of G and every pair of vertices $v, u \in N^k(x)$, that are in the same connected component of the graph $< V \setminus N^{k-1}(x) >$, we have*

$$N(v) \cap N^{k-1}(x) = N(u) \cap N^{k-1}(x).$$

(5) *(4-point condition)*
   For any four vertices u, v, w, x of G at least two of the following distance sums are equal:

$$dist(u,v) + dist(w,x); \ dist(u,w) + dist(v,x); \ dist(u,x) + dist(v,w).$$

   If the smaller sums are equal, then the largest one exceeds the smaller ones at most by 2.

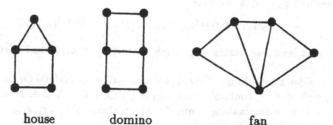

2   house   domino   fan
Figure 1: Forbidden induced subgraphs in a distance-hereditary graph.

Proposition 2 ([2],[3]) *Let G be a distance-hereditary graph.*

(1) *Every three pairwise intersecting in G disks have a nonempty intersection.*

(2) *For any vertex v of G there is no induced path of length 3 in the graph $< N^k(v) >$, i.e. every connected component of $< N^k(v) >$ is a cograph.*

(3) *For any vertex v of G if u, w are vertices in different components of $< N^k(v) >$, then $N(u) \cap N^{k-1}(v)$ and $N(w) \cap N^{k-1}(v)$ are either disjoint, or one of the two sets is contained in the other.*

3 Existence of r-Dominating Cliques

In this section we give a simple criterion for the existence of an r-dominating clique in a given distance-hereditary graph. This criterion is the same as in Helly or in chordal graphs [9].

We say that a subset M of V has an r-dominating clique C iff for every vertex $v \in M$ $dist(v, C) \leq r(v)$ holds, where $dist(v, C) = min\{dist(v, w) : w \in C\}$.

Theorem 1 *) *Let $G = (V, E)$ be a distance-hereditary graph with n-tuple $(r(v_1), \ldots, r(v_n))$ of non-negative integers and $M \subseteq V$ be any subset of V. Then M has an r-dominating clique C if and only if for every pair of vertices $v, u \in M$, the inequality*

$$dist(v, u) \leq r(v) + r(u) + 1$$

holds. Moreover such a clique C can be determined within time $O(|M| \cdot |E|)$.

Proof: " $\Longrightarrow$ " is obvious.

" $\Longleftarrow$ ": Assume that $v_1, \ldots, v_n$ is an ordering of V such that M consist of the first $|M|$ vertices of this ordering. Let i be the largest index such that there is a clique C with property $dist(v_j, C) \leq r(v_j)$ for all vertices $v_j \in M$, $1 \leq j \leq i$. If $i < |M|$ then for $v_{i+1} \in M$ $dist(v_{i+1}, C) \geq r(v_{i+1}) + 1$ holds.

Let $N^k(v, u) = N^k(v) \cap N^{dist(u,v)-k}(u)$. Consider the set

$$X = \bigcup_{x \in C} N^{r(v_{i+1})+1}(v_{i+1}, x).$$

Since $X \subseteq N^{r(v_{i+1})+1}(v_{i+1})$ and $C \subseteq V \setminus N^{r(v_{i+1})}(v_{i+1})$ vertices of X belong to the same connected component of graph $< V \setminus N^{r(v_{i+1})}(v_{i+1}) >$. By Proposition 1 (4) for any two vertices $v', v'' \in X$ we have

$$N(v') \cap N^{r(v_{i+1})}(v_{i+1}) = N(v'') \cap N^{r(v_{i+1})}(v_{i+1}).$$

So, there is at least one vertex u_{i+1} such that $X \subset N(u_{i+1})$ and $dist(u_{i+1}, v_{i+1}) = r(u_{i+1})$.

Now consider a new clique $C' \cup \{u_{i+1}\}$, where C' is a maximal (w.r.t. set inclusion) clique of graph $< X >$, containing the clique $C \cap X$ if $C \cap X \neq \emptyset$. Next we show that $C' \cup \{u_{i+1}\}$ is an r-dominating clique for the vertices $v_1, \ldots, v_i, v_{i+1}$. This will be a contradiction to the maximality of i.

Pick an arbitrary index j, $j \leq i$, and let x_j be a vertex from C with $dist(x_j, v_j) \leq r(v_j)$. The following cases may arise.

---

*) This theorem was independently proven by F. NIKOLAI [Algorithmische und strukturelle Aspekte distanz-erblicher Graphen, Dissertation Universität Duisburg, 1994]

Case (1). Either $dist(x_j, v_{i+1}) = r(v_{i+1}) + 2$ and $dist(x_j, v_j) < r(v_j)$ or $dist(x_j, v_{i+1}) > r(v_{i+1}) + 2$.

Since $dist(v_{i+1}, v_j) \leq r(v_{i+1}) + r(v_j) + 1$ the disks

$$D^{r(v_{i+1})+2}(v_{i+1}), \quad D^{r(v_j)-1}(v_j), \quad D^{dist(x_j, v_{i+1})-r(v_{i+1})-2}(x_j)$$

are pairwise intersecting. (Note that the disks $D^p(v)$ and $D^q(u)$ intersect if and only if $dist(v, u) \leq p + q$.) By Proposition 2 (1) there exists a vertex z with properties $dist(v_{i+1}, z) = r(v_{i+1}) + 2$, $dist(x_j, z) + dist(z, v_{i+1}) = dist(x_j, v_{i+1})$ and $dist(v_j, z) \leq r(v_j) - 1$. We claim that z is adjacent to all vertices from X.

Let u be an arbitrary vertex of X. Hence $u \in N^{r(v_{i+1})+1}(v_{i+1}, x)$ for some vertex $x \in C$. Evidently, vertex $u' \in N^1(u, x)$ and vertex z lie in the same connected component of graph $< V \setminus N^{r(v_{i+1})+1}(v_{i+1}) >$. Applying once more Proposition 1 (4) we obtain that $N(u') \cap N^{r(v_{i+1})+1}(v_{i+1}) = N(z) \cap N^{r(v_{i+1})+1}(v_{i+1})$, i. e. vertices u and z are adjacent. Thus we have $X \subset N(z)$ and so $dist(v_j, C') \leq dist(v_j, z) + 1 \leq r(v_j)$.

Case (2). $dist(x_j, v_{i+1}) = r(v_{i+1}) + 2$ and $dist(x_j, v_j) = r(v_j)$.

Now consider the disks

$$D^{r(v_{i+1})+1}(v_{i+1}), \quad D^{r(v_j)}(v_j), \quad D^1(x_j).$$

They are pairwise intersecting. Again by Proposition 2 (1) there is a vertex z such that $dist(v_{i+1}, z) = r(v_{i+1}) + 1$, $dist(x_j, z) + dist(z, v_{i+1}) = dist(x_j, v_{i+1})$ and $dist(v_j, z) \leq r(v_j)$. Since $z \in X$ vertices z and u_{i+1} are adjacent. Also as in case (1) we can show that $X \subset N(x_j)$.

If $dist(z, v_j) < r(v_j)$ then we immediately obtain $dist(v_j, u_{i+1}) \leq r(v_j)$. So, assume that $dist(x_j, v_j) = dist(z, v_j) = r(v_j)$ and $dist(v_j, C' \cup \{u_{i+1}\}) \geq r(v_j) + 1$. Since $X \subset N(x_j)$ in fact we have $C' \cup \{u_{i+1}\} \subseteq N^{r(v_j)+1}(v_j)$. By Proposition 1 (4) for any two vertices $u', u'' \in C' \cup \{u_{i+1}\}$ $N(u') \cap N^{r(v_j)}(v_j) = N(u'') \cap N^{r(v_j)}(v_j)$ holds. That is, $C' \cup \{u_{i+1}\} \subset N(z)$. This implies a contradiction with maximality of clique $C' \subset X$.

Case (3). $dist(x_j, v_{i+1}) = r(v_{i+1}) + 1$.

In this case the inequality $dist(v_j, C') \leq r(v_j)$ follows from the choice of clique C' (C' contains all vertices from $C \cap X$).

Time bound.
i-th step: The set X can be determined within time $O(|E|)$. A maximal clique $C' \subseteq X$ and vertex u_{i+1} can be found also within time $O(|E|)$.

There are at most $|M|$ such steps, and each step requires at most $O(|E|)$ time. $\square$

The *eccentricity* of vertex $v \in V$ is $e(v) = max\{dist(v, u) : u \in V\}$. The *diameter* $diam(G)$ of G is the maximum eccentricity, while the *radius* $radi(G)$ of G is the minimum eccentricity of vertices of G.

Corollary 1 *Let $r(v) = k$ for all $v \in V$. Then a distance-hereditary graph G has a k-dominating clique if and only if $diam(G) \leq 2k + 1$.*

Corollary 2 *Let G be a distance-hereditary graph. Then G has a dominating clique if and only if $diam(G) \leq 3$.*

Corollary 3 ([16]) *For every distance-hereditary graph G we have $diam(G) \geq 2radi(G) - 2$.*

For the case of chordal graphs Corollary 2 initially occurs in [15].

4 Minimum r-Dominating Cliques

Let $\gamma_r(G) = min\{|D| : D$ an r-dominating set in $G\}$, $\gamma_{r-clique}(G) = min\{|D| : D$ an r-dominating clique in $G\}$. The problem dual to the minimum r-dominating clique problem is: for a given graph G with n-tuple $(r(v_1), \ldots, r(v_n))$ of non-negative integers find a maximum cardinality vertex set P such that for all $v, u \in P$ $dist(u, v) = r(u) + r(v) + 1$ holds. We call such a set P a *strict r-packing set* of G and denote by $\pi_r(G)$ the cardinality of a maximum strict r-packing set. It is clear that for every graph G which has an r-dominating clique

$$\pi_r(G) \le \gamma_r(G) \le \gamma_{r-clique}(G)$$

holds but in general the parameters do not coincide. The next results show the coincidence of some of these parameters for distance-hereditary graphs that have an r-dominating clique.

We say that an r-dominating clique C is *minimal* iff for any vertex $v \in C$ the set $C \setminus \{v\}$ is not r-dominating.

Theorem 2 *Let G be a distance-hereditary graph which has an r-dominating clique for n-tuple $(r(v_1), \ldots, r(v_n))$ of non-negative integers. If $\gamma_{r-clique}(G) > 1$ then every minimal r-dominating clique of G is also a minimum one.*

Corollary 4 *Let G be a distance-hereditary graph which has an r-dominating clique for the n-tuple $(r(v_1), \ldots, r(v_n))$ of non-negative integers.*

(1) *If C is a minimal r-dom. clique of G and $|C| \ne 2$ then $\gamma_{r-clique}(G) = |C|$.*

(2) *If $\gamma_{r-clique}(G) \ne 2$ then $\gamma_{r-clique}(G) = \pi_r(G)$.*

(3) *If $\pi_r(G) > 1$ then $\gamma_{r-clique}(G) = \pi_r(G)$.*

(4) *$\gamma_r(G) = \gamma_{r-clique}(G)$.*

Corollary 5 *For every system of pairwise intersecting disks of a distance-hereditary graph G there exists either a single vertex or a pair of adjacent vertices that meet all disks of this system.*

Now we are able to present an $O(|V| \cdot |E|)$ algorithm for finding a minimum r-dominating clique in distance-hereditary graphs. For a distance-hereditary graph $G = (V, E)$ given together with n-tuple $(r(v_1), \ldots, r(v_n))$ first we can decide within time $O(|V| \cdot |E|)$ whether graph G has a single vertex r-dominating all other vertices of G. If such a vertex does not exist then we can find also within time $O(|V| \cdot |E|)$ (see Theorem 1) an r-dominating clique C of G (or decide that such a clique does not exist). Finally we can reduce this clique to a minimal one in time $O(|C| \cdot |E|)$. By Theorem 2 the obtained r-dominating clique is a minimum r-dominating clique of graph G.

Observe that a maximum strict r-packing set of a distance-hereditary graph G can be determined also within time $O(|V| \cdot |E|)$ (see Theorem 2).

As we will show in the next section, using the existence of special kinds of vertices (leaves and twins) in distance-hereditary graphs, one can construct a more efficient algorithm for these two problems.

5 Algorithm for Finding Minimum r-Dominating Cliques

As we already mentioned, every induced subgraph of distance-hereditary graph contains a leaf or a pair of twins. In the subsequent linear-time algorithm for the minimum r-dominating clique problem on distance-hereditary graphs these kinds of vertices turn out to be of importance.

Lemma 1 ([9]) *Let x be a leaf in graph G and let y be its neighbor. Let also $G - x = <V \setminus \{x\}>$ and $r(x) \geq 1$. A clique $C \subseteq V \setminus \{x\}$ is a minimum r-dominating clique of graph G if C is a minimum r'-dominating clique of graph $G - x$ with $r'(v) = r(v)$ when $v \neq y$ and $r'(y) = min\{r(y), r(x) - 1\}$. In particular, the graph G has an r-dominating clique if and only if the graph $G - x$ has an r'-dominating clique.*

Instead of twins next we consider some of their generalizations. A *homogeneous set* of a graph $G = (V, E)$ is a set of the vertices A such that every vertex in $V \setminus A$ is adjacent to either all or none of the vertices of A. A *proper homogeneous set* is a homogeneous set A such that $|A| \leq |V| - 2$. Observe that two vertices are twins if they form a homogeneous set of size 2. Evidently every vertex $v \in V \setminus A$ is equidistant from the vertices of a homogeneous set A.

Lemma 2 *Let $A \subset V$ be a proper homogeneous set of graph G, and x be a vertex of A with $r(x) = min\{r(y) : y \in A\}$. Assume that either $r(x) \geq 2$, or $r(x) = 1$ and there exists a vertex $v \in V \setminus A$ with $dist(v, x) > r(v)$. A clique $C \subseteq V \setminus A \cup \{x\}$ is a minimum r-dominating clique of graph G if C is a minimum r-dominating clique of graph $< V \setminus A \cup \{x\} >$. In particular, the graph G has an r-dominating clique if and only if the graph $< V \setminus A \cup \{x\} >$ has an r-dominating clique.*

Lemma 3 *Let x be a vertex of graph G such that $N(x)$ forms a proper homogeneous set in G, and y be a vertex of $N(x)$ with $r(y) = min\{r(z) : z \in N(x)\}$. Assume that $r(x) > r(y)$. A clique $C \subseteq V \setminus \{x\}$ is a minimum r-dominating clique of graph G if C is a minimum r-dominating clique of graph $G - x$. In particular, the graph G has an r-dominating clique if and only if the graph $G - x$ has an r-dominating clique.*

These lemmas will be used in the correctness proof of the subsequent algorithm which has a structure similar to the linear-time recognition algorithm presented in [11] for distance-hereditary graphs.

Algorithm RDC (Find a minimum r-dominating clique of a distance-hereditary graph if there is one, and answer NO otherwise)

Input: A distance-hereditary graph $G = (V, E)$ and an n-tuple $(r(v_1), \ldots, r(v_n))$ of non-negative integers.

Output: A minimum r-dominating clique DC of G if there is one and answer NO otherwise.

begin

(1) if for all $v \in V$ $r(v) > 0$ then

(2)   for arbitrary vertex $u \in V$ build its i-neighborhoods $N^1(u), N^2(u), \ldots, N^k(u)$;

(3)   for $i = k, k-1, \ldots, 2$ do

(4)   find the connected components $A_1, A_2, \ldots, A_p$ of $N^i(u) \cap V$;

(5)   in each component A_j pick a vertex x_j such that $r(x_j) = min\{r(y) : y \in A_j\}$;

(6)   order the vertices of $X = \{x_1, x_2, \ldots, x_p\}$ by increasing degree
$d'(x_j) = |N(x_j) \cap N^{i-1}(u)|$;

(7)   for all vertices $x_j \in X$ taken by increasing degree $d'(x_j)$ do

(8)   if $(r(x_j) \geq 2$ or $r(x_j) = 1$ and $\exists v \in V \setminus A_j$ $dist(x_j, v) > r(v))$ then

(9)   delete from V all vertices of $A_j \setminus \{x_j\}$;

(10)   in set $B = N(x_j) \cap N^{i-1}(u) \cap V$ pick a vertex y such that
$r(y) = min\{r(z) : z \in B\}$;

(11)   **do case**

(12)   case $r(y) = 1$ and $r(x_j) \geq 2$;

(13)   delete from V vertex x_j

(14)   case $r(y) \geq 2$ or $r(y) = r(x_j) = 1$ and $\exists v \in V \setminus B$ $dist(y, v) > r(v)$;

(15)   delete from V all vertices of $(B \cup \{x_j\}) \setminus \{y\}$;

(16)   put $r(y) := min\{r(x_j) - 1, r(y)\}$;

(17)   if $r(y) = 0$ then goto *outloop* endif

(18)   otherwise $(r(y) = r(x_j) = 1$ and vertex y r-dominates all vertices of
$V \setminus B)$

(19)   choose in set B a vertex w adjacent to all vertices $v \in B \setminus \{w\}$ with
$r(v) = 1$;

(20)   if (there is such a vertex) then $DC := \{w\}$ else $DC := \{x_j, y\}$
endif

(21)   **stop**

(22)   **endcase**

(23)   else $(r(x_j) = 1$ and vertex x_j r-dominates all vertices of $V \setminus A_j)$

(24)   choose in set A_j a vertex w adjacent to all vertices $v \in A_j \setminus \{w\}$ with
$r(v) = 1$;

(25)   if (there is such a vertex) then $DC := \{w\}$

(26)   else choose in set $B = N(x_j) \cap N^{i-1}(u) \cap V$ a vertex w adjacent to all
vertices $v \in B \setminus \{w\}$ with $r(v) = 1$;

(27)   if (there is such a vertex and it r-dominates also all vertices from $V \setminus B)$

(28)   then $DC := \{w\}$

(29)   else $DC := \{x_j, y\}$, where y is an arbitrary vertex from B;

(30)   endif

(31)   endif

(32)   **stop**

(33)   endif

(34)   endfor

(35)   endfor

(36)   stop with output $DC := \{u\}$

(37) else (now $r(u) = 0$ for some $u \in V$)

outloop:

(38)   in the rest of graph G build i-neighborhoods $N^1(u), \ldots, N^k(u)$ of vertex u with
$r(u) = 0$;

(39)   for $i = k, k-1, \ldots, 2$ do

(40)   repeat the steps (4),(5),(6);

```
(41)        for all vertices x_j ∈ X taken by increasing degree d'(x_j) do
(42)          if r(x_j) ≥ 1 then delete from V all vertices of A_j;
(43)            in set B = N(x_j) ∩ N^{i-1}(u) ∩ V pick a vertex y such that
                 r(y) = min{r(z) : z ∈ B};
(44)            if r(y) ≥ 1 then delete from V all vertices v ∈ B \ {y};
(45)              put r(y) := min{r(x_j) - 1, r(y)}
(46)            endif
(47)          else (r(x_j) = 0)
(48)            stop with output "there is no r-dominating clique in G"
(49)          endif
(50)        endfor
(51)      endfor      (now N[u] = V)
(52)      put DC := {v ∈ N[u] : r(v) = 0};
(53)      if DC is no clique then output "there is no r-dominating clique in G"
(54)      else DC is a minimum r-dominating clique of G
(55)      endif
(56) endif
end
```

Theorem 3 *Algorithm RDC is correct and works in linear time $O(|E|)$.*

The following algorithm solves in linear time the maximum strict r-packing problem on distance-hereditary graph having an r-dominating clique.

Algorithm RSP (Find a maximum strict r-packing set of a distance-hereditary graph which has an r-dominating clique)

Input: A distance-hereditary graph $G = (V, E)$ and an n-tuple $(r(v_1), \ldots, r(v_n))$ of non-negative integers.

Output: A maximum strict r-packing set SP of G.

```
begin
(1)  by Algorithm RDC find a minimum r-dominating clique DC of graph G;
(2)  if |DC| = 1 then SP := DC
(3)  else W := V; i := 1;
(4)    for all v ∈ DC do
(5)      num(v) := 1; l(v) := 0; pr(v) := v; W := W \ {v}
(6)    endfor
(7)    for all v ∈ V with dist(v, DC) = 1 do
(8)      num(v) := |N(v) ∩ DC|; l(v) := 1; W := W \ {v};
(9)      if num(v) := 1 then pr(v) := N(v) ∩ DC and g(v) := v endif
(10)   endfor
(11)   while W ≠ ∅ do
(12)     for all vertices x with l(x) = i do
(13)       for all vertices y from N(x) ∩ W do
(14)         num(y) := num(x); l(y) := i + 1; W := W \ {y};
(15)         if num(y) := 1 then pr(y) := pr(x) and g(y) := g(x) endif
(16)       endfor
(17)     endfor
```

```
(18)      i := i + 1;
(19)      enddo
(20)      for all v ∈ V do
(21)          if num(v) = 1 and l(v) = r(v) then pn(pr(v)) := v and pn(g(v)) := v endif
(22)      endfor
(23)      if |DC| ≥ 3 then SP := {pn(u) : u ∈ DC};
(24)      else (|DC| = 2)
(25)          let x and y be vertices from DC;
(26)          do case
(27)              case r(x) = 0;
(28)                  SP := {x, pn(y)}
(29)              case r(y) = 0;
(30)                  SP := {pn(x), y}
(31)              otherwise (r(y) ≥ 1, r(x) ≥ 1)
(32)                  X := {g(v) : num(v) = 1 and l(v) = r(v) and pr(v) = x};
(33)                  Y := {g(v) : num(v) = 1 and l(v) = r(v) and pr(v) = y};
(34)                  if (there is a pair of non-adjacent vertices u ∈ X and w ∈ Y) then
                              SP := {pn(u), pn(w)}
(35)                  else SP := {x}
(36)                  endif
(37)          endcase
(38)      endif
(39) endif
end
```

Theorem 4 *Algorithm RSP is correct and works in linear time $O(|E|)$.*

6 Related Problems

6.1 Central Vertex, Radius and Diameter

Recall that a vertex whose eccentricity is equal to $radi(G)$ is called *a central vertex* of graph G. Denote by $F(v)$ the set of all furthest from v vertices of G, i.e. $F(v) = \{u \in V : dist(v, u) = e(v)\}$.

Lemma 4 *For any vertex v of a distance-hereditary graph G and any furthest vertex $u \in F(v)$ we have $e(u) \geq 2radi(G) - 3$.*

Now we present linear-time algorithms for computing a central vertex, a central clique, the radius and diameter of a distance-hereditary graph G.

Let $G = (V, E)$ be a distance-hereditary graph. According to lemma 4 the value $k = [(e(u) + 1)/2]$ is an approximation of the radius of G, more precisely either $k = radi(G)$ or $k = radi(G) - 1$. Next we apply the algorithm RDC to graph G with $r(v) = k$ for all $v \in V$. If there exists an r-dominating clique in G then any vertex v from a minimum r-dominating clique DC is a central vertex of G, and $radi(G) = k$ when $|DC| = 1$ and $radi(G) = k + 1$ otherwise. If such a clique does not exist then $radi(G) = k + 1$ and a single vertex of a minimum r-dominating clique of G with $r(v) = k + 1$ for all $v \in V$ must be central.

For computing the diameter of G additionally to algorithm RDC we apply the algorithm RSP to graph G with $r(v) = radi(G) - 1$ for all $v \in V$. If there exists an r-dominating clique then $diam(G) = 2radi(G) - 1$ when the maximum strict r-packing set SP of G has at least two vertices and $diam(G) = 2radi(G) - 2$ otherwise. If such a clique does not exist then $diam(G) = 2radi(G)$.

The algorithm RDC can be used for finding a central clique of a distance-hereditary graph too. The *eccentricity* $e(C)$ of a clique C is the minimum distance from any vertex to C. A *clique center* of G is a clique with minimum eccentricity which is called the *clique radius* of G and is denoted by $cradi(G)$. For arbitrary graph G we have $radi(G) \geq cradi(G) \geq radi(G) - 1$, because the eccentricity of any clique containing a central vertex of G is at most $radi(G)$. So, it is sufficient to decide whether G contains an r-dominating clique with $r(v) = radi(G) - 1$ for all $v \in V$. Summarizing the results of this subsection we obtain

Theorem 5 *The radius, the diameter, a central vertex and a central clique of a distance-hereditary graph can be found in linear time.*

6.2 r-Dominating by Cliques and r-Dominant Cliques

Since a graph does not necessarily have a dominating clique in [9] we consider the following weaker *r-domination by cliques* problem: Given a graph $G = (V, E)$ with radius function $(r(v_1), \ldots, r(v_n))$ of non-negative integers find a minimum number of cliques $C_1, \ldots, C_k$ such that $\bigcup_{i=1}^{k} C_i$ r-dominates G. Note that for the special case $r(v_i) = 0$ for all $i \in \{1, \ldots, n\}$ this is the well-known problem *clique partition*.

Another problem closely related to that is to find a clique in G which r-dominates a maximum number of vertices. We call this the *r-dominant clique* problem. For the special case $r(v_i) = 0$ for all $i \in \{1, \ldots, n\}$ this is again a well-known problem namely the *maximum clique* problem.

It is obvious that these two problems are NP-complete. The following results show that for distance-hereditary graphs the problems are solvable in polynomial time.

For a graph $G = (V, E)$ with disks $\mathcal{D}(G) = \{D^k(v) : v \in V \text{ and } k \geq 0 \text{ a non-negative integer }\}$ let $\Gamma(G)$ be the following graph whose vertices are the disks of G and two disks $D^p(v), D^q(u)$ are adjacent iff $D^{p+1}(v) \cap D^q(u) \neq \emptyset$ (or, equivalently, $D^p(v) \cap D^{q+1}(u) \neq \emptyset$ i.e. $0 < dist(v, u) \leq p + q + 1$).

Lemma 5 *For each distance-hereditary graph G the graph $\Gamma(G)$ is weakly chordal.*

This lemma together with Theorem 1 will be used in the following.

Theorem 6 *The problem r-dominating by cliques is polynomial-time solvable on distance-hereditary graphs.*

Theorem 7 *The r-dominant clique problem is polynomial-time solvable on distance-hereditary graphs.*

References

[1] H.-J.BANDELT, A.HENKMANN, and F.NICOLAI Powers of distance-hereditary graphs, *Technical Report SM-DU-220*, University Duisburg 1993.

[2] H.-J.BANDELT and H.M.MULDER, Distance-hereditary graphs, *J. Comb. Theory* Ser.B, 41 (1986) 182–208.

[3] H.-J.BANDELT and H.M.MULDER, Pseudo-modular graphs, *Discrete Math.*, 62 (1986) 245–260.

[4] A.BRANDSTÄDT, Special graph classes - a survey, *Technical Report SM-DU-199*, University Duisburg 1993.

[5] A. BRANDSTÄDT, F.F. DRAGAN, V.D. CHEPOI, and V.I. VOLOSHIN, Dually chordal graphs, *Technical Report SM-DU-225*, University of Duisburg 1993, to appear as extended abstract in the proceedings of WG'93, (ed. J. van Leeuwen), Utrecht, The Netherlands, submitted to SIAM J. Discr. Math.

[6] A.BRANDSTÄDT and D.KRATSCH, Domination problems on permutation and other graphs, *Theoretical Computer Science*, 54 (1987) 181–198.

[7] G.J.CHANG , Centers of chordal graphs, *Graphs and Combinatorics*, 7 (1991) 305–313.

[8] A.D'ATRI and M.MOSCARINI, Distance-hereditary graphs, Steiner trees and connected domination, *SIAM J. Computing*, 17 (1988) 521–538.

[9] F.F.DRAGAN and A.BRANDSTÄDT, r-Dominating cliques in Helly graphs and chordal graphs, *Technical Report SM-DU-228*, University Duisburg 1993, *Proc. of the 11th STACS*, Caen, France, Springer, LNCS 775, 735 – 746, 1994

[10] M.C.GOLUMBIC, Algorithmic Graph Theory and Perfect Graphs, *Academic Press*, 1980.

[11] P.L.HAMMER and F.MAFFRAY, Completely separable graphs, *Discrete Appl. Math.*, 27 (1990) 85–99.

[12] S.C.HEDETNIEMI and R.LASKAR, (eds.), Topics on Domination, *Annals of Discr. Math.* 48, North-Holland, 1991.

[13] E.HOWORKA, A characterization of distance-hereditary graphs, *Quart. J. Math.* Oxford Ser. 2, 28 (1977) 417–420.

[14] D.KRATSCH, Finding dominating cliques efficiently in strongly chordal graphs and undirected path graphs, *Annals of Discr. Math.* 48, Topics on Domination, (S.C.Hedetniemi, R.Laskar, eds.), 225–238.

[15] D.KRATSCH, P.DAMASCHKE and A.LUBIW, Dominating cliques in chordal graphs, to appear in *Discrete Mathematics*.

[16] S.V.YUSHMANOV and V.D.CHEPOI, A general method of investigation of characteristic of graphs related with eccentricity *Math. Problems of Cybernetics, Moscow 3* 1991 217-232 (Russian, English transl.)

Author Index

Springer-Verlag
and the Environment

Lecture Notes in Computer Science

For information about Vols. 1–745
please contact your bookseller or Springer-Verlag

Vol. 783: C. G. Günther (Ed.), Mobile Communications. Proceedings, 1994. XVI, 564 pages. 1994.

Vol. 784: F. Bergadano, L. De Raedt (Eds.), Machine Learning: ECML-94. Proceedings, 1994. XI, 439 pages. 1994. (Subseries LNAI).

Vol. 785: H. Ehrig, F. Orejas (Eds.), Recent Trends in Data Type Specification. Proceedings, 1992. VIII, 350 pages. 1994.

Vol. 786: P. A. Fritzson (Ed.), Compiler Construction. Proceedings, 1994. XI, 451 pages. 1994.

Vol. 787: S. Tison (Ed.), Trees in Algebra and Programming – CAAP '94. Proceedings, 1994. X, 351 pages. 1994.

Vol. 788: D. Sannella (Ed.), Programming Languages and Systems – ESOP '94. Proceedings, 1994. VIII, 516 pages. 1994.

Vol. 789: M. Hagiya, J. C. Mitchell (Eds.), Theoretical Aspects of Computer Software. Proceedings, 1994. XI, 887 pages. 1994.

Vol. 790: J. van Leeuwen (Ed.), Graph-Theoretic Concepts in Computer Science. Proceedings, 1993. IX, 431 pages. 1994.

Vol. 791: R. Guerraoui, O. Nierstrasz, M. Riveill (Eds.), Object-Based Distributed Programming. Proceedings, 1993. VII, 262 pages. 1994.

Vol. 792: N. D. Jones, M. Hagiya, M. Sato (Eds.), Logic, Language and Computation. XII, 269 pages. 1994.

Vol. 793: T. A. Gulliver, N. P. Secord (Eds.), Information Theory and Applications. Proceedings, 1993. XI, 394 pages. 1994.

Vol. 794: G. Haring, G. Kotsis (Eds.), Computer Performance Evaluation. Proceedings, 1994. X, 464 pages. 1994.

Vol. 795: W. A. Hunt, Jr., FM8501: A Verified Microprocessor. XIII, 333 pages. 1994.

Vol. 796: W. Gentzsch, U. Harms (Eds.), High-Performance Computing and Networking. Proceedings, 1994, Vol. I. XXI, 453 pages. 1994.

Vol. 797: W. Gentzsch, U. Harms (Eds.), High-Performance Computing and Networking. Proceedings, 1994, Vol. II. XXII, 519 pages. 1994.

Vol. 798: R. Dyckhoff (Ed.), Extensions of Logic Programming. Proceedings, 1993. VIII, 362 pages. 1994.

Vol. 799: M. P. Singh, Multiagent Systems. XXIII, 168 pages. 1994. (Subseries LNAI).

Vol. 800: J.-O. Eklundh (Ed.), Computer Vision – ECCV '94. Proceedings 1994, Vol. I. XVIII, 603 pages. 1994.

Vol. 801: J.-O. Eklundh (Ed.), Computer Vision – ECCV '94. Proceedings 1994, Vol. II. XV, 485 pages. 1994.

Vol. 802: S. Brookes, M. Main, A. Melton, M. Mislove, D. Schmidt (Eds.), Mathematical Foundations of Programming Semantics. Proceedings, 1993. IX, 647 pages. 1994.

Vol. 803: J. W. de Bakker, W.-P. de Roever, G. Rozenberg (Eds.), A Decade of Concurrency. Proceedings, 1993. VII, 683 pages. 1994.

Vol. 804: D. Hernández, Qualitative Representation of Spatial Knowledge. IX, 202 pages. 1994. (Subseries LNAI).

Vol. 805: M. Cosnard, A. Ferreira, J. Peters (Eds.), Parallel and Distributed Computing. Proceedings, 1994. X, 280 pages. 1994.

Vol. 806: H. Barendregt, T. Nipkow (Eds.), Types for Proofs and Programs. VIII, 383 pages. 1994.

Vol. 807: M. Crochemore, D. Gusfield (Eds.), Combinatorial Pattern Matching. Proceedings, 1994. VIII, 326 pages. 1994.

Vol. 808: M. Masuch, L. Pólos (Eds.), Knowledge Representation and Reasoning Under Uncertainty. VII, 237 pages. 1994. (Subseries LNAI).

Vol. 809: R. Anderson (Ed.), Fast Software Encryption. Proceedings, 1993. IX, 223 pages. 1994.

Vol. 810: G. Lakemeyer, B. Nebel (Eds.), Foundations of Knowledge Representation and Reasoning. VIII, 355 pages. 1994. (Subseries LNAI).

Vol. 811: G. Wijers, S. Brinkkemper, T. Wasserman (Eds.), Advanced Information Systems Engineering. Proceedings, 1994. XI, 420 pages. 1994.

Vol. 812: J. Karhumäki, H. Maurer, G. Rozenberg (Eds.), Results and Trends in Theoretical Computer Science. Proceedings, 1994. X, 445 pages. 1994.

Vol. 813: A. Nerode, Yu. N. Matiyasevich (Eds.), Logical Foundations of Computer Science. Proceedings, 1994. IX, 392 pages. 1994.

Vol. 814: A. Bundy (Ed.), Automated Deduction—CADE-12. Proceedings, 1994. XVI, 848 pages. 1994. (Subseries LNAI).

Vol. 815: R. Valette (Ed.), Application and Theory of Petri Nets 1994. Proceedings. IX, 587 pages. 1994.

Vol. 816: J. Heering, K. Meinke, B. Möller, T. Nipkow (Eds.), Higher-Order Algebra, Logic, and Term Rewriting. Proceedings, 1993. VII, 344 pages. 1994.

Vol. 817: C. Halatsis, D. Maritsas, G. Philokyprou, S. Theodoridis (Eds.), PARLE '94. Parallel Architectures and Languages Europe. Proceedings, 1994. XV, 837 pages. 1994.

Vol. 818: D. L. Dill (Ed.), Computer Aided Verification. Proceedings, 1994. IX, 480 pages. 1994.

Vol. 819: W. Litwin, T. Risch (Eds.), Applications of Databases. Proceedings, 1994. XII, 471 pages. 1994.

Vol. 820: S. Abiteboul, E. Shamir (Eds.), Automata, Languages and Programming. Proceedings, 1994. XIII, 644 pages. 1994.

Vol. 821: M. Tokoro, R. Pareschi (Eds.), Object-Oriented Programming. Proceedings, 1994. XI, 535 pages. 1994.

Vol. 822: F. Pfenning (Ed.), Logic Programming and Automated Reasoning. Proceedings, 1994. X, 345 pages. 1994. (Subseries LNAI).

Vol. 823: R. A. Elmasri, V. Kouramajian, B. Thalheim (Eds.), Entity-Relationship Approach — ER '93. Proceedings, 1993. X, 531 pages. 1994.

Vol. 824: E. M. Schmidt, S. Skyum (Eds.), Algorithm Theory - SWAT '94. Proceedings. IX, 383 pages. 1994.

Vol. 826: D. S. Bowers (Ed.), Directions in Databases. Proceedings, 1994. X, 234 pages. 1994.

Vol. 827: D. M. Gabbay, H. J. Ohlbach (Eds.), Temporal Logic. Proceedings, 1994. XI, 546 pages. 1994.

Vol. 828: L. C. Paulson, Isabelle. XVII, 321 pages. 1994.